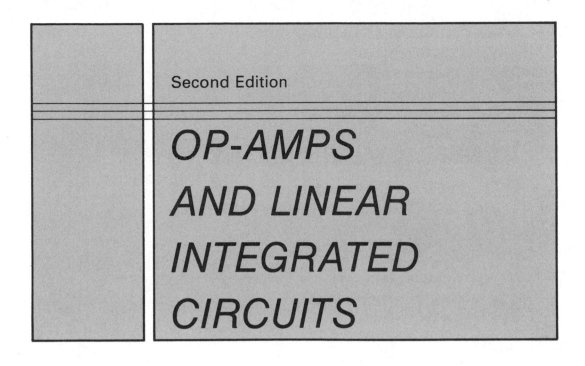

Second Edition

OP-AMPS
AND LINEAR
INTEGRATED
CIRCUITS

RAMAKANT A. GAYAKWAD

PRENTICE HALL
Englewood Cliffs, New Jersey 07632

Library of Congress Cataloging-in-Publication Data

Gayakwad, Ramakant A.
    Op-amps and linear integrated circuits/Ramakant A. Gayakwad. — 2nd ed.
      p.   cm.
    Rev. ed. of: Op-amps and linear integrated circuit technology.
    Bibliography: p.
    Includes index.
    ISBN 0-13-637174-4
    1. Operational amplifiers. 2. Linear integrated circuits.
I. Gayakwad, Ramakant A. Op-amps and linear integrated circuit
technology. II. Title.
TK7871.58.06G38   1988
621.381′735--dc19

87-20448
CIP

Editorial/production supervision: *Mary Carnis*
Cover design: *Diane Saxe*
Manufacturing buyer: *Lorraine Fumoso*

 © 1988 and 1983 by Prentice-Hall, Inc.
A Division of Simon & Schuster
Englewood Cliffs, New Jersey 07632

The first edition of this book was printed under the title
of *Op-Amps and Linear Integrated Circuit Technology,*
© 1983 by Prentice Hall, Inc., Englewood Cliffs, New Jersey.

Printed in the United States of America

10 9 8 7 6 5 4 3 2 1

ISBN 0-13-637174-4 025

Prentice-Hall International (UK) Limited, *London*
Prentice-Hall of Australia Pty. Limited, *Sydney*
Prentice-Hall Canada Inc., *Toronto*
Prentice-Hall Hispanoamericana, S.A., *Mexico*
Prentice-Hall of India Private Limited, *New Delhi*
Prentice-Hall of Japan, Inc., *Tokyo*
Simon & Schuster Asia Pte. Ltd., *Singapore*
Editora Prentice-Hall do Brasil, Ltda., *Rio de Janeiro*

To my father and mother

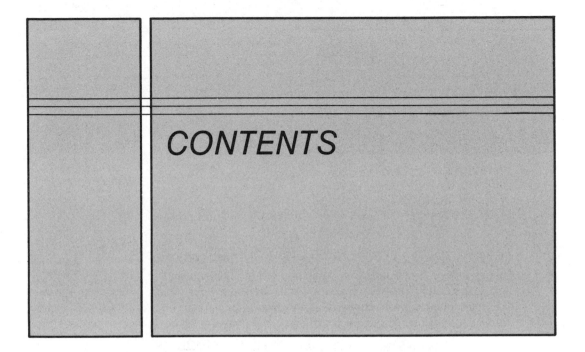

CONTENTS

Contents

5 THE PRACTICAL OP-AMP
145

5–1   Introduction   145

5–2   Input Offset Voltage   145
    5–2.1   Offset-Voltage Compensating Network Design, 147

5–3   Input Bias Current   157

5–4   Input Offset Current   163

5–5   Total Output Offset Voltage   165

5–6   Thermal Drift   167
    5–6.1   Error Voltage, 170

5–7   Effect of Variation in Power Supply Voltages on Offset Voltage   175

5–8   Change in Input Offset Voltage and Input Offset Current with Time   180

5–9   Other Temperature- and Supply Voltage-Sensitive Parameters   181

5–10   Noise   184

5–11   Common-Mode Configuration and Common-Mode Rejection Ratio   186

Summary   190

Questions   190

Problems   191

6 FREQUENCY RESPONSE OF AN OP-AMP
194

6–1   Introduction   194

6–2   Frequency Response   194

6–3   Compensating Networks   195

6–4   Frequency Response of Internally Compensated Op-Amps   197

6–5   Frequency Response of Noncompensated Op-Amps   197

6–6   High-Frequency Op-Amp Equivalent Circuit   198

6–7   Open-Loop Voltage Gain as a Function of Frequency   199

6–8   Closed-Loop Frequency Response   205

6–9   Circuit Stability   206

6–10   Slew Rate   209
    6–10.1   Causes of Slew Rate, 210
    6–10.2   Slew Rate Equation, 210
    6–10.3   Effect of Slew Rate in Applications, 213
    6–10.4   Difference between Bandwidth, Transient Response, and Slew Rate, 215

PREFACE

For both the publisher and myself, accuracy has been a top priority in the preparation of the second edition. In addition, the text has undergone a major revision that included removing lab experiments from the text to create a lab manual, while adding some most important topics, such as switched capacitor filters and switching regulators, expanding coverage in a number of vital areas, and highlighting examples. The input for these changes came from three major sources:

1. Most important has been the student response which continues to point out the topics in the text that need modification and improvement;
2. Timely suggestions and criticisms from teaching associates and users of the text are of paramount importance to, and are greatly appreciated by, the author;
3. The reviewers have provided detailed reviews to identify problem areas and recommended changes that will benefit the majority of users. Obviously, the credit for identifying such professional individuals and coordinating their inputs in a timely manner goes to the publisher. The publisher and I have acknowledged these individuals at the end of preface to express our gratitude for their continued input and support.

The primary objective of the second edition was to improve accuracy and to provide additional material that would clarify critical concepts and introduce subject areas that are currently popular. As in the first edition, this edition maintains the right blend of theory and practice and presents the theory in a practical way. The text is full of a variety of practical circuits which have been built and laboratory-tested to prove the theories behind them.

As indicated above, one major change was to remove the experiments from the text and create a separate laboratory manual, *Experiments in Op-Amps and Linear Integrated Circuits*. The labs have been heavily revised and updated to include data sheets of the ICs used in those experiments. The important feature of the lab manual which will be welcomed by users, students and faculty alike, is that the individual experiments can be removed from the manual for submission.

In the text, all eleven chapters have been thoroughly checked for accuracy. In addition, Chapters 3, 4, 6, 7, 8, and 10 have undergone changes with regard to additional material and clarification of critical concepts. Specifically, in Chapter 3, Table 3–1 is rearranged for easy understanding. In Chapter 4, discussion on inverting and noninverting amplifiers is expanded with reference to voltage gain, bandwidth, and total output offset voltage with feedback. The following sections are improved for accuracy and understanding in Chapter 6: 6–3, Compensating networks; 6–4, Frequency response of internally compensated op-amps; 6–5, Frequency response of noncompensated op-amps; 6–9, Circuit stability; and 6–10, Slew rate.

In Chapter 7, sections 7–2, dc and ac amplifiers and 7–5.3(b), summing amplifier, are improved for easy understanding. Also, the content of section 7–12, the integrator, is expanded to make it compatible to the differentiator. Chapter 8 is enhanced by the addition of derivations of oscillator equations. Finally, new topics that include the switched capacitor filter MF5, adjustable positive voltage regulator LM317, and the switching regulator 78S40 are added in Chapter 10 complete with theory and practical examples. In addition, Appendix C includes important derivations and appendix D includes the data sheets.

I am confident that the new cover, additions and deletions, modified sections, and the overall improved accuracy of the second edition will please most of the users.

ACKNOWLEDGMENTS

The author feels very fortunate to have the support of a number of key individuals at Prentice-Hall including Alice Barr, Acquisitions Editor, and Mary Carnis, Production Editor. Additionally, at Prentice-Hall, I thank Gregory Burnell, Executive Editor. Also, I would like to thank Paul McGuirk, President, DeVry Institute of Technology, Los Angeles, for his support and encouragement.

I wish to acknowledge the following reviewers for the quality, care, and depth of their comments and suggestions:

Nate Atun, *DeVry Institute of Technology*
Pankaj Patel, *DeVry Institute of Technology*
Stanley Hochman, *Mt. Hood Community College*
Carolyn A. Smith, *Howard Community College*
Gary Hecht, *Portland Community College*
Charlie Calkins, *Mt. Hood Community College*
Arthur Tobin, *Portland Community College*
Terence Dibble, *Portland Community College*

William Eaton, *Hinds Jr. College*

Andrew C. Woodson, *North Seattle Community College*

Sam Aiello, *DeVry Institute of Technology*

Carl Stancil, *Cabrillo College*

Doug Frey, *Lehigh University*

Russ Puckett, *Texas A&M University*

Steven Hunter, *Sierra College*

Carl S. Holzinger, *Lehigh University*

Harold L. Coomes, *Patterson State Technical College*

Greg Maninakis, *SUNY Agricultural & Technical College, Farmingdale*

Henry Peden, *Spokane Community College*

Paul B. Robinson, *New Hampshire Technical Institute*

Pravin M. Raghuwanshi, *DeVry Institute of Technology*

Omar Zia, *Oregon Institute of Technology*

Dr. S. E. Reyer, *Milwaukee School of Engineering*

Finally, I must express my deepest appreciation to my wife, Pratibha, and daughters, Leena and Neeta, for their patience and understanding while I spent many hours in the revision of this edition.

Ramakant A. Gayakwad

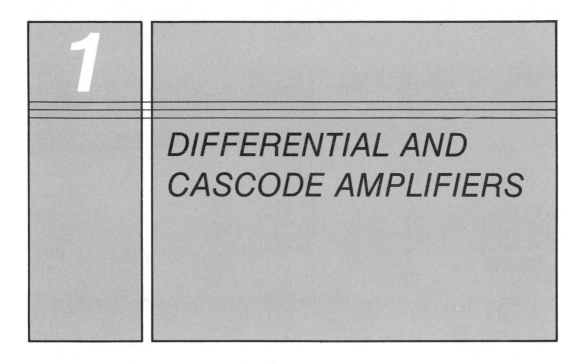

DIFFERENTIAL AND CASCODE AMPLIFIERS

1-1 INTRODUCTION

The first part of this chapter is devoted to an in-depth analysis of differential amplifier types using r-parameters. The cascaded differential amplifiers are presented in a stage-by-stage analysis. Detailed numerical examples are included in this analysis. The results of all types of differential amplifiers are summarized at the end of the first part as a quick and convenient cross-reference. The latter part of the chapter presents the cascode amplifier analysis.

Since the differential amplifier is the basic building block of operational amplifiers, the discussion of differential amplifiers in this chapter sets the groundwork for analysis and design procedures for the operational amplifiers. The analysis of differential amplifiers not only clarifies the operation of the operational amplifier but also makes the characteristics of the operational amplifier easy to understand. In addition, the analysis procedure developed in this chapter provides a means of checking operational amplifier parameters such as voltage gain and input and output resistances that are specified on the manufacturers' data sheets.

1-2 DIFFERENTIAL AMPLIFIER

Let us consider the emitter-biased circuit. Figure 1-1 shows two identical emitter-biased circuits in that transistor Q_1 has the same characteristics as transistor Q_2, $R_{E1} = R_{E2}$, $R_{C1} = R_{C2}$, and the magnitude of $+V_{CC}$ is equal to the magnitude of $-V_{EE}$. Remember that the supply voltages $+V_{CC}$ and $-V_{EE}$ are measured with respect to ground. To obtain a

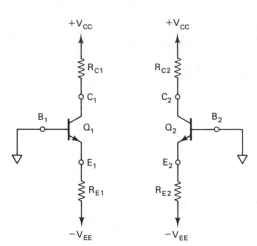

Figure 1–1 Two identical emitter-biased circuits.

single circuit such as the one in Figure 1-2, we should reconnect these two circuits as follows:

1. Reconnect $+V_{CC}$ supply voltages of the two circuits since the voltages are of the same polarity and amplitude. Similarly, reconnect $-V_{EE}$ supply voltages.
2. Reconnect the emitter E_1 of transistor Q_1 to the emitter E_2 of transistor Q_2. (This reconnection places R_{E1} in parallel with R_{E2}.)
3. Show the input signal $v_{in\,1}$ applied to the base B_1 of transistor Q_1 and $v_{in\,2}$ applied to the base B_2 of transistor Q_2.
4. Label the voltage between the collectors C_1 and C_2 as v_o. (The v_o is the output voltage.)

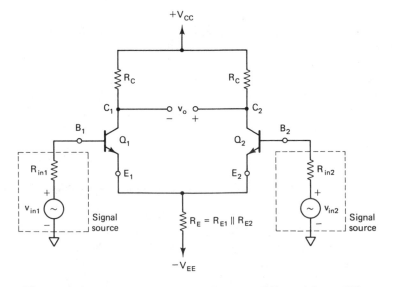

Figure 1–2 Dual-input, balanced-output differential amplifier.

The circuit of Figure 1-2 depicts these changes and is referred to as a *differential amplifier*. Note that in the circuit of Figure 1-2, $R_{C1} = R_{C2}$; hence their subscripts are omitted for simplicity. In addition, *the parallel combination of R_{E1} and R_{E2} is denoted by R_E.*

The differential amplifier of Figure 1-2, as its name implies, amplifies the difference between two input signals $v_{in\,1}$ and $v_{in\,2}$. (The differential amplifier is also referred to as a *difference amplifier*.) The circuit shown in Figure 1-2 is a *basic* differential amplifier circuit from which we can derive the remaining three differential amplifier circuit configurations.

1-3 DIFFERENTIAL AMPLIFIER CIRCUIT CONFIGURATIONS _____

The four differential amplifier configurations are the following:

1. Dual-input, balanced-output differential amplifier
2. Dual-input, unbalanced-output differential amplifier
3. Single-input, balanced-output differential amplifier
4. Single-input, unbalanced-output differential amplifier

The configurations listed are defined by the number of input signals used and the way an output voltage is measured. If we use two input signals, the configuration is said to be *dual input;* otherwise, it is a *single-input* configuration. On the other hand, if the output voltage is measured between two collectors, it is referred to as a *balanced output,* because both collectors are at the same dc potential with respect to ground. However, if the output is measured at one of the collectors with respect to ground, the configuration is called an *unbalanced output.* Before proceeding any further with the circuit analysis, some general observations on the differential amplifier are in order.

Two matched semiconductors of the same type (BJTs or FETs) are required for the differential amplifier. All the components in the two emitter-biased circuits, which constitute a differential amplifier, must be matched in all respects for proper operation. Furthermore, the magnitude of the supply voltages $+V_{CC}$ and $-V_{EE}$ must be equal. Fortunately, a variety of matched transistor pairs is readily available for differential amplifier applications. A multistage amplifier with a desired voltage gain can be formed using a direct connection between successive stages of differential amplifiers. The beauty of the direct connection between stages is that it removes the lower cutoff frequency imposed by the coupling capacitors. Therefore, the differential amplifier is capable of amplifying dc as well as ac input signals. In instrumentation systems, differential amplifiers are widely used to compare two input signals.

In the analysis to follow we use *r*-parameters instead of *h*-parameters for a number of reasons:

1. Because of the nature of *r*-parameters, ac analysis of differential amplifiers with *r*-parameters is simpler, more straightforward, and less cumbersome.
2. Unlike with *h*-parameters, there is no need to manipulate the *r*-parameters at different operating levels (different I_E and V_{CE} values) except for the r_e value.

3. The performance equations obtained are easy to remember since they are not as complex or as lengthy as *h*-parameter equations would be.

4. Above all, the results obtained using *r*-parameters compare favorably with the actual results.

1-4 DUAL-INPUT, BALANCED-OUTPUT DIFFERENTIAL AMPLIFIER

The circuit shown in Figure 1-2 is a dual-input, balanced-output differential amplifier. Notice that the two input signals (dual input), v_{in1} and v_{in2}, are applied to the bases B_1 and B_2 of transistors Q_1 and Q_2. The output v_o is measured between the two collectors, C_1 and C_2, which are at the same dc potential. Because of the equal dc potential at the two collectors with respect to ground, the output is referred to as a balanced output.

1-4.1 DC Analysis

To determine the operating point values (I_{CQ} and V_{CEQ}) for the differential amplifier of Figure 1-2, we need to obtain a dc equivalent circuit. The dc equivalent circuit can be obtained simply by reducing the input signals v_{in1} and v_{in2} to zero. The dc equivalent circuit thus obtained is shown in Figure 1-3. Note that the internal resistances of the input signals are denoted by R_{in} because $R_{in1} = R_{in2}$. Since both emitter-biased sections of the differential amplifier are symmetrical (matched in all respects), we need to determine the operating point collector current I_{CQ} and collector-to-emitter voltage V_{CEQ} for only one section. We shall determine the I_{CQ} and V_{CEQ} values for transistor Q_1 only. These I_{CQ} and V_{CEQ} values can then be used for transistor Q_2 also.

Applying Kirchhoff's voltage law to the base–emitter loop of the transistor Q_1 (see Figure 1-3),

$$-R_{in}I_B - V_{BE} - R_E(2I_E) + V_{EE} = 0 \qquad (1\text{-}1)$$

But

$$I_B = \frac{I_E}{\beta_{dc}} \qquad \text{since } I_C \cong I_E$$

Thus the emitter current through transistor Q_1 is determined directly from Equation (1-1) as follows:

$$I_E = \frac{V_{EE} - V_{BE}}{2R_E + R_{in}/\beta_{dc}} \qquad (1\text{-}2)$$

where $V_{BE} = 0.6$ V for silicon transistors
$V_{BE} = 0.2$ V for germanium transistors
Generally, $R_{in}/\beta_{dc} \ll 2R_E$. Therefore, Equation (1-2) can be rewritten as

$$I_E = \frac{V_{EE} - V_{BE}}{2R_E} \qquad (1\text{-}3)$$

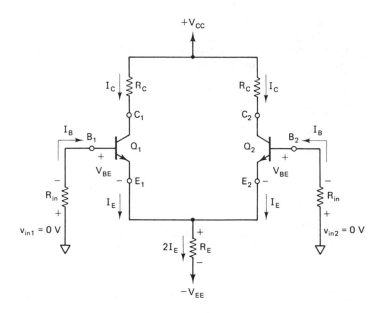

Figure 1–3 DC equivalent circuit of the dual-input, balanced-output differential amplifier.

From Equation (1-3) we see that the value of R_E sets up the emitter current in transistors Q_1 and Q_2 for a given value of V_{EE}. In other words, by selecting a proper value of R_E, we can obtain a desired value of emitter current for a known value of $-V_{EE}$. Notice that the emitter current in transistors Q_1 and Q_2 is independent of collector resistance R_C.

Next we shall determine the collector-to-emitter voltage V_{CE}. The voltage at the emitter of transistor Q_1 is approximately equal to $-V_{BE}$ if we assume the voltage drop across R_{in} to be negligibly small. Knowing the value of emitter current I_E $(\cong I_C)$, we can obtain the voltage at the collector V_C, as follows:

$$V_C = V_{CC} - R_C I_C$$

Thus the collector-to-emitter voltage V_{CE} is

$$V_{CE} = V_C - V_E$$
$$= (V_{CC} - R_C I_C) - (-V_{BE})$$
$$V_{CE} = V_{CC} + V_{BE} - R_C I_C \qquad (1\text{-}4)$$

Hence, for both transistors we can determine the I_{CQ} and V_{CEQ} by using Equations (1-2) and (1-4), respectively, because at the operating point $I_E = I_{CQ}$ and $V_{CE} = V_{CEQ}$.

Remember that the dc analysis equations, (1-2) and (1-4), are applicable for all four differential amplifier configurations as long as we use the same biasing arrangement (as discussed above) for each of them. That is, knowing the dc analysis procedure for one configuration (dual input, balanced output), we do not have to repeat the dc analysis procedure for the remaining three configurations.

1-4.2 AC Analysis

To perform ac analysis to derive the expression for the voltage gain A_d and the input resistance R_i of the differential amplifier shown in Figure 1-2:

1. Set the dc voltages $+V_{CC}$ and $-V_{EE}$ at zero.
2. Substitute the small-signal T-equivalent models for the transistors.

Figure 1-4(a) shows the resulting ac equivalent circuit of the dual-input, balanced-output differential amplifier.

(a)

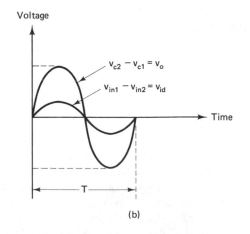

(b)

Figure 1–4   Dual-input, balanced-output differential amplifier. (a) AC equivalent circuit. (b) Input and output waveforms.

1–4.2(a) Voltage gain.   Before we proceed to derive the expression for the voltage gain A_d, the following should be noted about the circuit in Figure 1-4(a):

1. $I_{E1} = I_{E2}$; therefore, $r_{e1} = r_{e2}$. For this reason, the ac emitter resistance of transistor Q_1 and Q_2 is simply denoted by r_e.
2. The voltage across each collector resistor is shown out of phase by 180° with respect to the input voltages v_{in1} and v_{in2}. This polarity assignment is in accordance with the common-emitter configuration. As we have seen, the differential amplifier is basically constructed in a common-emitter configuration using two identical common-emitter circuits.
3. Note the assigned polarity of the output voltage v_o. This polarity simply indicates that the voltage at collector C_2 is assumed to be more positive with respect to that at collector C_1, even though both of them are negative with respect to ground.

Writing Kirchhoff's voltage equations for loops I and II in Figure 1-4(a) gives us

$$v_{in1} - R_{in1}i_{b1} - r_e i_{e1} - R_E(i_{e1} + i_{e2}) = 0 \qquad (1\text{-}5)$$

$$v_{in2} - R_{in2}i_{b2} - r_e i_{e2} - R_E(i_{e1} + i_{e2}) = 0 \qquad (1\text{-}6)$$

Substituting current relations $i_{b1} = i_{e1}/\beta_{ac}$ and $i_{b2} = i_{e2}/\beta_{ac}$ yields

$$v_{in1} - \frac{R_{in1}}{\beta_{ac}}i_{e1} - r_e i_{e1} - R_E(i_{e1} + i_{e2}) = 0$$

$$v_{in2} - \frac{R_{in2}}{\beta_{ac}}i_{e2} - r_e i_{e2} - R_E(i_{e1} + i_{e2}) = 0$$

Generally, R_{in1}/β_{ac} and R_{in2}/β_{ac} values are very small; therefore, we shall neglect them here for simplicity and rearrange these equations as follows:

$$(r_e + R_E)\,i_{e1} + (R_E)\,i_{e2} = v_{in1} \qquad (1\text{-}7)$$

$$(R_E)i_{e1} + (r_e + R_E)\,i_{e2} = v_{in2} \qquad (1\text{-}8)$$

Equations (1-7) and (1-8) can be solved simultaneously for i_{e1} and i_{e2} by using Cramer's rule:

$$i_{e1} = \frac{\begin{vmatrix} v_{in1} & R_E \\ v_{in2} & r_e + R_E \end{vmatrix}}{\begin{vmatrix} r_e + R_E & R_E \\ R_E & r_e + R_E \end{vmatrix}}$$

$$= \frac{(r_e + R_E)v_{in1} - (R_E)v_{in2}}{(r_e + R_E)^2 - (R_E)^2} \qquad (1\text{-}9a)$$

Similarly,

$$i_{e2} = \frac{\begin{vmatrix} r_e + R_E & v_{in\,1} \\ R_E & v_{in\,2} \end{vmatrix}}{\begin{vmatrix} r_e + R_E & R_E \\ R_E & r_e + R_E \end{vmatrix}}$$

$$= \frac{(r_e + R_E)v_{in\,2} - (R_E)v_{in\,1}}{(r_e + R_E) - (R_E)^2} \tag{1-9b}$$

The output voltage is

$$v_o = v_{c2} - v_{c1}$$
$$= -R_C i_{c2} - (-R_C i_{c1})$$
$$= R_C i_{c1} - R_C i_{c2}$$
$$= R_C(i_{e1} - i_{e2}) \quad \text{since } i_c \cong i_e \tag{1-10}$$

Substituting the current relations i_{e1} and i_{e2} in Equation (1-10), we get

$$v_0 = R_C \left[\frac{(r_e + R_E)v_{in\,1} - (R_E)v_{in\,2}}{(r_e + R_E)^2 - (R_E)^2} - \frac{(r_e + R_E)v_{in\,2} - (R_E)v_{in\,1}}{(r_e + R_E)^2 - (R_E)^2} \right]$$

$$= R_C \frac{(r_e + R_E)(v_{in\,1} - v_{in\,2}) + (R_E)(v_{in\,1} - v_{in\,2})}{(r_e + R_E)^2 - (R_E)^2}$$

$$= R_C \frac{(r_e + 2R_E)(v_{in\,1} - v_{in\,2})}{r_e^2 + 2r_e R_E + R_E^2 - R_E^2} \tag{1-11}$$

$$= R_C \frac{(r_e + 2R_E)(v_{in\,1} - v_{in\,2})}{r_e(r_e + 2R_E)}$$

$$= \frac{R_C}{r_e}(v_{in\,1} - v_{in\,2})$$

Thus a differential amplifier amplifies the *difference* between two input signals, as expected. Figure 1-4(b) shows the input and output waveforms of the dual-input, balanced-output differential amplifier. By defining $v_{id} = v_{in\,1} - v_{in\,2}$ as the difference in input voltages, we can write the voltage-gain equation of the dual-input, balanced-output differential amplifier as follows:

$$A_d = \frac{v_o}{v_{id}} = \frac{R_C}{r_e} \tag{1-12}$$

Notice that the voltage-gain equation of the differential amplifier is independent of R_E since R_E did not appear in the gain equation (1-12). Another point of interest is that this equation is identical to the voltage-gain equation of the common-emitter amplifier. The latter is obvious because the differential amplifier employs a common-emitter configuration.

1-4.2(b) Differential input resistance.   Differential input resistance is defined as the equivalent resistance that would be measured at either input terminal with the other terminal grounded. This means that the input resistance R_{i1} seen from the input signal source $v_{in\,1}$ is determined with the signal source $v_{in\,2}$ set at zero. Similarly, the input signal source $v_{\text{in}1}$ is set at zero to determine the input resistance R_{i2} seen from the input signal source $v_{\text{in}2}$. See Figure 1-4(a). Usually, the source resistances $R_{\text{in}1}$ and $R_{\text{in}2}$ are very small and hence will be ignored in the derivation of input resistances R_{i1} and R_{i2}. In equation form,

$$R_{i1} = \left. \left| \frac{v_{\text{in}1}}{i_{b1}} \right| \right|_{v_{\text{in}2}=0}$$

$$= \left. \left| \frac{v_{\text{in}1}}{i_{e1}/\beta_{\text{ac}}} \right| \right|_{v_{\text{in}2}=0}$$

Substituting the value of i_{e1} from Equation (1-9a), we get

$$R_{i1} = \frac{\beta_{\text{ac}}v_{\text{in}1}}{\dfrac{(r_e + R_E)v_{\text{in}1} - (R_E)(0)}{(r_e + R_E)^2 - (R_E)^2}}$$

$$= \frac{\beta_{\text{ac}}(r_e^2 + 2r_eR_E)}{(r_e + R_E)}$$

$$R_{i1} = \frac{\beta_{\text{ac}}r_e(r_e + 2R_E)}{(r_e + R_E)} \qquad (1\text{-}13)$$

Generally, $R_E \gg r_e$, which implies that $(r_e + 2R_E) \cong 2R_E$ and $(r_e + R_E) \cong R_E$. Therefore, Equation (1-13) can be rewritten as

$$R_{i1} = \frac{\beta_{\text{ac}}r_e(2R_E)}{(R_E)}$$

$$R_{i1} = 2\beta_{\text{ac}}r_e \qquad (1\text{-}14)$$

Similarly, the input resistance R_{i2} seen from the input signal source $v_{\text{in}2}$ is defined as

$$R_{i2} = \left. \left| \frac{v_{\text{in}2}}{i_{b2}} \right| \right|_{v_{\text{in}1}=0}$$

$$= \left. \left| \frac{v_{\text{in}2}}{i_{e2}/\beta_{\text{ac}}} \right| \right|_{v_{\text{in}1}=0}$$

Substituting the value of i_{e2} from Equation (1-9b), we obtain

$$R_{i2} = \frac{\beta_{\text{ac}}v_{\text{in}2}}{\dfrac{(r_e + R_E)v_{\text{in}2} - (R_E)(0)}{(r_e + R_E)^2 - (R_E)^2}}$$

$$R_{i2} = \frac{\beta_{\text{ac}}r_e(r_e + 2R_E)}{(r_e + R_E)} \qquad (1\text{-}15)$$

However, $(r_e + 2R_E) \cong 2R_E$, and $(r_e + R_E) \cong R_E$ if $R_E \gg r_e$. Therefore, Equation (1-15) can be rewritten as

$$R_{i2} = \frac{\beta_{\text{ac}}r_e(2R_E)}{R_E}$$

$$R_{i2} = 2\beta_{\text{ac}}r_e \qquad (1\text{-}16)$$

1-4.2(c) Output resistance.   Output resistance is defined as the equivalent resistance that would be measured at either output terminal with respect to ground. Therefore, the output resistance R_{o1} measured between collector C_1 and ground is equal to that of the collector resistor R_C [see Figure 1-4(a)]. Similarly, the output resistance R_{o2} measured at collector C_2 with respect to ground is equal to that of the collector resistor R_C. Thus

$$R_{o1} = R_{o2} = R_C \qquad\qquad (1\text{-}17)$$

The current gain of the differential amplifier is undefined; therefore, the current-gain equation will not be derived for any of the four differential amplifier configurations. Furthermore, like the common-emitter amplifier, the differential amplifier is a small-signal amplifier; therefore, it is generally used as a voltage amplifier and not as a current or power amplifier.

1-4.3 Inverting and Noninverting Inputs

In the differential amplifier circuit of Figure 1-2 the input voltage $v_{in\,1}$ is called the *noninverting input* because a positive voltage $v_{in\,1}$ acting alone produces a positive output voltage. This can be seen from voltage-gain equation (1-11). Similarly, the positive voltage $v_{in\,2}$ acting alone produces a negative output voltage; hence $v_{in\,2}$ is called the *inverting input* [see Equation (1-11)]. Consequently, the base terminal B_1 to which $v_{in\,1}$ is applied is referred to as the *noninverting input terminal,* and the base terminal B_2 is called the *inverting input terminal.*

1-4.4 Common-Mode Rejection Ratio

An important characteristic of the dual-input, balanced-output differential amplifier is its ability to suppress undesired disturbances that might be amplified along with the desired signal. When the matched pair of transistors is used in the differential amplifier, the unwanted signals, such as 60-Hz noise or hum pickup, would appear as common to both input bases, and therefore the net output would theoretically be zero. The practical effectiveness of rejecting the common signal, however, depends on the degree of matching between the two common-emitter stages forming the differential amplifier. In other words, the more closely equal are the currents in the input transistors Q_1 and Q_2, the better is the common-mode signal rejection (see Figure 1-5). When the same voltage is applied to both input terminals of a differential amplifier, the differential amplifier is said to operate in the common-mode configuration.

The ability of a differential amplifier to reject a common-mode signal is expressed by its common-mode rejection ratio (CMRR). It is the ratio of differential gain A_d to the common-mode gain A_{cm}. In equation form,

$$\text{CMRR} = \frac{A_d}{A_{cm}} \qquad\qquad (1\text{-}18)$$

The common-mode voltage gain A_{cm} can be determined as follows. Apply a known voltage v_{cm} to both input terminals of the differential amplifier as shown in Figure 1-5 and measure the resultant output voltage $v_{o\,cm}$. Then, using Equation (1-19), calculate A_{cm}.

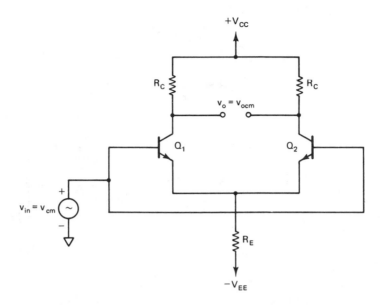

Figure 1–5   Differential amplifier in common-mode configuration.

$$A_{cm} = \frac{v_{o\,cm}}{v_{cm}} \tag{1-19}$$

Ideally, we expect A_{cm} to be zero, that is, $v_{o\,cm} = 0$ V. In other words, we would like to have CMRR of a differential amplifier ideally equal to infinity. Thus, in practice, it is advantageous to use a differential amplifier with higher CMRR since this amplifier is better able to reject common-mode signals. The significance of CMRR for op-amps is discussed in Section 3-2.

1-5 DUAL-INPUT, UNBALANCED-OUTPUT DIFFERENTIAL AMPLIFIER

In this configuration two input signals are used; however, the output is measured at only one of the two collectors with respect to ground. The output is referred to as an *unbalanced output* because the collector at which the output voltage is measured is at some finite dc potential with respect to ground. In other words, there is some dc voltage at the output terminal without any input signal applied. Let us assume that the output is measured at the collector of transistor Q_2 with respect to ground (see Figure 1-6).

1-5.1 DC Analysis

The dc analysis procedure for the dual-input, unbalanced-output differential amplifier is identical to that of the dual-input, balanced-output differential amplifier because both configurations use the same biasing arrangement. Therefore, the emitter current and emitter-to-collector voltage for the dual-input, unbalanced-output differential amplifier are

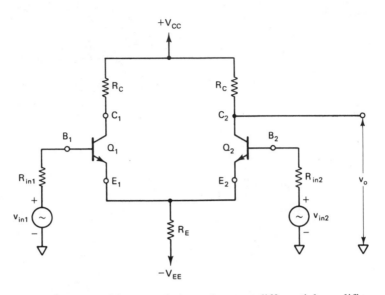

Figure 1–6 Dual-input, unbalanced-output differential amplifier.

determined from Equations (1-2) and (1-4), respectively. These equations are repeated here for convenience:

$$I_E = I_{CQ} = \frac{V_{EE} - V_{BE}}{2R_E + R_{in}/\beta_{dc}}$$

$$V_{CE} = V_{CEQ} = V_{CC} + V_{BE} - R_C I_{CQ}$$

1-5.2 AC Analysis

Figure 1-7(a) shows the ac equivalent circuit of the dual-input, unbalanced-output differential amplifier with small-signal T-equivalent models substituted for the transistors.

1-5.2(a) Voltage gain.   Writing Kirchhoff's voltage equations for loops I and II in Figure 1-7(a) gives us

$$v_{in1} - R_{in1}i_{b1} - r_e i_{e1} - R_E(i_{e1} + i_{e2}) = 0$$

$$v_{in2} - R_{in2}i_{b2} - r_e i_{e2} - R_E(i_{e1} + i_{e2}) = 0$$

Since these equations are the same as Equations (1-5) and (1-6), the expressions for i_{e1} and i_{e2} will be the same as in Equations (1-9a) and (1-9b), respectively. That is,

$$i_{e1} = \frac{(r_e + R_E)v_{in1} - (R_E)v_{in2}}{(r_e + R_E)^2 - (R_E)^2}$$

$$i_{e2} = \frac{(r_e + R_E)v_{in2} - (R_E)v_{in1}}{(r_e + R_E)^2 - (R_E)^2}$$

(a)

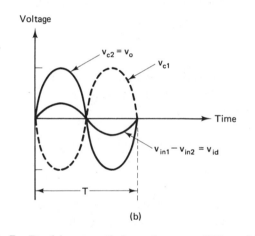

(b)

Figure 1–7 Dual-input, unbalanced-output differential amplifier. (a) AC equivalent circuit. (b) Input and output waveforms, assuming that the dc output voltage is zero.

The output voltage is

$$v_o = v_{c2} = -R_C i_{c2} = -R_C i_{e2} \quad \text{since } i_c \cong i_e$$

Substituting the value of i_{e2} yields

$$v_o = -R_C \frac{(r_e + R_E)v_{in2} - (R_E)v_{in1}}{(r_e + R_E)^2 - (R_E)^2}$$

$$= R_C \frac{(R_E)v_{in1} - (r_e + R_E)v_{in2}}{r_e(r_e + 2R_E)}$$

Generally, $R_E \gg r_e$; hence $(r_e + R_E) \cong R_E$ and $(r_e + 2R_E) \cong 2R_E$. Therefore,

$$v_o = R_C \frac{(R_E)v_{in1} - (R_E)v_{in2}}{2r_e R_E}$$

$$= R_C \frac{R_E(v_{in1} - v_{in2})}{2r_e R_E}$$

$$= \frac{R_C}{2r_e}(v_{in1} - v_{in2}) \qquad (1\text{-}20)$$

$$A_d = \frac{v_o}{v_{id}} = \frac{R_C}{2r_e} \qquad (1\text{-}21)$$

Thus the voltage gain of the dual-input, unbalanced-output differential amplifier is half the gain of the dual-input, balanced-output differential amplifier [see Equations (1-12) and (1-21)]. The input and output waveforms of a dual-input, unbalanced-output differential amplifier are shown in Figure 1-7(b). Notice also that in the dual-input, unbalanced-output differential amplifier the dc voltage at the output terminal is the error voltage in the desired output signal. Therefore, to reduce the undesired dc voltage to zero, this configuration is generally followed by a *level translator* circuit. (Refer to Section 1-13 for more information.)

1-5.2(b) Differential input resistance.   The only difference between the circuits in Figures 1-2 and 1-6 is in the way output voltage is measured. However, the input resistance seen from either input source does not depend on the way the output voltage is measured. Therefore, the input resistance R_i seen from either input source of the circuit of Figure 1-6 should be the same as that of the circuit in Figure 1-2 and is given by Equation (1-14). That is,

$$R_{i1} = R_{i2} = 2\beta_{ac}r_e$$

1-5.2(c) Output resistance.   The output resistance R_o measured at collector C_2 with respect to ground is equal to the collector resistor R_C (see Figure 1-7(a)). Thus

$$R_o = R_C$$

1-6 SINGLE-INPUT, BALANCED-OUTPUT DIFFERENTIAL AMPLIFIER

Figure 1-8 shows the single-input, balanced-output differential amplifier. In this figure the (single) input is applied to the base of transistor Q_1, and the output is measured between the two collectors, which are at the same dc potential. Therefore, the output is said to be a *balanced output*.

1-6.1 DC Analysis

The dc analysis procedure and bias equations for the single-input, balanced-output differential amplifier are identical to those of the two previous configurations because the dc

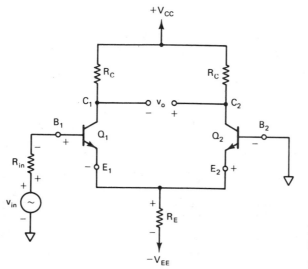

Figure 1–8   Single-input, balanced-output differential amplifier.

equivalent circuit for all three configurations is the same. Thus the bias equations [Equations (1-2) and (1-4)] are

$$I_E = I_{CQ} = \frac{V_{EE} - V_{BE}}{2R_E + R_{in}/\beta_{dc}}$$

$$V_{CE} = V_{CEQ} = V_{CC} + V_{BE} - R_c I_{CQ}$$

1-6.2 AC Analysis

The ac equivalent circuit of the single-input, balanced output differential amplifier with a small-signal T-equivalent model substituted for the transistors is shown in Figure 1-9(a).

1-6.2(a) Voltage gain.   Before we derive the expression for the voltage gain A_d, the following should be noted about the circuit in Figure 1-9(a).

1. During the positive half-cycle of the input signal (or when v_{in} is the positive voltage), the base–emitter voltage of the transistor Q_1 is positive and that of transistor Q_2 is negative, as shown in Figure 1-8. This means that the collector current in transistor Q_1 increases and that in transistor Q_2 decreases from the operating point value I_{CQ}. This change in collector currents during the positive half-cycle of the input signal is indicated in Figure 1-9(a), in which the currents of both sources i_{c1} and i_{c2} are shown to be in the same direction. In fact, during the negative half-cycle of the input signal (or when v_{in} is the negative voltage), the opposite action takes place; that is, the collector current of transistor Q_1 decreases and that in transistor Q_2 increases.

2. The polarity of voltage across each collector resistor is consistent with the direction of current sources i_{c1} and i_{c2}. That is, the voltage across the collector

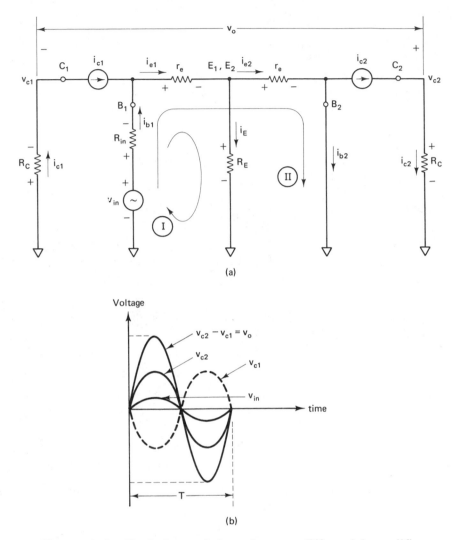

Figure 1–9 Single-input, balanced-output differential amplifier.
(a) AC equivalent circuit. (b) Input and output waveforms.

resistor of transistor Q_2 is positive, and that across the collector resistor of transistor Q_1 is negative with respect to ground [see Figure 1-9(a)].

3. Note the assigned polarity of the output voltage v_o. According to this polarity, the output voltage is equal to the voltage at collector C_2 minus the voltage at collector C_1.

Writing Kirchhoff's voltage equations for loops I and II of Figure 1-9(a) gives us

$$v_{in} - R_{in}i_{b1} - r_e i_{e1} - R_E i_E = 0 \qquad (1\text{-}22)$$
$$v_{in} - R_{in}i_{b1} - r_e i_{e1} - r_e i_{e2} = 0 \qquad (1\text{-}23)$$

Substituting current relations $i_E = (i_{e1} - i_{e2})$, $i_{b1} \cong i_{e1}/\beta_{ac}$, and $i_{b2} \cong i_{e2}/\beta_{ac}$, we get

$$v_{in} - \frac{R_{in}}{\beta_{ac}}i_{e1} - r_e i_{e1} - (R_E)(i_{e1} - i_{e2}) = 0$$

$$v_{in} - \frac{R_{in}}{\beta_{ac}}i_{e1} - r_e i_{e1} - r_e i_{e2} = 0$$

Generally, the R_{in}/β_{ac} value is very small; therefore, for simplicity we shall neglect it and rearrange the equations as follows:

$$(r_e + R_E)i_{e1} - (R_E)i_{e2} = v_{in} \tag{1-24}$$

$$(r_e)i_{e1} + (r_e)i_{e2} = v_{in} \tag{1-25}$$

Equations (1-24) and (1-25) can be solved simultaneously for i_{e1} and i_{e2} by using Cramer's rule:

$$i_{e1} = \frac{\begin{vmatrix} v_{in} & -R_E \\ v_{in} & r_e \end{vmatrix}}{\begin{vmatrix} r + R_E & -R_E \\ r_e & r_e \end{vmatrix}} = \frac{r_e v_{in} + R_E v_{in}}{(r_e + R_E)r_e + R_E r_e}$$

$$= \frac{r_e + R_E)v_{in}}{r_e(r_e + 2R_E)} \tag{1-26a}$$

Similarly,

$$i_{e2} = \frac{\begin{vmatrix} r_e + R_E & v_{in} \\ r_e & v_{in} \end{vmatrix}}{\begin{vmatrix} r_e + R_E & -R_E \\ r_e & r_e \end{vmatrix}} = \frac{(r_e + R_E)v_{in} - (r_e)v_{in}}{(r_e + R_E)r_e + R_E r_e}$$

$$= \frac{(R_E)v_{in}}{r_e(r_e + 2R_E)} \tag{1-26b}$$

The output voltage is

$$v_o = v_{c2} - v_{c1}$$
$$= R_C i_{c2} - (-R_C i_{c1})$$
$$= R_C(i_{c2} + i_{c1})$$
$$= R_C(i_{e2} + i_{e1}) \text{ since } i_e \cong i_c \tag{1-27}$$

Substituting the current relations i_{e1} and i_{e2} in Equation (1-27), we get

$$v_o = R_C \left[\frac{(R_E)v_{in}}{r_e(r_e + 2R_E)} + \frac{(r_e + R_E)v_{in}}{r_e(r_e + 2R_E)} \right]$$

$$= R_C \frac{(r_e + 2R_E)v_{in}}{r_e(r_e + 2R_E)}$$

$$= \frac{R_C}{r_e} v_{in} \tag{1-28}$$

Therefore,

$$A_d = \frac{v_o}{v_{in}} = \frac{R_C}{r_e} \tag{1-29}$$

Notice that the voltage gain of the single-input, balanced-output differential amplifier is equal to that of the dual-input, balanced-output differential amplifier [see Equations (1-12) and (1-29)]. The input and output waveforms of the single-input, balanced-output differential amplifier are shown in Figure 1-9(b).

1-6.2(b) Differential input resistance.   The input resistance R_i seen from the input signal source is determined as follows:

$$R_i = \frac{v_{in}}{i_{b1}} = \frac{v_{in}}{i_{e1}/\beta_{ac}} = \frac{\beta_{ac}v_{in}}{i_{e1}}$$

Substituting the value of i_{e1} from Equation (1-26a), we get

$$R_i = \frac{\beta_{ac}v_{in}}{\dfrac{(r_e + R_E)v_{in}}{r_e(r_e + 2R_E)}} = \frac{\beta_{ac}r_e(r_e + 2R_E)}{r_e + R_E}$$

$$R_i = 2\beta_{ac}r_e \text{ if } R_E \gg r_e \tag{1-30}$$

1-6.2(c) Output resistance.   Recall that the output resistance R_o is the equivalent resistance that would be measured at either output terminal with respect to ground. Therefore, the output resistance R_{o1} measured at the collector C_1 and the output resistance R_{o2} measured at the collector C_2 are given by Equation (1-17):

$$R_{o1} = R_{o2} = R_C$$

1-7 SINGLE-INPUT, UNBALANCED-OUTPUT DIFFERENTIAL AMPLIFIER

This differential amplifier configuration is not commonly used for several reasons:

1. Although identical to the common-emitter amplifier, it not only requires more components but also yields less voltage gain than the former one.
2. There exists a dc output voltage without any input signal applied.

Nevertheless, we shall include the analysis and derivations of the performance equations in order to make the differential amplifier discussion complete. In this configuration an input signal can be applied to the base of either transistor and the output can be measured at either collector. In Figure 1-10 the input is applied to the base of transistor Q_1, and the output is measured at the collector of transistor Q_2.

1-7.1 DC Analysis

The dc equivalent circuit for this configuration is identical to that of the dual-input, balanced-output configuration shown in Figure 1-3. Therefore, the I_{CQ} and V_{CEQ} equations

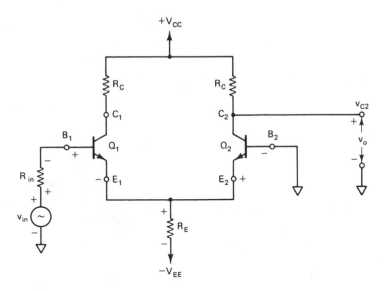

Figure 1–10   Single-input, unbalanced-output differential amplifier.

obtained from Figure 1-3 are applicable to this configuration also. These equations are repeated here for convenience:

$$I_E = I_{CQ} = \frac{V_{EE} - V_{BE}}{2R_E + R_{in}/\beta_{dc}}$$

$$V_{CE} = V_{CEQ} = V_{CC} + V_{BE} - R_C I_{CQ}$$

1-7.2 AC Analysis

The ac equivalent circuit of a single-input, unbalanced-output differential amplifier is shown in Figure 1-11(a).

1-7.2(a) Voltage gain.   The equivalent circuit shown in Figure 1-11(a) is identical to that shown in Figure 1-9(a), except in the way in which the output voltage is measured. Therefore, the current relations (1-26a) and (1-26b) derived for the circuit of Figure 1-9(a) are also valid for the circuit shown in Figure 1-11(a). The output voltage is

$$v_o = v_{c2} = R_C i_{c2} = R_C i_{e2} \text{ since } i_c \cong i_e$$

Substituting the i_{e2} value from Equation (1-26b) yields

$$v_o = R_C \frac{(R_E)v_{in}}{r_e(r_e + 2R_E)}$$

$$= R_C \frac{R_E v_{in}}{r_e(2R_E)} \text{ if } R_E >> r_e$$

$$= \frac{R_C}{2r_e} v_{in} \qquad (1\text{-}31)$$

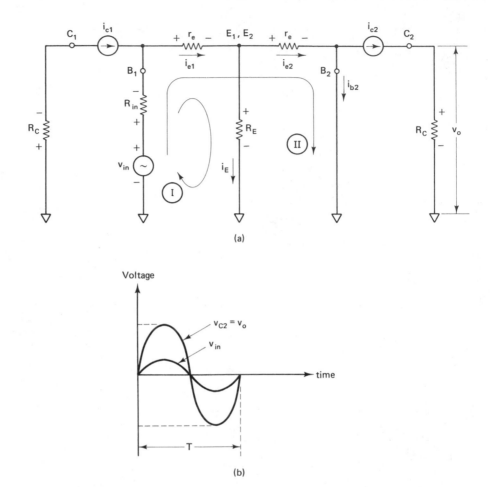

(a)

(b)

Figure 1–11 Single-input, unbalanced-output differential amplifier. (a) AC equivalent circuit. (b) Input and output waveforms, assuming that the dc output voltage is zero.

Therefore,

$$A_d = \frac{v_o}{v_{in}} = \frac{R_C}{2r_e} \tag{1-32}$$

The input and output waveforms of a single-input, unbalanced-output differential amplifier are shown in Figure 1-11(b).

1-7.2(b) Differential input resistance. The input resistance R_i of this configuration is identical to that of the previous one. That is,

$$R_i = 2\beta_{ac}r_e \qquad \text{if } R_E \gg r_e$$

1-7.2(c) Output resistance.   The output resistance R_o measured at collector C_2 with respect to ground is equal to the collector resistor R_C (see Figure 1-11(a)). Thus $R_o = R_C$.

The properties of the four types of differential amplifiers discussed are summarized in Table 1-1.

TABLE 1–1   PROPERTIES OF THE DIFFERENTIAL AMPLIFIER CIRCUIT CONFIGURATIONS

Configuration	Circuit	Voltage gain	Input resistance	Output resistance
1. Dual input, balanced output		$A_d = \dfrac{R_C}{r_e}$	$R_{i1} = 2\beta_{ac}r_e$ $R_{i2} = 2\beta_{ac}r_e$	$R_{o1} = R_C$ $R_{o2} = R_C$
2. Dual input, unbalanced output		$A_d = \dfrac{R_C}{2r_e}$	$R_{i1} = 2\beta_{ac}r_e$ $R_{i2} = 2\beta_{ac}r_e$	$R_o = R_C$
3. Single input, balanced output		$A_d = \dfrac{R_C}{r_e}$	$R_i = 2\beta_{ac}r_e$	$R_{o1} = R_C$ $R_{o2} = R_C$

TABLE 1-1 (Continued)

Configuration	Circuit	Voltage Gain	Input resistance	Output resistance
4. Single input, unbalanced output	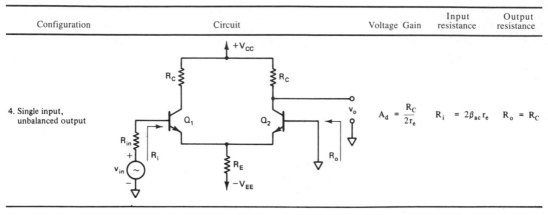	$A_d = \dfrac{R_C}{2r_e}$	$R_i = 2\beta_{ac}r_e$	$R_o = R_C$

1-8 FET DIFFERENTIAL AMPLIFIERS

In the differential amplifier configurations just discussed we have used BJTs. But if we require very high input resistance, we can use FETs instead. Fortunately, the voltage-gain equations derived for these configurations using BJTs can also be used for configurations using FETs, except for the following replacements:

$$R_C \rightarrow R_D$$

$$r_e \rightarrow \frac{1}{g_m}$$

For instance, the voltage gain of the JFET dual-input, balanced-output differential amplifier obtained from Equation (1-12) is

$$A_d = \frac{v_o}{v_{id}} = \frac{R_D}{1/g_m} = g_m R_D \qquad (1\text{-}33)$$

1-9 DIFFERENTIAL AMPLIFIER WITH SWAMPING RESISTORS

By using external resistors R_E' in series with each emitter, the dependence of the voltage gain of the differential amplifiers on variations in r_e can be reduced. Using R_E' also increases the linearity range of the differential amplifiers. Figure 1-12 shows the dual-input, balanced-output differential amplifier with emitter resistors R_E'. Generally, the value of R_E' is large enough to swamp the effect of r_e. For this reason R_E' is sometimes referred to as a *swamping resistor*.

The emitter current in each transistor can be determined by writing Kirchhoff's voltage equation around the base–emitter loop of transistor Q_1 with $v_{\text{in}\,1} = v_{\text{in}\,2} = 0$ V:

$$-R_{\text{in}}I_B - V_{BE} - R_E'I_E - R_E(2I_E) + V_{EE} = 0$$

(Note that we have omitted the subscripts because of symmetry between the two halves of the differential amplifier.)

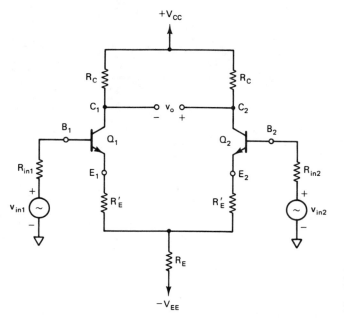

Figure 1–12 Dual-input, balanced-output differential amplifier with emitter resistors R'_E.

Substituting $I_B \cong I_E/\beta_{dc}$ and simplifying, we get

$$I_E = \frac{V_{EE} - V_{BE}}{2R_E + R'_E + R_{in}/\beta_{dc}} \qquad (1\text{-}34)$$

Remember that the emitter current equation (1-34) is also valid for the three remaining differential amplifier configurations with emitter resistor R'_E. The voltage between collector and emitter [Equation (1-4)] remains unchanged and will be repeated here for convenience:

$$V_{CE} = V_{CC} + V_{BE} - R_C I_C$$

When external resistors R'_E are added to each emitter lead, the new voltage gain and new input resistance can be obtained for any configuration simply by replacing r_e by $(r_e + R'_E)$. For instance, the voltage gain of the dual-input, balanced-output differential amplifier of Figure 1-12 becomes

$$A_d = \frac{v_o}{v_{id}} = \frac{R_C}{r_e + R'_E} \qquad (1\text{-}35)$$

and the new input resistance becomes

$$R_{i1} = R_{i2} = 2\beta_{ac}(r_e + R'_E) \qquad (1\text{-}36)$$

Keep in mind that the output resistance with or without R'_E is the same [see Equation (1-17)]. That is,

$$R_{o1} = R_{o2} = R_C$$

Thus the use of R'_E reduces the voltage gain but increases the input resistance, as can be seen from Equations (1-35) and (1-36), respectively.

EXAMPLE 1-1

The following specifications are given for the dual-input, balanced-output differential amplifier of Figure 1-2: $R_C = 2.2 \text{ k}\Omega$, $R_E = 4.7 \text{ k}\Omega$, $R_{in1} = R_{in2} = 50 \Omega$, $+V_{CC} = 10 \text{ V}$, $-V_{EE} = -10 \text{ V}$ and the transistor is the CA3086 with $\beta_{dc} = \beta_{ac} = 100$ and $V_{BE} = 0.715 \text{ V}$ typical.
(a) Determine the I_{CQ} and V_{CEQ} values.
(b) Determine the voltage gain.
(c) Determine the input and output resistances.

SOLUTION (a) To determine I_{CQ} and V_{CEQ} values, we have to substitute the known values in Equations (1-2) and (1-4) as follows:

$$I_{CQ} = I_E = \frac{V_{EE} - V_{BE}}{2R_E + R_{in}/\beta_{dc}} = \frac{10 - 0.715}{9400 + 50/100} = 0.988 \text{ mA}$$

$$V_{CEQ} = V_{CC} + V_{BE} - R_C I_{CQ}$$

$$= 10 + 0.715 - (2.2 \text{ k}\Omega)(0.988 \text{ mA})$$

$$= 8.54 \text{V}$$

(b) The ac emitter resistance

$$r_e = \frac{25 \text{ mV}}{I_E \text{ mA}} = \frac{25 \text{ mV}}{0.988 \text{ mA}} = 25.3 \ \Omega$$

Therefore, substituting the known values in the voltage-gain equation, (1-12), we obtain

$$A_d = \frac{v_o}{v_{id}} = \frac{R_C}{r_e} = \frac{2200}{25.3} = 86.96$$

(c) The input resistance seen from each input source is given by Equation (1-14):

$$R_{i1} = R_{i2} = 2\beta_{ac} r_e = (2)(100)(25.3) = 5.06 \text{ k}\Omega$$

The output resistance seen looking back into the circuit from each of the two output terminals is given by Equation (1-17):

$$R_{o1} = R_{o2} = 2.2 \text{ k}\Omega$$

EXAMPLE 1-2

Repeat Example 1-1 for the (a) dual-input, unbalanced-output and (b) single-input, balanced-output differential amplifiers.

SOLUTION (a) Since the component values remain unchanged and the biasing arrangement is same, the I_{CQ} and V_{CEQ} values as well as input and output resistance values for the dual-input, unbalanced-output configuration must be the same as those for the dual-input, balanced-output configuration (see Table 1-1). Thus

$$I_{CQ} = 0.988 \text{ mA}$$

$$V_{CEQ} = 8.54 \text{ V}$$

$$R_{i1} = R_{i2} = 5.06 \text{ k}\Omega$$

$$R_o = 2.2 \text{ k}\Omega$$

However, the voltage gain of the dual-input, unbalanced-output differential amplifier using Equation (1-21) is

$$A_d = \frac{R_C}{2r_e} = \frac{2200}{(2)(25.3)} = 43.48$$

(b) Because the same biasing arrangement and same component values are used in both configurations, the results obtained in Example 1-1 for the dual-input, balanced-output configuration are also valid for the single-input, balanced-output configuration. That is,

$$I_{CQ} = 0.988 \text{ mA}$$

$$V_{CEQ} = 8.54 \text{ V}$$

$$A_d = 86.96$$

$$R_i = 5.06 \text{ k}\Omega$$

$$R_{o1} = R_{o2} = 2.2 \text{ k}\Omega$$

EXAMPLE 1-3

For the dual-input, balanced-output differential amplifier of Example 1-1:
(a) Determine the output voltage (v_o) if $v_{in1} = 50$ mV peak to peak (pp) at 1 kHz and $v_{in2} = 20$ mV pp at 1 kHz.
(b) What is the maximum peak-to-peak output voltage without clipping?

SOLUTION   (a) In Example 1-1 we have determined the voltage gain of the dual-input, balanced-output differential amplifier. Substituting this voltage gain ($A_d = 86.96$) and the given values of input voltages in Equation (1-11), we get

$$v_o = \frac{R_C}{r_e}(v_{in1} - v_{in2}) = 86.96(50 \text{ mV} - 20 \text{ mV})$$

$$= 2.61 \text{ V pp}$$

(b) Note that in Figure 1-2 the output voltage v_o is measured across the collectors. Therefore, to calculate the maximum peak-to-peak output voltage, we need to determine the voltage drop across each collector resistor:

$$V_{RC} = R_C I_C$$

Substituting $I_C = I_{CQ} = 0.988$ mA, we get

$$V_{RC} = (2.2 \text{ k}\Omega)(0.988 \text{ mA}) = 2.17 \text{ V} < V_{CE} = 8.54 \text{ V} \quad \text{(see Example 1-1)}$$

This means that the maximum change in voltage across each collector resistor is ±2.17 (ideally) or 4.34 V pp. In other words, the maximum peak-to-peak output voltage without clipping is (2)(4.34) = 8.68 V pp.

Remember that the maximum peak-to-peak output voltage without clipping would have been 4.34 V pp if the same circuit were used as the dual-input, unbalanced-output differential amplifier.

EXAMPLE 1-4

Repeat Example 1-1 if 100-Ω resistors R'_E are placed externally in series with each emitter (see Figure 1-12).

SOLUTION (a) Substituting in Equations (1-34) and (1-4), we can determine the values of I_{CQ} and V_{CEQ}, respectively:

$$I_{CQ} = \frac{V_{EE} - V_{BE}}{2R_E + R'_E + R_{in}\beta_{dc}} = \frac{10 - 0.715}{9400 + 100 + 50/100}$$

$$= 0.977 \text{ mA}$$

$$V_{CEQ} = V_{CC} + V_{BE} - R_C I_{CQ}$$

$$= 10 + 0.715 - (2.2 \text{ k}\Omega)(0.977 \text{ mA})$$

$$= 8.57 \text{ V}$$

(b) The ac emitter resistance

$$r_e = \frac{25 \text{ mV}}{I_E \text{ mA}} = \frac{25}{0.977} = 25.59 \ \Omega$$

Therefore, substituting in Equation (1-35), the voltage gain is

$$A_d = \frac{R_C}{R'_E + r_e} = \frac{2200}{100 + 25.59} = 17.52$$

(c) The input resistance is given by Equation (1-36):

$$R_{i1} = R_{i2} = 2\beta_{ac}(R'_E + r_e)$$

$$= (2)(100)(100 + 25.59)$$

$$= 25.12 \text{ k}\Omega$$

The output resistance is independent of R'_E values and therefore is the same as in Example 1-1. That is,

$$R_{o1} = R_{o2} = 2.2 \text{ k}\Omega$$

Comparing the voltage gain and the input resistance values obtained in Examples 1-1 and 1-4, we can verify that the use of R'_E reduces the voltage gain but increases the input resistance significantly.

1-10 CONSTANT CURRENT BIAS

In the differential amplifiers discussed so far the combination of R_E and V_{EE} is used to set up the dc emitter current. We can also use constant current bias circuit to set up the dc emitter current if desired. In fact, the constant current bias circuit is better because it provides current stabilization and, in turn, assures a stable operating point for the differential amplifier. Figure 1-13 shows the dual-input, balanced-output differential amplifier using a resistive constant current bias. Notice that the resistor R_E is replaced by a constant

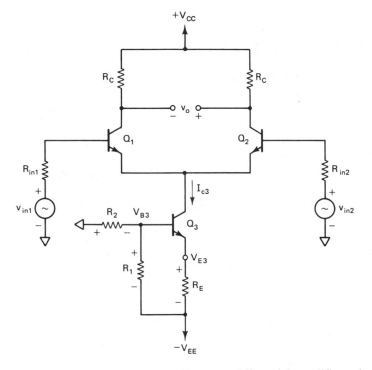

Figure 1–13   Dual-input, balanced-output differential amplifier using constant current bias.

current transistor (Q_3) circuit. The dc collector current in transistor Q_3 is established by resistors R_1, R_2, and R_E and can be determined as follows. Applying the voltage-divider rule, the voltage at the base of transistor Q_3 (neglecting base loading effect) is

$$V_{B3} = \frac{-R_2 V_{EE}}{R_1 + R_2}$$

$$V_{E3} = V_{B3} - V_{BE3} = -\frac{R_2 V_{EE}}{R_1 + R_2} - V_{BE3}$$

Therefore,

$$I_{E3} \cong I_{C3} = \frac{V_{E3} - (-V_{EE})}{R_E}$$

$$I_{C3} = \frac{V_{EE} - [R_2 V_{EE}/(R_1 + R_2)] - V_{BE3}}{R_E}$$

Because the two halves of the differential amplifier are symmetrical, each has half of the current I_{C3}. That is,

$$I_{E1} = I_{E2} \cong \frac{I_{C3}}{2} = \frac{V_{EE} - [R_2 V_{EE}/(R_1 + R_2)] - V_{BE3}}{2R_E} \qquad (1\text{-}37)$$

The collector current I_{C3} in transistor Q_3 is fixed and must be invariant because no signal is injected into either the emitter or the base of Q_3. Thus the transistor Q_3 is a source of constant emitter current for transistors Q_1 and Q_2 of the differential amplifier (see Figure 1-13).

Recall that, in the analysis of the differential amplifier circuits with emitter bias, we required that $R_E \gg r_e$. Besides supplying constant emitter current, the constant current bias also provides a very high source resistance since the ac equivalent of the dc current source is ideally an open circuit. Therefore, all the performance equations obtained for the differential amplifier configurations using emitter bias are also applicable to differential amplifiers with constant current bias. For instance, the voltage-gain equation, (1-12), is valid for the differential amplifier of Figure 1-2 as well as for that in Figure 1-13 with constant current bias.

To improve the thermal stability of the constant current transistor Q_3, resistor R_1 is replaced by diodes D_1 and D_2, as shown in Figure 1-14(a). The base of Q_3 is biased with the voltage divider containing components R_2, D_1, and D_2. Diodes D_1 and D_2 help to hold the emitter current I_{E3} constant even though the temperature changes. Note that I_2 flows to the node at the base of Q_3 and then divides into paths I_D and I_{B3}. If the temperature of Q_3 increases, its base–emitter voltage V_{BE3} decreases. In silicon units, V_{BE} decreases 2 mV/°C, and in germanium units, V_{BE} decreases 1.6 mV/°C. This reduced V_{BE3} tends to raise the voltage drop across R_E and, in turn, current I_{E3}. However, the voltage drops across D_1 and D_2 also decrease, causing a greater portion of I_2 to contribute to I_D, that is, causing I_D to increase. This causes I_{B3} to decrease, which prevents any significant increase in I_{E3}.

For better performance, a transistor array such as CA3086 may be used in Figure 1-14(a) as a constant current bias. Here an isolated transistor is used, and needed diodes are formed by using transistors connected for diode operation, although discrete diodes will work as well. The functional diagram of the CA3086 is shown in Figure 1-14(b).

The emitter current I_{E3} in Figure 1-14(a) is determined as follows. Assuming that the voltage drop across the diodes is equal and denoting it by V_D, the voltage at the base of transistor Q_3 is

$$V_{B3} = -V_{EE} + 2V_D$$

$$V_{E3} = V_{B3} - V_{BE3} = -V_{EE} + 2V_D - V_{BE3}$$

$$I_{E3} = \frac{V_{E3} - (-V_{EE})}{R_E} = \frac{2V_D - V_{BE3}}{R_E}$$

Assuming that the transistor Q_3 has the same characteristics as diodes D_1 and D_2, that is, if $V_D = V_{BE3}$, then

$$I_{E3} = \frac{V_D}{R_E} \tag{1-38}$$

Thus for a given value of R_E the emitter current I_{E3} depends on the voltage drop across the diodes. But the voltage drop across the diodes is a function of the current (I_D) through them (according to the diode characteristic curve). However, current I_D is a part of current I_2, which is determined by the value of R_2. This means that we can change the value of I_{E3} by varying either the R_2 or the R_E [see Equation (1-38)].

(a)

(b)

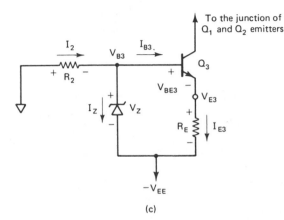

(c)

Figure 1–14   (a) Constant current bias with diode compensation for variations in V_{BE}. (b) Functional diagram of CA3086. (c) Constant current bias using zener.

To design the constant current bias circuit of Figure 1-14(a), we will use the following steps:

1. Choose a desired I_{E3} value.
2. Assuming that $V_D = 0.7$ V, determine the value of R_E using Equation (1-38).
3. Assuming that $I_2 = I_{E3}$ and $V_{D1} = V_{D2} = 0.7$ V, determine the value of R_2 from the following relation:

$$R_2 = \frac{V_{EE} - 1.4\ \text{V}}{I_{E3}} \qquad (1\text{-}39)$$

Often a zener diode is used in place of the diodes as shown in Figure 1-14(c). Zeners are available over a wide range of voltages and can have a matching temperature coefficient of voltage to those of transistors.

Referring to the circuit of Figure 1-14(c), the voltage at the base of transistor Q_3 is

$$V_{B3} = -V_{EE} + V_Z$$

Then the voltage at the emitter is

$$V_{E3} = -V_{EE} + V_Z - V_{BE3}$$

Therefore, the current I_{E3} is

$$I_{E3} = \frac{V_{E3} - (-V_{EE})}{R_E}$$

Substituting and simplifying, we obtain

$$I_{E3} = \frac{V_Z - V_{BE3}}{R_E} \qquad (1\text{-}40)$$

The value of R_2 should be selected so that $I_2 \cong 1.2 I_{Zt}$, where I_{Zt} is the minimum current required to cause the zener diode to conduct in the reverse region, that is, to block the rated voltage V_Z. The value of I_{Zt} is specified on the data sheet of a zener diode. Thus the value of R_2 can be calculated using the equation

$$R_2 = \frac{V_{EE} - V_Z}{I_2} \qquad (1\text{-}41)$$

where $I_2 \cong 1.2 I_{Zt}$. The conclusion is quite simple: The zener diode is useful for maintaining a constant base voltage and, in turn, the constant emitter current in a constant current bias circuit.

1-11 CURRENT MIRROR

The circuit in which the output current is forced to equal the input current is said to be a current mirror circuit. Thus in a current mirror circuit the output current is a mirror image of the input current. The block diagram and the circuit of the current mirror are shown in Figure 1-15(a) and (b), respectively. Once the current I_2 is set up, the current I_{C3} is automatically established to be nearly equal to I_2. The current mirror is a special case of constant current bias and therefore can be used to set up constant emitter currents in differential amplifier stages. Note that the current mirror circuit requires fewer compo-

Figure 1–15   (a) Block diagram of current mirror. (b) Current mirror circuit.

nents than constant current bias circuits. Because of its simplicity and ease of fabrication, the current mirror circuit is most commonly used in integrated amplifiers such as differential and operational amplifiers.

Remember that for the proper operation of the circuit it is necessary to construct a current mirror using a transistor array such as the CA3086 in which a needed diode is formed by using an adjacent transistor [see Figure 1-15(b)]. When used this way, the transistor array helps to achieve the required collector current thermal stability in a current mirror circuit.

Since Q_3 and Q_4 are identical transistors, their base–emitter voltages must be the same, and their base and collector currents must also be approximately equal:

$$V_{BE3} = V_{BE4}$$

$$I_{C3} = I_{C4}$$

$$I_{B3} = I_{B4}$$

Summing currents at node V_{B3}, we obtain

$$I_2 = I_{C4} + I$$

$$= I_{C4} + 2I_{B4} = I_{C3} + 2I_{B3}$$

$$= I_{C3} + 2\left(\frac{I_{C3}}{\beta_{dc}}\right)$$

$$I_2 = I_{C3}\left(1 + \frac{2}{\beta_{dc}}\right) \tag{1-42}$$

Generally, β_{dc} is large enough so that $2/\beta_{dc}$ is negligibly small. For instance, $\beta_{dc} = 100$ for the CA3086. Thus the collector current of Q_3 is nearly equal to the current I_2.

$$I_2 \cong I_{C3} \tag{1-43}$$

The current I_2 is obtained by writing the Kirchhoff's voltage equation for the base–emitter loop of transistor Q_3:

$$-R_2 I_2 - V_{BE3} + V_{EE} = 0$$

Rearranging, we get

$$I_2 = \frac{V_{EE} - V_{BE3}}{R_2} \tag{1-44}$$

Thus using Equation (1-44), an appropriate value of R_2 can be chosen to set a desired value of collector current in a current mirror circuit. Remember that, for satisfactory operation of the circuit, two identical transistors are necessary.

EXAMPLE 1-5

The following specifications are given for the JFET differential amplifier shown in Figure 1-16: $R_D = 3.9$ kΩ, $R_G = 1$ MΩ, $R_E = 1.5$ kΩ, $R_2 = 82$ Ω, $R_{in\,1} = R_{in\,2} = 50$ Ω, $V_{DD} = +10$ V, and $-V_{EE} = -10$ V. Assume that Q_1 and Q_2 are matched JFETs with $g_m = 8000$ μs, Q_3 is a BJT with $V_{BE} = 0.6$ V and $\beta_{ac} = \beta_{dc} = 100$, and D_1 is the 1N3826 zener with $V_Z = 5.1$ V and $I_{Zt} = 49$ mA. Determine:
(a) The emitter current in transistor Q_3.
(b) The voltage gain of the differential amplifier.
(c) The input resistance of the differential amplifier.

Figure 1–16 JFET differential amplifier with zener constant current bias.

SOLUTION   (a) Using Equation (1-40), we can determine the emitter current in transistor Q_3.

$$I_{E3} = \frac{V_Z - V_{BE3}}{R_E} = \frac{5.1 - 0.6}{1500} = 3 \text{ mA}$$

(b) The voltage gain of the JFET dual-input, balanced-output differential amplifier is given by

$$A_d = g_m R_D = (8000 \ \mu s)(3.9 \text{ k}\Omega) = 31.2$$

(c) The input resistance seen from either input source is equal to the gate resistance R_G. That is,

$$R_{i1} = R_{i2} = 1 \text{ M}\Omega$$

EXAMPLE 1-6

Design a dual-input, balanced-output differential amplifier with a constant current bias (using diodes) to satisfy the following requirements:

Differential voltage gain $A_d = 40 \pm 10$

Current supplied by the constant current bias circuit = 4 mA

Supply voltages $V_S = \pm 10$ V

SOLUTION   First, we design a constant current bias circuit following the steps outlined in Section 1-10.

1. The desired value of emitter current $I_{E3} = 4$ mA (see Figure 1-17).

2. We assume that the diodes D_1 and D_2 are identical and that $V_{D1} = V_{D2} = 0.7$ V. Rearranging Equation (1-38), we get

$$R_E = \frac{V_D}{I_{E3}}$$

Substituting yields

$$R_E = \frac{0.7 \text{ V}}{4 \text{ mA}} = 175 \ \Omega \quad (\text{use } 150 \ \Omega)$$

3. The value of R_2 can be calculated using Equation (1-39):

$$R_2 = \frac{V_{EE} - 1.4}{I_{E3}} = \frac{10 \text{ V} - 1.4 \text{ V}}{4 \text{ mA}} = 2.15 \text{ k}\Omega \quad (\text{use } 2 \text{ k}\Omega)$$

Next, we determine the value of the collector resistor R_C using Equation (1-12). Since $I_{E3} = 4$ mA,

$$I_{E1} = I_{E2} = 2 \text{ mA} \quad \text{and} \quad r_e = \frac{25 \text{ mV}}{2 \text{ mA}} = 12.5 \ \Omega$$

Figure 1–17 Differential amplifier of Example 1–6.

Therefore, to obtain the differential gain of 40, the required value of the collector resistor is

$$R_C = (A_d)r_e \quad \text{[from Equation (1-12)]}$$
$$= (40)(12.5) = 500 \ \Omega \quad (\text{use } 560 \ \Omega)$$

Figure 1-17 shows the dual-input, balanced-output differential amplifier with the designed component values. Note that for proper operation of the amplifier circuit the use of a transistor array such as CA3086 is recommended.

EXAMPLE 1-7

Design the single-input, balanced-output differential amplifier shown in Figure 1-18(a) to meet the following specifications:

Input resistance $R_i \geq 600 \ \text{k}\Omega$

Peak-to-peak output voltage swing $\leq 5 \ \text{V}$

Supply voltages $V_s = \pm 10 \ \text{V}$

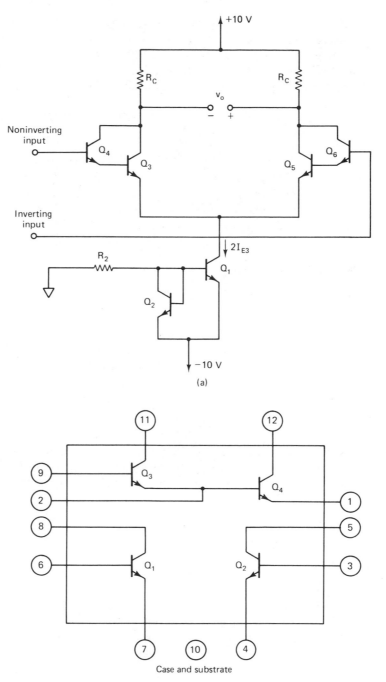

Figure 1–18   (a) Darlington-pair differential amplifier of Example 1–7.
(b) CA3018 transistor array (metal can).

SOLUTION   The circuit shown in Figure 1-18(a) requires two Darlington transistor pairs. We will choose the CA3018 or SK3542 silicon transistor array, which consists of a Darlington-connected transistor pair and two isolated transistors, as shown in Figure 1-18(b). The isolated transistors can be used to form the current mirror circuit. Remember that we need two transistor arrays. The following specifications apply for each transistor in the CA3018 transistor array: $h_{FE} = 100$, $h_{fe} = 110$, and $V_{BE} = 0.715$ V. The desired input resistance for the single-input, balanced-output configuration is 600 kΩ. Therefore, using Equation (1-30), for the Darlington pair

$$R_i = 2\beta_{acT}(2r_{e3})$$

where β_{acT} = total current gain of the Darlington pair
        = $(110)(110)$ = 12.1 k
    $2r_{e3}$ = ac emitter resistance of the Darlington pair $Q_3 Q_4 = Q_5 Q_6$
Substituting yields

$$600 \text{ k}\Omega = (2)(12.1 \text{ k})(2r_{e3})$$

Therefore, the ac emitter resistance of transistor $Q_3 (= Q_5)$ is

$$r_{e3} = \frac{600 \text{ k}\Omega}{48.2 \text{ k}} = 12.4 \ \Omega$$

But

$$r_{e3} = \frac{25 \text{ mV}}{I_{E3} \text{ mA}}$$

Therefore,

$$I_{E3} = \frac{25 \text{ mV}}{12.4 \ \Omega} = 2.02 \text{ mA}$$

Let us choose $I_{E3} = I_{E5} = 2$ mA. This means that the current mirror circuit must supply $2I_{E3} = 4$ mA. To form the current mirror circuit we can use isolated transistors in one of the two CA3018s, and the value of R_2 needed can be calculated by using Equation (1-44):

$$R_2 = \frac{V_{EE} - V_{BE3}}{I_2}$$

Substituting, we obtain

$$R_2 = \frac{10 \text{ V} - 0.715 \text{ V}}{4 \text{ mA}} = 2.32 \text{ k}\Omega \quad (\text{use } 2.2 \text{ k}\Omega)$$

Assume that the voltage drop across collector resistor R_C determines the peak-to-peak output voltage swing. The expected output voltage swing in the circuit of Figure 1-18(a) is 5 V pp. Since the output is "differential," the voltage drop across each collector resistor is 2.5 V pp or ± 1.25 V. In other words, $R_C I_C = 1.25$ V. Substituting, $I_C = 2$ mA, we get

$$R_C = 625 \ \Omega \quad (\text{use } 620 \ \Omega)$$

Thus for the single-input, balanced-output differential amplifier of Figure 1-18(a), we need the following components:

$$\text{Two transistor arrays:}\quad \text{CA3018}$$
$$R_C = 620\ \Omega$$
$$R_2 = 2.2\ \text{k}\Omega$$

1-12 CASCADED DIFFERENTIAL AMPLIFIER STAGES

Having gone through an analysis of the differential amplifier circuit configurations, let us see how differential amplifiers can be cascaded (connected in series). Figure 1-19 shows a typical two-stage differential amplifier. The first stage is a dual-input, balanced-output differential amplifier. The second stage is another differential amplifier driven by the output of the first differential amplifier. A single-ended (unbalanced) output is taken from this second stage. Both stages use the emitter biasing technique (the combination of R_E and V_{EE}) to set up the emitter currents in the differential pairs.

In many cases the matching of transistors as well as resistor values is essential for proper operation of the differential stages. For this reason the use of transistor arrays such

Figure 1–19   Cascaded differential amplifier of Example 1–8.

as CA3086 (or LM3146) is recommended. A transistor array such as CA3086 provides inherent advantages unique to integrated circuits: electrical and thermal matching, compactness, and ease of physical handling.

EXAMPLE 1-8

For the cascaded differential amplifier shown in Figure 1-19, determine:
(a) The collector current and collector-to-emitter voltage for each transistor.
(b) The overall voltage gain.
(c) The input resistance.
(d) The output resistance.
Assume that the transistor array CA3086 is used (for which $h_{FE} = 100$ typical, $h_{fe} = 100$, and $V_{BE} = 0.715$ V). The pin diagram of the CA3086 is shown in Figure 1-14(b).

SOLUTION

(a) To determine the collector current and collector-to-emitter voltage of transistors Q_1 and Q_2, we assume that the inverting and noninverting inputs are grounded. The collector current ($I_C \cong I_E$) in Q_1 and Q_2 is obtained by substituting the known values in Equation (1-2):

$$I_{E1} = \frac{V_{EE} - V_{BE}}{2R_{E1} + R_{in}/\beta_{dc}} = \frac{10 - 0.715}{2(4700) + 0} = 0.988 \text{ mA}$$

That is, $I_{C1} = I_{C2} = 0.988$ mA.

We can calculate the voltage between collector and emitter for Q_1 and Q_2 using the collector current as follows:

$$V_{C1} = V_{CC} - R_{C1}I_{C1} = 10\text{V} - (2.2 \text{ k}\Omega)(0.988 \text{ mA}) = 7.83 \text{ V} = V_{C2}$$

Since the voltage at the emitter of Q_1 and Q_2 is -0.715 V,

$$V_{CE1} = V_{CE2} = V_{C1} - V_{E1} = 7.83 + 0.715 = 8.545 \text{ V}$$

Next, we will determine the collector current in Q_3 and Q_4 by writing the Kirchhoff's voltage equation for the base–emitter loop of transistor Q_3:

$$V_{CC} - R_{C2}I_{C2} - V_{BE3} - R_E' I_{E3} - R_{E2}(2I_{E3}) + V_{EE} = 0$$

$$10 \text{ V} - (2.2 \text{ k}\Omega)(0.988 \text{ mA}) - 0.715 \text{ V} - (100)(I_{E3}) - (30 \text{ k}\Omega)I_{E3} + 10 \text{ V} = 0$$

$$10 \text{ V} - 2.17 \text{ V} - 0.715 \text{ V} + 10 \text{ V} - (30.1 \text{ k}\Omega)I_{E3} = 0$$

$$I_{E3} = \frac{17.12 \text{ V}}{30.1 \text{ k}\Omega} = 0.569 \text{ mA} = I_{E4}$$

Hence the voltage at the collector of Q_3 and Q_4 is

$$V_{C3} = V_{C4} = V_{CC} - R_{C3}I_{C3} = 10 \text{ V} - (1.2 \text{ k}\Omega)(0.569 \text{ mA})$$
$$= 9.32 \text{ V}$$

Therefore,

$$V_{CE3} = V_{CE4} = V_{C3} - V_{E3} = 9.32 - 7.12 = 2.2 \text{ V}$$

Thus, for Q_1 and Q_2,

$$I_{CQ} = 0.988 \text{ mA}$$
$$V_{CEQ} = 8.545 \text{ V}$$

and for Q_3 and Q_4,

$$I_{CQ} = 0.569 \text{ mA}$$
$$V_{CEQ} = 2.2 \text{ V}$$

Note that the output terminal (V_{C4}) is at 9.32 V and not at zero volts.

(b) First, we calculate the ac emitter resistance r_e of each stage and then its voltage gain.

$$r_{e1} = \frac{25 \text{ mV}}{I_{E1}} = \frac{25 \text{ mV}}{0.988 \text{ mA}} = 25.3 \ \Omega = r_{e2}$$

$$r_{e3} = \frac{25 \text{ mV}}{I_{E3}} = \frac{25 \text{ mV}}{0.569 \text{ mA}} = 43.94 \ \Omega = r_{e4}$$

The first stage is a dual-input, balanced-output differential amplifier; therefore, its voltage gain is

$$A_{d1} = \frac{v_{o1}}{v_{id}} = \frac{R_{C1} \| R_{i2}}{r_{e1}}$$

where

$$R_{i2} = \text{input resistance of the second stage}$$
$$= 2\beta_{ac}(r_{e3} + R_E')$$
$$= (200)(143.94) = 28.79 \text{ k}\Omega$$
$$A_{d1} = \frac{2.2 \text{ k}\Omega \ \| \ 28.79 \text{ k}\Omega}{25.3} = 80.78$$

The second stage is a dual-input, unbalanced-output differential amplifier with swamping resistor R_E', the voltage gain of which is

$$A_{d2} = \frac{v_o}{v_{o1}} = \frac{R_{C4}}{2(R_E' + r_{e4})} = \frac{1.2 \text{ k}\Omega}{287.88} = 4.17$$

Hence the overall voltage gain is

$$A_d = (A_{d1})(A_{d2}) = (80.78)(4.17) = 336.85$$

Thus we can obtain a higher voltage gain by cascading differential amplifier stages.

(c) The input resistance of the cascaded differential amplifier is the same as the input resistance of the first stage, that is,

$$R_{i1} = 2\beta_{ac}(r_{e1}) = (200)(25.3) = 5.06 \text{ k}\Omega$$

(d) The output resistance of the cascaded differential amplifier is the same as the output resistance of the last stage. Hence, using Equation (1-17),

$$R_{o2} = R_{C4} = 1.2 \text{ k}\Omega$$

1-13 LEVEL TRANSLATOR

From the results of the cascaded differential amplifier of Example 1-8, the following observations can be made:

1. Because of the direct coupling, the dc level at the emitters rises from stage to stage. This increase in dc level tends to shift the operating point of the succeeding stages and, therefore, limits the output voltage swing and may even distort the output signal. For example, in the circuit of Figure 1-19, the voltage at the emitters of Q_1 and Q_2 of the first stage is -0.715, whereas that at the emitters of Q_3 and Q_4 of the second stage is 7.12 V.

Figure 1–20   Level translator circuits. (a) Emitter follower with voltage divider. (b) Emitter follower with constant current bias. (c) Emitter follower with current mirror.

If used as a single stage, the voltage at the emitters of Q_3 and Q_4 would have been −0.715 V. Thus the increase in the emitter voltage of the second stage and in turn the change in its operating point are due to the cascading of differential amplifier stages.

2. The voltage at the output terminal of the second stage in Figure 1-19 is well above ground (0 V). This dc level is undesirable because it tends to limit the peak-to-peak output voltage swing without distortion and also contributes to the error in the dc output signal.

Therefore, in Figure 1-19 a final stage should be included to shift the output dc level at the second stage down to about zero volts to ground. Such a stage is referred to as a *level translator* or *shifter*. Thus, in the cascaded differential amplifier, to shift the output dc level down to zero volts, the final stage must be followed by a level translator circuit. Although there are a variety of level translator circuits, an emitter follower with a voltage divider is the simplest among them [see Figure 1-20(a)]. The output of the second differential stage in Figure 1-19 can work into the voltage-divider emitter-follower stage of Figure 1-20(a). Thus a positive 9.32 at the output terminal (V_{C4}) of the second stage in Figure 1-19 can produce zero volts at the junction of R_1 and R_2 with proper selection of components. Better results are obtained by using an emitter follower either with a diode constant current bias or a current mirror instead of the voltage divider, as shown in Figure 1-20(b) and (c), respectively. Note that the resistive or zener constant curent bias can also be used with the emitter follower in forming a level translator.

EXAMPLE 1-9
Design a level translator circuit as shown in Figure 1-20(b) for the cascaded differential amplifier in Figure 1-19.

SOLUTION We need to shift 9.32 V at the collector of Q_4 down to zero volts at the collector of Q_6 [see Figures 1-19 and 1-20(b)]. Following the design steps listed in Section 1-10, we first design the diode constant current bias circuit and then select an appropriate value for R so that the voltage at the collector of Q_6 is zero volts. Let us choose $I_{E6} = 3$ mA and $V_S = \pm 10$ V. Therefore, using Equation (1-38), the value of R_E is

$$R_E = \frac{V_D}{I_{E6}} = \frac{0.7}{3 \text{ mA}} = 233.3 \ \Omega \quad (\text{use } 270 \ \Omega)$$

Substituting the known values in Equation (1-39) yields

$$R_2 = \frac{V_{EE} - 1.4 \text{ V}}{I_{E6}} = \frac{10 \text{ V} - 1.4 \text{ V}}{3 \text{ mA}} = 2.87 \text{ k}\Omega \quad (\text{use } 2.7 \text{ k}\Omega)$$

Assuming that $V_{BE5} = 0.7$ V, the voltage at the emitter of Q_5 is

$$V_{E5} = V_{B5} - V_{BE5} = 9.32 - 0.7 = 8.62 \text{ V}$$

Since the voltage at the collector Q_6 is to be zero and $I_{E5} = I_{C6} \cong 3$ mA, the required value of R is

$$R = \frac{V_{E5}}{I_{E5}} = \frac{8.62 \text{ V}}{3 \text{ mA}} = 2.87 \text{ k}\Omega \quad (\text{use a 3-k}\Omega \text{ potentiometer})$$

Thus the desired level translator circuit components are

$$R = 3\text{-k}\Omega \text{ potentiometer}$$
$$R_E = 270\ \Omega$$
$$R_2 = 2.7\ \text{k}\Omega$$
$$D_1, D_2, Q_5, \text{ and } Q_6 = \text{CA3086}$$

EXAMPLE 1-10

For the cascaded differential amplifier of Figure 1-21, determine:
(a) The I_{CQ} and V_{CEQ} for each transistor.
(b) The overall voltage gain.
(c) The maximum peak-to-peak output voltage swing.
(Assume that the CA3086 transistor array is used for which $\beta_{ac} = \beta_{dc} = 100$ and $V_{BE} = 0.715$ V.)

SOLUTION   The circuit in Figure 1-21 consists of three stages:

1. Dual-input, balanced-output differential amplifier
2. Dual-input, unbalanced-output differential amplifier
3. Level translator circuit

All the stages use current mirror biasing. Especially notice the way current mirror transistors Q_4, Q_7, and Q_9 are connected. The bases of these transistors are tied together to the base of transistor Q_3. For this reason the collector currents in these transistors (Q_3, Q_4, Q_7, and Q_9) are equal and are set up by a 2.2-kΩ base resistor R_2 (see Figure 1-21).

Another interesting point about this circuit is that no collector resistor is used for Q_5 since the output is single-ended and is measured at the collector of transistor Q_6. To determine the dc conditions, we assume that the input terminals are shorted to ground.
(a) We can determine the collector currents in current mirror transistors Q_4, Q_7, and Q_9 using Equation (1-44):

$$I_2 = \frac{V_{EE} - V_{BE3}}{R_2}$$

Substituting gives us

$$I_2 = \frac{10\ \text{V} - 0.715\ \text{V}}{2.2\ \text{k}\Omega} = 4.22\ \text{mA}$$

This means that

$$I_{CQ1} = I_{CQ2} = I_{CQ5} = I_{CQ6} = 2.11\ \text{mA}$$

and

$$I_{CQ8} = 4.22\ \text{mA}\ \text{ since } I_{E8} = I_{C9}$$

Knowing these collector currents, we can calculate the collector-to-emitter voltages as follows:

$$V_{C1} = V_{C2} = V_{CC} - R_{C1}I_{C1} = 10\ \text{V} - (2.7\ \text{k}\Omega)(2.11\ \text{mA}) = 4.3\ \text{V}$$

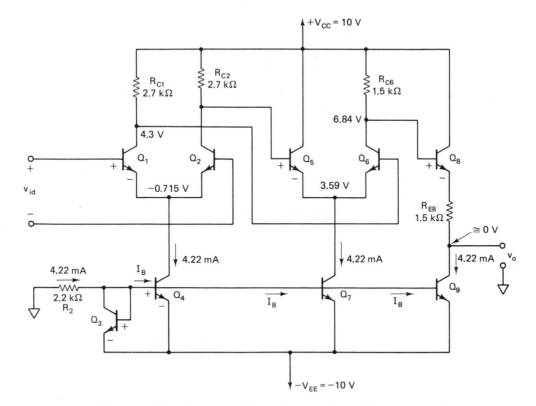

Figure 1–21   Cascaded differential amplifier of Example 1–10.

Therefore,

$$V_{CEQ1} = V_{CEQ2} = 4.3 - (-0.715) = 5.015 \text{ V}$$
$$V_{E5} = V_{E6} = (4.3 - 0.715) = 3.59 \text{ V}$$
$$V_{C6} = V_{CC} - R_{C6}I_{C6} = 10 \text{ V} - (1.5 \text{ k}\Omega)(2.11 \text{ mA}) = 6.84 \text{ V}$$

Hence,

$$V_{CEQ5} = 10 - 3.59 = 6.41 \text{ V}$$
$$V_{CEQ6} = 6.84 - 3.59 = 3.25 \text{ V}$$

Finally,

$$V_{E8} = 6.84 - 0.715 = 6.125 \text{ V} \quad \text{and} \quad V_{C8} = 10 \text{ V}$$

Therefore,

$$V_{CEQ8} = 10 - 6.125 = 3.875 \text{ V}$$

(b) The ac emitter resistances are

$$r_{e1} = r_{e6} = \frac{25 \text{ mV}}{I_{E1}} = \frac{25 \text{ mV}}{2.11 \text{ mA}} = 11.85 \text{ }\Omega$$

$$r_{e8} = \frac{25\ \text{mV}}{I_{E8}} = \frac{25\ \text{mV}}{4.22\ \text{mA}} = 5.92\ \Omega$$

and input resistance of the second stage is

$$R_{i2} = 2\beta_{ac}r_{e6} = (200)(11.85) = 2.37\ \text{k}\Omega$$

Therefore, the voltage gain of the first stage is

$$A_{d1} = \frac{R_{C1}\|R_{i2}}{r_{e1}} = \frac{2.7\ \text{k}\Omega\|2.37\ \text{k}\Omega}{11.85} = 106.5$$

The input resistance of the third stage (looking into the base of Q_8) is

$$R_{i3} = \beta_{ac}(r_{e8} + R_{E8}) = (100)(5.92\ \Omega + 1.5\ \text{k}\Omega) = 150.6\ \text{k}\Omega$$

Hence the voltage gain of the second stage is

$$A_{d2} = \frac{R_{C6}\|R_{i3}}{2r_{e6}} = \frac{1.5\ \text{k}\Omega\|150.6\ \text{k}\Omega}{(2)(11.85)\ \Omega} = 62.66$$

The voltage gain of the third stage is

$$A_{d3} \cong 1$$

Thus the overall gain of the circuit in Figure 1-21 is

$$A_d = (A_{d1})(A_{d2})(A_{d3}) = (106.5)(62.66)(1) = 6673.3$$

(c) The third stage in Figure 1-21 is the level translator circuit and, in addition, the voltage gain of this stage is unity. Therefore, the maximum peak-to-peak output voltage swing is determined by the second stage, specifically the voltage drop across the collector resistor R_{C6}. The voltage drop across collector resistor R_{C6} is

$$V_{RC6} = (1.5\ \text{k}\Omega)(2.11\ \text{mA}) = 3.17\ \text{V}$$

Hence the maximum peak-to-peak output voltage swing is

$$v_o = \pm 3.17\ \text{V} \quad \text{or} \quad 6.34\ \text{V pp}$$

1-14 CASCODE OR CE–CB CONFIGURATION

Each of the small-signal amplifier configurations, common emitter (CE), common base (CB), and common collector (CC), has some unique advantages as well as drawbacks. The specific property of the amplifier configuration can be enhanced, a specific drawback can be removed, or a special feature can be added by connecting two configurations. Because there are three configurations, we can form nine possible combinations, each having two configurations. But, unfortunately out of these nine possible combinations, some do not improve the amplifier's performance in any way and are, therefore, not practical. The most commonly used combinational configurations are the following:

1. Common collector–common collector (CC–CC) connection
2. Common collector–common emitter (CC–CE) connection
3. Common collector–common base (CC–CB) connection

4. Common emitter–common emitter (CE–CE) connection
5. Common emitter–common base (CE–CB) connection

The Darlington pair can be used to obtain a CC–CC configuration. The Darlington pair is a composite two-transistor device in which the collectors of two transistors are tied together and the base of the second transistor is driven by the emitter of the first. The principal advantages of this configuration are that the input resistance and the current gain are increased by a factor β over those in the standard CC configuration.

Another example of the CC–CC configuration is the double follower circuit in which complementary transistors are connected in the CC–CC configuration. The main features of this circuit are that the output voltage is equal to the input voltage and that a very high current gain is provided.

To obtain the CC–CE configuration, we can use the Darlington pair in place of a single transistor in CE configuration. In this configuration the current gain and the input resistance are improved by a factor β over that of the CE amplifier. Note that these increases are achieved at the sacrifice of a slight decrease in voltage gain.

The single-input, unbalanced-output differential amplifier of Figure 1-10 utilizes the CC–CB configuration. This configuration basically has the same input characteristics as the CC amplifier and the same output characteristics as the CB amplifier.

If the main objective is to obtain a very large voltage or power gain, we can use the CE–CE configuration. The complementary pair can also be utilized in forming the CE–CE configuration. Recall that in forming the dual-input, balanced output differential amplifier, we also have to use two symmetrical CE sections (see Figure 1-2). Thus the CE–CE is the most commonly used configuration.

The CE–CB configuration (usually referred to as the *cascode amplifier*) is shown in Figure 1-22. Note that the CE and CB stages are direct-coupled. This configuration, of course, has basically the same input characteristics as the CE amplifier and the same output characteristics as the CB amplifier. More specifically, it has high output resistance and is inherently more stable. The high output resistance attainable is useful in achieving large amounts of voltage gain. Also, in this configuration no high-frequency feedback occurs from the output back to the input, and the input Miller capacitance effect is at a minimum because the voltage gain of the CE configuration is very low. Therefore, a CE–CB configuration is inherently more stable and hence ideally suited for high-frequency applications.

1-14.1 DC Analysis

In the cascode amplifier of Figure 1-22, R_1, R_2 and R_3 form a bias network for Q_1 and Q_2; C_{S1} and C_{S2} provide ac signal ground paths from the Q_1 emitter and Q_2 base, respectively; and C_i is the coupling capacitor. For the proper operation of the circuit we require that transistors Q_1 and Q_2 be identical. Keep in mind that for dc conditions all capacitors are assumed to be open-circuit.

Because of identical transistors and the direct coupling between the CE–CB stages,

$$I_{E1} = I_{E2} \quad \text{or} \quad I_{C1} = I_{C2} \quad \text{and} \quad I_{B1} = I_{B2}$$

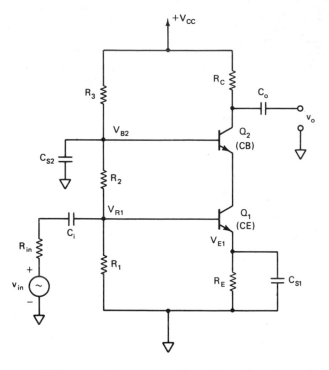

Figure 1–22 Cascode amplifier.

This means that we need to determine the emitter current of Q_1 only. Applying the voltage-divider rule, we get

$$V_{R1} = \frac{R_1 V_{CC}}{(R_1 + R_2 + R_3)} \tag{1-45}$$

$$V_{E1} = V_{R1} - V_{BE1}$$

Therefore,

$$I_{E1} = \frac{V_{E1}}{R_E}$$

Substituting yields

$$I_{E1} = \frac{[(R_1)V_{CC}/(R_1 + R_2 + R_3)] - V_{BE1}}{R_E} \tag{1-46}$$

Thus once we know the emitter current of Q_1 we can determine all the currents in and voltages at the three transistor terminals for Q_1 and Q_2 if desired.

1-14.2 AC Analysis

The small-signal T-equivalent circuit for the cascode amplifier is shown in Figure 1-23. Since we are considering only the low-frequency performance, we assume all capacitors as short-circuit elements. We will determine the low-frequency, small-signal properties of the cascode circuit: the voltage gain, current gain, and input resistance.

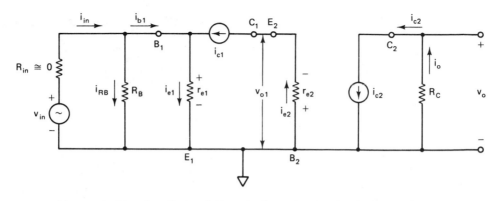

Figure 1–23   Small-signal T-equivalent circuit of cascode amplifier.

Note that, in Figure 1-23, $R_B = R_1 \| R_2$; $r_{e1} \cong r_{e2}$ since $I_{E1} \cong I_{E2}$; and v_{o1} is the output voltage of the CE stage. Since Q_1 and Q_2 are identical transistors, $\beta_{dc1} = \beta_{dc2}$ and $\beta_{ac1} = \beta_{ac2}$; therefore, we will omit subscripts in the calculations that follow.

1-14.2(a) Voltage gain.   For the input circuit in Figure 1-23,

$$v_{in} = (r_{e1})(i_{e1}) \text{ since } R_B \gg r_{e1}$$

$$v_{o1} = -(r_{e2})(i_{e2})$$

Since $i_{e2} = i_{c1}$ and $i_{c1} \cong i_{e1}$, implies that $i_{e2} \cong i_{e1}$. Therefore, the voltage gain A_{v1} of the CE stage is

$$A_{v1} = \frac{v_{o1}}{v_{in}} = -\frac{(r_{e2})(i_{e2})}{(r_{e1})(i_{e1})} = -1 \text{ since } r_{e1} = r_{e2} \qquad (1\text{-}47)$$

The output voltage

$$v_o = -(R_C)(i_o)$$

but

$$i_o = i_{c2} = i_{e2} \quad \text{and} \quad i_{e2} = -\frac{v_{o1}}{r_{e2}}$$

Therefore, substituting the value of i_{e2} gives us

$$v_o = -(R_C)\left(-\frac{v_{o1}}{r_{e2}}\right)$$

Then the voltage gain of the CB stage is

$$\frac{v_o}{v_{o1}} = \frac{R_C}{r_{e2}} \qquad (1\text{-}48)$$

Thus the overall voltage gain of the CE–CB configuration is

$$A_v = \frac{v_o}{v_{in}} = \frac{v_{o1}}{v_{in}}\frac{v_o}{v_{o1}} = -\frac{R_C}{r_{e2}} \qquad (1\text{-}49)$$

Note that $r_{e1} = r_{e2}$.

1-14.2(b) Current gain.   The current gain A_i of the CE–CB configuration can be expressed as

$$A_i = \frac{i_o}{i_{in}} = \frac{i_{b1}}{i_{in}}\left(\frac{i_{e1}}{i_{b1}}\right)\frac{i_{e2}}{i_{e1}}\left(\frac{i_{c2}}{i_{e2}}\right)\frac{i_o}{i_{c2}}$$

but

$$i_{e1} = i_{c1} = i_{e2} = i_{c2} = i_o$$

Therefore,

$$A_i = \frac{i_{b1}}{i_{in}}\frac{i_{e1}}{i_{b1}}$$

Applying the current-divider rule to the input circuit, we get

$$i_{b1} = \frac{R_B i_{in}}{R_B + \beta_{ac}r_{e1}}$$

where $R_B = R_1 \| R_2$
Also,

$$i_{e1} = \beta_{ac}i_{b1}$$

Therefore, the current-gain equation becomes

$$A_i = \frac{\beta_{ac}R_B}{R_B + \beta_{ac}r_{e1}} \tag{1-50}$$

1-14.2(c) Input resistance.   For the input circuit,

$$v_{in} = r_{e1}i_{e1} = \beta_{ac}r_{e1}i_{b1} \text{ since } i_{e1} = \beta_{ac}i_{b1}$$

Substituting

$$i_{b1} = \frac{R_B i_{in}}{R_B + \beta_{ac}r_{e1}}$$

we have

$$v_{in} = \beta_{ac}r_{e1}\frac{R_B i_{in}}{R_B + \beta_{ac}r_{e1}}$$

Solving for the ratio v_{in}/i_{in}, we get, for the input resistance,

$$R_i = \frac{v_{in}}{i_{in}} = \frac{\beta_{ac}r_{e1}R_B}{R_B + \beta_{ac}r_{e1}} \tag{1-51}$$

The cascode amplifier of Figure 1-22 is not encountered very often in discrete circuitry because of the following drawbacks:

1. The amplifier requires a large number of resistors and capacitors to provide bias and prevent undesirable feedback.
2. Unlike the integrated circuit (IC) cascode amplifier, it does not offer much flexibility. For instance, the CA3095E, a differential cascode amplifier, can

be used as a differential or a cascode amplifier. Some IC cascode amplifiers are even available with an automatic gain control (AGC) and have a provision for thermal stability. Obviously, such amplifiers are usually a little more complex in structure than the basic circuit shown in Figure 1-22.

EXAMPLE 1-11

The following specifications are given for the cascode amplifier of Figure 1-22.

$$R_1 = 27 \text{ k}\Omega \qquad R_C = 1.8 \text{ k}\Omega \qquad C_i = 10 \text{ } \mu F$$
$$R_2 = 39 \text{ k}\Omega \qquad R_E = 1 \text{ k}\Omega \qquad C_o = 10 \text{ } \mu F$$
$$R_3 = 47 \text{ k}\Omega \qquad\qquad\qquad C_{s1} = 100 \text{ } \mu F$$
$$V_{CC} = 12 \text{ V} \qquad\qquad\qquad C_{s2} = 100 \text{ } \mu F$$

Assume that 2N3393 transistors are used with $\beta_{dc} = 140$, $\beta_{ac} = 125$, and $V_{BE} = 0.63$ V. Determine:
(a) The dc conditions for both transistors.
(b) The overall voltage gain.
(c) The overall current gain.
(d) The input resistance.

SOLUTION   (a) Substituting in Equation (1-45), we get

$$V_{R1} = \frac{R_1 V_{CC}}{(R_1 + R_2 + R_3)} = \frac{(27 \text{ k}\Omega)(12)}{(27 \text{ k}\Omega + 39 \text{ k}\Omega + 47 \text{ k}\Omega)} = 2.87 \text{ V}$$

Therefore,

$$V_{E1} = V_{R1} - V_{BE1} = 2.87 - 0.63 = 2.24 \text{ V}$$
$$I_{E1} = \frac{V_{E1}}{R_E} = \frac{2.24 \text{ V}}{1 \text{ k}\Omega} = 2.24 \text{ mA}$$

Thus,

$$I_{E1} = I_{C1} = I_{E2} = I_{C2} = 2.24 \text{ mA}$$
$$V_{B2} = V_{CC} - \frac{R_3 V_{CC}}{(R_1 + R_2 + R_3)} = 12 - \frac{(47 \text{ k}\Omega)(12)}{(113 \text{ k}\Omega)} = 7.01 \text{ V}$$

Hence,

$$V_{E2} = V_{C1} = V_{B2} - V_{BE2} = 7.01 - 0.63 = 6.38 \text{ V}$$
$$V_{CE1} = V_{C1} - V_{E1} = 6.38 - 2.24 = 4.14 \text{ V}$$

Since

$$V_{C2} = V_{CC} - R_C I_{C2} = 12 - (1.8 \text{ k}\Omega)(2.24 \text{ mA}) = 7.97 \text{ V}$$

then

$$V_{CE2} = V_{C2} - V_{E2} = 7.97 - 6.38 = 1.59 \text{ V}$$

(b) The ac emitter resistances of Q_1 and Q_2 are

$$r_{e1} = r_{e2} = \frac{25 \text{ mV}}{I_{E1}} = \frac{25 \text{ mV}}{2.24 \text{ mA}} = 11.16 \ \Omega$$

Substituting in Equation (1-49), we have the overall voltage gain

$$A_v = -\frac{R_C}{r_{e2}} = -\frac{1.8 \text{ k}\Omega}{11.16 \ \Omega} = 161.3$$

(c) $R_B = R_1 \| R_2 = 27 \text{ k}\Omega \| 39 \text{ k}\Omega = 15.95 \text{ k}\Omega$ and $\beta_{ac} = 125$. Substituting these values in Equation (1-50), we get

$$A_i = \frac{\beta_{ac} R_B}{R_B + \beta_{ac} r_{e1}} = \frac{(125)(15.95 \text{ k}\Omega)}{15.95 \text{ k}\Omega + (125)(11.16 \ \Omega)} = 114.9$$

(d) Input resistance is given by Equation (1-51):

$$R_i = \frac{\beta_{ac} r_{e1} R_B}{R_B + \beta_{ac} r_{e1}} = \frac{(125)(11.16 \ \Omega)(15.95 \text{ k}\Omega)}{15.95 \text{ k}\Omega + (125)(11.16 \ \Omega)} = 1.28 \text{ k}\Omega$$

SUMMARY

1. The differential amplifier consists of two symmetrical common-emitter sections and is capable of amplifying the difference between two input signals. The differential amplifier can amplify ac as well as dc input signals because it employs direct coupling.

2. There are four types of differential amplifier configurations:
   (a) The dual input, balanced output
   (b) The dual input, unbalanced output
   (c) The single input, balanced output
   (d) The single input, unbalanced output
   These configurations are defined according to the number of input signals used and the way output is measured. When we use two inputs, the configuration is said to be dual input; otherwise, it is a single input. If the output voltage is measured between two collectors, the configuration is referred to as a balanced output. On the other hand, if the output is measured at either collector with respect to ground, the configuration is said to be unbalanced output.

3. For proper operation of the differential amplifier, a transistor array and matched components must be used. The differential amplifier can be biased by using emitter bias (a combination of R_E and V_{EE}), constant current bias, or current mirror techniques.

4. In a cascaded differential amplifier, the output of the first stage is used as an input for the second stage, the output of the second stage is applied as an input to the third stage, and so on. Because of direct coupling between the stages, the operating point of succeeding stages changes.

5. Generally, in a cascaded differential amplifier, the last dual-input, unbalanced-output stage is followed by a level translator. The function of the level translator is to make the dc voltage at the output terminal zero.

6. The cascode amplifier is composed of direct-coupled common-emitter and common-base configurations. It provides a large amount of voltage gain at higher frequencies.

QUESTIONS ───

1-1. What is a differential amplifier?

1-2. What are the four differential amplifier configurations? Which one is not commonly used, and why?

1-3. What does the term "balanced output" mean?

1-4. What is the main advantage of constant current bias over emitter bias?

1-5. Explain the differences between constant current bias and current mirror.

1-6. What is a level translator circuit? Why is it used with the cascaded differential amplifier?

1-7. What is a cascode amplifier? List the characteristics of the cascode amplifier.

PROBLEMS ───

1-1. An emitter-biased dual-input, balanced-output differential amplifier has the following specifications: $|V_{CC}| = |-V_{EE}| = 10$ V, $R_{C1} = R_{C2} = 2.7$ kΩ, and $R_E = 3.9$ kΩ; the transistor array is CA3086 with $\beta_{ac} = \beta_{dc} = 100$ and $V_{BE} = 0.715$ V. Calculate:
(a) The operating current and voltage values for each transistor.
(b) The voltage gain.
(c) The input resistance of the circuit.

1-2. Repeat Problem 1-1 for the dual-input, unbalanced-output differential amplifier.

1-3. Repeat Problem 1-1 with $R_E = 5.6$ kΩ.

1-4. Repeat Problem 1-1 with $R_{C1} = R_{C2} = 5.6$ kΩ.

1-5. Repeat Problem 1-3 for the single-input, balanced-output configuration.

1-6. Repeat Problem 1-4 for the dual-input, unbalanced-output configuration.

1-7. Repeat Problem 1-4 for the single-input, balanced-output configuration.

1-8. For the dual-input, balanced-output differential amplifier of Problem 1-1:
(a) Draw the output voltage waveform if $v_{in\,1} = 5$ mV pp and $v_{in\,2} = 2$ mV pp at 1 kHz. Assume that the collector C_1 is negative with respect to collector C_2 (see Figure 1-2).
(b) Determine the maximum peak-to-peak output voltage swing without clipping.

1-9. Repeat Problem 1-8 for the dual-input, unbalanced-output configuration.

1-10. Repeat Problem 1-8 if 180-Ω resistors R'_E are placed externally in series with each emitter (see Figure 1-12).

1-11. Design a zener constant current bias circuit as shown in Figure 1-14(c) according to the following specifications:
(a) Emitter current I_{E3} = 5 mA.
(b) 1N3825 zener diode with V_Z = 4.7 V and I_{Zt} = 53 mA.
(c) CA3086 transistor with β_{ac} = β_{dc} = 100 and V_{BE} = 0.715 V.
(d) Supply voltage $-V_{EE}$ = -9 V.

1-12. The JFET dual-input, balanced-output differential amplifier uses the zener constant current bias circuit designed in Problem 1-11. Determine the voltage gain and the input resistance of the amplifier if R_D = 3.9 kΩ, R_G = 1 MΩ, g_m = 4000 μs and V_s = $\pm$9 V.

1-13. Design the dual-input, balanced-output differential amplifier using diode constant current bias (Figure 1-14(a)) to meet the following specifications:
(a) Supply voltages V_s = $\pm$12 V.
(b) Emitter current I_E in each differential amplifier transistor = 1.5 mA.
(c) Voltage gain $\leq$ 60.

1-14. Design the dual-input, balanced-output differential amplifier with the current mirror bias (Figure 1-15(b)) according to the following specifications:
(a) Supply voltages V_s = $\pm$10 V.
(b) Maximum output voltage swing = 4 V pp.

1-15. Design the single-input, balanced-output differential amplifier with diode constant current bias (Figure 1-14(a)) to meet the following specifications:
(a) Voltage gain $\leq$ 10.
(b) Input resistance R_i $\geq$ 20 kΩ.

1-16. For the modified current mirror circuit shown in Figure 1-24, determine the emitter current in transistor Q_3. Assume that the CA3086 transistor array is used. For each transistor in the array, β_{dc} = β_{ac} = 100 and V_{BE} = 0.715 V.

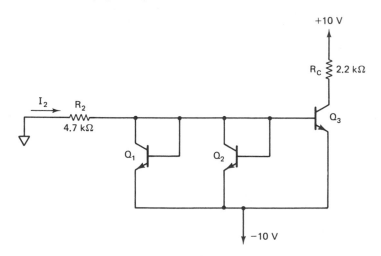

Figure 1–24   Modified current mirror circuit of Problem 1–16.

1-17. Referring to the circuit of Figure 1-25, determine:
(a) The current through collector resistor R_C.
(b) The collector current in each transistor.
Assume that the CA3086 transistor array is used.

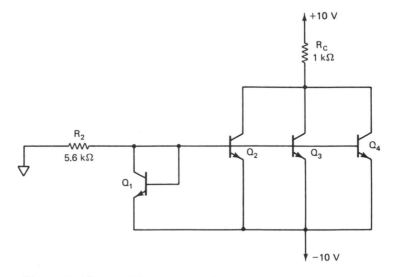

Figure 1–25   Modified current mirror circuit of Problem 1–17.

1-18. For the single-input, balanced-output differential amplifier shown in Figure 1-26, determine:
(a) The operating point values (I_{CQ} and V_{CEQ}).
(b) The voltage gain.
(c) The input resistance.
(d) The maximum peak-to-peak output voltage.
Assume that $\beta_{ac} = \beta_{dc} = 100$, $V_{BE} = V_{D1} = 0.715$ V, $V_Z = 6.2$ V, and $I_{Zt} = 41$ mA.

1-19. For the cascaded differential amplifier of Figure 1-27, determine:
(a) The dc conditions for each stage.
(b) The overall voltage gain.
(c) The maximum peak-to-peak output voltage swing.
Assume that the CA3086 transistor array is used.

1-20. Repeat Problem 1-19 for the circuit shown in Figure 1-28, page 55.

1-21. Repeat Problem 1-19 for the circuit shown in Figure 1-29, page 56.

1-22. Repeat Problem 1-19 for the circuit shown in Figure 1-30, page 56.

1-23. Repeat Problem 1-19 for the circuit shown in Figure 1-31, page 57.

1-24. Repeat Problem 1-19 for the circuit shown in Figure 1-32, page 57. Assume that the input common-mode voltage V_{CM} is 3.5 V.

1-25. Repeat Problem 1-19 for the circuit shown in Figure 1-33, page 58.

Figure 1–26 Differential amplifier of Problem 1–18.

Figure 1–27 Cascaded differential amplifier of Problem 1–19.

Figure 1-28 Cascaded differential amplifier of Problem 1–20.

1-26. Design the level translator circuit using diode constant current bias for the circuit in Problem 1-19. Refer to Figure 1-20(b).

1-27. Design the level translator circuit using zener constant current bias for the circuit in Problem 1-20. Use Figure 1-20(b) and replace diodes by a zener.

1-28. For the circuit in Problem 1-21, design the level translator circuit using current mirror bias. Refer to Figure 1-20(c).

1-29. For the cascode amplifier shown in Figure 1-22, determine:
(a) The dc conditions for each transistor.
(b) The overall voltage gain.
(c) The overall current gain.
(d) The input resistance.
Assume that $R_1 = 12$ kΩ, $R_2 = 20$ kΩ, $R_3 = 24$ kΩ, $R_C = 1.8$ kΩ, $R_E = 1$ kΩ, $C_i = C_o = 10$ μF, $C_{S1} = C_{S2} = 100$ μF, and $V_{CC} = 12$ V. The transistors are 2N3393 with $\beta_{ac} = 125$ and $V_{BE} = 0.63$ V.

1-30. Repeat Example 1-11 with $R_E = 2.2$ kΩ.

1-31. Repeat Problem 1-29 with $R_C = 1$ kΩ.

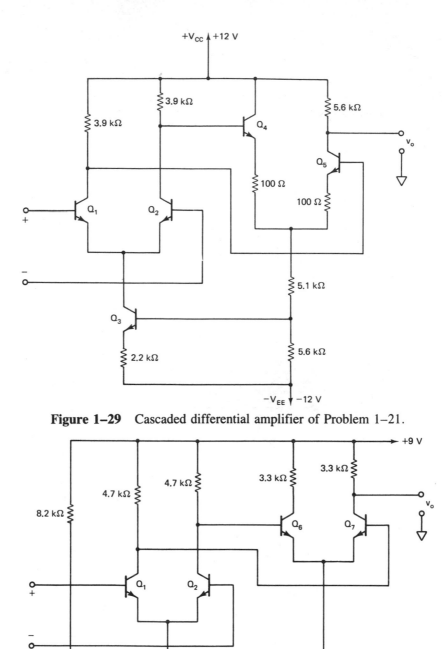

Figure 1–29 Cascaded differential amplifier of Problem 1–21.

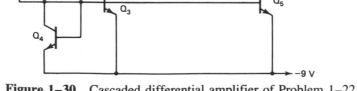

Figure 1–30 Cascaded differential amplifier of Problem 1–22.

Figure 1–31 Cascaded differential amplifier of Problem 1–23.

Figure 1–32 Schematic diagram of μA730 differential amplifier.

Figure 1–33 Cascaded differential amplifier of Problem 1–25.

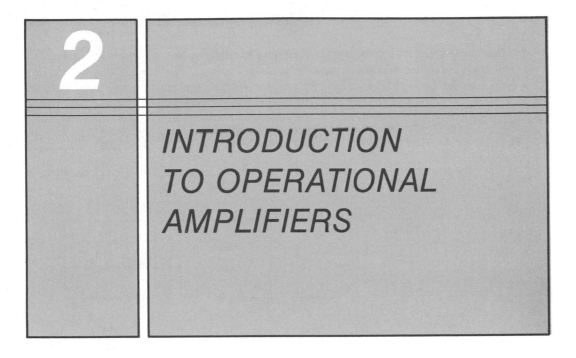

2

INTRODUCTION TO OPERATIONAL AMPLIFIERS

2-1 INTRODUCTION

In Chapter 1 you studied the two important types of direct-coupled amplifiers—differential and cascode. You saw that, when cascaded, the differential amplifier yields a very high voltage gain. However, in differential amplifiers the matching of resistors and transistors is essential for the proper operation of the circuit. For this reason the differential amplifier and the cascode amplifier are manufactured as individual integrated circuits, and the differential amplifiers, in particular, are used as "building blocks" for more elaborate circuits called operational amplifiers. The cost of integrated amplifiers is currently so low that they have effectively displaced the discrete amplifiers.

In this chapter we study the basic structure of the operational amplifier (op-amp) and then analyze some typical operational amplifier circuits. We will also investigate the op-amp symbol, types of op-amps, different grades of op-amps, and op-amp development—among other things.

2-2 THE OPERATIONAL AMPLIFIER

An operational amplifier is a direct-coupled high-gain amplifier usually consisting of one or more differential amplifiers and usually followed by a level translator and an output stage. The output stage is generally a push-pull or push-pull complementary-symmetry pair. An operational amplifier is available as a single integrated circuit package.

The operational amplifier is a versatile device that can be used to amplify dc as well as ac input signals and was originally designed for computing such mathematical functions as addition, subtraction, multiplication, and integration. Thus the name *operational ampli-*

fier stems from its original use for these mathematical operations and is abbreviated to *op-amp*. With the addition of suitable external feedback components, the modern day op-amp can be used for a variety of applications, such as ac and dc signal amplification, active filters, oscillators, comparators, regulators, and others.

2-3 BLOCK DIAGRAM REPRESENTATION OF A TYPICAL OP-AMP ───

Since an op-amp is a multistage amplifier, it can be represented by a block diagram as shown in Figure 2-1.

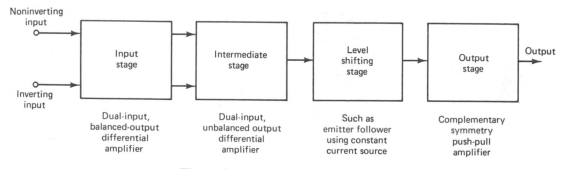

Figure 2–1   Block diagram of a typical op-amp.

The input stage is the dual-input, balanced-output differential amplifier. This stage generally provides most of the voltage gain of the amplifier and also establishes the input resistance of the op-amp. The intermediate stage is usually another differential amplifier, which is driven by the output of the first stage. In most amplifiers the intermediate stage is dual input, unbalanced (single-ended) output. Because direct coupling is used, the dc voltage at the output of the intermediate stage is well above ground potential. Therefore, generally, the level translator (shifting) circuit is used after the intermediate stage to shift the dc level at the output of the intermediate stage downward to zero volts with respect to ground. The final stage is usually a push-pull complementary amplifier output stage. The output stage increases the output voltage swing and raises the current supplying capability of the op-amp. A well-designed output stage also provides low output resistance.

2-4 ANALYSIS OF TYPICAL OP-AMP EQUIVALENT CIRCUITS ───

First we consider a specific op-amp with a simpler equivalent circuit and perform the analysis of this circuit to determine the input resistance and voltage gain. Then the op-amp with a more complex equivalent circuit will be presented. However, instead of a quantitative analysis, a qualitative analysis will be used for this complex op-amp equivalent circuit.

While there are a variety of op-amps, each with specific inner design features such as internal frequency compensation, FET inputs, Darlington inputs, current sources as active loads, input voltage and output current limiter, and many others, the analysis of specific op-amp equivalent circuits will provide a good basis for understanding the inner operation and construction of the op-amp and aid in selecting a proper op-amp for a desired application.

EXAMPLE 2-1

The equivalent circuit of the Motorola op-amp MC 1435 is shown in Figure 2-2.
(a) Determine the collector current in each transistor and the dc voltage at the output terminal.
(b) Calculate the voltage gain of the op-amp.
(c) Determine the input resistance of the op-amp.
Assume that the $\beta_{ac} = \beta_{dc} = 150$ and $V_{BE} = 0.7$ V for each transistor.

SOLUTION   Figure 2-2 should give you an idea of the internal structure of the op-amp. A close examination of this equivalent circuit reveals that it consists of four stages. These stages are labeled in Figure 2-2 and are separated by dashed lines. Transistors Q_1 and Q_2 form the first differential amplifier stage, which uses a constant current bias provided by transistor Q_3 and associated resistors. This first stage has inverting and noninverting inputs and hence can be driven with two inputs or a single input. The second differential amplifier formed by transistors Q_4 and Q_5 uses an emitter bias and is driven by the outputs of the first differential amplifer. The single-ended output of the second differential amplifier drives an emitter follower, which is composed of transistor Q_6 and 15-kΩ resistor. The output of the emitter follower drives a complementary transistor pair Q_7 and Q_8, which produces the final output voltage.

(a) To determine the collector current and the collector-to-emitter voltage of the first differential amplifer stage, the inverting and noninverting input terminals are assumed to be connected to ground. Referring to Figure 2-2, the collector current in transistor Q_1 and Q_2 is then determined as follows:

$$V_{C1} = V_{CC} - (6.7 \text{ k}\Omega)I_{C1}$$
$$= 6 - (6.7 \text{ k}\Omega)I_{C1}$$
$$V_{E4} = V_{C1} - V_{BE5}$$
$$= 5.3 - (6.7 \text{ k}\Omega)I_{C1}$$

Since I_{B3} is negligibly small,

$$I_4 = \frac{V_{E4} + V_{EE}}{9.1 \text{ k}\Omega + 5.5 \text{ k}\Omega}$$
$$= \frac{11.3 - (6.7 \text{ k}\Omega)I_{C1}}{14.6 \text{ k}\Omega}$$
$$V_{B3} = (5.5 \text{ k}\Omega)I_4 - V_{EE}$$
$$= 5.5 \text{ k}\Omega\left[\frac{11.3 - (6.7 \text{ k}\Omega)I_{C1}}{14.6 \text{ k}\Omega}\right] - 6$$

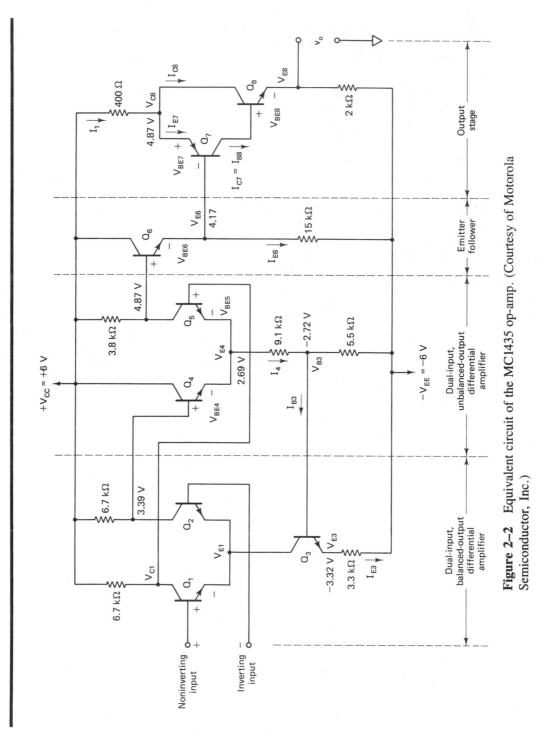

Figure 2–2 Equivalent circuit of the MC1435 op-amp. (Courtesy of Motorola Semiconductor, Inc.)

$$V_{E3} = V_{B3} - V_{BE3}$$

$$= \frac{5.5 \text{ k}\Omega}{14.6 \text{ k}\Omega}[11.3 - (6.7 \text{ k}\Omega)I_{C1}] - 6.7$$

$$I_{E3} = \frac{V_{E3} + V_{EE}}{3.3 \text{ k}\Omega}$$

$$= \frac{(5.5 \text{ k}\Omega/14.6 \text{ k}\Omega)[11.3 - (6.7 \text{ k}\Omega)I_{C1}] - 6.7 + 6}{3.3 \text{ k}\Omega}$$

But $I_{E3} = 2I_{C1}$. Therefore,

$$2I_{C1} = \frac{(5.5 \text{ k}\Omega)(11.3)}{(14.6 \text{ k}\Omega)(3.3 \text{ k}\Omega)} - \frac{(5.5 \text{ k}\Omega)(6.7 \text{ k}\Omega)I_{C1}}{(14.6 \text{ k}\Omega)(3.3 \text{ k}\Omega)} - \frac{0.7}{3.3 \text{ k}\Omega}$$

$$2I_{C1} + 0.765I_{C1} = 1.29 \text{ mA} - 0.212 \text{ mA}$$

$$I_{C1} = \frac{1.08 \text{ mA}}{2.765} \cong 0.39 \text{ mA}$$

This collector current can be used to compute the collector currents of the remaining stages as follows:

$$V_{C1} = V_{CC} - (6.7 \text{ k}\Omega)I_{C1}$$

$$= 6 - (6.7 \text{ k}\Omega)(0.39 \text{ mA}) = 3.39 \text{ V}$$

$$V_{E4} = V_{C1} - V_{BE5}$$

$$= 2.69 \text{ V}$$

Hence

$$2I_{E4} = \frac{V_{E4} + V_{EE}}{14.6 \text{ k}\Omega}$$

$$I_{E4} = \frac{2.69 + 6}{29.2 \text{ k}\Omega} = 0.298 \text{ mA} = I_{C5}$$

$$V_{C5} = V_{CC} - (3.8 \text{ k}\Omega)I_{C5}$$

$$= 6 - (3.8 \text{ k}\Omega)(0.298 \text{ mA}) = 4.87 \text{ V}$$

$$V_{E6} = V_{C5} - V_{BE6}$$

$$= 4.87 - 0.7 = 4.17 \text{ V}$$

Therefore,

$$I_{E6} = \frac{V_{E6} + V_{EE}}{15 \text{ k}\Omega}$$

$$= \frac{4.17 + 6}{15 \text{ k}\Omega} = 0.678 \text{ mA}$$

$$V_{E7} = V_{C8} = V_{E6} + V_{BE7}$$

$$= 4.17 + 0.7 = 4.87 \text{ V}$$

Hence

$$I_1 = \frac{V_{CC} - V_{E7}}{400}$$

$$= \frac{6 - 4.87}{400} = 2.83 \text{ mA}$$

Since $I_{C7} = I_{B8}$ and $I_{C7} \cong I_{E7}$,

$$I_1 = I_{C8} + I_{B8}$$

$$= I_{E8} = 2.83 \text{ mA}$$

The voltage V_{E8} at the output terminal is

$$V_{E8} = -V_{EE} + (2 \text{ k}\Omega)I_{E8}$$

$$= -6 + (2 \text{ k}\Omega)(2.83 \text{ mA})$$

$$= -0.34 \text{ V} \cong 0 \text{ V} \text{ as expected}$$

Thus

$$I_{C1} = I_{C2} \cong 0.39 \text{ mA}$$

$$I_{C3} \cong 0.78 \text{ mA}$$

$$I_{C4} = I_{C5} \cong 0.298 \text{ mA}$$

$$I_{C6} \cong 0.678 \text{ mA}$$

$$I_{C7} = I_{B8} = 18.8 \text{ } \mu\text{A}$$

$$I_{E8} = 2.83 \text{ mA}$$

$$V_{E8} = -0.34 \text{ V}$$

(b) The voltage gain of the emitter follower as well as that of the output stage is approximately 1. To calculate the voltage gains of the differential amplifier stages, we need to know their ac emitter resistances. The ac emitter resistance of each stage can be calculated using the collector current obtained in part (a):

$$r_{e1} = r_{e2} = \frac{25 \text{ mV}}{I_{E1}} = \frac{25 \text{ mV}}{0.39 \text{ mA}} \cong 64.1 \text{ } \Omega$$

$$r_{e4} = r_{e5} = \frac{25 \text{ mV}}{I_{E4}} = \frac{25 \text{ mV}}{0.298 \text{ mA}} \cong 83.89 \text{ } \Omega$$

$$r_{e6} = \frac{25 \text{ mV}}{I_{E6}} = \frac{25 \text{ mV}}{0.678 \text{ mA}} \cong 36.87 \text{ } \Omega$$

Therefore, the voltage gain (A_{d1}) of the dual-input, balanced-output differential amplifier is

$$A_{d1} = \frac{R_{C1} \| 2\beta_{ac}(r_{e4})}{r_{e1}} = \frac{6.7 \text{ k}\Omega \| (2)(150)(83.89 \text{ } \Omega)}{64.1 \text{ } \Omega}$$

$$= 82.55$$

And the voltage gain (A_{d2}) of the dual-input, unbalanced output-differential amplifier is

$$A_{d2} = \frac{R_{C5}\|\beta_{ac}(r_{e6} + 15\ \text{k}\Omega)}{2r_{e5}}$$

$$= \frac{3.8\ \text{k}\Omega\|(150)(15.04\ \text{k}\Omega)}{(2)(83.89\ \Omega)}$$

$$= 22.6$$

Thus the overall voltage gain of the op-amp is

$$A_d = (A_{d1})(A_{d2}) = (82.55)(22.6)$$

$$= 1865.63$$

(c) The input resistance of the op-amp is the same as the input resistance of the first differential amplifier stage and is given by Equation (1-14):

$$R_i = 2\beta_{ac}r_{e1}$$

$$= (2)(150)(64.1) = 19.23\ \text{k}\Omega$$

This completes the analysis of MC1435 op-amp.

   Figure 2-3 shows the schematic diagram of the 741 op-amp that has become an industry standard. It is a high-gain, stable amplifier that features internal compensation. A desired phase control at high frequencies is provided by a 30-pF capacitor that is fabricated on the chip. Since the equivalent circuit of the 741 op-amp is more complex than that of the 1435, we will perform a qualitative (rather than quantitative) analysis of the circuit.

   The input differential amplifier consists of transistors Q_1–Q_3 and Q_2–Q_4. A differential input signal is applied between the bases of transistors Q_1 and Q_2. The maximum input signal capacity of the input differential stage is increased by PNP transistors Q_3 and Q_4, since the base–emitter junction of Q_3 is in series with the base–emitter junction of Q_1 and the base–emitter junction of Q_4 is in series with that of Q_2. Since the signal levels normally encountered do not exceed the voltage across both of the series junctions, the input differential stage is practically indestructible.

   The constant base current in Q_3 and Q_4 is supplied by a constant current source consisting of transistors Q_9 and Q_{10} and diode-connected transistors Q_8 and Q_{11}. This constant base current in transistors Q_3 and Q_4 in turn establishes the operating current for the input differential stage. The diode-connected transistors Q_8 and Q_{11} stabilize the base potentials of transistors Q_9 and Q_{10}, respectively.

   A controlled current source is formed by transistors Q_5 and Q_6 and resistors R_1, R_2, and R_3. The emitter current of transistors Q_5 and Q_6 can, however, be controlled by varying a 10-kΩ potentiometer that is externally connected between the offset null terminals and by connecting the wiper to the negative supply. (For more information on offset null capabilities of the 741, refer to Section 5-2). Transistor Q_7 supplies the base currents to transistors Q_5 and Q_6. Since transistors Q_5 and Q_6 are identical (and $R_1 = R_2$), their base currents must be equal, and in turn their collector currents must also be equal at all times. The collector voltage of transistor Q_5 controls the operation of transistor Q_7; that is, any

Figure 2–3 Equivalent circuit of the 741 op-amp. (Courtesy of Fairchild Semi-conductor Corporation.)

change in Q_5 collector voltage produces identical changes in transistor Q_5 and Q_6 collector currents via transistor Q_7.

The output of the differential stage is taken at the junction of transistors Q_4 and Q_6, which act as a complementary-symmetry amplifier. For example, suppose that a differential input signal applied between Q_1 and Q_2 causes the collector currents of Q_1, Q_3, and Q_5 to increase and the collector currents of Q_2 and Q_4 to decrease. This differential input signal also causes the collector voltage of Q_5 to increase, which in turn causes the collector currents of Q_5 and Q_6 to increase. Thus, at the junction of transistors Q_4 and Q_6, the net change in current is twice that of Q_4 alone. This action verifies that transistors Q_4 and Q_6 act as a complementary-symmetry amplifier. The output at the junction of Q_4 and Q_6 is proportional to the differential input signal.

The output of the differential stage is applied to the common collector amplifier formed by Q_{16} and R_9. The output of Q_{16} drives the common-emitter amplifier, which is a bias source for transistors Q_{18} and Q_{19}. The CE amplifier is composed of Q_{17}, R_8, and constant current load formed by Q_{13}. Transistors Q_{13} and Q_{12} form a current mirror that establishes the current through transistors Q_{17}, Q_{18}, and Q_{19}. In fact, the currents in the collectors of Q_{13} are equal to the current through R_5. The network consisting of transistor Q_{18}, Q_{19}, and R_{10} is designed to bias the output complementary-symmetry stage formed by transistors Q_{14}, Q_{20} and resistors R_6, R_7 in the linear region. These transistors also separate the bases of Q_{14} and Q_{20} by two diode drops and in turn temperature-compensate currents in Q_{14} and Q_{20}. The output is measured at the junction of R_6 and R_7, which are in the emitter circuits of Q_{14} and Q_{20}, respectively.

Transistor Q_{22} performs two functions. It serves as a buffer between Q_{17} and Q_{20} and also provides a negative feedback to Q_{16}. The current limiting for the output complementary-symmetry stage is provided by transistors Q_{15}, Q_{21}, and Q_{23}. If the output (load) current exceeds the safe limit, the voltage drops across R_6 and R_7 increase, causing Q_{15} and Q_{21} to turn on, which in turn causes Q_{23} to turn on. This action however, turns off Q_{16} and in turn Q_{17}, which reduces the emitter currents in Q_{22}. The reduction in the currents of Q_{22} causes the currents in Q_{18} and Q_{19} to decrease, which in turn lowers the currents in Q_{14} and Q_{20}. The diode-connected transistor Q_{24} is a temperature-compensating diode for transistor Q_{23}.

2-5 SCHEMATIC SYMBOL

Given an op-amp schematic diagram like the one in Figure 2-2 or 2-3, we can save time by using a schematic symbol for the entire circuit. Figure 2-4 shows the most widely used of such symbols for a circuit with two inputs and one output. For simplicity, power supply and other pin connections are omitted. Since the input differential amplifier stage of the op-amp is designed to be operated in the differential mode, the differential inputs are designated by the (+) and (−) notations. The (+) input is the noninverting input. An ac signal (or dc voltage) applied to this input produces an inphase (or same polarity) signal at the output. On the other hand, the (−) input is the inverting input because an ac signal (or dc voltage) applied to this input produces an 180° out-of-phase (or opposite polarity) signal at the output.

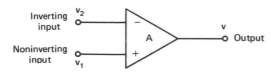

Figure 2–4   Schematic symbol for the op-amp.

In Figure 2-4,

$$v_1 = \text{voltage at the noninverting input (volts)}$$

$$v_2 = \text{voltage at the inverting input (volts)}$$

$$v_o = \text{output voltage (volts)}$$

All these voltages are measured with respect to ground.

$$A = \text{large-signal voltage gain, which is specified}$$
$$\text{on the data sheet for an op-amp}$$

2-6 INTEGRATED CIRCUITS

The circuits shown in Figures 2-2 and 2-3 are *integrated* circuits (ICs), meaning that all the components in each circuit are fabricated on the same "chip." ICs have become a vital part of modern electronic circuit design. They are used in the computer industry, automobile industry, home appliances, communication, and control systems, where they permit miniaturization and superior performance not possible with discrete components. ICs are now being used in all types of electronic equipment because of the long, trouble-free service they provide. In addition, they are economical because they are mass produced.

Classified according to their mode of operation, ICs are of two basic types: digital or linear.

Digital ICs are complete functioning logic networks that are equivalents of basic transistor logic circuits. They are used to form such circuits as gates, counters, multiplexers, demultiplexers, shift registers, and others. Since a digital IC is a complete predesigned package, it usually requires nothing more than a power supply, input, and output.

Digital circuits are primarily concerned with only two levels of voltage (or current): "high" and "low." Therefore, accurate control of operating-region characteristics is not required in digital circuits, unlike in linear circuits. For this reason, digital circuits are easy to design and are produced in large quantitites as low-cost devices.

Linear ICs are equivalents of discrete transistor networks, such as amplifiers, filters, frequency multipliers, and modulators, that often require additional external components for satisfactory operation. For example, external resistors are necessary to control the voltage gain and frequency response of an op-amp. In linear circuits the output electrical signals vary in proportion to the input signals applied or the physical quantities (parameters) they represent. Since the electrical signals are *analogous* to the physical quantities, linear circuits are also referred to as *analog circuits*.

Of all presently available linear ICs, the majority are operational amplifiers. With suitable external components, the op-amp is used in amplifiers, active filters, integrators, differentiators, and in countless other applications. A wide variety of special-purpose linear ICs is available for uses in comparators, voltage regulators, digital-interface circuits, and in radio-frequency and power amplifiers.

In addition to the op-amp and special-purpose linear ICs, component arrays are also available in IC form. Such arrays may consist of groups of isolated transistors, diodes, and resistors as well as of Darlington pairs or individual stages such as differential and cascode amplifiers.

Op-amps are further classified into two groups: general purpose and special purpose. *General-purpose* op-amps may be used for a variety of applications, such as integrator, differentiator, summing amplifier, and others. An example of a widely used general-purpose op-amp is the 741/351. On the other hand, *special-purpose* op-amps are used only for the specific applications they are designed for. For example, the LM380 op-amp can be used *only* for audio power applications.

2-7 TYPES OF INTEGRATED CIRCUITS

Integrated circuits may be classified as either monolithic or hybrid. Most linear ICs are produced by the monolithic process in that all transistors and passive elements (resistors and capacitors) are fabricated on a single piece of semiconductor material, usually silicon. "Monolithic" is a Greek-based word meaning "one stone."

In *monolithic ICs* all components (active and passive) are formed simultaneously by a diffusion process. Then a metallization process is used in interconnecting these components to form the desired circuit. Electrical isolation between the components in monolithic ICs can be achieved by any one of the three isolation techniques: dielectric, beam-lead, or PN-junction. However, the PN-junction isolation is most economical and is, therefore, commonly used.

The monolithic process makes low-cost mass production of ICs possible. Also, monolithic ICs exhibit good thermal stability because all the components are integrated on the same chip very close to each other.

However, the large values of resistance and capacitance that are required in some linear circuits cannot be formed using the monolithic process. Moreover, there is no method available to fabricate transformers or to form large values of inductors in integrated-circuit form. However, if these components are required in a given application, external discrete components can be used with the IC.

In *hybrid ICs,* passive components (such as resistors and capacitors) and the interconnections between them are formed on an insulating substrate. The substrate is used as a chassis for the integrated components. Active components such as transistors and diodes, as well as monolithic integrated circuits, are then connected to form a complete circuit. For this reason, low-volume production methods are best suited to hybrid IC technology.

Hybrid ICs are further classed as thin film or thick film, depending on the method used to form the resistors, capacitors, and related interconnections on the substrate. When a suitable material is evaporated on a substrate in forming resistors, capacitors, and interconnections, a *thin-film* hybrid IC is obtained. On the other hand, in a *thick-film* hybrid IC the resistors, capacitors, and interconnections are etched on the substrate by silk screening.

2-8 MANUFACTURERS' DESIGNATIONS FOR INTEGRATED CIRCUITS

In the United States alone there are well over 30 IC manufacturers producing millions of ICs per year. Each manufacturer uses a specific code and assigns a specific type number to the ICs it produces. That is, each manufacturer uses its own identifying initials followed by its own type number. For example, the 741 type of internally compensated op-amp was originally manufactured by Fairchild and is sold as the μA741, where "μA" represents the identifying initials used by Fairchild. Initials used by some of the well-known manufacturers of *linear* ICs are as follows:

Fairchild	μA
	μAF
National Semiconductor	LM
	LH
	LF
	TBA
Motorola	MC
	MFC
RCA	CA
	CD
Texas Instruments	SN
Sprague	ULN
	ULS
	ULX
Intersil	ICL
	IH
Siliconix, Inc.	L
Signetics	N/S
	NE/SE
	SU
Burr-Brown	BB

Remember that the initials used by manufacturers in designating digital ICs may differ from those used for linear ICs. For example, DM and CD are the initials used for digital monolithic and CMOS (complementary metal-oxide semiconductor) digital ICs, respectively, by National Semiconductor.

In addition to producing their own ICs, a number of manufacturers also produce one another's popular ICs. In second-sourcing such ICs, the manufacturers usually retain the original type number of the IC in their own IC designation. For example, Fairchild's original μA741 is also manufactured by various other manufacturers under their own designations, as follows:

National Semiconductor	LM741
Motorola	MC1741
RCA	CA3741
Texas Instruments	SN52741
Signetics	N5741

Note that the last three digits in each manufacturer's designation are 741. All these op-amps have the same specifications and, therefore, behave the same.

More information is available in the linear industry's cross reference guides on the types of ICs manufactured by different manufacturers. An industry cross-reference guide is generally included in the manufacturer's data book.

Since a number of manufacturers produce the same IC, for convenience we shall refer to such ICs by their type numbers and delete manufacturers' identifying initials. For example, instead of referring to an op-amp as a μA741 or MC1741, we shall refer to it simply as a 741.

Some linear ICs are available in different classes, such as A, C, E, S, and SC. For example, the 741, 741A, 741C, 741E, 741S, and 741SC are different versions of the same op-amp. The 741 is a military-grade op-amp (operating temperature range: $-55°$ to $125°C$) and the 741C is a commercial-grade op-amp (operating temperature range: $0°$ to $70°/75°C$). On the other hand, the 741A and 741E are improved versions of the 741 and 741C, respectively, in that they have improved electrical specifications over their counterparts. The 741C and 741E are identical to the 741 and 741A except that the former have their performance guaranteed over a $0°$ to $70°/75°C$ temperature range, instead of $-55°$ to $125°C$. The 741S and 741SC are military- and commercial-grade op-amps, respectively, with higher slew rate (rate of change of output voltage per unit of time) than the 741 and 741C.

2-9 DEVELOPMENT OF INTEGRATED CIRCUITS

The development of linear ICs can be traced back to early 1960s, when arrays were first fabricated on a single silicon chip. The arrays are combinations of isolated components such as diodes and transistors or individual stages such as differential amplifiers and Darlington pairs. The use of such IC arrays helped minimize the temperature-drift problem inherent in discrete transistor and diode circuits. It also greatly reduced the size of discrete electronic circuits.

In 1963 Fairchild Semiconductor introduced the first IC op-amp, its μA702, which set the stage for the development of other IC op-amps. The unequal supply voltages, such as $+V_{CC} = +12$ V and $-V_{EE} = -6$ V, relatively low input resistance (40 kΩ typically), and low voltage gain (3600 V/V) were the major drawbacks of the μA702 op-amp. For these reasons the μA702 was not universally accepted.

In 1965 Fairchild introduced the μA709, an improved op-amp compared to the μA702. More specifically, the μA709 had symmetrical supply voltages such as $+V_{CC} = +15$ V and $-V_{EE} = -15$ V, much higher input resistance (400 kΩ, typically), and a voltage gain of 45,000 V/V. The μA709 was the first quality op-amp and is, therefore, remembered for its historical significance. The μA709 is also regarded as a *first-generation* op-amp. Another example of a first-generation op-amp is the MC1537. The disadvantages of first-generation op-amps are as follows:

1. *No short-circuit protection.* The op-amp is susceptible to burnout if output is accidentally shorted to ground.
2. *A possible latch-up problem.* Output voltage can be latched up to some value and then fails to respond to changes in input signal applied.
3. *Requires an external frequency-compensating network (two capacitors and a resistor) for stable operation.* This means that when an external compensating network is not used the op-amp may oscillate. Ideally, these compensating components should be internally integrated so that no extra work of connecting them to the op-amp is required.

The next major advancement in IC op-amp technology came in 1968 with the introduction of the Fairchild μA741, an internally compensated op-amp. Unlike the μA709, it has short-circuit protection, has no latch-up problem, and is inherently stable. Besides that, it has very high input resistance (2 MΩ, typically), extremely high voltage gain (200,000 V/V), and offset null capability. It is regarded as a classic among IC op-amps because of its performance, versatility, and economy. It is one of the most widely used general-purpose op-amps in industry even a decade after its introduction. The 741 is an example of *second-generation* op-amps. Other examples of second-generation op-amps are the LM101, LM307, μA748, and MC1558. General-purpose op-amps such as the 741 and 307 are used in the greatest percentage of applications.

The design philosophy and technology used in the 741 were implemented in later-generation op-amps to enhance and optimize their performance characteristics. This trend gave birth to a variety of special-purpose op-amps. These devices had one or more of the following distinguishing features: high slew rate, that is, high rate of change of output voltage per unit of time; wide bandwidth; high input resistance; very low input currents and voltage offsets or drifts; programmability; and other characteristics. For example, the LM318 is a high-speed op-amp with 15-MHz small-signal bandwidth and guaranteed 50-V/μs slew rate. The key features of the μA771 op-amp are a low input bias current of 200 pA and a high slew rate of 13 V/μs. The μA714 is a precision op-amp with low offset voltage (75 μV), low input offset current (2.8 nA), low offset voltage drift (1.3 μV/1°C), and low noise. The μA791 op-amp is a power amplifier with output current capability of 1 A, whereas the μA776 is a multipurpose programmable op-amp. Thus most later-generation ICs have improved electrical specifications and hence exhibit superior performance over preceding-generation ICs.

New directions in linear ICs are aimed at the development of building-block circuits that not only include consumer circuits (audio, radio, and TV), but also industrial circuits (timers, regulators, etc.) The prime example of this concept is the emergence of cost-saving quad devices such as the LM324 and MC4741. The trend is to increase component density and yield while keeping the cost down through improved technology. Today's state of the art developments are concerned with BI-FET and laser trimming technology. National Semiconductor is the developer of BI-FET technology, which has introduced wide bandwidths and fast slew rates, together with low bias currents and low input offset currents. For example, the LF351 and LF353 BI-FET op-amps have low input bias and offset currents (200 and 100 pA, respectively), a large unity gain bandwidth (4 MHz), and a fast slew rate (13 V/μs). In addition, these op-amps have the same pin-outs as the 741 and 1458 (dual 741), respectively. This means that designers can immediately upgrade the overall performance of existing designs by simply placing LF351/LF353 op-amps into sockets presently hosting the 741 types. Laser trimming technology, at the same time, has achieved lower input offset voltages. *Trim*ming and BI-*FET* technologies are combined (TRIMFET, for short) by Motorola to provide precision input characteristics at a low cost. Motorola's MC34001/MC34002/MC34004 single, dual, and quad op-amps are the best examples of TRIMFET technologies. These devices are pincompatible with the 741, 1458, and 324 (quad) op-amps and have nearly identical electrical characteristics to those of the LF351/LF353.

The growth of IC technology and the implementation of new techniques is more evident in the development of digital ICs. Advances in semiconductor technology have had a remarkable effect on the development of digital ICs. The best example of such developments is the MC68000 microprocessor, which has 70,000 active components integrated on a single chip. The fabrication of such a complex and powerful digital IC was made possible through dry plasma etching, projection printing, and HMOS (high-density, short-channel MOS) circuit design techniques. In short, advances in technology have made ICs easy to use, more reliable, and more flexible in application, while maximizing performance.

2-9.1 SSI, MSI, LSI, and VLSI Packages

ICs are classified according to the number of components (or gates, in the case of digital ICs) integrated on the same chip, as follows:

Small-scale integration	SSI $<$ 10 components
Medium-scale integration	MSI $<$ 100 components
Large-scale integration	LSI $>$ 100 components
Very large scale integration	VLSI $>$ 1000 components

In the SSI package, the number of components integrated on the same chip is typically $<$ 10. Most of the arrays fall into this group. In the MSI IC, the number of components is $<$ 100, whereas the LSI package includes $>$ 100 components. Almost all the linear integrated circuits and integrated combination logic circuits are MSI packages. Most of the sequential logic circuits are of the LSI type. In the VLSI package, the number of components formed on the same chip is typically $>$ 1000. A classic example of a VLSI package is Motorola's MC68000 microprocessor IC, which houses some 70,000 components on the same chip.

As IC technology progresses and more efficient manufacturing techniques allow the manufacture of more complex yet better quality ICs, the price of an IC package continues to fall. A typical example would be the hand-held calculator. Current calculator prices are remarkably below the levels established when these devices were introduced in the late 1960s.

2-10 INTEGRATED CIRCUIT PACKAGE TYPES, PIN IDENTIFICATION, AND TEMPERATURE RANGES

2-10.1 Package Types

Figure 2-5 shows the three basic types of linear IC packages:

1. The flat pack
2. The metal can or transistor pack
3. The dual-in-line package (for short, DIP)

Ceramic flat package

14-lead version

(a)

TO-5 Style package with straight leads

8, 10, and 12-lead versions

(b)

TO-5 Style package with dual-in-line formed leads (DIL-CAN)

8-lead version

(c)

TO-5 Style package with radial formed leads

(d)

Dual-in-line plastic package

8, 14, and 16-lead versions

(e)

Dual-in-line welded-seal ceramic package

14, and 16-lead versions

(f)

Figure 2–5   Types of IC packages. (a) Flat pack. (b)–(d) Metal can. (e) and (f) Dual-in-line-package. (Courtesy of RCA Corporation.)

In the *flat pack,* the chip is enclosed in a rectangular ceramic case with terminal leads extending through the sides and ends as shown in Figure 2-5(a). The flat pack comes with 8, 10, 14, or 16 leads. These leads accommodate the power supplies, inputs, outputs, and several special connections required to complete the circuit.

In the *metal can* or *transistor pack,* the chip is encapsulated in a metal or plastic case [see Figure 2-5(b)–(d)]. The transistor pack is available with 3, 5, 8, 10, or 12 pins. Most of the voltage-regulator ICs, such as the LM117, have 3 pins. Power op-amps and audio power amplifiers are usually available in 5-pin packages. The metal can package is best suited for power amplifiers because metal is a good heat conductor and consequently has better dissipation capability than the flat-pack or dual-in-line package. In addition, the metal can package permits the use of external heat sinks. Most of the general-purpose op-amps come in 8-, 10-, or 12-pin packages.

In the *dual-in-line package* (DIP), the chip is mounted inside a plastic or ceramic case, as shown in Figure 2-5(e) and (f). The DIP is the most widely used package type because it can be mounted easily. The 8-pin dual-in-line packages are referred to as mini DIPs. Dual-in-line packages are also available with 12, 14, 16, and 20 pins. In general, as the density of components integrated on the same chip increases, the number of pins also goes up. This is especially true in digital ICs. For example, there are 64 pins on the MC68000 microprocessor chip, compared to 40 pins on MC6800 microprocessor.

Table 2-1 is a summary of the number of pins and types of packages in which linear integrated circuits are presently available. On the other hand, almost all digital ICs are DIP packages. Metal can packages are also available with dual-in-line formed leads (DIL-CAN) and with radial formed leads as shown in Figure 2-5(c) and (d). Different outlines exist within each package style to accommodate various die sizes and number of pins (leads). For example, TO-99, TO-100, and TO-101 are some of the outlines available in a transistor pack.

TABLE 2-1   LINEAR INTEGRATED CIRCUIT PACKAGES AND PIN NUMBERS

Number of pins	Type of package		
	Flat pack (ceramic)	Metal can or transistor pack (metal or plastic)	Dual-in-line (DIP) package (ceramic or plastic)
10	Yes	Yes	Yes (mini DIP)
3	No	Yes	No
5	No	Yes	No
8	Yes	Yes	Yes (mini DIP)
10	Yes	Yes	No
12	No	Yes	Yes
14	Yes	No	Yes
16	No	No	Yes
20	No	No	Yes

2-10.1(a) Selecting an IC package.   If the IC is used for experimentation/ breadboarding purposes, the best choice is the DIP package, because it is easy to mount. The mounting does not require bending or soldering of the leads. The DIP is also suitable for mounting on printed circuit boards because of its lead construction and more spacing between the leads. Generally, the ceramic DIP is more expensive than the plastic DIP, but the ceramic dissipates more heat.

The flat pack is more reliable and lighter than a comparable DIP package and is, therefore, suited for airborne applications. On the other hand, the metal can is the best choice if the IC is to be operated at relatively high power and expected to dissipate considerable heat.

When all three packages are available for a specific application, the choice can be made based on the relative cost and ease of fabrication/breadboarding the IC. Some applications may not allow a choice of package because the IC is available in only one package style.

2-10.2 Pin Identification

Although there has been a tendency to standardize IC terminal connections, the various manufacturers still use their own systems. Most manufacturers use the pin designations depicted in Figure 2-6 for the three types of packages.

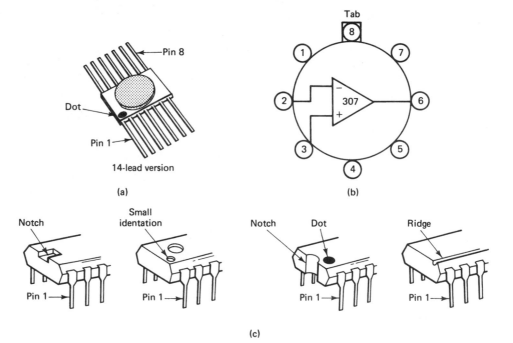

Figure 2–6 Typical pin designations for different packages. (a) Flat pack. (b) Metal can. (c) DIP.

2-10.3 Temperature Ranges

All ICs manufactured fall into one of the three basic temperature grades:

1. Military temperature range: $-55°$ to $+125°C$ (or $-55°$ to $+85°C$)
2. Industrial temperature range: $-20°$ to $+85°C$ (or $-40°$ to $+85°C$)
3. Commercial temperature range: $0°$ to $+70°C$ (or $0°$ to $+75°C$)

Individual data sheets specify exact values of the IC parameters and conditions under which these parameters are determined. The standard practice is to specify IC parameters at room temperature, that is, 25°C.

The military- and commercial-grade ICs differ in specifications for supply voltages, input current and voltage offsets and drifts, voltage gains, and others. The military-grade devices are almost always of superior quality, with tightly controlled parameters, and consequently cost more. Commercial-grade ICs have the worst tolerances among the three types but are the cheapest. In short, performance and cost are the important factors in selecting an IC.

2-11 ORDERING INFORMATION

Generally, in ordering an IC the following information must be specified: device type, package type, and temperature range. The device type is a group of alphanumeric characters such as μA741, LM741, and MC1741. These characters refer to the data sheet that specifies the device's functional and electrical characteristics. The basic package type — flat pack, transistor pack, or DIP — is represented by one letter. The military, industrial, or commercial temperature range is either numerically specified or included in the device type number or represented by a letter.

Unfortunately, the manufacturers do not have a consistent format by which the ordering information should be specified. For example, the ordering information for Fairchild's 741 mini DIP with commercial temperature range is as follows:

On the other hand, the ordering information format for a typical Motorola IC is as follows:

In National Semiconductor ICs the temperature range is denoted in the device number itself:

In National's linear ICs, a 1–2–3 numbering system is used to represent the temperature range. For example, in LM101/LM201/LM301 the 1 denotes a military temperature range device, the 2 denotes an industrial temperature range device, and the 3 denotes a commercial temperature range device.

Note that the package letter designations and temperature range nomenclatures are independently defined by each manufacturer. Therefore, the best source of information for ordering an IC is the data book.

2-12 DEVICE IDENTIFICATION

The IC is identified by marking the device type number on the face of the IC. This number is usually accompanied by the date code, indicating the year and week the device was manufactured. Figure 2-7 shows an example of Fairchild's device identification method.

Figure 2–7 Device identification.

2-13 POWER SUPPLIES FOR INTEGRATED CIRCUITS

Most linear ICs (particularly op-amps) use one or more differential amplifier stages, and differential amplifiers require both a positive and negative power supply for proper operation of the circuit. This means that most linear ICs need both a positive and a negative power supply. A few linear ICs (especially earlier op-amps) use unequal power supplies, and some ICs require only a positive supply. For example, the 702 op-amp requires unequal power supplies, whereas the 324 requires only a positive supply. When a single supply is used, it is normally necessary to connect an extra circuit to the IC. Some dual-supply op-amp ICs can also be operated from a single supply voltage, provided that a special external circuit is used with it. Digital ICs, on the other hand, generally require only one positive supply voltage. [An exception is the emitter-coupled logic (ECL)IC.]

The two power supplies required for a linear IC are usually equal in magnitude, +15 V and −15 V, for example. These power supply voltages must be referenced to a

common point or ground. Unfortunately, as in the case of ordering information, manufacturers do not agree on power supply labeling. For example, Fairchild uses $V+$ to
indicate the positive voltage and $V-$ to indicate the negative voltage or V_S to indicate both
positive and negative voltages. On the other hand, Motorola uses the symbols $+V_{CC}$ and
$-V_{EE}$ (as in discrete transistor circuits) to represent positive and negative voltages,
respectively. We will follow Motorola's notation for the power supplies.

Figure 2-8 shows power supply connections for the 741/351 op-amp. The numerals
adjacent to terminals are pin numbers. Thus, for the 741/351 op-amp, pin 7 is a positive
supply pin and pin 4 is a negative supply pin. The remaining pins on the 741/351 are
omitted for the sake of simplicity.

Figure 2-8 Typical 8-pin mini-
DIP linear IC op-amp power supply
connections.

Instead of using two separate power supplies, we can use a single power supply to
obtain $+V_{CC}$ and $-V_{EE}$, as shown in Figure 2-9. In Figure 2-9(a) the value of the total
resistance $(2R)$ should be $\geqslant 10$ kΩ so that it does not draw much current from the supply

Figure 2-9 Different arrangements for obtaining positive and negative supply voltages for an op-amp.

V_S. The two capacitors provide for decoupling (bypass) of the power supply; they range in value from 0.01 to 10 μF. In Figure 2-9(b) zener diodes are used to obtain symmetrical supply voltages. The value of R_S should be chosen such that it supplies sufficient current for the diodes to operate in the avalanche mode. The potentiometer is used in Figure 2-9(c) to assure equality between $+V_{CC}$ and $-V_{EE}$ values. Diodes D_1 and D_2 are intended to protect the IC if the positive and negative leads of the supply voltage V_S are accidentally reversed. Note that these diodes could also be connected in Figure 2-9(a) and (b).

SUMMARY

1. An operational amplifier is a direct-coupled high-gain amplifier consisting of one or more differential amplifiers, usually followed by a level translator and an output stage. The block diagram representation of the op-amp helps to visualize its internal construction.
2. In integrated circuits (ICs) all the components are fabricated on the same chip (chassis). Classified according to their mode of operation, ICs are of two basic types: digital or linear. Digital ICs are complete functioning logic networks that are equivalents of basic transistor logic circuits. On the other hand, linear ICs are equivalents of discrete transistor networks such as amplifiers, filters, frequency multipliers, and modulators. Most linear ICs are produced by a monolithic process in that all transistors and passive elements (R and C) are fabricated on a single piece of semiconductor material. In hybrid ICs, passive components and the interconnections between them are formed on an insulating substrate. Active components and monolithic ICs are then connected to form a complete circuit.
3. Each manufacturer uses a specific code and assigns a specific type number to the ICs it produces. Since the invention of the op-amp in the early 1960s, there have been a number of op-amp (IC) generations. Later-generation ICs have improved specifications and hence exhibit superior performance.
4. Linear ICs are available in three basic types: the flat pack, the metal can, and the dual-in-line (DIP) package. Almost all digital ICs are DIPs.
5. Most linear ICs require both a positive and a negative supply. The standard supply voltages for op-amps are ± 15 V. On the other hand, most digital ICs require only one positive supply voltage.

QUESTIONS

2-1. What is an op-amp?
2-2. List four basic building blocks of an op-amp.
2-3. What is the advantage of using a schematic symbol for an op-amp?
2-4. Explain briefly the difference between digital and linear ICs.
2-5. What is the difference between monolithic and hybrid ICs?

2-6. List the disadvantages of first-generation op-amps, such as the 709.

2-7. What is the major difference among SSI, MSI, LSI, and VLSI ICs?

2-8. List three types of linear IC packages.

2-9. What are the three operating temperature ranges of the IC?

2-10. What is the major difference between the power supply requirements of linear and digital ICs?

2-11. Name two recent op-amps (1980 technology) that are direct replacements for the 741 op-amp (1960 technology).

PROBLEMS

2-1. Referring to the LH0005 op-amp equivalent circuit shown in Figure 2-10, determine (a) the collector current through each transistor and (b) the dc voltage at the output terminal. (Assume that each transistor has $\beta_{dc} = \beta_{ac} = 100$ and $V_{BE} = 0.7$ V.)

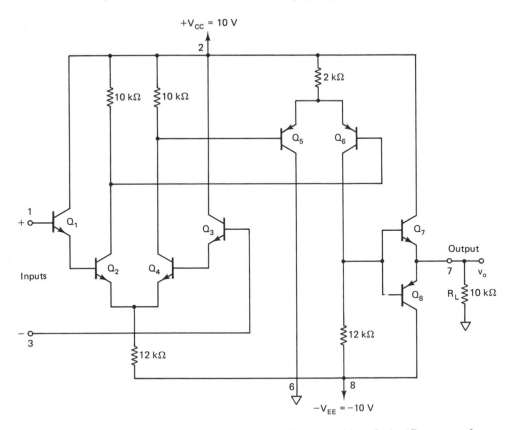

Figure 2–10 Equivalent circuit of the LH0005 for Problem 2–1. (Courtesy of National Semiconductor.)

(c) Draw the block diagram for the circuit shown in Figure 2-10 and explain the function of each block.

2-2. For the op-amp of Problem 2-1, determine:
   (a) The overall voltage gain
   (b) The input resistance
   (c) The maximum output voltage swing

3

INTERPRETATION OF DATA SHEETS AND CHARACTERISTICS OF AN OP-AMP

3-1 INTRODUCTION

In present-day electronic systems a basic building block is the op-amp. Therefore, a better understanding of the characteristics of op-amps and of their importance to overall circuit operation is essential. The manufacturers supply data sheets for the ICs they produce. These data sheets provide a wealth of information: absolute maximum ratings, intended applications, electrical characteristics, performance limitations, pin diagrams, equivalent circuits of the devices, and more. To get the most use out of these data sheets, we must be able to interpret properly the information presented in them. The lack of standardization of the IC specifications, which includes package type designations, device nomenclature, pin configuration, and ordering information, also makes it imperative to refer to the data sheets.

　　The purpose of this chapter is to explain how to read the typical op-amp data sheet, to define most electrical parameters given there, and to evaluate their significance. Proper interpretation of the data sheet should not only help you to understand the characteristics of the op-amp but should also help you to select a proper op-amp for a desired application. After these considerations, the chapter discusses equivalent circuit and open-loop op-amp configurations.

3-2 INTERPRETING A TYPICAL SET OF DATA SHEETS

Figure 3-1 shows data sheets for a Fairchild μA741 op-amp. Although the series includes the 741, 741A, 741C, and 741E models, the schematic diagrams and electrical parameters for all of them are the same, with only the values of the parameters differing from

μA741
FREQUENCY-COMPENSATED OPERATIONAL AMPLIFIER
FAIRCHILD LINEAR INTEGRATED CIRCUITS

GENERAL DESCRIPTION — The μA741 is a high performance monolithic Operational Amplifier constructed using the Fairchild Planar* epitaxial process. It is intended for a wide range of analog applications. High common mode voltage range and absence of latch-up tendencies make the μA741 ideal for use as a voltage follower. The high gain and wide range of operating voltage provides superior performance in integrator, summing amplifier, and general feedback applications.

- NO FREQUENCY COMPENSATION REQUIRED
- SHORT CIRCUIT PROTECTION
- OFFSET VOLTAGE NULL CAPABILITY
- LARGE COMMON MODE AND DIFFERENTIAL VOLTAGE RANGES
- LOW POWER CONSUMPTION
- NO LATCH-UP

ABSOLUTE MAXIMUM RATINGS

Supply Voltage	
μA741A, μA741, μA741E	±22 V
μA741C	±18 V
Internal Power Dissipation (Note 1)	
Metal Can	500 mW
Molded and Hermetic DIP	670 mW
Mini DIP	310 mW
Flatpak	570 mW
Differential Input Voltage	±30 V
Input Voltage (Note 2)	±15 V
Storage Temperature Range	
Metal Can, Hermetic DIP, and Flatpak	−65°C to +150°C
Mini DIP, Molded DIP	−55°C to +125°C
Operating Temperature Range	
Military (μA741A, μA741)	−55°C to +125°C
Commercial (μA741E, μA741C)	0°C to +70°C
Pin Temperature (Soldering)	
Metal Can, Hermetic DIPs, and Flatpak (60 s)	300°C
Molded DIPs (10 s)	260°C
Output Short Circuit Duration (Note 3)	Indefinite

CONNECTION DIAGRAMS

8-PIN METAL CAN
(TOP VIEW)
PACKAGE OUTLINE 5B
PACKAGE CODE H

Note: Pin 4 connected to case

ORDER INFORMATION

TYPE	PART NO.
μA741A	μA741AHM
μA741	μA741HM
μA741E	μA741EHC
μA741C	μA741HC

14-PIN DIP
(TOP VIEW)
PACKAGE OUTLINES 6A, 9A
PACKAGE CODES D P

ORDER INFORMATION

TYPE	PART NO.
μA741A	μA741ADM
μA741	μA741DM
μA741E	μA741EDC
μA741C	μA741DC
μA741C	μA741PC

8-PIN MINI DIP
(TOP VIEW)
PACKAGE OUTLINES 6T 9T
PACKAGE CODES R T

ORDER INFORMATION

TYPE	PART NO.
μA741C	μA741TC
μA741C	μA741RC

10-PIN FLATPAK
(TOP VIEW)
PACKAGE OUTLINE 3F
PACKAGE CODE F

ORDER INFORMATION

TYPE	PART NO.
μA741A	μA741AFM
μA741	μA741FM

Notes on following pages.

*Planar is a patented Fairchild process.

Figure 3–1 Data sheets for the μA741. (Courtesy of Fairchild Semiconductor Corporation.)

ELECTRICAL CHARACTERISTICS: $V_S = \pm15$ V, $T_A = 25°C$ unless otherwise specified.

CHARACTERISTICS (see definitions)		CONDITIONS	MIN	TYP	MAX	UNITS
Input Offset Voltage		$R_S \leqslant 50\Omega$		0.8	3.0	mV
Average Input Offset Voltage Drift					15	$\mu V/°C$
Input Offset Current				3.0	30	nA
Average Input Offset Current Drift					0.5	$nA/°C$
Input Bias Current				30	80	nA
Power Supply Rejection Ratio		$V_S = +20, -20; V_S = -20, +10V, R_S = 50\Omega$		15	50	$\mu V/V$
Output Short Circuit Current			10	25	40	mA
Power Dissipation		$V_S = \pm20V$		80	150	mW
Input Impedance		$V_S = \pm20V$	1.0	6.0		$M\Omega$
Large Signal Voltage Gain		$V_S = \pm20V, R_L = 2k\Omega, V_{OUT} = \pm15V$	50			V/mV
Transient Response	Rise Time			0.25	0.8	μs
(Unity Gain)	Overshoot			6.0	20	%
Bandwidth (Note 4)			.437	1.5		MHz
Slew Rate (Unity Gain)		$V_{IN} = \pm10V$	0.3	0.7		$V/\mu s$
The following specifications apply for $-55°C \leqslant T_A \leqslant +125°C$						
Input Offset Voltage					4.0	mV
Input Offset Current					70	nA
Input Bias Current					210	nA
Common Mode Rejection Ratio		$V_S = \pm20V, V_{IN} = \pm15V, R_S = 50\Omega$	80	95		dB
Adjustment For Input Offset Voltage		$V_S = \pm20V$	10			mV
Output Short Circuit Current			10		40	mA
Power Dissipation	$V_S = \pm20V$	$-55°C$			165	mW
		$+125°C$			135	mW
Input Impedance		$V_S = \pm20V$	0.5			$M\Omega$
Output Voltage Swing	$V_S = \pm20V,$	$R_L = 10k\Omega$	±16			V
		$R_L = 2k\Omega$	±15			V
Large Signal Voltage Gain		$V_S = \pm20V, R_L = 2k\Omega, V_{OUT} = \pm15V$	32			V/mV
		$V_S = \pm5V, R_L = 2k\Omega, V_{OUT} = \pm2$ V	10			V/mV

NOTES
1. Rating applies to ambient temperatures up to 70°C. Above 70°C ambient derate linearly at 6.3mW/°C for the metal can, 8.3mW/°C for the DIP and 7.1mW/°C for the Flatpak.
2. For supply voltages less than ±15V, the absolute maximum input voltage is equal to the supply voltage.
3. Short circuit may be to ground or either supply. Rating applies to +125°C case temperature or 75°C ambient temperature.
4. Calculated value from: $BW(MHz) = \dfrac{0.35}{Rise\ Time\ (\mu s)}$

Figure 3–1 (Continued)

μA741

ELECTRICAL CHARACTERISTICS: $V_S = \pm 15$ V, $T_A = 25°C$ unless otherwise specified.

CHARACTERISTICS (see definitions)		CONDITIONS	MIN	TYP	MAX	UNITS
Input Offset Voltage		$R_S \leqslant 10$ kΩ		1.0	5.0	mV
Input Offset Current				20	200	nA
Input Bias Current				80	500	nA
Input Resistance			0.3	2.0		MΩ
Input Capacitance				1.4		pF
Offset Voltage Adjustment Range				±15		mV
Large Signal Voltage Gain		$R_L \geqslant 2$ kΩ, $V_{OUT} = \pm 10$ V	50,000	200,000		
Output Resistance				75		Ω
Output Short Circuit Current				25		mA
Supply Current				1.7	2.8	mA
Power Consumption				50	85	mW
Transient Response (Unity Gain)	Rise time	$V_{IN} = 20$ mV, $R_L = 2$ kΩ, $C_L \leqslant 100$ pF		0.3		μs
	Overshoot			5.0		%
Slew Rate		$R_L \geqslant 2$ kΩ		0.5		V/μs

The following specifications apply for $-55°C \leqslant T_A \leqslant +125°C$:

			MIN	TYP	MAX	UNITS
Input Offset Voltage		$R_S \leqslant 10$ kΩ		1.0	6.0	mV
Input Offset Current	$T_A = +125°C$			7.0	200	nA
	$T_A = -55°C$			85	500	nA
Input Bias Current	$T_A = +125°C$			0.03	0.5	μA
	$T_A = -55°C$			0.3	1.5	μA
Input Voltage Range			±12	±13		V
Common Mode Rejection Ratio		$R_S \leqslant 10$ kΩ	70	90		dB
Supply Voltage Rejection Ratio		$R_S \leqslant 10$ kΩ		30	150	μV/V
Large Signal Voltage Gain		$R_L \geqslant 2$ kΩ, $V_{OUT} = \pm 10$ V	25,000			
Output Voltage Swing	$R_L \geqslant 10$ kΩ		±12	±14		V
	$R_L \geqslant 2$ kΩ		±10	±13		V
Supply Current	$T_A = +125°C$			1.5	2.5	mA
	$T_A = -55°C$			2.0	3.3	mA
Power Consumption	$T_A = +125°C$			45	75	mW
	$T_A = -55°C$			60	100	mW

TYPICAL PERFORMANCE CURVES FOR μA741A AND μA741

Figure 3–1 (Continued)

ELECTRICAL CHARACTERISTICS: $V_S = \pm 15$ V, $T_A = 25°C$ unless otherwise specified.

CHARACTERISTICS (see definitions)		CONDITIONS	MIN	TYP.	MAX	UNITS
Input Offset Voltage		$R_S \leqslant 50\Omega$		0.8	3.0	mV
Average Input Offset Voltage Drift					15	$\mu V/°C$
Input Offset Current				3.0	30	nA
Average Input Offset Current Drift					0.5	nA/°C
Input Bias Current				30	80	nA
Power Supply Rejection Ratio		$V_S = +10, -20$; $V_S = +20, -10V$, $R_S = 50\Omega$		15	50	$\mu V/V$
Output Short Circuit Current			10	25	40	mA
Power Dissipation		$V_S = \pm 20V$		80	150	mW
Input Impedance		$V_S = \pm 20V$	1.0	6.0		$M\Omega$
Large Signal Voltage Gain		$V_S = \pm 20V$, $R_L = 2k\Omega$, $V_{OUT} = \pm 15V$	50			V/mV
Transient Response	Rise Time			0.25	0.8	μs
(Unity Gain)	Overshoot			6.0	20	%
Bandwidth (Note 4)			.437	1.5		MHz
Slew Rate (Unity Gain)		$V_{IN} = \pm 10V$	0.3	0.7		$V/\mu s$
The following specifications apply for $0°C \leqslant T_A \leqslant 70°C$						
Input Offset Voltage					4.0	mV
Input Offset Current					70	nA
Input Bias Current					210	nA
Common Mode Rejection Ratio		$V_S = \pm 20V$, $V_{IN} = \pm 15V$, $R_S = 50\Omega$	80	95		dB
Adjustment For Input Offset Voltage		$V_S = \pm 20V$	10			mV
Output Short Circuit Current			10		40	mA
Power Dissipation		$V_S = \pm 20V$			150	mW
Input Impedance		$V_S = \pm 20V$	0.5			$M\Omega$
Output Voltage Swing	$R_L = 10k\Omega$	$V_S = \pm 20V$,	± 16			V
	$R_L = 2k\Omega$		± 15			V
Large Signal Voltage Gain		$V_S = \pm 20V$, $R_L = 2k\Omega$, $V_{OUT} = \pm 15V$	32			V/mV
		$V_S = \pm 5V$, $R_L = 2k\Omega$, $V_{OUT} = \pm 2$ V	10			V/mV

EQUIVALENT CIRCUIT

Figure 3–1 (Continued)

μA741C

ELECTRICAL CHARACTERISTICS: V_S = ±15 V, T_A = 25°C unless otherwise specified.

CHARACTERISTICS (see definitions)		CONDITIONS	MIN	TYP	MAX	UNITS
Input Offset Voltage		$R_S \leqslant 10$ kΩ		2.0	6.0	mV
Input Offset Current				20	200	nA
Input Bias Current				80	500	nA
Input Resistance			0.3	2.0		MΩ
Input Capacitance				1.4		pF
Offset Voltage Adjustment Range				±15		mV
Input Voltage Range			±12	±13		V
Common Mode Rejection Ratio		$R_S \leqslant 10$ kΩ	70	90		dB
Supply Voltage Rejection Ratio		$R_S \leqslant 10$ kΩ		30	150	μV/V
Large Signal Voltage Gain		$R_L \geqslant 2$ kΩ, V_{OUT} = ±10 V	20,000	200,000		
Output Voltage Swing		$R_L \geqslant 10$ kΩ	±12	±14		V
		$R_L \geqslant 2$ kΩ	±10	±13		V
Output Resistance				75		Ω
Output Short Circuit Current				25		mA
Supply Current				1.7	2.8	mA
Power Consumption				50	85	mW
Transient Response (Unity Gain)	Rise time	V_{IN} = 20 mV, R_L = 2 kΩ, $C_L \leqslant$ 100 pF		0.3		μs
	Overshoot			5.0		%
Slew Rate		$R_L \geqslant 2$ kΩ		0.5		V/μs

The following specifications apply for $0°C \leqslant T_A \leqslant +70°C$:

Input Offset Voltage					7.5	mV
Input Offset Current					300	nA
Input Bias Current					800	nA
Large Signal Voltage Gain		$R_L \geqslant 2$ kΩ, V_{OUT} = ±10 V	15,000			
Output Voltage Swing		$R_L \geqslant 2$ kΩ	±10	±13		V

TYPICAL PERFORMANCE CURVES FOR μA741E AND μA741C

Figure 3-1 (Continued)

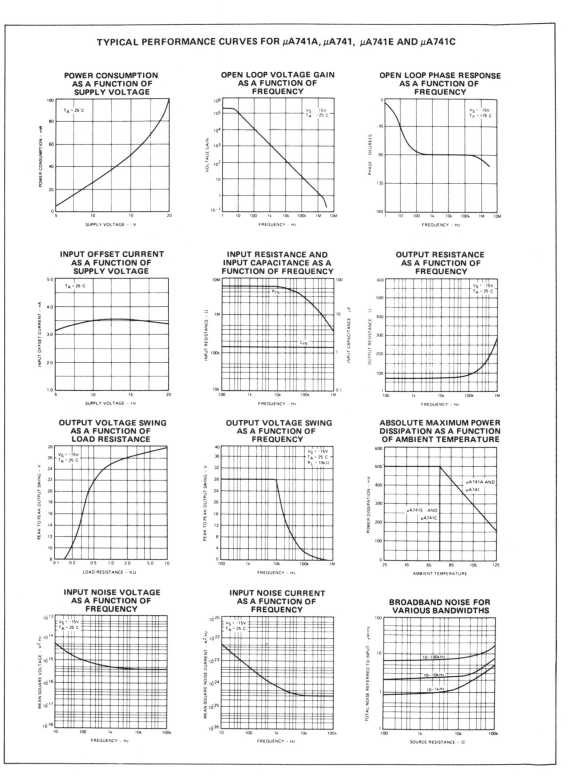

TYPICAL PERFORMANCE CURVES FOR μA741A, μA741, μA741E AND μA741C

Figure 3–1 (Continued)

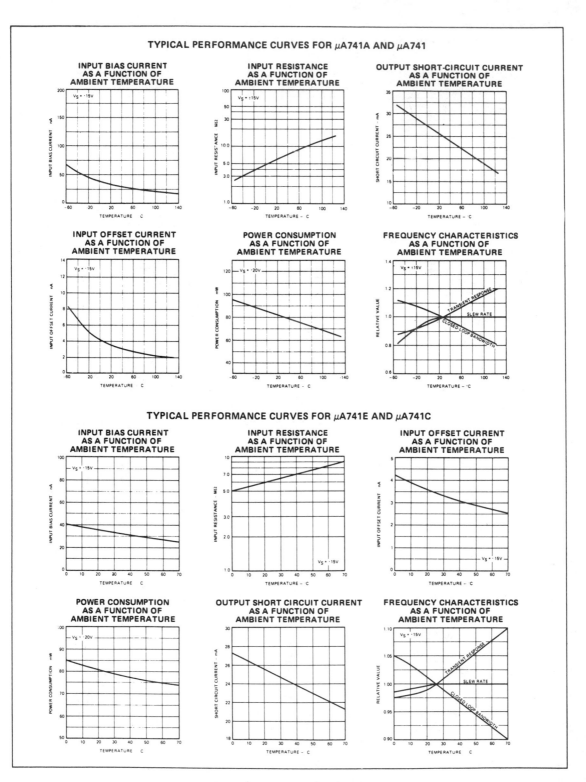

TYPICAL PERFORMANCE CURVES FOR μA741A AND μA741

Figure 3–1 (Continued)

Figure 3–1 (Continued)

TYPICAL APPLICATIONS (Cont'd)

SIMPLE INTEGRATOR

$$E_{OUT} = -\frac{1}{R_1 C_1} \int E_{IN} dt$$

SIMPLE DIFFERENTIATOR

$$E_{OUT} = -R2C \frac{dE_{IN}}{dt}$$

LOW DRIFT LOW NOISE AMPLIFIER

Voltage Gain = 10^3
Input Offset Voltage Drift = 0.6 μV/°C
Input Offset Current Drift = 2.0 pA/°C

HIGH SLEW RATE POWER AMPLIFIER

NOTCH FILTER USING THE μA741 AS A GYRATOR

Trim R3 such that
$$\frac{R1}{R2} = \frac{R3}{2\ R4}$$

NOTCH FREQUENCY AS A FUNCTION OF C1

Figure 3–1 (Continued)

one model to another. For instance, the 741A has the best tolerance (tightly controlled parameters) and costs the most. At the other extreme, the 741C has the worst tolerance and consequently costs the least. In this section we consider only the 741C op-amp specifications.

Generally, the information found on data sheets can be broken down into the following groups:

1. At the top of the data sheet is a device number and a brief description of the basic type of the device, such as *frequency-compensated op-amp, low-power op-amp,* or *low-cost programmable op-amp*.

2. A general description is given that includes the construction process of the device, intended applications, and a list of the main features.

3. Absolute maximum ratings for the proper operation of the device are then specified. These ratings are limiting values of operating and environmental conditions applicable to the device and should not be exceeded.

4. The pin configuration (connection diagram), package types, and order information are given.

5. The internal schematic diagram (equivalent circuit) is shown.

6. Electrical characteristics and parameter values under specific conditions are also given.

7. Typical performance curves such as voltage gain versus supply voltage, output voltage swing as a function of frequency, and power consumption as a function of temperature are provided.

8. Finally, typical applications and test circuits for the device are illustrated.

Let us now take a close look at the μA741C data sheets with reference to the information just outlined.

1. The Fairchild 741 is an internally frequency compensated operational amplifier.

2. The 741 is a monolithic IC constructed by a special process called "planar epitaxial." It is suited for integrator, summing amplifier, voltage follower, and other general feedback applications. The features of the 741 are as follows:

a. No external frequency compensation required

b. Short-circuit protection

c. Offset null capability

d. Large common-mode and differential voltage ranges

e. Low power consumption

f. No latch-up problem

3. Absolute maximum ratings are specified for supply voltage, input and differential input voltages, storage and operating temperature ranges, soldering pin temperature, and output short-circuit duration. Supply voltage and operating temperature range are given for Models 741, 741A, 741C, and 741E, whereas internal power dissipation, storage temperature range, and soldering pin temperature are listed for all package types: flat pack,

metal can, and DIP. For safe and proper operation of the device, these ratings should not be exceeded even under the worst operating conditions.

4. The 741 is available in all *three* package types: 8-pin metal can, 10-pin flat pack, and 8- or 14-pin DIP. The pin diagram for all these packages is shown on the data sheets. Order information is included for each package as well.

5. The equivalent circuit diagram illustrates the internal structure of the 741 op-amp. This diagram helps to clarify the capabilities and limitations of the op-amp. Note that all the 741 models have the same equivalent circuit.

6. For the 741C, two sets of electrical specifications are given. One set of specifications applies at room temperature (T_A) of 25°C, whereas the other set applies to the commercial temperature range from 0° to +70°C. Since our main interest at this point is in the definition and significance of such parameters rather than their values, we shall discuss the parameters at 25°C. These parameters are applicable at supply voltages $+V_{CC} = +15$ V and $-V_{EE} = -15$ V. The electrical parameters are defined in the following paragraphs.

Input Offset Voltage.   Input offset voltage is the voltage that must be applied between the two input terminals of an op-amp to null the output, as shown in Figure 3-2. In the figure V_{dc1} and V_{dc2} are dc voltages and R_s represents the source resistance. We denote input offset voltage by V_{io}. This voltage V_{io} could be positive or negative; therefore, its absolute value is listed on the data sheet. For a 741C the maximum value of V_{io} is 6 mV dc. The smaller the value of V_{io}, the better the input terminals are matched. For instance, the 714C *precision op-amp* has $V_{io} = 150$ μV maximum. Refer to the data sheet in Appendix D.

Figure 3–2   Defining input offset voltage V_{io}. (Pin numbers for 8-pin mini DIP.)

Input Offset Current.   The algebraic difference between the currents into the inverting and noninverting terminals is referred to as *input offset current* I_{io} (see Figure 3-3). In the form of an equation,

$$I_{io} = |I_{B1} - I_{B2}| \qquad (3\text{-}1)$$

where I_{B1} is the current into the noninverting input and I_{B2} the current into the inverting input.

The input offset current for the 741C is 200 nA maximum. As the matching between two input terminals is improved, the difference between I_{B1} and I_{B2} becomes smaller; that

$+V_{CC} = +15$ V

I_{B1}

3

7

+

6

741C

Output

2

4

I_{B2}

$I_{io} = |I_{B1} - I_{B2}|$ $-V_{EE} = -15$ V

Figure 3–3   Defining input offset current I_{io}.

is, the I_{io} value decreases further. For instance, the precision op-amp 714C has a maximum value of I_{io} equal to 6 nA, a dramatic improvement over older technology. Refer to the data sheet of the 714C in Appendix D.

   Input Bias Current.   Input bias current I_B is the average of the currents that flow into the inverting and noninverting input terminals of the op-amp. In equation form,

$$I_B = \frac{I_{B1} + I_{B2}}{2} \tag{3-2}$$

$I_B = 500$ nA maximum for the 741C, whereas I_B for the precision 714C is ± 7 nA. Note that the two input currents I_{B1} and I_{B2} are actually the base currents of the first differential amplifier stage.

   Differential Input Resistance.   Differential input resistance R_i (often referred to as input resistance) is the equivalent resistance that can be measured at either the inverting or noninverting input terminal with the other terminal connected to ground. For the 741C the input resistance is a relatively high 2 MΩ. However, for FET input op-amps this value is amazingly large. For example, $R_i = 1000$ GΩ ($10^{12}\Omega$) for the μAF771 FET input op-amp.

   Input Capacitance.   Input capacitance C_i is the equivalent capacitance that can be measured at either the inverting or noninverting terminal with the other terminal connected to ground. A typical value of C_i is 1.4 pF for the 741C. This parameter is not listed on all op-amp data sheets.

   Offset Voltage Adjustment Range.   One of the features of the 741 family op-amps is an offset voltage null capability. The 741 op-amps have pins 1 and 5 marked as *offset null* for this purpose. As shown in Figure 3-4, a 10-kΩ potentiometer can be connected

$+V_{CC} = +15$ V

3

7

+

6

741C

Output

5

2

4

1

$-V_{EE} = -15$ V

10 kΩ

Figure 3–4   Offset voltage adjustment range.

between offset null pins 1 and 5, and the wiper of the potentiometer can be connected to the negative supply $-V_{EE}$. By varying the potentiometer, the *output offset voltage* (output voltage without any input applied) can be reduced to zero volts. Thus the offset voltage adjustment range is the range through which the input offset voltage can be adjusted by varying the 10-kΩ potentiometer. For the 741C the offset voltage adjustment range is ± 15 mV. Very few op-amps have the offset voltage null capability, some of these being the 301, 748, and 777. This means that for most op-amps we have to design an offset voltage compensating network in order to reduce the output offset voltage to zero. The design of such a network is presented in Section 5-2.1.

Input Voltage Range.   When the same voltage is applied to both input terminals, the voltage is called a common-mode voltage V_{cm} and the op-amp is said to be operating in the common-mode configuration (see Figure 3-5). For the 741C the range of the input common-mode voltage is ± 13 V maximum. This means that the common-mode voltage applied to both input terminals can be as high as $+13$ V or as low as -13 V without disturbing proper functioning of the op-amp. In other words, the input voltage range is the range of common-mode voltages over which the offset specifications apply. Obviously, the common-mode configuration is used only for test purposes to determine the degree of matching between the inverting and noninverting input terminals.

Figure 3–5   Common-mode configuration.

Common-Mode Rejection Ratio.   The common-mode rejection ratio (CMRR) is defined in several essentially equivalent ways by various manufacturers. Generally, it can be defined as the ratio of the differential voltage gain A_d to the common-mode voltage gain A_{cm}; that is,

$$\text{CMRR} = \frac{A_d}{A_{cm}} \tag{3-3}$$

The differential voltage gain A_d is the same as the large-signal voltage gain A, which is specified on the data sheets; however, the common-mode voltage gain can be determined from the circuit of Figure 3-5 using the equation

$$A_{cm} = \frac{V_{ocm}}{V_{cm}} \tag{3-4}$$

where V_{ocm} = output common-mode voltage
V_{cm} = input common-mode voltage
A_{cm} = common-mode voltage gain

Generally, the A_{cm} is very small and $A_d = A$ is very large; therefore, the CMRR is very large. Being a large value, CMRR is most often expressed in decibels (dB). For the 741C, CMRR is 90 dB typically. Note that this value of CMRR is determined under the test condition that the input source resistance $R_s \leq 10$ kΩ. In Figure 3-5, R_s is assumed to be zero because most of the practical voltage sources have negligible source resistances.

The higher the value of CMRR, the better is the matching between two input terminals and the smaller is the output common-mode voltage. For the 714C precision op-amp, CMRR=120 dB. Refer to the data sheet in Appendix D. This means that the 714C has a better ability to reject common-mode voltages such as electrical noise than the 741C and is preferred in noise environments.

Supply Voltage Rejection Ratio.   The change in an op-amp's input offset voltage V_{io} caused by variations in supply voltages is called the *supply voltage rejection ratio* (SVRR). A variety of terms equivalent to SVRR are used by different manufacturers, such as the *power supply rejection ratio* (PSRR) and the *power supply sensitivity* (PSS). These terms are expressed either in microvolts per volt or in decibels. If we denote the change in supply voltages by ΔV and the corresponding change in input offset voltage by ΔV_{io}, SVRR can be defined as follows:

$$SVRR = \frac{\Delta V_{io}}{\Delta V} \qquad (3\text{-}5)$$

For the 741C, SVRR=150 μV/V. On the other hand, for the 714C (see Appendix D)

$$SVRR = 20 \log \left(\frac{\Delta V}{\Delta V_{io}}\right) = 104 \text{ dB} \qquad (3\text{-}6)$$

or, equivalently,

$$SVRR = 6.31 \ \mu V/V$$

This means that the lower the value of SVRR in microvolts/volt, the better for op-amp performance.

Note that for the 741C (and most other op-amps) SVRR is measured for both supply magnitudes increasing or decreasing simultaneously, with $R_s \leq 10$ kΩ. However, for some op-amps SVRR is separately specified as *positive SVRR* for positive voltage and *negative* SVRR for negative supply voltage.

Large-Signal Voltage Gain.   Since the op-amp amplifies difference voltage between two input terminals, the voltage gain of the amplifier is defined as

$$\text{voltage gain} = \frac{\text{output voltage}}{\text{differential input voltage}}$$

that is,

$$A = \frac{V_o}{V_{id}} \qquad (3\text{-}7)$$

Because output signal amplitude is much larger than the input signal, the voltage gain is commonly called *large-signal voltage gain*. Under the test conditions $R_L \geq 2$ kΩ and

$V_o = \pm 10$ V (or 20 V peak to peak), the large-signal voltage gain of the 741C is 200,000 typically (see Figure 3-6).

Figure 3–6 Determining voltage gain.

Output Voltage Swing.   The output voltage swing $V_{o\,max}$ of the 741C is guaranteed to be between -13 and $+13$ V for $R_L \geq 2$ kΩ, that is, giving a 26-V peak-to-peak undistorted sine wave for ac input signals. In fact, the output voltage swing indicates the values of positive and negative saturation voltages of the op-amp. The output voltage never exceeds these limits for given supply voltages $+V_{CC}$ and $-V_{EE}$.

Output Resistance.   Output resistance R_o is the equivalent resistance that can be measured between the output terminal of the op-amp and the ground (or common point). It is 75 Ω for the 741C op-amp.

Output Short-Circuit Current.   Nobody would deliberately want to short the output terminal of the op-amp to ground, but if such an event were to happen accidentally, the current through the short would certainly be much higher in value than either I_B or I_{io}. This high current may damage the op-amp if it does not have output short-circuit protection. Recall, however, that the 741 family op-amps do have short-circuit protection circuitry built in.

The short-circuit current I_{sc} = 25 mA for the 741C op-amp. This means that the built-in short-circuit protection is guaranteed to withstand 25 mA of current in protecting the op-amp.

Supply Current.   Supply current I_s is the current drawn by the op-amp from the power supply. This parameter is not given on most of the op-amp data sheets. For the 741C op-amp the supply current I_s = 2.8 mA.

Power Consumption.   Power consumption P_c is the amount of quiescent power (v_{in} = 0 V) that must be consumed by the op-amp in order to operate properly. The amount of power consumed by the 741C is 85 mW.

Transient Response.   The response of any practically useful network to a given input is composed of two parts: the *transient* and *steady-state* response. The transient response is that portion of the complete response before the output attains some fixed value. Once reached, this fixed value remains at that level and is, therefore, referred to as a *steady-state value*. The response of the network after it attains a fixed value is independent of time and is called the *steady-state response*. Unlike the steady-state response, the transient response is time variant. The rise time and the percent of overshoot are the characteristics of the transient response. The time required by the output to go from 10% to 90% of its final value is called the *rise time*. Conversely, *overshoot* is the maximum amount by which the output deviates from the steady-state value. Overshoot is generally expressed as a percentage.

The transient response test circuit for the 741C as well as the response of this test circuit for V_{in} = 20 mV dc is included in the data sheets. The rise time is 0.3 μs and overshoot is 5% for the 741C op-amp.

The transient response is one of the important considerations in selecting an op-amp in ac applications. In fact, the rise time is inversely proportional to the unity gain bandwidth of the op-amp. This means that the smaller the value of rise time, the higher is the bandwidth.

Slew Rate.   Slew rate (SR) is defined as the maximum rate of change of output voltage per unit of time and is expressed in volts per microseconds. In equation form,

$$SR = \frac{dV_o}{dt}\bigg|_{maximum} \quad V/\mu s \qquad (3\text{-}8)$$

Slew rate indicates how rapidly the output of an op-amp can change in response to changes in the input frequency. The slew rate changes with change in voltage gain and is normally specified at unity (+1) gain. The slew rate of an op-amp is fixed; therefore, if the slope requirements of the output signal are greater than the slew rate, then distortion occurs. Thus slew rate is one of the important factors in selecting the op-amp for ac applications, particularly at relatively high frequencies.

One of the drawbacks of the 741C is its low slew rate (0.5 V/μs), which limits its use in relatively high frequency applications, especially in oscillators, comparators, and filters. The newer op-amps—LF351, μAF771, and MC34001—which are direct replacements for 741, have a slew rate of 13 V/μs. In high-speed op-amps especially, the slew rate is significantly improved. For instance, the LM318 has a slew rate of 70 V/μs.

Some Additional Electrical Parameters.   On the data sheets of some op-amps you may find additional electrical parameters that are related to their intended applications. These electrical parameters include gain–bandwidth product, average temperature coefficients of input offset voltage and current, long-term input offset voltage and current stability, and equivalent noise voltage and current. Besides these, one more parameter is applicable to only dual and quad op-amps: channel separation.

Gain–Bandwidth Product.   The gain–bandwidth product (GB) is the bandwidth of the op-amp when the voltage gain is 1. Although for the 741 op-amp it is not listed under electrical characteristics, from the open-loop voltage gain versus frequency graph it can be found to be approximately 1 MHz. Equivalent terms for gain–bandwidth product are *closed-loop bandwidth, unity gain bandwidth,* and *small-signal bandwidth.* The newer op-amps LF351 and MC34001 have a gain–bandwidth product of 4 MHz.

Average Temperature Coefficient of Input Offset Voltage (and Current).   These parameters are also referred to as *average input offset voltage* or *current drift.* The average temperature coefficient of input offset voltage is the average rate of change in input offset voltage per unit change in temperature expressed as μV/°C. Similarly, the average temperature coefficient of input offset current is the average rate of change in input offset current per unit change in temperature and is usually expressed as pA/°C. Both of these parameters are generally given for the instrumentation and precision type op-amps. For example, for the precision op-amp 714C, the average temperature coefficient of input offset voltage $\Delta V_{io}/\Delta T = 0.5\ \mu V/°C$ typically, and the average temperature coefficient of input offset current $\Delta I_{io}/\Delta T = 12$ pA/°C.

Long-Term Input Offset Voltage (and Current) Stability. $\Delta V_{io}/\Delta t$ is the average rate of change in input offset voltage per unit of time and is generally expressed as μV/week. It is also referred to as *input offset voltage drift with time*. Similarly, $\Delta I_{io}/\Delta t$ is the average rate of change in input offset current per unit of time in pA/week. Again, these parameters are normally given for instrumentation and precision-type op-amps. For instance, the precision-type op-amp 714C has $\Delta V_{io}/\Delta t = 0.1\ \mu V$/week typically. The value of $\Delta I_{io}/\Delta t$ is negligible for the 714C; hence it is not listed on the data sheet. See Appendix D.

Equivalent Input Noise Voltage and Current.   Since electrical noise is random in nature, it is expressed as a root-mean-square value. Standard industry practice is to express the noise as a power density. The equivalent input noise voltage is, therefore, expressed as square voltage (V^2/Hz) and the equivalent input noise current as square noise current A^2/Hz). Using input noise voltage and input noise current versus frequency curves given on the data sheets (see Figure 3-1), the minimum amount of signal power that is necessary to overcome the noise signal and produce a measurable output signal can be determined. As a general rule, the signal-to-noise ratio must be at least larger than by a factor of 10.

Channel Separation.   This parameter is specified in the data sheets of dual and quad (four) op-amps such as the μAF772 and μAF774, respectively. It is a measure of the amount of electrical coupling between op-amps that are integrated on the same chip. Because of the physical closeness of op-amps in dual and quad packages, when a signal is applied to the input of only one op-amp, some signal will appear at the output of other op-amps. The amplitude of these output signals is approximately the same and can be calculated using channel separation and a given input signal. Channel separation is also called *amplifier-to-amplifier coupling*.

The 774/348 is a true quad 741 and has a channel separation of -120 dB. This means that if a signal is applied to one of the op-amps, the signal at the outputs of the undriven op-amps will be at least 120 dB (equivalent to a ratio of 10^6) below the signal output of the driven op-amp.

Thus, as more and more data are made available on op-amp characteristics, it becomes easier to select a proper op-amp for a desired application.

A list of parameters that should be considered for ac and dc applications is given in Table 3-1. Depending on the application, a designer sets priorities among these parameters and then selects a suitable op-amp for his or her needs.

7. Typical performance curves included on the data sheets of the 741C are the graphs of electrical parameters that are affected by mainly three factors:

    a. Change in supply voltage

    b. Change in operating frequency

    c. Change in temperature

TABLE 3-1   PARAMETER EVALUATION FOR AC AND DC
APPLICATIONS

For ac applications, consider:	For dc applications, consider:
Input resistance	Input resistance
Output resistance	Output resistance
Large-signal voltage gain	Large-signal voltage gain
Output voltage swing	Output voltage swing
Average input offset voltage and current drifts	Input offset voltage and input offset current
Long-term input offset voltage stability	Average input offset voltage and current drifts
Gain–bandwidth product	Long-term input offset voltage stability
Transient response	
Slew rate	
Equivalent input noise voltage and current	

The graphs of parameters that are affected by each of these factors are as follows:

Supply-voltage-dependent parameters. Voltage gain, output voltage swing, input common-mode voltage range, power consumption, and input offset current.

Frequency-dependent parameters. Voltage gain, input resistance, output resistance, output voltage swing, input noise voltage and noise current, and CMRR.

Temperature-dependent parameters. Absolute maximum power dissipation, input bias current, input offset current, power consumption, input resistance, output short-circuit current, transient response, and gain–bandwidth product.

In short, the information obtained from the performance curves can be used to improve the op-amp's performance.

8. Finally, a collection of amplifier applications with circuit diagrams is included on the data sheets. These are the applications in which the op-amp is guaranteed to perform satisfactorily. Such applications of the 741C include voltage follower, inverting and noninverting amplifiers, clipping amplifier, simple integrator, differentiator, and others.

We have seen that a wide range of important information, including a pin diagram, electrical characteristics, and applications, is included on the data sheets of an op-amp. To get the most out of this device, it is essential to refer to the data sheets.

3-3 THE IDEAL OP-AMP

An *ideal* op-amp would exhibit the following electrical characteristics:

1. Infinite voltage gain A.

2. Infinite input resistance R_i so that almost any signal source can drive it and there is no loading of the preceding stage.

3. Zero output resistance R_o so that output can drive an infinite number of other devices.

4. Zero output voltage when input voltage is zero.

5. Infinite bandwidth so that any frequency signal from 0 to ∞ Hz can be amplified without attenuation.

6. Infinite common-mode rejection ratio so that the output common-mode noise voltage is zero.

7. Infinite slew rate so that output voltage changes occur simultaneously with input voltage changes.

There are practical op-amps that can be made to approximate some of these characteristics using a negative feedback arrangement. In particular, the input resistance, output resistance, and bandwidth can be brought close to ideal values by this method. Chapter 4 includes a discussion of negative feedback configurations.

3-4 EQUIVALENT CIRCUIT OF AN OP-AMP

Figure 3-7 shows an equivalent circuit of an op-amp. This circuit includes important values from the data sheets: A, R_i, and R_o. Note that Av_{id} is an equivalent Thévenin voltage source, and R_o is the Thévenin equivalent resistance looking back into the output terminal of an op-amp.

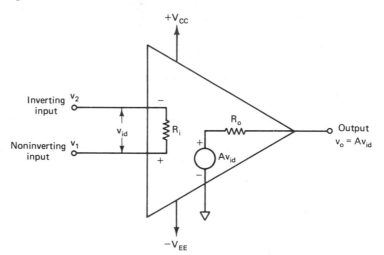

Figure 3–7 Equivalent circuit of an op-amp.

The equivalent circuit is useful in analyzing the basic operating principles of op-amps and in observing the effects of feedback arrangements. For the circuit shown in Figure 3-7, the output voltage is

$$v_o = Av_{id} = A(v_1 - v_2) \tag{3-9}$$

where A = large-signal voltage gain

v_{id} = difference input voltage

v_1 = voltage at the noninverting input terminal with respect to ground

v_2 = voltage at the inverting terminal with respect to ground

Equation (3-9) indicates that the output voltage v_o is directly proportional to the algebraic difference between the two input voltages. In other words, the op-amp amplifies the difference between the two input voltages; it does not amplify the input voltages themselves. For this reason the polarity of the output voltage depends on the polarity of the difference voltage.

3-5 IDEAL VOLTAGE TRANSFER CURVE

Equation (3-9) is the basic op-amp equation, in which the output offset voltage is assumed to be zero. This equation is useful in studying the op-amp's characteristics and in analyzing different circuit configurations that employ feedback. The graphic representation of this equation is shown in Figure 3-8, where the output voltage v_o is plotted against input difference voltage v_{id}, keeping gain A constant.

Figure 3–8 Ideal voltage transfer curve.

Note, however, that the output voltage cannot exceed the positive and negative saturation voltages. These saturation voltages are specified by an output voltage swing rating of the op-amp for given values of supply voltages. This means that the output voltage is directly proportional to the input difference voltage only until it reaches the saturation voltages and that thereafter output voltage remains constant, as shown in Figure 3-8.

The curve shown in Figure 3-8 is called an *ideal voltage transfer curve*, ideal because output offset voltage is assumed to be zero. In normal op-amp use (with negative feedback), this voltage is near zero and is ignored for simplicity of calculation. Notice that the curve is not drawn to scale. If drawn to scale, the curve would be almost vertical because of the very large values of A.

3-6 OPEN-LOOP OP-AMP CONFIGURATIONS

In the case of amplifiers the term *open loop* indicates that no connection, either direct or via another network, exists between the output and input terminals. That is, the output signal is not fed back in any form as part of the input signal, and the loop that would have been formed with feedback is open.

When connected in open-loop configuration, the op-amp simply functions as a high-gain amplifier. There are three open-loop op-amp configurations:

1. Differential amplifier
2. Inverting amplifier
3. Noninverting amplifier

These configurations are classed according to the number of inputs used and the terminal to which the input is applied when a single input is used.

3-6.1 The Differential Amplifier

Figure 3-9 shows the open-loop differential amplifier in which input signals v_{in1} and v_{in2} are applied to the positive and negative input terminals. Since the op-amp amplifies the difference between the two input signals, this configuration is called the *differential amplifier*.

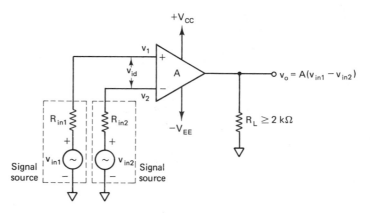

Figure 3–9   Open-loop differential amplifier.

The op-amp is a versatile device because it amplifies both ac and dc input signals. This means that v_{in1} and v_{in2} could be either ac or dc voltages. The source resistances R_{in1} and R_{in2} are normally negligible compared to the input resistance R_i. Therefore, the voltage drops across these resistors can be assumed to be zero, which then implies that $v_1 = v_{in1}$ and $v_2 = v_{in2}$. Substituting these values of v_1 and v_2 in Equation (3-9), we get

$$v_o = A(v_{in1} - v_{in2})$$

Thus, as expected, the output voltage is equal to the voltage gain A times the difference between the two input voltages. Also, notice that the polarity of the output voltage is dependent on the polarity of the input difference voltage $(v_{in1} - v_{in2})$. In open-loop configurations, gain A is commonly referred to as *open-loop gain*.

3-6.2 The Inverting Amplifier

In the inverting amplifier only one input is applied and that is to the inverting input terminal. The noninverting input terminal is grounded (refer to Figure 3-10). Since $v_1 = 0$ V, and $v_2 = v_{in}$ from Equation (3-9),

$$v_o = -Av_{in}$$

The negative sign indicates that the output voltage is out of phase with respect to input by 180° or is of opposite polarity. Thus in the inverting amplifier the input signal is amplified by gain A and is also inverted at the output.

Figure 3–10   Inverting amplifier.

3-6.3 The Noninverting Amplifier

Figure 3-11 shows the open-loop noninverting amplifier. In this configuration the input is applied to the noninverting input terminal, and the inverting terminal is connected to ground.

Figure 3–11   Noninverting amplifier.

In the circuit of Figure 3-11, $v_1 = v_{in}$ and $v_2 = 0$ V. Therefore, according to Equation (3-9),

$$v_o = Av_{in}$$

This means that the output voltage is larger than the input voltage by gain A and is in phase with the input signal.

In all three open-loop configurations any input signal (differential or single) that is only slightly greater than zero drives the output to saturation level. This results from the very high gain (A) of the op-amp. Thus, when operated open-loop, the output of the op-amp is either negative or positive saturation or switches between positive and negative saturation levels. For this reason, open-loop op-amp configurations are not used in linear applications.

EXAMPLE 3-1

Determine the output voltage in each of the following cases for the open-loop differential amplifier of Figure 3-9:

(a) $v_{in\,1} = 5$ μV dc, $v_{in\,2} = -7$ μV dc
(b) $v_{in\,1} = 10$ mV rms, $v_{in\,2} = 20$ mV rms

The op-amp is a 741 with the following specifications: $A = 200{,}000$, $R_i = 2$ MΩ, $R_o = 75$ Ω, $+V_{CC} = +15$ V, $-V_{EE} = -15$ V. and output voltage swing $= \pm14$ V.

SOLUTION   (a) By Equation (3-9),

$$v_o = 200{,}000[(5)(10^{-6}) - (-7)(10^{-6})] = 2.4 \text{ V dc}$$

Remember that $v_o = 2.4$ V dc with the assumption that the dc output voltage is zero when the input signals are zero.

(b) Equation (3-9) is valid for both ac and dc input signals. However, the restriction on ac input signals is that they must be of the same frequency (see Figure 3-12). By Equation (3-9),

$$v_o = 200{,}000[(10)(10^{-3}) - (20)(10^{-3})] = -2000 \text{ V rms}$$

Thus the theoretical value of output voltage $v_o = -2000$ V rms. However, the op-amp saturates at ±14 V. Therefore, the actual output waveform will be *clipped* as shown in Figure 3-12. This nonsinusoidal waveform is unacceptable in amplifier applications. The normal solution to this problem is to use a negative feedback, which is discussed in Chapter 4.

EXAMPLE 3-2

Determine the output voltage for the inverting amplifier shown in Figure 3-10, if

(a) $v_{in} = 20$ mV dc
(b) $v_{in} = -50$ μV peak sine wave

Assume that the op-amp is a 741.

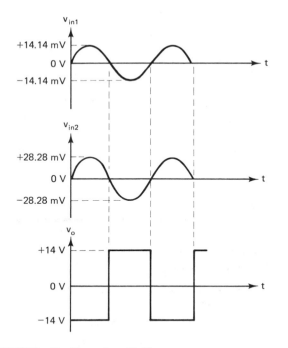

Figure 3–12 Waveforms for Example 3–1.

SOLUTION   By Equation (3-9),

$$\text{(a)} \quad v_o = -Av_{in} = -(2)(10^5)(20)(10^{-3}) = -4000 \text{ V}$$

This is the theoretical value; the actual value will be a negative saturation voltage of −14 V.

$$\text{(b)} \quad v_o = -Av_{in} = -(2)(10^5)(-50)(10^{-6}) = 10 \text{ V peak sine wave}$$

This means that the output is a sine wave, since it is less than the output voltage swing of ±14 V or 28 V peak to peak.

SUMMARY

1. Manufacturers supply data sheets for the ICs they produce. These data sheets include absolute maximum ratings, electrical characteristics, pin diagrams, equivalent circuits, intended applications of the devices, and more. To get the most use out of these sheets, proper interpretation of them is essential.

2. Generally, the information found on the op-amp data sheets can be broken down into the following groups: device type, absolute maximum ratings, connection diagram, equivalent circuit, electrical characteristics, and typical applications.

3. An ideal op-amp has infinity voltage gain, input resistance, CMRR, and slew rate, together with zero output resistance and output offset voltage.

4. The equivalent circuit is useful in analyzing the basic operating principles of an op-amp and in observing the effects of feedback arrangements.
5. The voltage transfer characteristic curve of an op-amp is the graph of output voltage versus differential input voltage.
6. Differential, inverting, and noninverting amplifiers are the three open-loop op-amp configurations in which the output signal is not fed back in any form as part of the input signal. When operated open-loop, generally the op-amp's output is either positive or negative saturation or switches between positive and negative saturation levels. This action is undesirable in linear applications; hence open-loop configurations are rarely used in linear applications.

QUESTIONS

3-1. What information is contained in the typical op-amp data sheet?

3-2. Explain why proper interpretation of op-amp data sheets is important.

3-3. Define the following electrical parameters: input offset voltage, input resistance, CMRR, output voltage swing, and slew rate.

3-4. List the parameters that should be considered for ac and dc applications.

3-5. What are the three factors that affect the electrical parameters of an op-amp?

3-6. What are the characteristics of an ideal op-amp?

3-7. What is a voltage transfer curve of an op-amp?

3-8. List three open-loop op-amp configurations.

3-9. Explain why open-loop op-amp configurations are not used in linear applications.

PROBLEMS

3-1. For the μAF771 op-amp, PSRR = 70 dB minimum. What is the numerical value of the PSRR?

3-2. For a given op-amp, CMRR = 10^5 and differential gain $A_d = 10^5$. Determine the common-mode gain A_{cm} of the op-amp.

3-3. The output voltage of a certain op-amp circuit changes by 20 V in 4 μs. What is its slew rate?

3-4. In the differential amplifier of Figure 3-9, $v_{in\,1} = 2.1$ V dc and $v_{in\,2} = 2.0$ V dc. Determine the output voltage v_o. Assume that the op-amp is 741, with supply voltages = ± 15 V.

3-5. Repeat Problem 3-4 for the following values of inputs: $v_{in\,1} = -25\ \mu$V rms and $v_{in\,2} = 20\ \mu$V rms.

3-6. Referring to the circuit of Figure 3-10, determine v_o if $v_{in} = -15\ \mu$V dc. Assume that the op-amp is a 741 and that supply voltages = ± 15 V.

3-7. Repeat Problem 3-6 for $v_{in} = -10\ \mu$V rms.

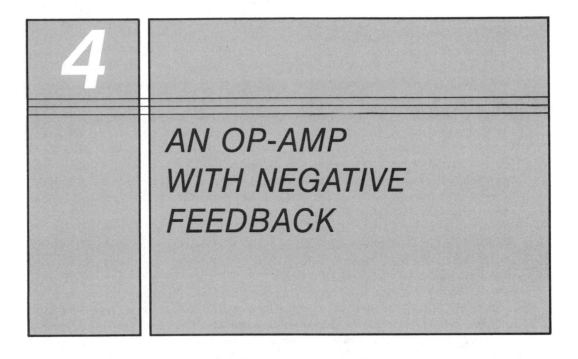

AN OP-AMP WITH NEGATIVE FEEDBACK

4-1 INTRODUCTION

In Chapter 3 the characteristics of the op-amp and interpretation of its parameters are discussed. Recall that clipping occurs in open-loop configurations when the output attempts to exceed the saturation levels of the op-amp. In other words, because the open-loop gain of the op-amp is very high, only the smaller signals (of the order of microvolts or less) having very low frequency may be amplified accurately without distortion. However, signals this small are very susceptible to noise and almost impossible to obtain in the laboratory.

Besides being large, the open-loop voltage gain of the op-amp is not a constant. The voltage gain varies with changes in temperature and power supply as well as mass production techniques. The variations in voltage gain are relatively large in open-loop op-amps in particular, which makes the open-loop op-amp unsuitable for many linear applications. In most linear applications the output is proportional to the input and is of the same type.

In addition, the *bandwidth* (band of frequencies for which the gain remains constant) of most open-loop op-amps is negligibly small—almost zero. For this reason the open-loop op-amp is impractical in ac applications. For instance, the open-loop bandwidth of the 741C is approximately 5 Hz. (Refer to the voltage gain versus frequency graph supplied on the data sheet.) However, in almost all ac applications a bandwidth larger than 5 Hz is needed.

For the reasons stated, the open-loop op-amp is generally not used in linear applications. Nevertheless, in certain applications the open-loop op-amp is purposely used as a nonlinear device; that is, a square-wave output is obtained by deliberately applying a relatively large input signal. Open-loop op-amp configurations are most suitable in such applications.

We can select as well as control the gain of the op-amp if we introduce a modification in the basic circuit. This modification involves the use of *feedback;* that is, an output signal is fed back to the input either directly or via another network. If the signal fed back is of opposite polarity or out of phase by 180° (or odd integer multiples of 180°) with respect to the input signal, the feedback is called *negative feedback.* An amplifier with negative feedback has a self-correcting ability against any change in output voltage caused by changes in environmental conditions. Negative feedback is also known as *degenerative* feedback because when used it degenerates (reduces) the output voltage amplitude and in turn reduces the voltage gain.

On the other hand, if the signal fed back is in phase with the input signal, the feedback is called *positive feedback.* In positive feedback the feedback signal aids the input signal. For this reason it is also referred to as *regenerative* feedback. Positive feedback is necessary in oscillator circuits.

When used in amplifiers, negative feedback stabilizes the gain, increases the bandwidth, and changes the input and output resistances. Of course, the price paid for these improvements is reduced voltage gain. Other benefits of negative feedback include a decrease in harmonic or nonlinear distortion and reduction in the effect of input offset voltage at the output. Negative feedback also reduces the effect of variations in temperature and supply voltages on the output of the op-amp.

Negative feedback and its effect on the performance of an op-amp are the subject of this chapter. First the block diagrams for different op-amp configurations using negative feedback are presented. Then specific arrangements of these feedback configurations are studied in more depth.

4-2 BLOCK DIAGRAM REPRESENTATION OF FEEDBACK CONFIGURATIONS

An op-amp that uses feedback is called a *feedback amplifier.* A feedback amplifier is sometimes referred to as a *closed-loop amplifier* because the feedback forms a closed loop between the input and the output. A feedback amplifier essentially consists of two parts: an op-amp and a *feedback circuit.* The feedback circuit can take any form whatsoever, depending on the intended application of the amplifier. This means that the feedback circuit may be made up of either passive components, active components, or combinations of both. This chapter, in order to develop the basic feedback concepts, presents only purely resistive feedback circuits.

A closed-loop amplifier can be represented by using two blocks, one for an op-amp and another for a feedback circuit. There are four ways to connect these two blocks. These connections are classified according to whether the voltage or current is fed back to the input in *series* or in *parallel,* as follows:

1.   Voltage-series feedback
2.   Voltage-shunt feedback

3.   Current-series feedback

4.   Current-shunt feedback

The four types of configurations are illustrated in Figure 4-1. In Figure 4-1(a) and (b) the voltage across load resistor R_L is the input voltage to the feedback circuit. The feedback quantity (either voltage or current) is the output of the feedback circuit and is proportional to the output voltage. On the other hand, in the current-series and current-shunt feedback circuits of Figure 4-1(c) and (d), the load current i_L flows into the feedback circuit. The output of the feedback circuit (either voltage or current) is proportional to the load current i_L.

Note that in all four of these configurations the signal direction through the op-amp is from the input to output. On the other hand, in the ideal case the signal direction through the feedback circuit is exactly opposite: from output to input.

The voltage-series and voltage-shunt feedback configurations are important because they are most commonly used. An in-depth analysis of these two configurations is presented here, computing voltage gain, input resistance, output resistance, and bandwidth for each. The other two configurations are not discussed in this chapter. However,

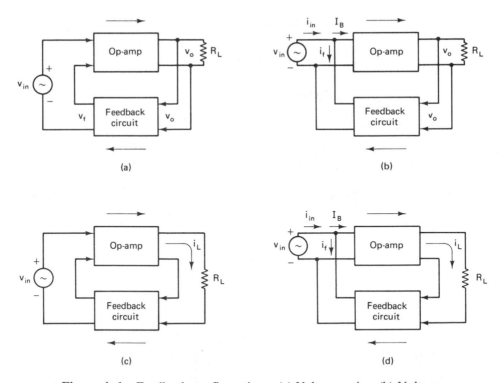

Figure 4–1   Feedback configurations. (a) Voltage-series. (b) Voltage-shunt. (c) Current-series. (d) Current-shunt. Arrows indicate the signal flow directions.

the interested reader may analyze the latter using the same procedures as illustrated for the first two configurations.

4-3 VOLTAGE-SERIES FEEDBACK AMPLIFIER

The schematic diagram of the voltage-series feedback amplifier is shown in Figure 4-2. The op-amp is represented by its schematic symbol, including its large-signal voltage gain A, and the feedback circuit is composed of two resistors, R_1 and R_F.

The circuit shown in Figure 4-2 is commonly known as a *noninverting amplifier with feedback* (or *closed-loop noninverting amplifier*) because it uses feedback, and the input signal is applied to the noninverting input terminal of the op-amp.

Before proceeding, it is necessary to define some important terms for the voltage-series feedback amplifier of Figure 4-2. Specifically, the voltage gain of the op-amp with and without feedback, and the gain of the feedback circuit are defined as follows:

$$\text{open-loop voltage gain (or gain without feedback)}\quad A = \frac{v_o}{v_{id}}$$

$$\text{closed-loop voltage gain (or gain with feedback)}\quad A_F = \frac{v_o}{v_{\text{in}}}$$

$$\text{gain of the feedback circuit}\qquad\qquad\qquad B = \frac{v_f}{v_o}$$

4-3.1 Negative Feedback

Referring to the circuit of Figure 4-2, Kirchhoff's voltage equation for the input loop is

$$v_{id} = v_{\text{in}} - v_f \tag{4-1}$$

where $v_{\text{in}} = $ input voltage
$\quad\ \ v_f = $ feedback voltage
$\quad\ \ v_{id} = $ difference input voltage

Figure 4–2 Voltage-series feedback amplifier (or noninverting amplifier with feedback).

Recall, however, that an op-amp always amplifies the difference input voltage v_{id}. From Equation (4-1), this difference voltage is equal to the input voltage v_{in} minus the feedback voltage v_f. In other words, the feedback voltage always opposes the input voltage (or is out of phase by 180° with respect to the input voltage); hence the feedback is said to be *negative*.

Returning now to the analysis of the voltage-series feedback amplifier, we should note that it will be performed by computing closed-loop voltage gain, input and output resistances, and the bandwidth.

4-3.2 Closed-Loop Voltage Gain

As defined previously, the closed-loop voltage gain

$$A_F = \frac{v_o}{v_{in}}$$

However, by Equation (3-9),

$$v_o = A(v_1 - v_2)$$

Referring to Figure 4-2, we see that

$$v_1 = v_{in}$$

$$v_2 = v_f = \frac{R_1 v_o}{R_1 + R_F} \qquad \text{since } R_i >> R_1$$

Therefore,

$$v_o = A\left(v_{in} - \frac{R_1 v_o}{R_1 + R_F}\right)$$

Rearranging, we get

$$v_o = \frac{A(R_1 + R_F)v_{in}}{R_1 + R_F + AR_1}$$

Thus

$$A_F = \frac{v_o}{v_{in}} = \frac{A(R_1 + R_F)}{R_1 + R_F + AR_1} \quad \text{(exact)} \tag{4-2}$$

Generally, A is very large (typically 10^5). Therefore,

$$AR_1 >> (R_1 + R_F) \quad \text{and} \quad (R_1 + R_F + AR_1) \cong AR_1$$

Thus

$$A_F = \frac{v_o}{v_{in}} = 1 + \frac{R_F}{R_1} \quad \text{(ideal)} \tag{4-3}$$

Equation (4-3) is important because it shows that the gain of the voltage-series feedback amplifier is determined by the ratio of two resistors, R_1 and R_F. For instance, if a gain of 11 is desired, we can then choose $R_1 = 1\ k\Omega$ and $R_F = 10\ k\Omega$ or $R_1 = 100\ \Omega$ and $R_F = 1\ k\Omega$. In other words, in setting the gain the ratio of R_1 and R_F is important and not the absolute values of these resistors. As a general rule, however, all external component values should be less than 1 MΩ so that they do not adversely affect the internal circuitry of the op-amp. This is especially true for older-generation ICs such as the 709 and 741.

Another interesting result can be obtained from Equation (4-3). As defined previously, the gain of the feedback circuit (B) is the ratio of v_f and v_o. Referring to Figure 4-2, this gain is

$$B = \frac{v_f}{v_o}$$

$$= \frac{R_1}{R_1 + R_F} \tag{4-4}$$

Comparing Equations (4-3) and (4-4), we can conclude that

$$A_F = \frac{1}{B} \quad \text{(ideal)} \tag{4-5}$$

This means that the gain of the feedback circuit is the reciprocal of the closed-loop voltage gain. In other words, for given R_1 and R_F the values of A_F and B are fixed. Besides that, Equation (4-5) is an alternative to Equation (4-3), and its simpler form makes it easier to remember.

Finally, the closed-loop voltage gain A_F can be expressed in terms of open-loop gain A and feedback circuit gain B as follows. Rearranging Equation (4-2), we get

$$A_F = \frac{A\left(\dfrac{R_1 + R_F}{R_1 + R_F}\right)}{\dfrac{R_1 + R_F}{R_1 + R_F} + \dfrac{AR_1}{R_1 + R_F}}$$

Using Equation (4-4) yields

$$A_F = \frac{A}{1 + AB} \tag{4-6}$$

where A_F = closed-loop voltage gain
     A = open-loop voltage gain
     B = gain of the feedback circuit
   AB = loop gain

A *one-line* block diagram of Equation (4-6) is shown in Figure 4-3. This block diagram illustrates a standard form for representing a system with feedback and also indicates the

Figure 4–3 Block diagram representation of noninverting amplifier with feedback.

relationship between different variables of the system. The block-diagram approach helps to simplify the analysis of complex closed-loop networks, particularly if they are composed of nonresistive feedback circuits.

4-3.3 Difference Input Voltage Ideally Zero

Let us reconsider Equation (3-9), which can be rewritten as

$$v_{id} = \frac{v_o}{A}$$

Since A is very large (ideally infinite),

$$v_{id} \cong 0 \tag{4-7a}$$

That is,

$$v_1 \cong v_2 \quad \text{(ideal)} \tag{4-7b}$$

Equation (4-7b) says that the voltage at the noninverting input terminal of an op-amp is approximately equal to that at the inverting input terminal provided that A is very large. This concept is useful in the analysis of *closed-loop* op-amp circuits. For example, ideal closed-loop voltage gain [Equation (4-3)] can be obtained using the preceding results as follows. In the circuit of Figure 4-2,

$$v_1 = v_{in}$$
$$v_2 = v_f$$
$$= \frac{R_1 v_o}{R_1 + R_F}$$

Substituting these values of v_1 and v_2 in Equation (4-7b), we get

$$v_{in} = \frac{R_1 v_o}{R_1 + R_F}$$

That is,

$$A_F = \frac{v_o}{v_{in}} = 1 + \frac{R_F}{R_1}$$

4-3.4 Input Resistance with Feedback

Figure 4-4 shows a voltage-series feedback amplifier with the op-amp equivalent circuit. In this circuit R_i is the input resistance (open loop) of the op-amp, and R_{iF} is the input resistance of the amplifier with feedback. The input resistance with feedback is defined as

$$R_{iF} = \frac{v_{in}}{i_{in}}$$
$$= \frac{v_{in}}{v_{id}/R_i}$$

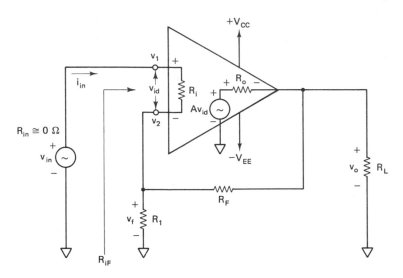

Figure 4–4   Derivation of input resistance with feedback.

However,

$$v_{id} = \frac{v_o}{A} \quad \text{and} \quad v_o = \frac{A}{1 + AB}v_{in}$$

Therefore,

$$R_{iF} = R_i \frac{v_{in}}{v_o/A}$$

$$= AR_i \frac{v_{in}}{Av_{in}/(1 + AB)}$$

$$= R_i(1 + AB) \tag{4-8}$$

This means that the input resistance of the op-amp with feedback is $(1 + AB)$ times that without feedback.

4-3.5 Output Resistance with Feedback

Output resistance is the resistance determined looking back into the feedback amplifier from the output terminal as shown in Figure 4-5. This resistance can be obtained by using Thévenin's theorem for *dependent* sources. Specifically, to find output resistance with feedback R_{oF}, reduce independent source v_{in} to zero, apply an external voltage v_o, and then calculate the resulting current i_o. In short, the R_{oF} is defined as follows:

$$R_{oF} = \frac{v_o}{i_o} \tag{4-9a}$$

Writing Kirchhoff's current equation at output node N, we get

$$i_o = i_a + i_b$$

Figure 4–5   Derivation of output resistance with feedback.

since $[(R_F + R_1)\|R_i] \gg R_o$ and $i_a \gg i_b$. Therefore,

$$i_o \cong i_a$$

The current i_o can be found by writing Kirchhoff's voltage equation for the output loop:

$$v_o - R_o i_o - A v_{id} = 0$$

$$i_o = \frac{v_o - A v_{id}}{R_o}$$

However,

$$v_{id} = v_1 - v_2$$

$$= 0 - v_f$$

$$= -\frac{R_1 v_o}{R_1 + R_F} = -B v_o$$

Therefore,

$$i_o = \frac{v_o + A B v_o}{R_o}$$

Substituting the value of i_o in Equation (4-9a), we get

$$R_{oF} = \frac{v_o}{(v_o + A B v_o)/R_o}$$

$$= \frac{R_o}{1 + AB} \tag{4-9b}$$

This result shows that the output resistance of the voltage-series feedback amplifier is $1/(1 + AB)$ times the output resistance R_o of the op-amp. That is, the output resistance of the op-amp with feedback is much smaller than the output resistance without feedback.

4-3.6 Bandwidth with Feedback

The bandwidth of an amplifier is defined as the band (range) of frequencies for which the gain remains constant. Manufacturers generally specify either the gain–bandwidth product or supply open-loop gain versus frequency curve for the op-amp. For the 741 op-amp the latter is typical.

Figure 4-6 shows the open-loop gain versus frequency curve of the 741C op-amp. From this curve for a gain of 200,000, the bandwidth is approximately 5 Hz; or the gain–bandwidth product is (200,000 × 5 Hz) = 1 MHz. On the other extreme, the bandwidth is approximately 1 MHz when the gain is 1. Thus, the gain–bandwidth product is constant. However, this holds true only for op-amps, like the 741 that have just one break frequency below unity gain–bandwidth. For the 741, 5 Hz is the *break frequency;* the frequency at which the gain A is 3 dB down from its value at 0 Hz. We will denote it by f_o. On the other hand, the frequency at which the gain equals 1 is known as the *unity gain–bandwidth* (UGB). The relationship between the break frequency f_o, open-loop voltage gain A, bandwidth with feedback f_F, and the closed-loop gain A_F can be established as follows. Since for an op-amp with a single break frequency f_o, the gain–bandwidth product is constant, and equal to the *unity gain–bandwidth* (UGB), we can write,

$$\text{UGB} = (A)(f_o) \tag{4-10a}$$

where A = open-loop voltage gain
$\quad f_o$ = break-frequency of an op-amp

or, alternatively, only for a single break frequency op-amp.

$$\text{UGB} = (A_F)(f_F) \tag{4-10b}$$

where A_F = closed-loop voltage gain
$\quad f_F$ = bandwidth with feedback

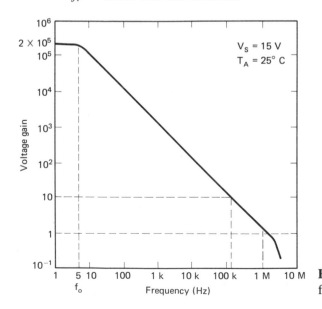

Figure 4–6 Open-loop gain versus frequency curve of the 741C.

Therefore, equating Equations (4-10a) and (4-10b),

$$(A)(f_o) = (A_F)(f_F)$$

or

$$f_F = \frac{(A)(f_o)}{A_F} \tag{4-10c}$$

However, for the noninverting amplifier with feedback,

$$A_F = \frac{A}{1 + AB}$$

Therefore, substituting the value of A_F in Equation (4-10c), we get

$$f_F = \frac{(A)(f_o)}{A/(1 + AB)}$$

or

$$f_F = f_o(1 + AB) \tag{4-10d}$$

Equation (4-10d) indicates that the bandwidth of the noninverting amplifier with feedback, f_F, is equal to its bandwidth without feedback, f_o, times $(1 + AB)$. In other words, if negative feedback is used as shown in Figure 4-2, gain A decreases to $A/(1 + AB)$; consequently, the open-loop bandwidth f_o (the break frequency) should increase to $f_o(1 + AB)$. For instance, let us assume that the 741C is used in the circuit of Figure 4-2 and that the desired voltage gain A_F is 10; then the closed-loop bandwidth f_F [using Equation (4-10c) or (4-10d)] will be approximately 100 kHz.

The closed-loop bandwidth can also be determined from the open-loop gain versus frequency plot. To do this we locate the closed-loop voltage gain value on the gain axis and draw a line through this value parallel to the frequency axis. Then we project the point of intersection of the line with the curve on the frequency axis and read the value of the closed-loop bandwidth. Using this procedure in Figure 4-6, the bandwidth is approximately 100 kHz for a closed-loop gain of 10.

4-3.7 Total Output Offset Voltage with Feedback

In an op-amp when the input is zero, the output is also expected to be zero. However, because of the effect of input offset voltage and current, the output is significantly larger, a result in large part of very high open-loop gain. That is, the high gain aggravates the effect of input offset voltage and current at the output. We call this enhanced output voltage the *total output offset voltage* V_{ooT}. In an open-loop op-amp the total output offset voltage is equal to either the positive or negative saturation voltage. The saturation voltages are specified on the data sheets as *output voltage swing*.

Since with feedback the gain of the noninverting amplifier changes from A to $A/(1 + AB)$ [Equation (4-6)], the total output offset voltage with feedback must also be $1/(1 + AB)$ times the voltage without feedback. That is,

$$\left(\begin{array}{c} \text{total output offset} \\ \text{voltage with feedback} \end{array} \right) = \frac{\text{total output offset voltage without feedback}}{1 + AB}$$

or

$$V_{ooT} = \frac{\pm V_{sat}}{1 + AB}$$ (4-11)

where $1/(1 + AB)$ is always less than 1 and $\pm V_{sat}$ = saturation voltages, the maximum voltages the output of an op-amp can reach.

Remember that in an open-loop configuration even a very small voltage at the input of an op-amp can cause the output to reach maximum value ($+V_{sat}$ or $-V_{sat}$) because of its very high voltage gain. Therefore, according to Equation (4-11), for a given op-amp circuit the V_{ooT} is either positive or negative voltage because V_{sat} can be either positive or negative.

Negative feedback can also be used to reduce significantly the effect of noise, variations in supply voltages, and changes in temperature on the output voltage of a noninverting amplifier. In fact, the higher the value of $(1 + AB)$, the smaller is the effect of noise and of variations in supply voltages and changes in temperature on the output voltage of a noninverting amplifier.

From this analysis it is clear that the noninverting amplifier with feedback exhibits the characteristics of the *perfect voltage amplifier*. That is, it has very high input resistance, very low output resistance, stable voltage gain, large bandwidth, and very little (ideally zero) output offset voltage.

4-3.8 Voltage Follower

The lowest gain that can be obtained from a noninverting amplifier with feedback is 1. When the noninverting amplifier is configured for unity gain, it is called a *voltage follower* because the output voltage is equal to and in phase with the input. In other words, in the voltage follower the output follows the input.

Although it is similar to the discrete emitter follower, the voltage follower is preferred because it has much higher input resistance, and the output amplitude is exactly equal to the input.

To obtain the voltage follower from the noninverting amplifier of Figure 4-2, simply *open* R_1 and *short* R_F. The resulting circuit is shown in Figure 4-7. In this figure all the output voltage is fed back into the inverting terminal of the op-amp; consequently, the gain of the feedback circuit is 1 ($B = A_F = 1$).

Since the voltage follower is a special case of the noninverting amplifier, all the formulas developed for the latter are indeed applicable to the former except that the gain of the feedback circuit is 1 ($B = 1$). The applicable formulas are

Figure 4–7 Voltage follower.

$$A_F = 1$$

$$R_{iF} = AR_i$$

$$R_{oF} = \frac{R_o}{A}$$

$$f_F = Af_o$$

$$V_{ooT} = \frac{\pm V_{\text{sat}}}{A}$$

since $(1 + A) \cong A$.

The voltage follower is also called a *noninverting buffer* because, when placed between two networks, it removes the loading on the first network.

EXAMPLE 4-1

The 741C op-amp having the following parameters is connected as a noninverting amplifier (Figure 4-2) with $R_1 = 1 \text{ k}\Omega$ and $R_F = 10 \text{ k}\Omega$:

$$A = 200,000$$

$$R_i = 2 \text{ M}\Omega$$

$$R_o = 75 \ \Omega$$

$$f_o \cong 5 \text{ Hz}$$

$$\text{supply voltages} = \pm 15 \text{ V}$$

$$\text{output voltage swing} = \pm 13 \text{ V}$$

Compute the values of A_F, R_{iF}, R_{oF}, f_F, and V_{ooT}.

SOLUTION   Let us first calculate the value of B. Then the closed-loop parameters A_F, R_{iF}, R_{oF}, f_F, and V_{ooT} can be obtained by using Equations (4-6), (4-8), (4-9b), (4-10d), and (4-11), respectively.

$$B = \frac{R_1}{R_1 + R_F} = \frac{1 \text{ k}\Omega}{1 \text{ k}\Omega + 10 \text{ k}\Omega} = \frac{1}{11}$$

$$1 + AB = 1 + \frac{200,000}{11} = 18,182.8$$

$$A_F = \frac{200,000}{18,182.8} = 10.99$$

$$R_{iF} = 2 \text{ M}\Omega \, (18,182.8) = 36.4 \text{ G}\Omega$$

$$R_{oF} = \frac{75 \ \Omega}{18,182.8} = 4.12 \text{ m}\Omega$$

$$f_F = (5 \text{ Hz})(18,182.8) = 90.9 \text{ kHz}$$

$$V_{ooT} = \frac{\pm 13 \text{ V}}{18,182.8} = \pm 0.715 \text{ mV}$$

Note that the $(\pm)$ sign indicates that V_{ooT} could be of either polarity.

In the example above, the voltage gain calculated using the exact equation [Equation (4-2)] is 10.99. The gain would have been 11 if we had used the ideal voltage-gain equation [Equation (4-3)]. Thus the difference error is very small (0.09%) and can be ignored. That is, for all practical purposes we may use the ideal voltage-gain equation, provided that $A >> A_F$. Remember that as A_F approaches A, the difference error also increases.

EXAMPLE 4-2

Repeat Example 4-1 for the voltage follower of Figure 4-7.

SOLUTION   For the voltage follower, $B = 1$; therefore, $1 + AB = 200,000$. To compute the closed-loop parameters, we merely substitute the known values into Equations (4-6), (4-8), (4-9b), (4-10d), and (4-11).

$$A_F = 1$$

$$R_{iF} = 2 \text{ M}\Omega(200,000) = 400 \text{ G}\Omega$$

$$R_{oF} = \frac{75 \text{ }\Omega}{200,000} = 0.375 \text{ m}\Omega$$

$$f_F = (5 \text{ Hz})(200,000) = 1 \text{ MHz}$$

$$V_{ooT} = \frac{\pm 13 \text{ V}}{200,000} = \pm 65 \text{ }\mu\text{V}$$

Thus the input and output resistances of the voltage follower approach ideal values, and the bandwidth is equal to the maximum operating frequency of the op-amp. In addition, since $(1 + AB) = A$, the smallest possible value for V_{ooT} is possible.

4-4 VOLTAGE-SHUNT FEEDBACK AMPLIFIER

Figure 4-8 shows the voltage-shunt feedback amplifier using an op-amp. The input voltage drives the inverting terminal, and the amplified as well as inverted output signal is also applied to the inverting input via the feedback resistor R_F. This arrangement forms a negative feedback because any increase in the output signal results in a feedback signal into the inverting input, causing a decrease in the output signal.

Note that the noninverting terminal is grounded, and the feedback circuit has only one resistor R_F. However, an extra resistor R_1 is connected in series with the input signal source v_{in}.

First we derive the formulas for the voltage gain, input and output resistances, bandwidth, and the total output offset voltage. Then we study the special cases of this configuration.

4-4.1 Closed-Loop Voltage Gain

The closed-loop voltage gain A_F of the voltage-shunt feedback amplifier can be obtained by writing Kirchhoff's current equation at the input node v_2 [see Figure 4-8] as follows:

$$i_{in} = i_F + I_B \tag{4-12a}$$

Figure 4–8   Voltage-shunt feedback amplifier (or inverting amplifier with feedback).

Since R_i is very large, the input bias current I_B is negligibly small. For instance, $R_i = 2\ M\Omega$ and $I_B = 0.5\ \mu A$ for the 741C. Therefore,

$$i_{in} \cong i_F$$

That is,

$$\frac{v_{in} - v_2}{R_1} = \frac{v_2 - v_o}{R_F} \qquad (4\text{-}12b)$$

However, from Equation (3-9),

$$v_1 - v_2 = \frac{v_o}{A}$$

Since $v_1 = 0$ V,

$$v_2 = -\frac{v_o}{A}$$

Substituting this value of v_2 in Equation (4-12b) and rearranging, we get

$$\frac{v_{in} + v_o/A}{R_1} = \frac{-(v_o/A) - v_o}{R_F}$$

$$A_F = \frac{v_o}{v_{in}} = -\frac{AR_F}{R_1 + R_F + AR_1} \qquad \text{(exact)} \qquad (4\text{-}13)$$

The negative sign in Equation (4-13) indicates that the input and output signals are out of phase by 180° (or of opposite polarities). In fact, because of this phase inversion, the configuration in Figure 4-8 is commonly called an *inverting amplifier* with feedback.

Since the internal gain A of the op-amp is very large (ideally infinity), $AR_1 \gg R_1 + R_F$. This means that Equation (4-13) can be rewritten as

$$A_F = \frac{v_o}{v_{in}} = -\frac{R_F}{R_1} \qquad \text{(ideal)} \qquad (4\text{-}14)$$

This equation shows that the gain of the inverting amplifier is set by selecting a ratio of feedback resistance R_F to the input resistance R_1. In fact, the ratio R_F/R_1 can be set to any value whatsoever, even to less than 1. Because of this property of the gain equation, the inverting amplifier configuration with feedback lends itself to a majority of applications as against those of the noninverting amplifier.

Let us now rewrite Equation (4-13) in the feedback form of Equation (4-6), for a couple of reasons. First, it facilitates analysis of the inverting amplifier with feedback. Second, it helps compare and contrast inverting and noninverting amplifier configurations, as we shall soon see. However, to express Equation (4-13) in the form of Equation (4-6), we must represent the current-summing junction at the input terminals of an amplifier as a voltage-summing junction (see Figures 4-8 and 4-9).

To begin with, we divide both numerator and denominator of Equation (4-13) by $(R_1 + R_F)$:

$$A_F = -\frac{AR_F/R_1 + R_F}{1 + \dfrac{AR_1}{R_1 + R_F}}$$

$$= -\frac{AK}{1 + AB} \qquad (4\text{-}15)$$

where $K = \dfrac{R_F}{R_1 + R_F}$, a voltage attenuation factor

$B = \dfrac{R_1}{R_1 + R_F}$, gain of the feedback circuit

A comparison of Equation (4-15) with the feedback Equation (4-6) indicates that, in addition to the phase inversion ($-$ sign), the closed-loop gain of the inverting amplifier is K times the closed-loop gain of the noninverting amplifier, where $K < 1$.

The one-line block diagram of the inverting amplifier with feedback is shown in Figure 4-9. The reason for the block diagram is twofold: (1) to facilitate the analysis of the inverting amplifier, and (2) to express the performance equations in the same form as those for the noninverting amplifier.

The block diagram in Figure 4-3 for the noninverting amplifier and the block diagram in Figure 4-9 for the inverting amplifier are identical, except for the K block.

Figure 4–9   Block diagram of inverting amplifier with feedback using a voltage-summing junction as a model for current summing.

However, the major difference is that in Figure 4-9 a voltage-summing junction is being used as a model for what is actually current summing.

To derive the ideal closed-loop gain, we can use Equation (4-15) as follows. If $AB \gg 1$, then $(1 + AB) \cong AB$ and

$$A_F = -\frac{K}{B}$$

$$= -\frac{R_F}{R_1} \tag{4-16}$$

4-4.2 Inverting Input Terminal at Virtual Ground

Refer again to the inverting amplifier of Figure 4-8. In this figure, the noninverting terminal is grounded, and the input signal is applied to the inverting terminal via resistor R_1. However, as discussed in Section 4-3.3, the difference input voltage is ideally zero; that is, the voltage at the inverting terminal (v_2) is approximately equal to that at the noninverting terminal (v_1). In other words, the inverting terminal voltage v_2 is approximately at ground potential. Therefore, the inverting terminal is said to be at *virtual ground*. This concept is extremely useful in the analysis of closed-loop inverting amplifier circuits. For example, ideal closed-loop gain [Equation (4-14)] can be obtained using the virtual-ground concept as follows:

In the circuit of Figure 4-8,

$$i_{\text{in}} \cong i_F \tag{4-17}$$

That is,

$$\frac{v_{\text{in}} - v_2}{R_1} = \frac{v_2 - v_o}{R_F}$$

However,

$$v_1 = v_2 = 0 \text{ V}$$

Therefore,

$$\frac{v_{\text{in}}}{R_1} = -\frac{v_o}{R_F}$$

or

$$A_F = \frac{v_o}{v_{\text{in}}} = -\frac{R_F}{R_1}$$

This is the same result obtained in Equation (4-14).

4-4.3 Input Resistance with Feedback

The easiest method of finding the input resistance is to Millerize the feedback resistor R_F; that is, split R_F into its two Miller components, as shown in Figure 4-10.

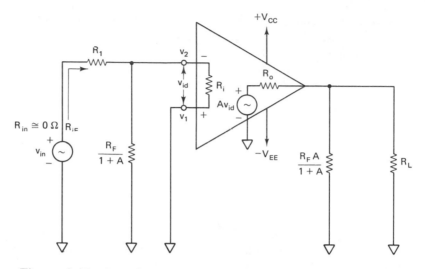

Figure 4–10   Inverting amplifier with Millerized feedback resistor.

In the circuit of Figure 4-10, the input resistance with feedback R_{iF} is then

$$R_{iF} = R_1 + \left. \frac{R_F}{1 + A} \right\| (R_i) \quad \text{(exact)} \tag{4-18}$$

Since R_i and A are very large,

$$\left. \frac{R_F}{1 + A} \right\| R_i \cong 0 \ \Omega$$

Hence

$$R_{iF} = R_1 \quad \text{(ideal)} \tag{4-19}$$

4-4.4 Output Resistance with Feedback

The output resistance with feedback R_{oF} is the resistance measured at the output terminal of the feedback amplifier. The output resistance of the noninverting amplifier was obtained by using Thévenin's theorem, and we can do the same for the inverting amplifier. Thévenin's equivalent circuit for R_{oF} of the inverting amplifier is shown in Figure 4-11. Note that this Thévenin's equivalent circuit is exactly the same as that for noninverting amplifier (Figure 4-5) because the output connections in both amplifiers are the same. This means that the equation for the output resistance R_{oF} of the inverting amplifier must be identical to that of the noninverting amplifier [Equation (4-9b)]. Specifically,

$$R_{oF} = \frac{R_o}{1 + AB} \tag{4-20}$$

where R_o = output resistance of the op-amp
A = open-loop voltage gain of the op-amp
B = gain of the feedback circuit

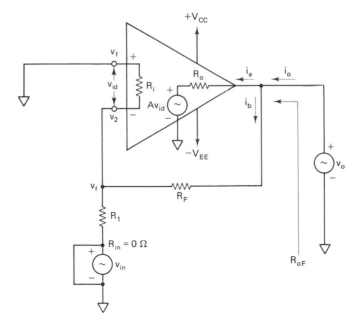

Figure 4–11   Thévenin's equivalent circuit for R_{oF} of the inverting amplifier.

4-4.5 Bandwidth with Feedback

As mentioned previously, the gain–bandwidth product of a single break frequency op-amp is always constant. We also saw that the gain of the amplifier with feedback is always less than the gain without feedback. Therefore, the bandwidth of the amplifier with feedback f_F must be larger than that without feedback.

$$f_F = f_o(1 + AB) \tag{4-21a}$$

where f_o = break frequency of the op-amp

$\qquad = \dfrac{\text{unity gain bandwidth}}{\text{open-loop voltage gain}}$

$\qquad = \dfrac{\text{UGB}}{A}$ (true only for the single break frequency op-amp such as the 741).

Substituting the value of f_o in Equation (4-21a), we get

$$f_F = \frac{\text{UGB}}{A}(1 + AB)$$

$$f_F = \frac{(\text{UGB})\,(K)}{A_F} \tag{4-21b}$$

where $K = \dfrac{R_F}{R_1 + R_F}$

$$A_F = \frac{AK}{1 + AB}$$

To find the closed-loop bandwidth of the inverting amplifier, use Equation (4-21a) if f_o is known and use Equation (4-21b) if unity gain–bandwidth (UGB) is given. However, to calculate the bandwidth graphically from the gain versus frequency curve, we can use the same procedure that was outlined for the noninverting amplifier. From Equations (4-10b) and (4-21b), it is obvious that for the *same* closed-loop gain the closed-loop bandwidth for the inverting amplifier is lower than that for the noninverting amplifier by a factor of K (<1). For example, when the closed-loop gain is equal to 1, the bandwidths will be

$$f_F = \text{UGB} \quad \text{for the noninverting amplifier}$$

and

$$f_F = \frac{\text{UGB}}{2} \quad \text{for the inverting amplifier, since } R_1 = R_F.$$

However, as the closed-loop gain A_F approaches the open-loop gain A, the difference between the noninverting and inverting amplifier bandwidths approaches zero. As an extreme limit, when $K \cong 1$, the value of f_F for both the noninverting and inverting amplifiers is approximately the same.

4-4.6 Total Output Offset Voltage with Feedback

When the temperature and power supply voltages are fixed, the output offset voltage is a function of the gain of an op-amp. However, we saw that the gain of the op-amp with feedback is always less than that without feedback. Therefore, the output offset voltage with feedback V_{ooT} must always be smaller than that without feedback. Specifically,

$$\left(\begin{array}{c} \text{total output offset} \\ \text{voltage with feedback} \end{array} \right) = \frac{\text{total output offset voltage without feedback}}{1 + AB}$$

That is,

$$V_{ooT} = \frac{\pm V_{\text{sat}}}{1 + AB} \tag{4-22}$$

where $\pm V_{\text{sat}}$ = saturation voltages
A = open-loop voltage gain of the op-amp
B = gain of the feedback circuit
$B = \dfrac{R_F}{R_1 + R_F}$

The output voltage of the op-amp without feedback can be either $+V_{\text{sat}}$ or $-V_{\text{sat}}$ because of its very high voltage gain A, which is typically on the order of 10^5. Note that the V_{ooT}

equation for the inverting amplifier is the same as that for the noninverting amplifier. This is because, when the input signal v_{in} is reduced to zero, both inverting and noninverting amplifiers result in the *same* circuit. For more information on the total output offset voltage, refer to Chapter 5.

In addition, because of the negative feedback, the effect of noise, variations in supply voltages, and changes in temperature on the output voltage of the inverting amplifiers are significantly reduced.

Finally, the two special cases of the inverting amplifier with feedback are the current-to-voltage converter and the inverter.

4-4.7 Current-to-Voltage Converter

Let us reconsider the ideal voltage-gain Equation (4-14) of the inverting amplifier,

$$\frac{v_o}{v_{in}} = -\frac{R_F}{R_1}$$

Therefore,

$$v_o = -\left(\frac{v_{in}}{R_1}\right)R_F$$

However, since $v_1 = 0$ V and $v_1 = v_2$,

$$\frac{v_{in}}{R_1} = i_{in}$$

and

$$v_o = -i_{in}R_F \tag{4-23}$$

This means that if we replace the v_{in} and R_1 combination by a current source i_{in} as shown in Figure 4-12, the output voltage v_o becomes proportional to the input current i_{in}. In other words, the circuit of Figure 4-12 *converts* the input current into a proportional output voltage.

One of the most common uses of the current-to-voltage converter is in sensing current from photodetectors and in digital-to-analog converter applications, which are discussed in more depth in Chapter 7.

Figure 4–12   Current-to-voltage converter.

4-4.8 Inverter

If we need an output signal equal in amplitude but opposite in phase to that of the input signal, we can use the inverter. The inverting amplifier of Figure 4-8 works as an inverter if $R_1 = R_F$. Since the inverter is a special case of the inverting amplifier, all the equations developed for the inverting amplifier are also applicable here. The equations can be applied by merely substituting $(A/2)$ for $(1+AB)$, since $B=1/2$.

EXAMPLE 4-3

For the inverting amplifier of Figure 4-8, $R_1 = 470\ \Omega$ and $R_F = 4.7\ \text{k}\Omega$. Assume that the op-amp is the 741 having the same specifications as those given in Example 4-1. Calculate the values of A_F, R_{iF}, R_{oF}, f_F, and V_{ooT}.

SOLUTION   Using the given values of R_1 and R_F,

$$K = \frac{R_F}{R_1 + R_F} = \frac{4700}{470 + 4700} = \frac{1}{1.1}$$

$$B = \frac{R_1}{R_1 + R_F} = \frac{470}{470 + 4700} = \frac{1}{11}$$

and

$$1 + AB = \left[1 + (2 \times 10^5)\left(\frac{1}{11}\right)\right] = 18{,}182.8$$

Therefore, using Equations (4-15), (4-18), (4-20), (4-21), and (4-22), the values of the closed-loop parameters are

$$A_F = -\frac{(200{,}000)(1/1.1)}{18{,}182.8} = -10$$

$$R_{iF} = 470\ \Omega + \left[\frac{4700\ \Omega}{200{,}000}\middle\| (2 \times 10^6)\right] = 470\ \Omega$$

$$R_{oF} = \frac{75\ \Omega}{18{,}182.8} = 4.12\ \text{m}\Omega$$

$$f_F = \frac{(5\ \text{Hz})(18{,}182.8)}{(1/1.1)} \cong 100\ \text{kHz}$$

$$V_{ooT} = \frac{\pm 13\ \text{V}}{18{,}182.8} = \pm 0.715\ \text{mV}$$

EXAMPLE 4-4

For the inverting amplifier of Example 4-3, determine the value of output voltage if the input is a 1-V peak-to-peak sine wave at 1 kHz. Also sketch the output waveform. Assume that V_{ooT} is zero.

SOLUTION   Using the value of the gain calculated in Example 4-3, the output voltage is
$$v_o = -(10)(1) = -10 \text{ V peak to peak}$$
The input and output waveforms are shown in Figure 4-13.

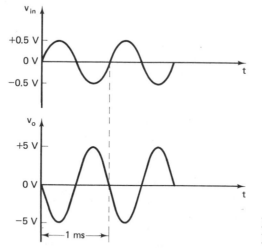

Figure 4–13   Waveforms for Example 4–4.

For quick review, the results of noninverting and inverting amplifiers are summarized in Table 4-1.

TABLE 4-1   SUMMARY OF RESULTS OBTAINED FOR NONINVERTING AND INVERTING AMPLIFIERS

Parameter	Noninverting amplifier	Inverting amplifier
1. Voltage gain	$A_F = \dfrac{A(R_1 + R_F)}{R_1 + R_F + AR_1}$ (exact)	$A_F = \dfrac{-AR_F}{R_1 + R_F + AR_1}$ (exact)
	$= \dfrac{A}{1 + AB}$	$= \dfrac{-AK}{1 + AB}$, where $K = \dfrac{R_F}{R_1 + R_F}$
	$= 1 + \dfrac{R_F}{R_1}$ (ideal)	$= -\dfrac{R_F}{R_1}$ (ideal)
2. Gain of the feedback circuit	$B = \dfrac{R_1}{R_1 + R_F}$	$B = \dfrac{R_1}{R_1 + R_F}$
3. Input resistance	$R_{iF} = R_i(1 + AB)$	$R_{iF} = R_1 + \left(\dfrac{R_F}{1 + A} \parallel R_i\right)$
4. Output resistance	$R_{oF} = \dfrac{R_o}{1 + AB}$	$R_{oF} = \dfrac{R_o}{1 + AB}$
5. Bandwidth	$f_F = f_o(1 + AB)$	$f_F = f_o(1 + AB)$
	$f_F = \dfrac{\text{UGB}}{A_F}$	$f_F = \dfrac{(\text{UGB})(K)}{A_F}$
6. Total output offset voltage	$V_{ooT} = \dfrac{\pm V_{\text{sat}}}{1 + AB}$	$V_{ooT} = \dfrac{\pm V_{\text{sat}}}{1 + AB}$

As mentioned earlier, current-shunt and current-series feedback configurations seldom find use in practice; therefore, they will not be discussed here. However, interested readers may analyze them following the same procedure as used for voltage-series and voltage-shunt feedback configurations.

4-5 DIFFERENTIAL AMPLIFIERS

We will now study the closed-loop differential amplifier configurations. Specifically, we evaluate three different arrangements of the differential amplifier with negative feedback. We classify these arrangements according to the number of op-amps used. That is,

1. Differential amplifier with one op-amp
2. Differential amplifier with two op-amps
3. Differential amplifier with three op-amps

Generally, the differential amplifiers are used in instrumentation and industrial applications to amplify differences between two input signals, such as the outputs of the *Wheatstone bridge* circuits. Differential amplifiers are preferred in these applications because they are better able to reject common-mode (noise) voltages than single-input circuits such as inverting and noninverting amplifiers. They also present a balanced input impedance.

4-5.1 Differential Amplifier with One Op-Amp

Figure 4-14 shows the differential amplifier with one op-amp. We will analyze this circuit by deriving voltage gain and input resistance. A close examination of Figure 4-14 reveals that the differential amplifier is a combination of inverting and noninverting amplifiers. That is, when v_x is reduced to zero the circuit is a noninverting amplifier, whereas the circuit is an inverting amplifier when input v_y is reduced to zero.

4-5.1 (a) Voltage gain.   The circuit in Figure 4-14 has two inputs, v_x and v_y; we will, therefore, use the superposition theorem in order to establish the relationship between inputs and output. When $v_y = 0$ V, the configuration becomes an inverting amplifier; hence the output due to v_x only is

$$v_{ox} = -\frac{R_F(v_x)}{R_1} \tag{4-24a}$$

Similarly, when $v_x = 0$ V, the configuration is a noninverting amplifier having a voltage-divider network composed of R_2 and R_3 at the noninverting input. Therefore,

$$v_1 = \frac{R_3(v_y)}{R_2 + R_3}$$

and the output due to v_y then is

$$v_{oy} = \left(1 + \frac{R_F}{R_1}\right)v_1$$

Figure 4–14 Differential amplifier with one op-amp. $R_1 = R_2$ and $R_F = R_3$.

That is,

$$v_{oy} = \frac{R_3}{R_2 + R_3}\left(\frac{R_1 + R_F}{R_1}\right)v_y$$

Since $R_1 = R_2$ and $R_F = R_3$,

$$v_{oy} = \frac{R_F(v_y)}{R_1} \tag{4-24b}$$

Thus, from Equations (4-24a) and (4-24b), the net output voltage is

$$v_o = v_{ox} + v_{oy}$$

$$v_o = -\frac{R_F}{R_1}(v_x - v_y) = -\frac{R_F(v_{xy})}{R_1}$$

or the voltage gain

$$A_D = \frac{v_o}{v_{xy}} = -\frac{R_F}{R_1} \tag{4-25}$$

Note that the gain of the differential amplifier is the same as that of the inverting amplifier.

4-5.1(b) Input Resistance.   The input resistance R_{iF} of the differential amplifier is the resistance determined looking into either one of the two input terminals with the other grounded. Therefore, with $v_y = 0$ V, the circuit in Figure 4-14 is an inverting amplifier the input resistance of which is

$$R_{iFx} \cong R_1 \tag{4-26a}$$

Similarly, with $v_y = 0$ V, the differential amplifier of Figure 4-14 becomes a noninverting amplifier whose input resistance can then be written as

$$R_{iFy} \cong (R_2 + R_3) \tag{4-26b}$$

From Equations (4-26a) and (4-26b), it is obvious that the input resistances seen by the signal sources v_x and v_y are not the same. This inequality can be corrected and both input resistances can be made equal if we modify the basic differential amplifier of Figure 4-14

Chap. 4 An Op-Amp with Negative Feedback">

as shown in Section 4-5.3. However, for the differential amplifier of Figure 4-14 to perform properly, both R_1 and $(R_2 + R_3)$ can be made much larger than the source resistances so that the loading of the signal sources does not occur.

If we need a variable gain, we can use the differential amplifier of Figure 4-15. In this circuit $R_1 = R_2$, $R_F = R_3$, and the potentiometer $R_P = R_4$. Therefore, depending on the position of the wiper in R_P, voltage gain can be varied from the closed-loop gain of $-2R_F/R_1$ to the open-loop gain of A.

Figure 4–15 Differential amplifier with variable gain.

EXAMPLE 4-5

In the circuit of Figure 4-14, $R_1 = R_2 = 1$ kΩ, $R_F = R_3 = 10$ kΩ, and the op-amp is a 741C.
(a) What are the gain and input resistance of the amplifier?
(b) Calculate the output voltage v_o if $v_x = 2.7$ V pp and $v_y = 3$ V pp sine waves at 100 Hz.

SOLUTION (a) Substituting known values in Equations (4-25) and (4-26), we obtain A_D and R_{iF}:

$$A_D = -\frac{10 \text{ k}\Omega}{1 \text{ k}\Omega} = -10$$

$$R_{iFx} = 1 \text{ k}\Omega$$

$$R_{iFy} = 1 \text{ k}\Omega + 10 \text{ k}\Omega = 11 \text{ k}\Omega$$

(b) Rearranging Equation (4-25) yields

$$v_o = -A_D v_{xy}$$

$$= -(10)(2.7 - 3) = 3 \text{ V pp sine wave at 100 Hz}$$

4-5.2 Differential Amplifier with Two Op-Amps

Recall that the gain expression for the differential amplifier of Figure 4-14 is the same as that of the inverting amplifier of Figure 4-8 [see Equations (4-14) and (4-25)]. We can increase the gain of the differential amplifier and also increase the input resistance R_{iF} if

Figure 4-16   Differential amplifier with two op-amps.

we use two op-amps. Such a circuit is shown in Figure 4-16. In fact, the characteristics of this amplifier are identical to those of the noninverting amplifier.

4-5.2(a) Voltage gain.   A close examination of the circuit of Figure 4-16 shows that it is composed of two stages: (1) the noninverting amplifier, and (2) the differential amplifier with unequal gains. By finding the gain of these two stages, we can obtain the overall gain of the circuit as follows:

The output v_z of the first stage is

$$v_z = \left(1 + \frac{R_3}{R_2}\right)v_y \qquad (4\text{-}27a)$$

By applying the superposition theorem to the second stage, we can obtain the output voltage:

$$v_o = -\frac{R_F(v_z)}{R_1} + \left(1 + \frac{R_F}{R_1}\right)v_x \qquad (4\text{-}27b)$$

Substituting the value of v_z from Equation (4-27a), we get

$$v_o = -\left(\frac{R_F}{R_1}\right)\left(1 + \frac{R_3}{R_2}\right)v_y + \left(1 + \frac{R_F}{R_1}\right)v_x$$

Since $R_1 = R_3$ and $R_F = R_2$,

$$v_o = \left(1 + \frac{R_F}{R_1}\right)(v_x - v_y)$$

Therefore,

$$A_D = \frac{v_o}{v_{xy}} = 1 + \frac{R_F}{R_1} \qquad (4\text{-}28)$$

where $v_{xy} = v_x - v_y$.

4-5.2(b) Input resistance. The input resistance R_{iF} of the differential amplifier is the resistance determined looking into either one of the two noninverting input terminals with the other grounded (see Figure 4-16). Note, however, that the first stage (A_1) is a noninverting amplifier; therefore [from Equation (4-8)], its input resistance is

$$R_{iFy} = R_i(1 + AB) \qquad (4\text{-}29a)$$

where R_i = open-loop input resistance of the op-amp

$$B = \frac{R_2}{R_2 + R_3}$$

Similarly, with v_y shorted to ground ($v_y = 0$ V), the second stage (A_2) also becomes a noninverting amplifier whose input resistance can then be written as

$$R_{iFx} = R_i(1 + AB) \qquad (4\text{-}29b)$$

where R_i = open-loop input resistance of the op-amp

$$B = \frac{R_1}{R_1 + R_F}$$

However, since $R_1 = R_3$ and $R_F = R_2$, the $R_{iFy} \neq R_{iFx}$. This is the drawback of the differential amplifier of Figure 4-16. Nevertheless, with proper selection of components, both R_{iFy} and R_{iFx} can be made much larger than the source resistances so that the loading of the input sources does not occur.

EXAMPLE 4-6

The following specifications are given for the differential amplifier of Figure 4-16: $R_1 = R_3 = 680 \ \Omega$, $R_F = R_2 = 6.8 \ \text{k}\Omega$, $v_x = -1.5$ V pp, and $v_y = -2$ V pp sine waves at 1 kHz. The op-amp is a 741C. Calculate (a) the voltage gain and the input resistance and (b) the output voltage of the amplifier. Assume that the output is initially nulled ($V_{ooT} = 0$ V).

SOLUTION (a) From Equation (4-28),

$$A_D = 1 + \frac{6.8 \ \text{k}\Omega}{680 \ \Omega} = 11$$

The input resistance can be calculated using Equations (4-29a) and (4-29b).

$$R_{iFy} = (2 \ \text{M}\Omega)\left[1 + \frac{(2)(10^5)(6.8 \ \text{k}\Omega)}{6.8 \ \text{k}\Omega + 680 \ \Omega}\right] = 364 \ \text{G}\Omega$$

$$R_{iFx} = (2 \ \text{M}\Omega)\left[1 + \frac{(2)(10^5)(680) \ \Omega}{6.8 \ \text{k}\Omega + 680 \ \Omega}\right] = 36.4 \ \text{G}\Omega$$

(b) Output voltage can be calculated by rearranging Equation (4-28):

$$v_o = \left(1 + \frac{R_F}{R_1}\right)v_{xy}$$

$$= (11)(-1.5 + 2) = 5.5 \ \text{V pp sine wave at 1 kHz}$$

4-5.3 Differential Amplifier with Three Op-Amps

The advantages of the two differential amplifiers just discussed can be combined in a single differential amplifier having variable voltage gain and very high input resistance. However, to achieve this desirable combination, three op-amps must be used, as shown in Figure 4-17.

4-5.3(a) Voltage gain.   The differential op-amp of Figure 4-17 consists of two stages. The first stage is composed of op-amps A_1 and A_2, while the second stage is formed by op-amp A_3. Therefore, to find the overall voltage gain A_D of the amplifier, the voltage gain of each stage must be determined. To begin with, the first stage can be viewed as two separate differential amplifiers, as shown in Figure 4-18. The output voltages of these differential amplifiers can be found by applying the superposition theorem. For Figure 4-18(a),

$$v_z = \frac{2R_4 + R_5}{R_4 + R_5}v_x - \frac{R_4}{R_4 + R_5}v_t \tag{4-30a}$$

and for Figure 4-18(b),

$$v_t = \frac{2R_4 + R_5}{R_4 + R_5}v_y - \frac{R_4}{R_4 + R_5}v_z \tag{4-30b}$$

However, the output voltage of the first stage is

$$v_{zt} = v_z - v_t$$

Figure 4-17   Differential amplifier with three op-amps.

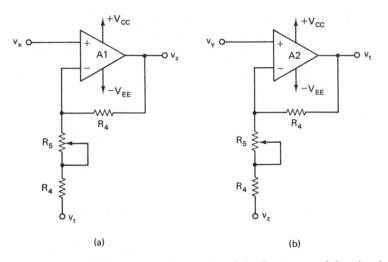

Figure 4–18   Deriving the voltage gain of the first stage of the circuit in Figure 4–17.

Therefore, from Equations (4-30a) and (4-30b),

$$v_{zt} = \frac{2R_4 + R_5}{R_4 + R_5}(v_x - v_y) + \frac{R_4}{R_4 + R_5}(v_z - v_t)$$

$$= \frac{2R_4 + R_5}{R_4 + R_5}v_{xy} + \frac{R_4}{R_4 + R_5}v_{zt}$$

Simplifying and rearranging, the voltage gain of the first stage is

$$\frac{v_{zt}}{v_{xy}} = \frac{2R_4 + R_5}{R_5} \tag{4-30c}$$

Next, using Equation (4-25), the gain of the second stage is

$$\frac{v_o}{v_{zt}} = -\frac{R_F}{R_1} \tag{4-30d}$$

Thus, from Equations (4-30c) and (4-30d), the overall voltage gain is

$$A_D = \frac{v_{zt}}{v_{xy}}\frac{v_o}{v_{zt}} = -\left(\frac{2R_4 + R_5}{R_5}\right)\frac{R_F}{R_1}$$

or

$$A_D = \frac{v_o}{v_{xy}} = -\left(1 + \frac{2R_4}{R_5}\right)\frac{R_F}{R_1} \tag{4-31}$$

Remember that the gain can be changed by varying potentiometer R_5. Of course, R_5 should never be set to zero or infinity. If R_5 is set to zero, the voltage gain equation (4-31) becomes indeterminent, whereas if R_5 is set to infinity (open R_5), the first stage acts as a voltage follower.

4-5.3(b) Input resistance.   The input resistance R_{iF} of the differential amplifier in Figure 4-17 is the same as the input resistance of the first stage, that is, the resistance determined at input v_x and v_y looking into the circuit with the other terminal grounded.

In Figure 4-18a, for instance, when v_t is reduced to zero, that is, when v_y is grounded, the circuit is a noninverting amplifier. Applying the concepts developed for the noninverting amplifier, the input resistance determined at input v_x is

$$R_{iF} = R_i\left(1 + A\frac{R_4 + R_5}{2R_4 + R_5}\right) \tag{4-32}$$

Similarly, the input resistance determined at input v_y will be the same as that given in Equation (4-32).

A simple modification in the differential amplifier of Figure 4-17 is to replace the first stage by voltage followers, as shown in Figure 4-19. The main advantage of this circuit is that it has the highest input resistance of the three amplifiers; this resistance is given by

$$R_{iF} = R_i(1 + A) \tag{4-33a}$$

where R_i = input resistance of the op-amp
$\quad A$ = open-loop voltage gain of the op-amp

However, the voltage gain of the circuit in Figure 4-19 is equal to the voltage gain of the differential amplifier with one op-amp; that is,

$$A_D = \frac{v_o}{v_{xy}} = -\frac{R_F}{R_1} \tag{4-33b}$$

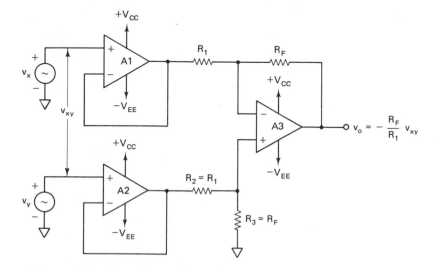

Figure 4–19   Special case of differential amplifier with three op-amps.

4-5.4 Output Resistance and Bandwidth of Differential Amplifiers with Feedback

Recall that the closed-loop op-amp configurations—noninverting, inverting, and differential amplifier—employ negative feedback and use the same type of output connections. Therefore, the output resistance of all three configurations must be identical. In other words, the output resistance of the differential amplifier should be the same as that of the noninverting or inverting amplifier, except that $B = 1/A_D$. That is,

$$R_{oF} = \frac{R_o}{1 + A/A_D} \qquad (4\text{-}34)$$

where A_D = closed-loop gain of the differential amplifier
$\quad R_o$ = output resistance of the op-amp
$\quad A$ = open-loop voltage gain of the op-amp

Remember that A_D is different for each of the differential amplifier configurations.

As in the case of noninverting or inverting amplifiers, the bandwidth of differential amplifiers also depends on the closed-loop gain of the amplifier and is given by

$$f_F = \frac{\text{unity gain bandwidth}}{\text{closed-loop gain } A_D} \qquad (4\text{-}35a)$$

or,

$$f_F = \frac{(A)(f_o)}{A_D} \qquad (4\text{-}35b)$$

where f_o is the open-loop break frequency of the op-amp.

EXAMPLE 4-7

The following specifications are given for the differential amplifier of Figure 4-17: $R_4 = 3.9 \text{ k}\Omega$, R_5 is a 5-kΩ potentiometer, $R_1 = R_2 = 1 \text{ k}\Omega$, and $R_F = R_3 = 4.7 \text{ k}\Omega$. The op-amps are 741Cs and the supply voltages are ± 15 V. If R_5 is set at 3 kΩ, calculate (a) the voltage gain, (b) the input resistance, (c) the output resistance, and (d) the bandwidth of the differential amplifier.

SOLUTION   (a) Substituting the given values in Equation (4-31), the voltage gain is

$$A_D = -\left[1 + \frac{(2)(3.9 \text{ k}\Omega)}{3 \text{ k}\Omega}\right]\frac{4.7 \text{ k}\Omega}{1 \text{ k}\Omega} = -16.92$$

Note that the negative sign indicates the phase reversal between differential input and output.
(b) Using Equation (4-32),

$$R_{iF} = (2 \times 10^6)\left[1 + \frac{(200,000)(3.9 \text{ k}\Omega + 3 \text{ k}\Omega)}{(2)(3.9 \text{ k}\Omega) + 3 \text{ k}\Omega}\right] = 256 \text{ G}\Omega$$

(c) Output resistance can be obtained substituting the values in Equation (4-34):

$$R_{oF} = \frac{75 \ \Omega}{1 + 200,000/16.92} = 6.3 \ \text{m}\Omega$$

(d) Finally, the bandwidth of the amplifier is given by Equation (4-35b):

$$f_F = \frac{(200,000) \ (5)}{16.92} = 59.1 \ \text{kHz}$$

SUMMARY

1. The very high voltage gain of the open-loop op-amp as well as its variation with temperature, power supply, and production yield makes the open-loop op-amp configuration unsuitable for linear applications.
2. The introduction of negative feedback stabilizes the gain; however, it is also smaller than the open-loop gain. On the other hand, positive feedback is necessary in oscillator circuits.
3. There are four closed-loop configurations using negative feedback: voltage series, voltage shunt, current series, and current shunt. These configurations are labeled according to whether the voltage or current is fed back to the input in series or in parallel.
4. The voltage-series negative feedback configuration is commonly called a noninverting amplifier with feedback, while the voltage-shunt feedback configuration is called an inverting amplifier with feedback. The ideal closed-loop gain of these two configurations depends only on the feedback network (components) and is independent of the internal gain A of the op-amp.
5. The use of negative feedback in a noninverting amplifier increases the input impedance and bandwidth, and decreases the output resistance, total output-offset voltage, and the effect of varying environmental conditions on the gain. The same results are obtained for the inverting amplifier using negative feedback, except for input resistance. The input resistance of the inverting amplifier is relatively smaller because it depends on the resistance connected in series with the input signal source.
6. A special case of the noninverting amplifier is the voltage follower, while the current-to-voltage converter and the inverter are two special cases of the inverting amplifier. The voltage follower has a gain of unity. The current-to-voltage converter produces output voltage proportional to input current, and the inverter has an output equal in amplitude but opposite in phase to that of the input.
7. Negative feedback is also used in differential amplifiers. According to the number of op-amps used, the differential amplifier configurations are classified as differential amplifier with one op-amp, differential amplifier with two op-amps, and differential amplifier with three op-amps. The differential amplifier with one op-amp

has the same characteristics as the inverting amplifier, while the amplifier with two op-amps has the same characteristics as the noninverting amplifier. Very high input resistance and variable gain are the features of the differential amplifier with three op-amps.

QUESTIONS

4-1. Give two reasons why an open-loop op-amp is unsuitable for linear applications.

4-2. What is a feedback? List two types of feedback. Which type is used in linear applications?

4-3. List the four negative feedback configurations. Which two configurations are most commonly used?

4-4. Explain briefly why negative feedback is desirable in amplifier applications.

4-5. What is the effect of negative feedback in noninverting amplifiers?

4-6. How does negative feedback affect the performance of an inverting amplifier?

4-7. In what way is the voltage follower a special case of the noninverting amplifier?

4-8. List two special cases of inverting amplifiers.

4-9. What are the three differential amplifier configurations? Briefly compare and contrast these configurations.

PROBLEMS

4-1. The open-loop gain A of a particular op-amp varies from 15,000 to 20,000 over the operating temperature range from 25° to 35°C. This op-amp is used in the circuit of Figure 4-2 with $R_1 = 470\ \Omega$ and $R_F = 47\ \text{k}\Omega$.
  (a) Calculate the closed-loop gain at both temperature extremes.
  (b) Compute the percentage of variation in the open-loop gain and in the closed-loop gain; explain the results.

4-2. The following specifications apply to the noninverting amplifier of Figure 4-2:

Op-amp is the 741C with:	Feedback circuit
$A = 200,000$	$R_1 = 1.8\ \text{k}\Omega$
$R_i = 2\ \text{M}\Omega$	$R_F = 18\ \text{k}\Omega$
$R_o = 75\ \Omega$	
$f_o \cong 5\ \text{Hz}$	
UGB $\cong 1\ \text{MHz}$	
Supply voltages $= \pm 15\ \text{V}$	
Maximum output voltage swing $= \pm 13\ \text{V}$	

(a) Calculate the *exact* closed-loop gain.
(b) Calculate the *ideal* closed-loop gain.
(c) Explain the results obtained in parts (a) and (b).

4-3. Repeat Problem 4-2 with $R_1 = 470 \ \Omega$ and $R_F = 47 \ k\Omega$.

4-4. The 714C is configured as a noninverting amplifier as shown in Figure 4-2. The following data are given for the circuit:

$$A = 400,000 \qquad R_1 = 470 \ \Omega$$
$$R_i = 33 \ M\Omega \qquad R_F = 4.7 \ k\Omega$$
$$R_o = 60 \ \Omega$$

Supply voltages $= \pm 15$ V

Maximum output voltage swing $= \pm 13$ V

Unity gain bandwidth $= 0.6$ MHz

Compute the closed-loop parameters A_F, R_{iF}, R_{oF}, f_F, and V_{ooT}.

4-5. Repeat Problem 4-4 with $R_1 = 100 \ \Omega$.

4-6. For the noninverting amplifier of Problem 4-4, compute the output voltage v_o and then sketch it if $v_{in} = 100$ mV pp sine wave at 1 kHz. Assume that the op-amp is initially nulled.

4-7. The op-amp of Problem 4-4 is used in a voltage follower. Compute the closed-loop parameters A_F, R_{iF}, R_{oF}, f_F, and V_{ooT}.

4-8. The 741C is configured as an inverting amplifier as shown in Figure 4-8, with $R_1 = 1 \ k\Omega$ and $R_F = 10 \ k\Omega$.
(a) Calculate the *exact* closed-loop gain.
(b) Caculate the *ideal* closed-loop gain.
(c) Compare and explain the results of parts (a) and (b).
(For op-amp specifications, refer to Problem 4-2.)

4-9. Repeat Problem 4-8 with $R_1 = 100 \ \Omega$.

4-10. The 714C is connected as an inverting amplifier with $R_1 = 1 \ k\Omega$ and $R_F = 4.7 \ k\Omega$. Compute the closed-loop parameters A_F, R_{iF}, R_{oF}, f_F, and V_{ooT}. (Refer to Problem 4-4 for the op-amp parameters.)

4-11. Repeat Problem 4-10 with $R_F = 1 \ k\Omega$. What is the name of this circuit?

4-12. For the inverting amplifier of Problem 4-10, $v_{in} = 1$ V pp sine wave at 100 Hz.
(a) Compute the output voltage v_o.
(b) Draw the output voltage waveform, assuming that the output is initially nulled.

4-13. The input current into the current-to-voltage converter of Figure 4-12 varies from 500 μA to 1 mA. Calculate the variations in the output voltage if $R_F = 4.7 \ k\Omega$. Assume that the op-amp is a 714C with its output initially nulled.

4-14. The 741C is used for the differential amplifier of Figure 4-14 with $R_1 = R_2 = 100 \ \Omega$ and $R_F = R_3 = 3.9 \ k\Omega$.

(a) Calculate the closed-loop voltage gain.
(b) Calculate the input resistances R_{iFx} and R_{iFy} assuming that the circuit is initially nulled.
(The 741C parameters are given in Problem 4-2.)

4-15. For the differential amplifier of Problem 4-14, compute the output voltage if $v_x = -3.2$ V and $v_y = -3.0$ V. Assume that the circuit is initially nulled.

4-16. In the variable-gain differential amplifier of Figure 4-15, $R_1 = R_2 = 100\ \Omega$, $R_F = R_3 = 2.2\ \text{k}\Omega$, $R_4 = 10\ \text{k}\Omega$, and R_P is a 20-kΩ potentiometer. The op-amp is a 714C. Compute the gain of the differential amplifier if (a) $R_P = 10\ \text{k}\Omega$ and (b) $R_P = 5\ \text{k}\Omega$.

4-17. The following specifications apply to the differential amplifier of Figure 4-16: $R_2 = R_F = 2.2\ \text{k}\Omega$, $R_3 = R_1 = 1\ \text{k}\Omega$, $v_x = 700$ mV pp, and $v_y = 500$ mV pp sine waves at 500 Hz; the op-amp is a 741C. Assume that the op-amp is initially nulled. Calculate:
(a) The voltage gain.
(b) The bandwidth.
(c) The input resistance seen by each signal source.
(d) The output resistance.
(e) The output voltage of the amplifier.
(For the electrical parameters of the 741C, refer to Problem 4-2.)

4-18. For the differential amplifier of Figure 4-17, $R_4 = 1\ \text{k}\Omega$, R_5 is a 10-kΩ potentiometer, $R_1 = R_2 = 1.5\ \text{k}\Omega$, and $R_F = R_3 = 15\ \text{k}\Omega$. The op-amps are 741Cs with supply voltages $= \pm 15$ V. If R_5 is set at 5 kΩ, calculate:
(a) The voltage gain.
(b) The input resistance seen by each signal source.
(c) The output resistance.
(d) The bandwidth of the differential amplifier.

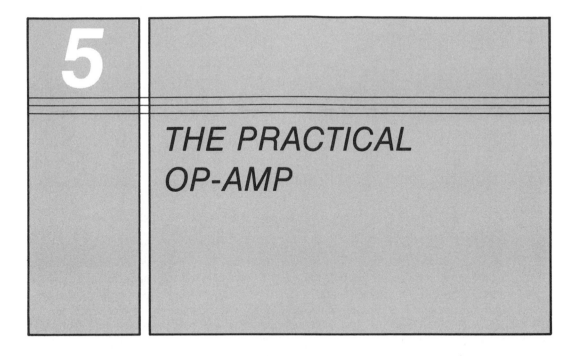

THE PRACTICAL OP-AMP

5-1 INTRODUCTION

In preceding chapters we have seen that the op-amp may be used as an inverting, non-inverting, or differential amplifier, and that the negative feedback can be used to stabilize the voltage gain and increase the bandwidth of the op-amp circuit. However, up to this point in our discussion we have treated the op-amp as an ideal device that gives us the desired results, such as high input impedance, low output impedance, high voltage gain, and broader bandwidth, if we use appropriate external components. We have also treated the op-amp as though it responds equally to both ac and dc input voltages. However, these ideal characteristics are not fully present in the practical op-amp circuits. The practical op-amp has some dc output voltage, called *output offset voltage,* even though both inverting and noninverting input terminals are grounded. Such an output offset voltage is an error voltage and is therefore undesirable. In this chapter we discuss the properties of the practical op-amp that produce the output offset voltage. We also see that it is necessary to incorporate additional circuitry to increase the effectiveness of the op-amp, especially if it is used as a dc amplifier.

5-2 INPUT OFFSET VOLTAGE

Input offset voltage V_{io} is the differential input voltage that exists between two input terminals of an op-amp without any external inputs applied. In other words, it is the amount of the input voltage that should be applied between two input terminals in order to force the output voltage to zero [see Figure 5-1(a)]. Let us denote the output offset voltage due to input offset voltage V_{io} as V_{oo}. The output offset voltage V_{oo} is caused by

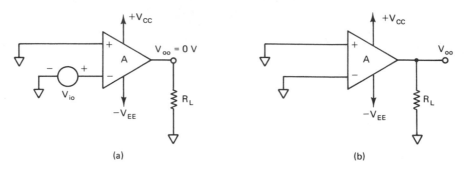

Figure 5–1 (a) Input offset voltage in an op-amp. (b) Output offset voltage in an op-amp.

mismatching between two input terminals. Even though all the components are integrated on the same chip, it is not possible to have two transistors in the input differential amplifier stage with exactly the same characteristics. This means that the collector currents in these two transistors are not equal, which causes a differential output voltage from the first stage. The output of first stage is amplified by following stages and possibly aggravated by more mismatching in them. Thus the output voltage caused by mismatching between two input terminals is the output offset V_{oo}. Figure 5-1(b) shows the output offset voltage in an op-amp without feedback.

The output offset voltage V_{oo} is a dc voltage; it may be positive or negative in polarity depending on whether the potential difference between two input terminals is positive or negative. It is impossible to predict the polarity of the input offset voltage since it is dependent on mismatching between two input terminals. Therefore, on the data sheets the value of the input offset voltage V_{io} is listed as an absolute maximum value. For example, for a 741, $V_{io} = 6$ mV maximum, whereas for a 740, $V_{io} = 20$ mV maximum, which means that the maximum potential difference between two input terminals in a 741 op-amp can be as large as 6 mV dc; that is, voltage at the noninverting input terminal may differ from that at the inverting input terminal by as much as 6 mV dc, or vice versa. This input offset voltage gives rise to an output offset voltage V_{oo}. Thus we need to apply a differential input voltage of specific amplitude and correct polarity in order to reduce the output offset voltage V_{oo} to zero. This voltage is referred to as input offset voltage V_{io}.

The V_{io} value for op-amps of the same type may not be the same in amplitude and polarity because of mass production, but it will always be less than the maximum value given on the specification sheets. For example, if we take three different 741s and use them one at a time as in Figure 5-1(b) and measure the corresponding output offset voltage for each of them, we find that the output offset voltage in these three op-amps is not of the same amplitude and polarity, which means that input offset V_{io} is not of the same amplitude and polarity even though the op-amps are of the same type.

To reduce V_{oo} to zero, we need to have a circuit at the input terminals of the op-amp that will give us the flexibility of obtaining V_{io} of proper amplitude and polarity. Such a circuit is called an input offset voltage-compensating network. Before we apply external input to the op-amp, with the help of an offset voltage-compensating network we reduce

Figure 5–2 Voltage offset null circuit for the 741.

the output offset voltage V_{oo} to zero; the op-amp is then said to be *nulled* or *balanced*. Before we get into the design of a compensating network, it is worth noting that the compensating network is not needed for those op-amps that have offset null pins, such as 741, 748, 777, and 301. For the 741-type op-amp the manufacturer recommends that a 10-kΩ potentiometer be placed across offset null pins 1 and 5 and a wiper be connected to the negative supply pin 4 as shown in Figure 5-2. Adjustment of this pot will null the output.

Refer to the 741 equivalent circuit in Figure 2-3. By varying the position of the wiper on the 10-kΩ potentiometer, we are trying to remove the mismatch between inverting and noninverting input terminals of the op-amp. Adjust the wiper until the output offset voltage is reduced to zero. See the analysis of the 741 op-amp circuit in Section 2-4.

5-2.1 Offset-Voltage Compensating Network Design

The op-amp with offset-voltage compensating network is shown in Figure 5-3. The compensating network consists of potentiometer R_a and resistors R_b and R_c. If we are planning to make use of the op-amp as an inverting amplifier, the compensating network should be connected to the noninverting input terminal of the op-amp. The circuit in Figure 5-3 can be used as a noninverting amplifier since the compensating network is connected to the inverting input terminal of the op-amp.

Figure 5–3 Op-amp with offset voltage-compensating network.

147

Note that the potentiometer R_a is connected to $+V_{CC}$ and $-V_{EE}$ and the voltage across R_c is the voltage at the inverting input terminal V_2. By adjusting the wiper on R_a, voltage V_2 can be made to equal voltage V_1: that is, V_{io} can be made zero, which forces V_{oo} to go to zero. Say, for example, that V_{oo} is positive, which implies that $V_1 > V_2$. This means that V_2 should be increased until it equals V_1; the increase can be accomplished by moving the wiper toward V_{CC}. On the other hand, if V_{oo} is negative, that is, $V_2 > V_1$, the wiper can be moved toward $-V_{EE}$ until V_{oo} is reduced to zero.

To establish a relationship between V_{io}, supply voltages, and the compensating components, we first have to Thévenize the circuit, looking back into R_a from point T. The maximum Thévenin's equivalent resistance R_{max} occurs when the wiper is at the center of the potentiometer, as shown in Figure 5-4(a). Thus

$$R_{max} = \frac{R_a}{2} \left\| \frac{R_a}{2} = \frac{R_a}{4}.$$

The maximum Thévenin's equivalent voltage V_{max} is equal to either V_{CC} or $-V_{EE}$ when the wiper is uppermost in the potentiometer or lowest in the potentiometer [see Figure 5-4(b) and (c)].

Supply voltages V_{CC} and $-V_{EE}$ are equal in magnitude; therefore, let us denote their magnitude by voltage V. Thus $V_{max} = V$.

Next we redraw the compensating network using the maximum Thévenin's voltage and resistance as shown in Figure 5-5. Applying the voltage-divider rule to the circuit in Figure 5-5, we get

$$V_2 = \frac{R_c}{R_{max} + R_b + R_c} V_{max} \tag{5-1}$$

where V_2 has been expressed as a function of maximum Thévenin's voltage V_{max} and maximum Thévenin's resistance R_{max}. But the maximum value of V_2 can be equal to V_{io}, since $|V_1 - V_2| = V_{io}$. Thus Equation (5-1) becomes

$$V_{io} = \frac{R_c}{R_{max} + R_b + R_c} V_{max} \tag{5-2}$$

(a)   (b)   (c)

Figure 5–4 (a) Thévenin's equivalent for resistance R_{max}. (b) and (c) Thévenin's equivalent for voltage V_{max}.

Figure 5–5 Compensating network with maximum Thévenin's equivalent resistance and voltage.

There are too many unknowns in Equation (5-2). To simplify Equation (5-2), let us make the assumption that $R_b > R_{max} > R_c$, where $R_{max} = R_a/4$. An explanation for this assumption is in order. The bias currents in the op-amp are fixed and very small, in the range of nanoamperes. Therefore, the bias current in an inverting terminal of the op-amp in Figure 5-3 will also be very small and fixed. In no case will the bias current be larger than the value listed on the data sheets. To null the op-amp the required variation in the voltage across R_c is also small, on the order of millivolts; in the case of some op-amps it is even on the order of microvolts. This means that not only the current through R_c should be small, but the value of R_c should also be very small. The bias current in an inverting terminal and the current through R_c are derived from the current through R_b. Since two of these currents will be very small, the current through R_b will be significantly smaller than the current through R_a. To accomplish this we needed $R_b > R_a$, which implies that $R_b > R_{max}$ since $R_{max} = R_a/4$. Thus the assumption $R_b > R_{max} > R_c$ is valid. Using this assumption, we can say that $R_{max} + R_b + R_c \cong R_b$. Therefore, Equation (5-2) can be rewritten as

$$V_{io} = \frac{R_c V_{max}}{R_b} \qquad (5\text{-}3)$$

where

$$V_{max} = V = |V_{CC}| = |-V_{EE}|$$

$$V_{io} = \frac{R_c V}{R_b} \qquad (5\text{-}4)$$

Note that V_{io} depends on the magnitude of the supply voltages $+V_{CC}$ and $-V_{EE}$.

Equation (5-4) will be used to design the compensating network. We will obtain the value of V_{io} from the data sheet of a given op-amp, whereas the value of V will be fixed according to the supply voltages chosen; once these two values are known, we will establish the relationship between R_b and R_c. Then we will select the value for R_c to compute the value of R_b. The value for R_c will be selected to be less than 100 Ω so that R_b and R_a values will not be too large.

EXAMPLE 5-1

Design a compensating network for the LM307 op-amp. The op-amp uses ±10-V supply voltages.

SOLUTION   The value of V_{io} specified on the data sheets of the LM307 is 10 mV maximum. The value of $V = |V_{CC}| = |-V_{EE}| = 10$ V. Substituting these values in Equation (5-4), we get

$$10 \text{ mV} = \frac{R_c(10 \text{ V})}{R_b}$$

$$R_b = \frac{10 \text{ V}}{10 \text{ mV}} R_c$$

If we select $R_c = 10$ Ω, the value of R_b should be

$$R_b = (1000)R_c = 10{,}000 \text{ } \Omega$$

Since $R_b > R_{max}$, let us choose $R_b = 10R_{max}$, where $R_{max} = R_a/4$. Therefore,

$$R_b = (10)\frac{R_a}{4}$$

or

$$R_a = \frac{R_b}{2.5} = \frac{10 \text{ k}\Omega}{2.5} = 4\text{-k}\Omega \text{ potentiometer}$$

If a 4-kΩ potentiometer is not available, we may prefer to use the next lower value available, such as 3 kΩ, so that the value of R_a will be larger than R_b by a factor of 10. If we select a 3-kΩ potentiometer as the R_a value, R_b is 13.3 times larger than R_a. Thus

$$R_a = 3\text{-k}\Omega \text{ potentiometer}$$
$$R_b = 10 \text{ k}\Omega$$
$$R_c = 10 \text{ } \Omega$$

The final circuit, which also includes the pin connections for the LM307, is shown in Figure 5-6.

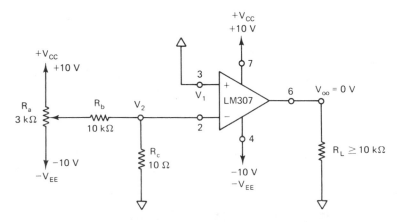

Figure 5–6   Offset voltage-compensating network for LM307.

After the circuit in Figure 5-6 is breadboarded, we will adjust the potentiometer R_a until the output is reduced to zero. This nulled op-amp circuit now can be used as a noninverting amplifier by applying an external voltage to the noninverting input terminal. But the magnitude of the input voltage applied has to be very small to avoid any distortion and clipping of the output signal. A small input voltage is needed because the voltage gain of the op-amp without feedback is extremely high, ideally infinity. Because of the high risk of distortion and clipping of the output signal, an op-amp in open-loop configuration is not used in linear circuit applications.

Let us now examine the effect of V_{io} in amplifiers with feedback. The noninverting and inverting amplifiers with feedback are shown in Figure 5-7. To determine the effect of V_{io} in each case, we have to reduce the input voltage v_{in} to zero.

(a)

(b)

Figure 5–7   (a) Noninverting amplifier with feedback. (b) Inverting amplifier with feedback.

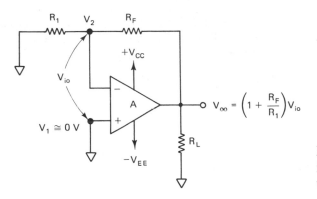

Figure 5–8 Closed-loop non-inverting or inverting amplifier with $v_{in} = 0$ V.

With v_{in} reduced to zero, the circuits of both noninverting and inverting amplifiers are the same as the circuit in Figure 5-8. We assume here and in future cases that the internal resistance R_{in} of the input signal voltage is negligibly small.

We wish to express V_{oo} in terms of external components R_1, R_F, and the specified input offset voltage V_{io} for a given op-amp. In the figure, the noninverting input terminal is connected to ground; therefore, we assume voltage V_1 at the noninverting input terminal to be zero. The voltage V_2 at the inverting input terminal can be determined by applying the voltage-divider rule:

$$V_2 = \frac{R_1 V_{oo}}{R_1 + R_F} \tag{5-5}$$

Therefore,

$$V_{oo} = \frac{R_1 + R_F}{R_1} V_2 \tag{5-6}$$

Since $V_{io} = |V_1 - V_2|$ and $V_1 = 0$ V,

$$V_{io} = |0 - V_2| = V_2 \tag{5-7}$$

Therefore,

$$V_{oo} = \left(1 + \frac{R_F}{R_1}\right) V_{io} = (A_{oo}) V_{io} \tag{5-8}$$

where $A_{oo} = 1 + R_F/R_1$. Thus, when an op-amp is used in a closed-loop configuration as either a noninverting or inverting amplifier, we can compute the maximum possible output offset voltage V_{oo} caused by the input offset voltage V_{io} by using Equation (5-8). According to this equation, for a given op-amp the amount of V_{oo} depends on the values of external components R_1 and R_F, that is, A_{oo}. Therefore, the smaller the value of A_{oo}, the smaller will be the value of V_{oo} for a given value of V_{io}. In the extreme case when $R_1 \gg R_F$, $A_{oo} \cong 1$ and $V_{oo} \cong V_{io}$. Thus, in practice, all op-amp circuits will have some output offset voltage V_{oo}. To null the output offset voltage V_{oo}, we can use a compensating network identical to the one that was used to null the open-loop op-amp circuit. The closed-loop noninverting and inverting amplifiers with their compensating networks are shown in Figure 5-9.

(a)

(b)

Figure 5–9   (a) Compensated noninverting amplifier with feedback.
(b) Compensated inverting amplifier with feedback.

The offset-voltage compensating networks in Figure 5-9 will be designed by using Equation (5-4). The compensating network is connected in the noninverting terminal for the inverting amplifier and in the inverting terminal for the noninverting amplifier. Note that the voltage gain of the noninverting amplifier with compensating network is $A_F = 1 + [R_F/(R_1 + R_c)]$. The gain changes because the Thévenin equivalent resistance of the compensating network, which is approximately equal to R_c, is in series with R_1. In fact, this result is based on the assumption that $R_b > R_a > R_c$. Before an external input v_{in} is applied, the op-amp in Figure 5-9 should be nulled by adjusting the wiper in pot R_a.

The closed-loop differential amplifier is somewhat more difficult to null because the use of the compensating network can change the common-mode rejection ratio. (See

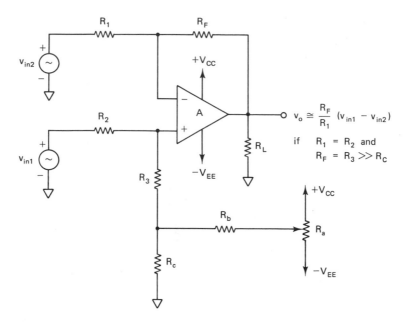

Figure 5–10 Differential amplifier with offset voltage-compensating network.

Section 5-11 for further information on the common-mode rejection ratio.) Figure 5-10 shows the differential amplifier with its input offset voltage compensating network. To maximum CMRR, we should use $R_1 = R_2$ and $R_F = R_3 + R_c$.

The compensated differential amplifier shown in Figure 5-11 uses the op-amp with the offset voltage null pins. Here $R_1 = R_2$ and $R_F = R_3$. This amplifier circuit has two

Figure 5–11 Differential amplifier with offset voltage null circuit (pins 1 and 5 are offset null pins for the 741 op-amp.)

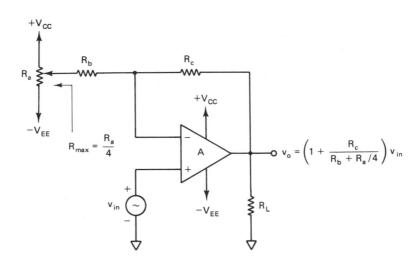

Figure 5–12   Voltage follower with offset voltage-compensating network.

advantages over the one in Figure 5-10. One advantage is that it is simpler, since it uses fewer components, and the other is that the offset null circuit does not affect the CMRR.

A voltage follower may be nulled using a similar compensating network, as shown in Figure 5-12. The voltage drop across R_c is used to cancel the offset voltage. Even though $R_b > R_a$ and R_c, their finite values cause the gain of the voltage follower to increase slightly (see the v_o expression in Figure 5-12). This insignificant error in gain usually can be ignored, since $R_c < (R_b + R_a/4)$.

The offset voltage-compensating network just discussed is a versatile circuit. It is applicable to all op-amp types and can be used in different operating modes without any modifications. It also allows nulling without interfering with the internal circuitry of an op-amp.

EXAMPLE 5-2

The op-amp in the circuit of Figure 5-13 is the LM307 with $V_{io} = 10$ mV dc maximum. What is the maximum possible output offset voltage, V_{oo}, caused by the input offset voltage V_{io}?

SOLUTION   To find the maximum possible value of V_{oo}, we reduce the input voltage v_{in} to zero. The closed-loop gain of the amplifier is

$$A_{oo} = 1 + \frac{R_F}{R_1} = 1 + \frac{10 \text{ k}\Omega}{1 \text{ k}\Omega} = 11$$

Since $V_{io} = 10$ mV dc maximum, the V_{oo} might be as large as

$$V_{oo} = A_{oo}V_{io} = (11)(10 \text{ mV}) = 110 \text{ mV dc}$$

Figure 5–13   Noninverting amplifier with feedback.

This means that the output terminal can be either at a negative or a positive 110 mV dc with respect to ground even though $v_{in} = 0$ V.

EXAMPLE 5-3

Design an input offset voltage-compensating network for the circuit in Figure 5-13.

SOLUTION   We have designed the compensating network for the LM307 in Example 5-1. We can use the same circuit and connect it in the inverting terminal as shown in Figure 5-14.

Applying Thévenin's theorem across R_c, the equivalent resistance $R_{TH} = R_c$ because $R_c < R_a < R_b$. Also, the value of R_1 is much larger than R_c. This means that the closed-

Figure 5–14   Compensated noninverting amplifier with feedback.

loop gain of the noninverting amplifier will not be affected by the use of the compensating network, that is,

$$A_F = 1 + \frac{R_F}{R_1 + R_c} = 1 + \frac{10\text{ k}\Omega}{1\text{ k}\Omega + 10\text{ }\Omega} = 11$$

Note that we have distinguished between the voltage gains A_{oo} and A_F. The voltage gain $A_{oo} = 1 + R_F/R_1$ is used to calculate the maximum possible output offset voltage V_{oo} due to the input offset voltage V_{io} both in the inverting and noninverting amplifiers (see Figure 5-8). On the other hand, the voltage gain $A_F = 1 + R_F/R_1$ of the noninverting amplifier of Figure 5-7(a) or the voltage gain $A_F = -R_F/R_1$ of the inverting amplifier of Figure 5-7(b) is used to compute the output voltage v_o due to the input voltage v_{in}.

5-3 INPUT BIAS CURRENT

An input bias current I_B is defined as the average of the two input bias currents, I_{B1} and I_{B2}, as shown in Figure 5-15; that is,

$$I_B = \frac{I_{B1} + I_{B2}}{2} \tag{5-9}$$

where I_{B1} = dc bias current flowing into the noninverting input
$\quad\quad I_{B2}$ = dc bias current flowing into the inverting input

In Figure 5-15, both input terminals are grounded so that no input voltage is applied to the op-amp. But the plus–minus supply voltages are necessary to bias the op-amp properly.

Actually, the input bias currents I_{B1} and I_{B2} are the base bias currents of the two transistors in the input (first) differential amplifier stage of the op-amp. Even though both of the input transistors are identical, it is not possible to have I_{B1} and I_{B2} exactly equal to each other because of the internal imbalance between the two inputs. In this section we use the specified input bias current I_B as being equal to either one of the two input currents I_{B1} and I_{B2}; that is,

$$I_B = I_{B1} = I_{B2} \tag{5-10}$$

The value of input bias current I_B is very small, in the range of a few to a few hundred nanoamperes. It is a dc current and is specified on the op-amp data sheets. For example,

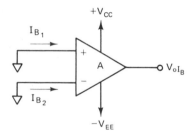

Figure 5–15 Input bias currents in an op-amp.

I_B is 500 nA maximum at supply voltages $= \pm 15$ V dc for the μA741, whereas it is 75 nA maximum for the MLM101A. The value of I_B has been significantly reduced in later-generation op-amps.

Even though very small, the input bias current I_B can cause a significant output offset voltage in circuits using relatively large feedback resistors. This output offset voltage may not be as large as that caused by the input offset voltage, but certain precautions must be taken to minimize it.

First we obtain the expression for the output offset voltage caused by the input bias current I_B in the inverting and noninverting amplifiers and then devise some scheme to eliminate or minimize it. The noninverting or inverting amplifier with $v_{in} = 0$ V is redrawn in Figure 5-16. We assume for the duration of this section that the input offset voltage V_{io} is zero; that is, there is no output offset voltage due to V_{io}. Let V_{ol_B} be the output offset voltage due to input bias current I_B.

In the figure, the input bias currents I_{B1} and I_{B2} are flowing into the noninverting and inverting input leads, respectively. The noninverting terminal is connected to ground; therefore, the voltage $V_1 = 0$ V. The controlled voltage source $AV_{io} = 0$ V since $V_{io} = 0$ V is assumed. With output resistance R_o negligibly small, the right end of R_F is essentially at ground potential; that is, resistors R_1 and R_F are in parallel and the bias current I_{B2} flows through them. Therefore, the voltage at the inverting terminal is

$$V_2 = (R_1 \| R_F) I_{B2} \tag{5-11}$$

$$V_2 = \frac{R_1 R_F}{R_1 + R_F} I_{B2} \tag{5-12}$$

Writing a node voltage equation for node V_2, we get

$$I_1 + I_2 = I_{B2}$$

$$\frac{0 - V_2}{R_1} + \frac{V_{ol_B} - V_2}{R_F} = \frac{V_2}{R_i} \tag{5-13}$$

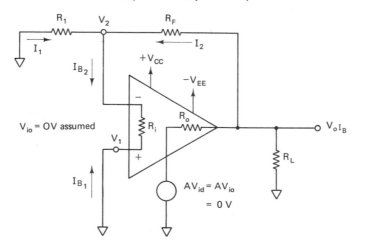

Figure 5–16   Output offset voltage due to input bias current in a noninverting or inverting amplifier.

where V_{ol_B} = output offset voltage due to input bias current
  R_i = input resistance of the op-amp (see Figure 5-16)

Rearranging Equation (5-13), we get

$$\frac{V_{ol_B}}{R_F} = V_2\left(\frac{1}{R_1} + \frac{1}{R_F} + \frac{1}{R_i}\right)$$

Since R_i is extremely high (ideally ∞), $1/R_i \cong 0$ siemens. Therefore,

$$\frac{V_{ol_B}}{R_F} = V_2\frac{R_1 + R_F}{R_1 R_F} \qquad (5\text{-}14)$$

Substituting in Equation (5-14) the value of V_2 from Equation (5-12), we get

$$V_{ol_B} = \frac{R_1 R_F I_{B2}}{R_1 + R_F}\left(\frac{R_1 + R_F}{R_1}\right)$$

$$V_{ol_B} = R_F I_{B2} \qquad (5\text{-}15)$$

From Equation (5-10),

$$V_{ol_B} = R_F I_B \qquad (5\text{-}16)$$

According to Equation (5-16), the amount of output offset voltage V_{ol_B} is a function of feedback resistor R_F for a specified value of input bias current I_B. The amount of V_{ol_B} can be increased by the use of relatively large feedback resistors. Therefore, the use of small feedback resistors is recommended.

To eliminate or reduce the output offset voltage V_{ol_B} due to input bias current I_B, we have to devise some scheme at the input by which voltage V_1 can be made equal to V_2. In other words, if voltages V_1 and V_2 caused by the currents I_{B1} and I_{B2} can be made equal, there will be no output offset voltage V_{ol_B}. From Equation (5-12), we have

$$V_2 = R_p I_{B2} \qquad (5\text{-}17)$$

where

$$R_p = \frac{R_1 R_F}{R_1 + R_F}$$

Equation (5-17) implies that we must express voltage V_1 at the noninverting input terminal as a function of I_{B1} and some specific resistor R_{OM}. This can be accomplished as follows. The input bias current I_{B1} does not produce any voltage at the noninverting input terminal because this terminal is directly connected to ground (see Figure 5-16). If we could connect the proper value of resistor R_{OM} in the noninverting terminal, the voltage V_1 would be

$$V_1 = R_{OM} I_{B1} \qquad (5\text{-}18)$$

To have voltage V_1 equal to V_2, the right-hand sides of Equations (5-17) and (5-18) must be equal; that is,

$$R_p I_{B2} = R_{OM} I_{B1} \qquad (5\text{-}19)$$

Now if the currents I_{B1} and I_{B2} are equal, Equation (5-19) implies that

$$R_p = R_{OM} \qquad (5\text{-}20)$$

or

$$\frac{R_1 R_F}{R_1 + R_F} = R_{OM} \tag{5-21}$$

Thus the proper value required of an R_{OM} resistor connected in the noninverting terminal is the parallel combination of resistors R_1 and R_F. However, the use of R_{OM} may not completely eliminate the output offset voltage V_{oI_B} because the currents I_{B1} and I_{B2} are not exactly equal. Nevertheless, the use of R_{OM} will minimize the amount of output offset voltage V_{oI_B}; therefore, the R_{OM} resistor is referred to as the *offset minimizing resistor* (see Figure 5-17).

Figure 5–17 R_{OM} reduces the output offset voltage V_{oI_B} caused by the input bias current I_B.

Note that if we reduce both the inputs to zero (that is, $v_{in1} = v_{in2} = 0$ V) in the closed-loop differential amplifier, the resulting circuit becomes the same as in Figure 5-17. This means that there is no need to use a separate resistor R_{OM} in the differential amplifier circuits. Figure 5-18 shows the noninverting and inverting amplifiers with offset minimizing resistor R_{OM}.

(a)

Figure 5–18 (a) Noninverting amplifier with offset minimizing resistor R_{OM}. (b) Inverting amplifier with offset minimizing resistor R_{OM}.

(b)

Figure 5–18   (Continued)

EXAMPLE 5-4

(a) For the inverting amplifier of Figure 5-19, determine the maximum possible output offset voltage due to: 1. Input offset voltage V_{io}. 2. Input bias current I_B. The op-amp is a type 741.

(b) What value of R_{OM} is needed to reduce the effect of input bias current I_B?

Figure 5–19   Inverting amplifier of Example 5–4.

SOLUTION   (a) From the 741 op-amp data sheets we have

$$V_{io}\,\text{max} = 6\text{ mV dc}$$

$$I_B\text{max} = 500\text{ nA dc at } T_A = 25°C \quad\text{and}\quad V_S = \pm 15\text{ V}$$

1. Since $R_1 = 470\ \Omega$ and $R_2 = 47\ k\Omega$, the output voltage offset due to V_{io} might be as large as

$$V_{oo} = \left(1 + \frac{R_F}{R_1}\right) V_{io}$$

$$= \left(1 + \frac{47 \text{ k}\Omega}{470 \text{ }\Omega}\right)(6 \text{ mV})$$

$$= 606 \text{ mV dc}$$

Remember that this means that the voltage at the output terminal may be below or above the ground potential by 606 mV dc without any external input signal v_{in}.

2. The value of feedback resistor $R_F = 47$ kΩ; therefore,

$$V_{ol_B} = R_F I_B$$

$$= (47 \text{ k}\Omega)(500 \text{ nA})$$

$$= 23.5 \text{ mV dc}$$

Again, this means that the output offset voltage V_{ol_B} could be positive or negative with respect to ground by as much as 23.5 mV dc.

From the results in parts 1 and 2, it is obvious that $V_{oo} \gg V_{ol_B}$; that is, the output offset voltage due to V_{io} is potentially a greater problem than the output offset voltage due to I_B.

(b) The value of R_{OM} to be used is

$$R_{OM} = \frac{R_1 R_F}{R_1 + R_F}$$

$$= \frac{(470 \text{ }\Omega)(47 \text{ k}\Omega)}{470 \text{ }\Omega + 47 \text{ k}\Omega} \cong 470 \text{ }\Omega$$

EXAMPLE 5-5

Repeat Example 5-4 if R_1 is replaced by 1 kΩ and R_F by 100 kΩ.

SOLUTION

(a)

1. The closed-loop voltage gain

$$A_{oo} = 1 + \frac{R_F}{R_1} = 1 + \frac{100 \text{ k}\Omega}{1 \text{ k}\Omega} = 101$$

is unchanged. Therefore, the output offset voltage V_{oo} due to V_{io} is also unchanged; that is, $V_{oo} = 606$ mV dc as in Example 5-4.

2. With $R_F = 100$ kΩ, the V_{ol_B} is

$$V_{ol_B} = (100 \text{ k}\Omega)(500 \text{ nA}) = 50 \text{ mV dc}$$

This is relatively larger output offset voltage than in Example 5-4. Thus V_{ol_B} becomes larger and therefore potentially troublesome, with larger values of feedback resistor R_F.

(b) The value of R_{OM} we can use is

$$R_{OM} = 1 \text{ k}\Omega \| 100 \text{ k}\Omega = 990 \ \Omega$$

Since a 990-Ω resistor is not a standard value, in practice we may use a 1-kΩ resistor as R_{OM}.

5-4 INPUT OFFSET CURRENT

We have seen in Section 5-3 that the use of R_{OM} in series with the noninverting terminal reduces the output offset voltage V_{ol_B} due to I_B. However, the value of R_{OM} was derived based on the assumption that the input bias currents I_{B1} and I_{B2} are equal. In practice, these currents are not equal because of the internal imbalances in the op-amp's circuitry. The input offset current I_{io} is used as an indicator of the degree of mismatching between these two currents. The value of I_{io} specified on the data sheets indicates the maximum amount by which the two input bias currents may differ. In fact, the smaller this value, the better.

The input offset current I_{io} is defined as the algebraic difference between two input bias currents I_{B1} and I_{B2}. In equation form,

$$I_{io} = |I_{B1} - I_{B2}| \tag{5-22}$$

For a 741-type op-amp, maximum $I_{io} = 200$ nA dc. This means that I_{B1} may be larger than I_{B2} or I_{B2} may be larger than I_{B1}, at the most, by 200 nA. In other words, the maximum difference between I_{B1} and I_{B2} can be as large as 200 nA. In FET input op-amps, the value of I_{io} is extremely small. For example, the maximum I_{io} is 0.3 nA for the μA740C op-amp.

In a circuit like that in Figure 5-17, there will be an output offset voltage due to the input bias currents I_{B1} and I_{B2}. In other words, the output offset voltage in a case like this can be expressed as a function of input offset current I_{io}. Let $V_{ol_{io}}$ be the output offset voltage caused by the input offset current I_{io}. To separate the effect of input offset current from that of input offset voltage, let us again assume that $V_{io} = 0$ V.

Referring to Figure 5-20, we will express the voltages V_1 and V_2 as a function of I_{B1} and I_{B2}, for given values of R_1 and R_F, as follows:

$$V_1 = R_{OM} I_{B1} \tag{5-18}$$
$$V_2 = R_p I_{B2} \tag{5-17}$$

where

$$R_{OM} = R_p = \frac{R_1 R_F}{R_1 + R_F}$$

Figure 5-20 Output offset voltage $V_{ol_{io}}$ caused by the input offset current I_{io} in an inverting or noninverting amplifier.

      Applying the superposition theorem, we will now find the output offset voltage due to V_1 and V_2 in terms of I_{B1}, I_{B2}, and R_F. We know, from Equation (5-15), that

$$V_{ol_{B2}} = -R_F I_{B2}$$

Here the negative sign is used because V_2 is the voltage at the inverting input terminal. This output offset voltage $V_{ol_{B2}}$ is due to voltage V_2 only in terms of I_{B2} and R_F. Similarly, the output offset voltage $V_{ol_{B1}}$ due to V_1 only in terms of I_{B1} and R_F can be obtained as follows:

$$V_{ol_{B1}} = V_1 \left(1 + \frac{R_F}{R_1} \right) \tag{5-23}$$

where V_1 = voltage at the noninverting input terminal

$\left(1 + \dfrac{R_F}{R_1} \right)$ = gain of the noninverting amplifier

Substituting in Equation (5-23) the value of V_1 from Equation (5-18), we get

$$V_{ol_{B1}} = R_{OM} I_{B1} \left(1 + \frac{R_F}{R_1} \right)$$

$$= \frac{R_1 R_F}{R_1 + R_F} I_{B1} \frac{R_1 + R_F}{R_1} \tag{5-24}$$

$$V_{ol_{B1}} = R_F I_{B1} \tag{5-25}$$

Therefore, the maximum magnitude of the output offset voltage due to I_{B1} and I_{B2} is

$$V_{ol_{B1}} + V_{ol_{B2}} = R_F I_{B1} - R_F I_{B2}$$

$$= R_F (I_{B1} - I_{B2})$$

$$V_{ol_{io}} = R_F (I_{io}) \tag{5-26}$$

where $V_{ol_{B1}} + V_{ol_{B2}} = V_{ol_{io}}$ is the output offset voltage due to I_{io} and $I_{io} = |I_{B1} - I_{B2}|$, the input offset current. Thus for a given value of input offset current I_{io}, the amount of output offset voltage $V_{ol_{io}}$ depends on the value of feedback resistor R_F. As in the two previous cases, the $V_{ol_{io}}$ is also a dc voltage and could be positive or negative with respect to ground. Since I_{io} is generally much smaller than I_B, the output offset caused by I_{io} is always smaller than that caused by I_B.

Figure 5–21 Inverting amplifier of Example 5–6.

EXAMPLE 5-6

For the inverting amplifier in Figure 5-21, determine the maximum output offset voltage $V_{ol_{io}}$ caused by the input offset current I_{io}. The op-amp is a type 741.

SOLUTION   For the 741 op-amp, I_{io} = 200 nA max. Therefore,

$$V_{ol_{io}} = (R_F)I_{io}$$
$$= (100\ k\Omega)(200\ nA)$$
$$= 20\ mV\ dc$$

Example 5-6 illustrates that we may not be able to eliminate completely the output offset voltage due to unequal input bias currents, even though a proper value of R_{OM} is being used. However, the output offset voltage has been reduced from 50 mV (in Example 5-5) to 20 mV (in Example 5-6) with the use of R_{OM}, which is a significant change.

5-5 TOTAL OUTPUT OFFSET VOLTAGE ━━━━━━━━━━━━━━━━━━━━━━

We know that in a circuit like the one in Figure 5-19, the output offset voltage V_{oo} caused by V_{io} could be either positive or negative with respect to ground. Similarly, the output offset voltage V_{ol_B} caused by I_B could also be either positive or negative with respect to ground. If these output offset voltages are of different polarities, the resultant output offset will be very little. On the other hand, if both of these output offset voltages are of the same polarity, the maximum amplitude of the total output offset would be

$$V_{ooT} = V_{oo} + V_{ol_B}$$
$$\bar{V}_{ooT} = \left(1 + \frac{R_F}{R_1}\right)V_{io} + (R_F)I_B \qquad (5\text{-}27)$$

By the same token, in a circuit such as that in Figure 5-21, the total output offset voltage V_{ooT} can be given by the expression

$$V_{ooT} = V_{oo} + V_{ol_{io}}$$

$$V_{ooT} = \left(1 + \frac{R_F}{R_1}\right)V_{io} + (R_F)I_{io} \tag{5-28}$$

Note the difference between Equations (5-27) and (5-28). The only difference in the circuits of Figures 5-19 and 5-21 is the R_{OM} resistor. Therefore, since $I_{io} < I_B$, the use of the R_{OM} resistor in the noninverting or inverting amplifier assures a reduction in the current-generated output offset voltage.

EXAMPLE 5-7

Compute the maximum possible total output offset voltages in the amplifier circuits shown in Figure 5-22. The op-amp is the MC1536 with the following specifications:

$$V_{io} = 7.5 \text{ mV maximum}$$

$$I_{io} = 50 \text{ nA maximum}$$

$$I_B = 250 \text{ nA maximum at } T_A = 25°C$$

SOLUTION   In the circuit shown in Figure 5-22(a), the current-generated output offset voltage is due to the input bias current I_B in the inverting lead. Therefore, to compute the total output offset voltage, we have to use Equation (5-27):

$$V_{ooT} = \left(1 + \frac{R_F}{R_1}\right)V_{io} + (R_F)I_B$$

$$= \left(1 + \frac{10 \text{ k}\Omega}{1 \text{ k}\Omega}\right)(7.5 \text{ mV}) + (10 \text{ k}\Omega)(250 \text{ nA})$$

$$= 82.5 \text{ mV} + 2.5 \text{ mV} = 85 \text{ mV}$$

The R_{OM} is used in the circuit of Figure 5-22(b), which means that the current-generated output offset is caused by the input offset current I_{io}. Therefore, using Equation (5-28), we get

$$V_{ooT} = \left(1 + \frac{R_F}{R_1}\right)V_{io} + (R_F)I_{io}$$

$$= \left(1 + \frac{10 \text{ k}\Omega}{1 \text{ k}\Omega}\right)(7.5 \text{ mV}) + (10 \text{ k}\Omega)(50 \text{ nA})$$

$$= 82.5 \text{ mV} + 0.5 \text{ mV} = 83 \text{ mV}$$

Thus it is obvious that the output offset voltage due to the input offset voltage V_{io} is potentially a more troublesome problem than the output offset voltage due to the input bias current I_B or the input offset current I_{io}.

(a)

(b)

Figure 5–22 (a) MC1556 inverting amplifier of Example 5–7.
(b) MC1556 inverting amplifier of Example 5–7 with R_{OM}.

5-6 THERMAL DRIFT

In previous sections we learned to compensate for the effects of input offset voltage and input bias currents. In our discussion so far, we have assumed that the parameters V_{io}, I_B, and I_{io} are constant for a given op-amp. However, in practice, the values of V_{io}, I_B, and I_{io} vary with:

1. Change in temperature
2. Change in supply voltages: $+V_{CC}$ and $-V_{EE}$
3. Time

In this section we discuss the effect of these factors on the performance of a given op-amp circuit.

The most serious variation in the values of V_{io}, I_B, and I_{io} is due to the change in temperature. Before we proceed further to find the change in output offset voltage due to the change in temperature, let us define the term *thermal drift*. The average rate of change of input offset voltage per unit change in temperature is called *thermal voltage drift* and is denoted by $\Delta V_{io}/\Delta T$. It is expressed in $\mu V/°C$. By the same concept, we can also define the thermal drift in the input offset current and input bias current as follows:

$$\frac{\Delta I_{io}}{\Delta T} = \text{thermal drift in the input offset current (pA/°C)}$$

$$\frac{\Delta I_B}{\Delta T} = \text{thermal drift in the input bias current (pA/°C)}$$

It is important to note that the drift is not a constant value; that is, it is not uniform over a specified operating temperature range. Furthermore, the value of the input offset voltage or current may increase or decrease with increasing temperature, as shown in Figure 5-23(a) and (b). Referring to the graph in Figure 5-23(c), we can see that the value of the input bias current is not decreasing at a constant rate with increasing temperature. The curves shown in Figure 5-23 are for the MC1741 op-amp.

Note that in Figure 5-23(a) and (b) the normalized values of the input offset voltage and input offset current are plotted versus temperature. This means that the input offset voltage and current are assumed to be zero at room temperature (25°C). If the change in temperature from 25°C is denoted by ΔT, the changes in input offset voltage and input offset current corresponding to this change in temperature ΔT are represented by ΔV_{io} and ΔI_{io}, respectively. In fact, once we calculate these values from the graphs in Figure 5-23, for a desired operating temperature range, we can obtain

$$\frac{\Delta V_{io}}{\Delta T} = \text{thermal voltage drift}$$

$$\frac{\Delta I_{io}}{\Delta T} = \text{thermal current drift}$$

But this procedure is time consuming and also involves mathematical manipulation. Therefore, some manufacturers specify the average value of the drift in offset voltage and offset current over an entire operating temperature range instead. On the data sheet these values are listed as the average temperature coefficients of input offset voltage and input offset current. It is worth mentioning that the average temperature coefficient of input bias current is not listed on the data sheets. Instead, a graph of I_B versus temperature may be furnished by some manufacturers [see Figure 5-23(c)]. For example, for the LM101A op-amp the average temperature coefficient of input offset voltage (that is, thermal voltage drift) is

$$\frac{\Delta V_{io}}{\Delta T} = 15 \ \mu V/°C \text{ maximum}$$

and the average temperature coefficient of input offset current is

$$\frac{\Delta I_{io}}{\Delta T} = 200 \, pA/°C \text{ maximum}$$

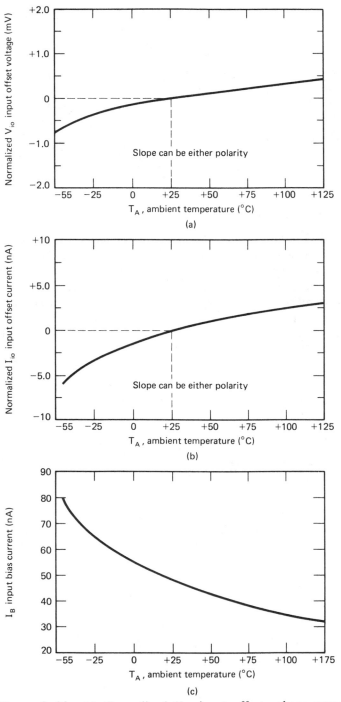

Figure 5–23 (a) Normalized V_{io}, input offset voltage versus temperature curve of the MC1741. (b) Normalized I_{io}, input offset current versus temperature curve of the MC1741. (c) Input bias current I_B versus temperature curve of the MC1741. (Courtesy of Motorola Semiconductor, Inc.)

No information is available on the change in input bias current versus temperature. In fact, when we use an R_{OM} resistor, we are not concerned about the change in input bias current as a function of change in temperature. Note again that these values are denoted as absolute values on the data sheets because they may be positive in one temperature range and negative in another. More specifically, for the LM101A this means that the input offset voltage may change at most by ± 15 $\mu V/°C$ change in temperature about 25°C. Similarly, the change in input offset current can be at most ± 200 pA/°C change in temperature from 25°C.

5-6.1 Error Voltage

Next we find the change in the total output offset voltage caused by the input offset voltage and input offset current drifts. This is important in evaluating the amplifier performance. To get a clear idea of the point, we consider a specific case.

Let us consider the inverting amplifier with R_{OM} and compensating network as shown in Figure 5-24. We assume that the amplifier has been nulled at room temperature (25°C); that is, the effects of V_{io} and I_{io} have been reduced to zero by use of the potentiometer arrangement shown in Figure 5-24. We have seen that the values of V_{io} and I_{io} drift (change) with temperature. According to Equation (5-28), any change in the values of V_{io} and I_{io} results in a change in the total output offset voltage. Therefore, the total output offset voltage will not be zero at any temperature other than room temperature. This means that the desired output accuracy will be affected by the presence of the output offsets at all temperatures other than 25°C. We can conclude that thermal drifts are the major problem and not the output offsets.

To determine the effect of voltage and current drifts on the performance of an amplifier, we first need to find the average change in total output offset voltage per unit change in temperature. This value is obtained from Equation (5-28) and is given by Equation (5-29):

Figure 5–24   Completely compensated inverting amplifier.

$$\frac{\Delta V_{ooT}}{\Delta T} = \left(1 + \frac{R_F}{R_1}\right)\frac{\Delta V_{io}}{\Delta T} + (R_F)\frac{\Delta I_{io}}{\Delta T} \tag{5-29}$$

where $\Delta V_{ooT}/\Delta T$ is the average change in total output offset voltage per unit change in temperature, in $\mu V/°C$. We know that the voltage and current drifts can be either positive or negative since they are specified as absolute values on the data sheets. Therefore, to obtain the maximum possible average change in total output offset voltage per unit change in temperature, we shall assume that the two drift effects are cumulative (additive). In other words, the $\Delta V_{ooT}/\Delta T$ value obtained from Equation (5-29) will also be an absolute value, meaning that it could be either positive or negative. We can now obtain the maximum possible change in total output offset voltage ΔV_{ooT} resulting from a change in temperature ΔT by multiplying both sides of Equation (5-29) by ΔT. Let us define this maximum possible change in total output offset voltage ΔV_{ooT} as the *error voltage* and denote it by E_v. Therefore, from Equation (5-29), we have

$$\Delta V_{ooT} = \left(1 + \frac{R_F}{R_1}\right)\left(\frac{\Delta V_{io}}{\Delta T}\right)\Delta T + (R_F)\left(\frac{\Delta I_{io}}{\Delta T}\right)\Delta T$$

$$E_v = \left(1 + \frac{R_F}{R_1}\right)\left(\frac{\Delta V_{io}}{\Delta T}\right)\Delta T + (R_F)\left(\frac{\Delta I_{io}}{\Delta T}\right)\Delta T \tag{5-30}$$

As we have mentioned before, E_v could be either positive or negative. Thus the expression for the output voltage for the inverting amplifier in Figure 5-24 may be written as

$$v_o = \left(-\frac{R_F}{R_1}\right)v_{in} \pm E_v \tag{5-31}$$

where E_v is the error voltage.

But v_o is expected to be directly proportional to v_{in}, which means that we would like to have E_v at zero. According to Equation (5-31), $E_v = 0$ V is possible only if either voltage and current drifts were zero or change in temperature ΔT was zero. Practically, neither of these two conditions is possible. In other words, in practice all amplifiers will exhibit some finite error E_v resulting from the change in temperature. However, if the amount of error is excessive for the application we have in mind, we can then consider a precision-type op-amp. High-performance precision op-amps are available that feature extremely low thermal voltage and current drifts. For example, the LH0044A has $\Delta V_{io}/\Delta T = 0.1 \ \mu V/°C$ and $\Delta I_{io}/\Delta T = 5 \ pA/°C$. But remember that we cannot get something for nothing; we also have to pay a higher price for these precision-type op-amps than for the general-purpose types. All in all, it is a trade-off between performance and cost.

It should be noted that the effect of voltage and current drifts is most pronounced in dc amplifiers, especially when the amplitude of the input to be amplified is relatively small. In fact, voltage and current drifts can also impose a serious problem in ac amplifiers, particularly if relatively large input voltage is to be amplified by an amplifier operating at its maximum capacity. The amplifier is said to be operating at its maximum capacity when the amplitude of the output voltage is equal to the saturation voltages. In any given amplifier circuit the saturation voltages are assumed to be slightly smaller than the supply voltages $+V_{CC}$ and $-V_{EE}$.

EXAMPLE 5-8

Refer to the inverting amplifier in Figure 5-24. The op-amp is the LM307 with the following specifications:

$$\frac{\Delta V_{io}}{\Delta T} = 30 \ \mu\text{V/°C maximum}$$

$$\frac{\Delta I_{io}}{\Delta T} \ \text{maximum} = 300 \ \text{pA/°C}$$

$$V_S = \pm 15 \ \text{V}$$
$$R_1 = 1 \ \text{k}\Omega, \qquad R_F = 100 \ \text{k}\Omega, \qquad R_L = 10 \ \text{k}\Omega$$

Assume that the amplifier is nulled at 25°C. Calculate the value of the error voltage E_v and the output voltage at 35°C if:
(a) $V_{in} = 1 \ \text{mV dc}$
(b) $V_{in} = 10 \ \text{mV dc}$

SOLUTION   The change in temperature $\Delta T = 35° - 25° = 10°\text{C}$. Using this change in temperature, we can calculate E_v and then V_o values as follows.
(a) Substituting known values in Equation (5-30), we get

$$E_v = \left(1 + \frac{R_F}{R_1}\right)\left(\frac{\Delta V_{io}}{\Delta T}\right)\Delta T + (R_F)\left(\frac{\Delta I_{io}}{\Delta T}\right)\Delta T$$

$$= \left(1 + \frac{100 \ \text{k}\Omega}{1 \ \text{k}\Omega}\right)\left(\frac{30 \ \mu\text{V}}{1°\text{C}}\right)(10°\text{C}) + (100 \ \text{k}\Omega)\left(\frac{300 \ \text{pA}}{1°\text{C}}\right)(10°\text{C})$$

$$= 30.3 \ \text{mV} + 0.3 \ \text{mV} = 30.6 \ \text{mV}$$

For $V_{in} = 1 \ \text{mV dc}$, the output voltage V_o using Equation (5-31) is

$$V_o = \left(-\frac{R_F}{R_1}\right)V_{in} \pm E_v$$

$$= \left(-\frac{100 \ \text{k}\Omega}{1 \ \text{k}\Omega}\right)(1 \ \text{mV}) \pm 30.6 \ \text{mV}$$

$$= -100 \ \text{mV} \pm 30.6 \ \text{mV}$$

$$= -130.6 \ \text{mV} \quad \text{or} \quad -69.4 \ \text{mV}$$

Thus the error voltage of ± 30.6 mV dc may cause the output voltage to equal -69.4 mV dc or -130.6 mV dc as the temperature varies from 25°C to 35°C.

(b) All the other factors are unchanged except the V_{in} value; therefore, the value of error voltage is still equal to ± 30.6 mV. Substituting in Equation (5-31) the known values, we have

$$V_o = \left(-\frac{R_F}{R_1}\right)V_{in} \pm E_v$$

$$= \left(\frac{-100\ k\Omega}{1\ k\Omega}\right)(10\ mV) \pm 30.6\ mV$$

$$= -1000\ mV \pm 30.6\ mV = -1030.6\ mV \quad or \quad -969.4\ mV$$

This means that V_o can be equal to -969.4 mV or -1030.6 mV as the temperature changes between 25° and 35°C.

   Let us now take a close look at the results in parts (a) and (b). Even though the amount of error is the same in both cases, the percentage error in V_o is larger in part (a) because the amplitude of the input is smaller. Thus, for a given op-amp circuit and a given temperature change, the percentage of error in the output voltage increases as the amplitude of the input signal is decreased. The percentage of error in output voltage can be reduced significantly if we use a precision-type op-amp.

EXAMPLE 5-9

Refer again to the amplifier circuit in Figure 5-24. Use the same circuit specifications that are given in Example 5-8. Assume that the amplifier is nulled at 25°C. If v_{in} is a 10-mV peak sine wave at 1 kHz:
(a) Calculate E_v and v_o values at 55°C.
(b) Draw the output voltage waveform at 55°C.

SOLUTION   The change in temperature ΔT is $(55° - 25°) = 30°C$.
   (a) First we calculate the error voltage and then the output voltage:

$$E_v = \left(1 + \frac{R_F}{R_1}\right)\left(\frac{\Delta V_{io}}{\Delta T}\right)\Delta T + (R_F)\left(\frac{\Delta I_{io}}{\Delta T}\right)\Delta T$$

$$= \left(1 + \frac{100\ k\Omega}{1\ k\Omega}\right)\left(\frac{30\ \mu V}{1°C}\right)(30°C) + (100\ k\Omega)\left(\frac{300\ pA}{1°C}\right)(30°C)$$

$$= 90.9\ mV + 0.9\ mV$$

$$= 91.8\ mV\ dc$$

$$v_o = \left(-\frac{R_2}{R_1}\right)v_{in} \pm E_v$$

$$= \left(\frac{-100\ k\Omega}{1\ k\Omega}\right)(10\ mV) \pm 91.8\ mV$$

$$= -1000\ mV\ peak \pm 91.8\ mV\ dc$$

This means that the 1000-mV peak ac voltage signal rides either on a $+91.8$-mV or a -91.8-mV dc level. When the signal rides on a positive dc level, it is said to be shifted up, whereas if it rides on a negative dc level, it is said to be shifted down. Thus the 1000-mV peak ac voltage signal is either shifted up or down at most by 91.8 mV.

Figure 5-25   Waveform of Example 5-9.

(b) The waveform in Figure 5-25 is drawn assuming that the ac voltage signal is shifted up by 91.8 mV.

It is clear from this example that the error voltage causes a dc level shift in ac amplifiers. A comparatively small dc level shift may be tolerated. However, a relatively large dc level shift in an ac amplifier can cause the output waveform to distort. That is, a dc level shift, depending on its polarity, can clip off either a positive peak or a negative peak portion of the ac output waveform. Again, to reduce the amount of dc level shift in an ac amplifier, we may use precision-type op-amps.

Thus far we have discussed the effect of drifts on the performance of an inverting amplifier. The same problems arise in a noninverting amplifier. Figure 5-26 shows the completely compensated noninverting amplifier.

Recall that the total output offset voltage Equation (5-28) is applicable to inverting as well as noninverting amplifiers. Therefore, the error voltage equation, (5-30), must also

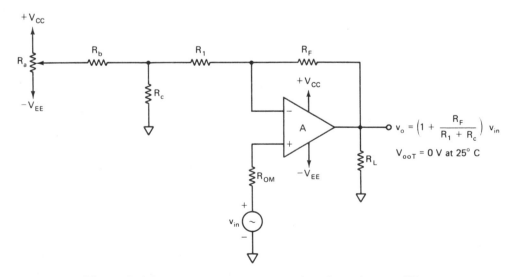

Figure 5-26   Completely compensated noninverting amplifier.

be applicable to both amplifiers, since it is derived from Equation (5-28). We have seen that the gain of the noninverting amplifier with an offset voltage-compensating network is equal to $1 + [R_F/(R_1 + R_c)]$ (see Figure 5-26). Therefore, the expression for the output voltage of a noninverting amplifier may be written as

$$v_o = \left(1 + \frac{R_F}{R_1 + R_c}\right)v_{\text{in}} \pm E_v \tag{5-32}$$

where

$$E_v = \left(1 + \frac{R_F}{R_1}\right)\left(\frac{\Delta V_{io}}{\Delta T}\right)\Delta T + (R_F)\left(\frac{\Delta I_{io}}{\Delta T}\right)\Delta T$$

Remember that if $R_c < R_1$ the voltage gain of the noninverting amplifier becomes equal to $(1 + R_F/R_1)$.

EXAMPLE 5-10

Repeat Example 5-8 for the noninverting amplifier shown in Figure 5-26. Assume that $R_c \ll R_1$.

SOLUTION   The error voltage E_v is the same as in Example 5-8; that is, $E_v = 30.6$ mV.
   (a) If $V_{\text{in}} = 1$ mV dc, then

$$V_o = \left(1 + \frac{R_F}{R_1}\right)V_{\text{in}} \pm E_v$$

$$= (101)(1 \text{ mV}) \pm 30.6 \text{ mV}$$

$$= +70.4 \text{ mV} \text{ or } +131.6 \text{ mV dc}$$

   (b) If $V_{\text{in}} = 10$ mV dc, then

$$V_o = (101)(10 \text{ mV}) \pm 30.6 \text{ mV}$$

$$= +979.4 \text{ mV} \text{ or } +1040.6 \text{ mV dc}$$

5-7 EFFECT OF VARIATION IN POWER SUPPLY VOLTAGES ON OFFSET VOLTAGE

In the preceding section we studied the effect of input offset voltage and input offset current thermal drifts on the output voltage of inverting as well as noninverting amplifiers. As we have mentioned before, the V_{io}, I_{io}, and I_B values are also susceptible to the changes in the supply voltages $+V_{CC}$ and $-V_{EE}$. Obviously, because the op-amp is capable of amplifying dc inputs, it is sensitive to changes in its supply voltages. This section is concerned with the effect of variation in supply voltages $+V_{CC}$ and $-V_{EE}$ on the values of V_{io}, I_{io}, and I_B and, in turn, the effect of changes in V_{io}, I_{io}, and I_B on output offset voltage.

Once we select the specific values for supply voltages $+V_{CC}$ and $-V_{EE}$ in a given op-amp amplifier, we do not change them deliberately. However, sometimes these voltages may change as a result of poor regulation and filtering. A poorly regulated power supply gives different values depending on the size and type of load connected to it. On the other hand, a poorly filtered power supply has a ripple voltage riding on some specific dc level.

Figure 5-27(a) shows the input bias current versus supply voltage curve for the LH0001 op-amp. A glance at output offset voltage V_{oI_B} [Equation (5-16)] reveals that for a given value of R_F any change in I_B causes a change in V_{oI_B}. Of course, you will recall that we can use the R_{OM} resistor to minimize the effect of I_B or of changes in it on the output offset voltage V_{oI_B}.

Even though input bias currents change due to the change in supply voltages, the input offset current should remain relatively constant because it is the absolute value of the difference between two input bias currents [see Figure 5-27(b)]. Thus, in practice, if we use a proper value of the R_{OM} resistor in a given amplifier circuit, there will be a negligible change in the current-generated output offset voltage due to the change in supply voltages. Therefore, manufacturers do not furnish curves like those in Figure 5-27 for all op-amps.

Recall that the supply voltages change because of poor regulation and filtering. For a given op-amp any change in the values of the supply voltages results in a change in the input offset voltage, which in turn causes a change in the output offset voltage. The change in op-amp's input offset voltage caused by variations in the supply voltages is generally specified on the data sheets by a variety of terms: The input offset voltage sensitivity, the power supply rejection ratio, the power supply sensitivity, and the supply voltage rejection ratio are some of them. All these terms are equivalent since they convey the same information. These terms are expressed either in microvolts per volt or in decibels. For example, the supply voltage rejection ratio (SVRR) for the μA741 is $\Delta V_{io}/\Delta V = 150\ \mu V/V$ maximum, and it is typically $20 \log (\Delta V/\Delta V_{io}) = 96$ dB for the LM307,

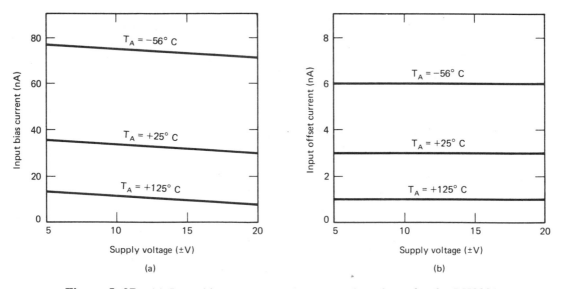

Figure 5–27 (a) Input bias current I_B versus supply voltage for the LH0001 op-amp. (b) Input offset current I_{io} versus supply voltage for the LH0001 op-amp. (Courtesy of National Semiconductor.)

where ΔV is the change in supply voltages $+V_{CC}$ and $-V_{EE}$, and ΔV_{io} is the resulting change in the input offset voltage.

Given a supply voltage rejection ratio in microvolts per volt, we can obtain an equivalent value in decibels (dB), or vice versa. For instance, SVRR $(\Delta V_{io}/\Delta V)$ of 150 μV/V is equivalent to

$$20 \log\left(\frac{1}{\text{SVRR}}\right) = 20 \log\left(\frac{1}{\Delta V_{io}/\Delta V}\right) = 20 \log\left(\frac{1}{150 \ \mu\text{V/V}}\right) = 20 \log\left(\frac{10^6}{150}\right)$$

$$= 76.48 \text{ dB}$$

Similarly, an SVRR of 96 dB is equivalent to 15.85 μV/V as follows.

Since $20 \log(1/\text{SVRR}) = 96$ dB,

$$\log\left(\frac{1}{\text{SVRR}}\right) = \frac{96}{20}$$

$$\frac{1}{\text{SVRR}} = 10^{4.8}$$

$$\text{SVRR} = \frac{1}{10^{4.8}}$$

$$= 15.85 \ \mu\text{V/V}$$

Note that the higher the value of SVRR in decibels, the lower is the change in input offset voltage due to the change in supply voltages or, in other words, the lower the value of SVRR in μV/V, the better for op-amp performance. In fact, ideally the value of SVRR in μV/V should be zero.

Now we find for a given amplifier the change in output offset voltage due to the supply voltage rejection ratio. Refer again to the completely compensated inverting amplifier circuit shown in Figure 5-24. Let us assume that the amplifier is nulled initially. Suppose that, after the circuit is in operation for a while, the supply voltages change in value due to poor regulation. We know that any change in the supply voltages results in a change in the input offset voltage. And, according to Equation (5-8), any change in input offset voltage results in a change in the output offset voltage. Therefore, using Equation (5-8) we can establish a relationship between the change in output offset voltage and SVRR as follows:

$$V_{oo} = \left(1 + \frac{R_F}{R_1}\right) V_{io} \tag{5-8}$$

where $(1 + R_F/R_1)$ is a constant for given values of R_1 and R_F. Therefore, the average change in V_{oo} per unit change in supply voltages can be

$$\frac{\Delta V_{oo}}{\Delta V} = \left(1 + \frac{R_F}{R_1}\right)\left(\frac{\Delta V_{io}}{\Delta V}\right) \tag{5-33}$$

Multiplying both sides of Equation (5-33) by ΔV, we get

$$\Delta V_{oo} = \left(1 + \frac{R_F}{R_1}\right)\left(\frac{\Delta V_{io}}{\Delta V}\right)\Delta V \tag{5-34}$$

where ΔV_{oo} = change in output offset voltage (volts)
ΔV = change in supply voltages $+V_{CC}$ and $-V_{EE}$

$$\frac{\Delta V_{io}}{\Delta V} = \text{supply voltage rejection ratio } (\mu V/V)$$

Remember that ΔV_{oo} is a dc voltage, and it could be either positive or negative. Thus all practical op-amps are affected by changes in the supply voltages, and therefore regulated supplies are recommended.

EXAMPLE 5-11

The amplifier in Figure 5-28(b) is nulled when the low dc supply is 20 V [see Figure 5-28(a)]. Because of poor regulation, low dc voltage varies with time from 18 V to 22 V. Determine (a) the change in the output offset voltage caused by the change in supply voltages, and (b) the output voltage V_o if V_{in} = 10 mV dc. The op-amp is the LM307 with SVRR = 96 dB.

(a)

(b)

Figure 5–28 (a) Supply voltages for the inverting amplifier. (b) Inverting amplifier of Examples 5–11 and 5–12.

SOLUTION   (a) The variation in low dc voltage from 18 V to 22 V, compared to its desired value of 20 V ($+V_{CC} = +10$ V and $-V_{EE} = -10$ V) implies that the change in supply voltages $\Delta V = 2$ V. The supply voltage rejection ratio (SVRR) equivalent to 96 dB is 15.85 μV/V. That is,

$$\frac{\Delta V_{io}}{\Delta V} = 15.85 \ \mu\text{V/V}$$

Substituting known values in Equation (5-34),

$$\Delta V_{oo} = \left(1 + \frac{100 \ \text{k}\Omega}{1 \ \text{k}\Omega}\right)(15.8 \ \mu\text{V/V})(2 \ \text{V})$$

$$= (101)(31.6)(10^{-6})$$

$$= 3.20 \ \text{mV}$$

It is worth noting that the ΔV value is the same regardless of whether it is computed from the change in low dc supply or change in $+V_{CC}$ or $-V_{EE}$. In this example, as a worst-case situation, suppose that $-V_{EE}$ remains constant at -10 V; then the $+V_{CC}$ has to vary from 8 to 12 V as a result of change in low dc voltage. This means that the change ΔV in supply voltage $+V_{CC}$ is 2 V in either direction from 10 V. In fact, for some op-amps the positive ($-V_{EE}$ constant) and negative ($+V_{CC}$ constant) supply voltage rejection ratios are specified on the data sheets. For example, the μA1458C op-amp has positive power supply sensitivity $= 30 \ \mu$V/V and negative power supply sensitivity $= 30 \ \mu$V/V.

(b) The total output voltage, including the change in output offset voltage ΔV_{oo}, is given by

$$V_o = \left(-\frac{R_F}{R_1}\right)V_{\text{in}} \pm \Delta V_{oo}$$

$$= \left(-\frac{100 \ \text{k}\Omega}{1 \ \text{k}\Omega}\right)(10 \ \text{mV}) \pm 3.2 \ \text{mV}$$

$$= -1000 \ \text{mV} \pm 3.2 \ \text{mV}$$

$$= -1003.2 \ \text{mV or} \ -996.8 \ \text{mV}$$

It can be seen from Example 5-11 that the variation in supply voltages could be troublesome in small-signal op-amp amplifiers.

EXAMPLE 5-12

Referring to Figure 5-28(b), suppose that the circuit is nulled when the voltage across terminals $+V_{CC}$ and $-V_{EE}$ measures 20 V dc. Also suppose that, because of poor filtering, 10-mV rms ac ripple is measured across the terminals $+V_{CC}$ and $-V_{EE}$. While the input signal $v_{\text{in}} = 0$ V, how much ripple voltage can we expect at the output if the op-amp is the LM307?

SOLUTION   The SVRR $= 15.85$ μV/V for the LM307, and because of poor filtering $\Delta V = 10$ mV rms. Substituting known values in Equation (5-34), we get an output ripple voltage of

$$\Delta V_{oo} = \left(1 + \frac{100 \text{ k}\Omega}{1 \text{ k}\Omega}\right)\left(\frac{15.85 \text{ }\mu\text{V}}{V}\right)(10 \text{ mV})$$

$$= 16 \text{ }\mu\text{V rms}$$

5-8 CHANGE IN INPUT OFFSET VOLTAGE AND INPUT OFFSET CURRENT WITH TIME

Under operating conditions the characteristics of all semiconductor devices, such as diodes and transistors, change to some extent with time. The amount of change in semiconductor characteristics then affects the performance of a circuit, in particular its linearity and accuracy. Since an op-amp is composed of transistors and diodes among other things, the op-amp's input offset voltage and input offset current change (drift) with time. For long-term stability the amount of change in input offset voltage and input offset current with time is crucial. On some op-amp data sheets the input offset voltage and input offset current drifts with time are specified and are denoted by $\Delta V_{io}/\Delta t$ and $\Delta I_{io}/\Delta t$, respectively. Generally, $\Delta V_{io}/\Delta t$ is expressed in microvolts per week and $\Delta I_{io}/\Delta t$ in nanoamperes per week. For example, for the LH0041C the typical values of $\Delta V_{io}/\Delta t =$ 5 μV/week and $\Delta I_{io}/\Delta t = 2$ nA/week. Remember that the drifts in V_{io} and I_{io} with time are the absolute values, meaning that V_{io} may increase or decrease typically by 5 μV/week in the case of the LH0041C. Similarly, the typical change in I_{io} could be as much as either $+2$ nA or -2 nA per week for the LH0041C.

Thus, with the help of Equation (5-28), we should be able to determine the maximum possible change in output offset voltage as a function of input offset voltage and input offset current drifts with time over a certain time period.

$$V_{ooT} = \left(1 + \frac{R_F}{R_1}\right)V_{io} + (R_F)I_{io} \tag{5-28}$$

Therefore, the average change in V_{ooT} per unit time is

$$\frac{\Delta V_{ooT}}{\Delta t} = \left(1 + \frac{R_F}{R_1}\right)\frac{\Delta V_{io}}{\Delta t} + (R_F)\frac{\Delta I_{io}}{\Delta t} \tag{5-35}$$

Multiplying both the sides of Equation (5-35) by Δt, we get

$$\Delta V_{ooT} = \left(1 + \frac{R_F}{R_1}\right)\left(\frac{\Delta V_{io}}{\Delta t}\right)\Delta t + (R_F)\left(\frac{\Delta I_{io}}{\Delta t}\right)\Delta t \tag{5-36}$$

where $\Delta t =$ time elapsed (weeks)

$\dfrac{\Delta V_{io}}{\Delta t} =$ input offset voltage drift with time (volts/week)

$\dfrac{\Delta I_{io}}{\Delta t} =$ input offset current drift with time (amperes/week)

$\Delta V_{ooT} =$ change in output offset voltage (volts)

Note that Equation (5-36) is applicable to noninverting as well as inverting amplifiers. Thus, in practice, the output offset voltage in all op-amp circuits will change with time. To maintain the desired accuracy and linearity of a system it is necessary to calibrate all op-amps in that system periodically.

EXAMPLE 5-13

Suppose that the circuit in Figure 5-24 is initially nulled. Assume also that room temperature and the voltage across terminals $+V_{CC}$ and $-V_{EE}$ remain constant. Determine the maximum possible change in output offset voltage after one month if the op-amps is the LH0041C. Assume that $R_1 = 1$ kΩ, $R_F = 100$ kΩ and $R_L = 10$ kΩ.

SOLUTION   For the LH0041C, $\Delta V_{io}/\Delta t = 5$ μV/week and $\Delta I_{io}/\Delta t = 2$ nA/week. Due to the time drift, $\Delta t = 4$ weeks,

$$\Delta V_{ooT} = \left(1 + \frac{R_F}{R_1}\right)\left(\frac{\Delta V_{io}}{\Delta t}\right)\Delta t + (R_F)\left(\frac{\Delta I_{io}}{\Delta t}\right)\Delta t$$

$$= \left(1 + \frac{100 \text{ k}\Omega}{1 \text{ k}\Omega}\right)(5 \times 10^{-6})(4) + (100 \text{ k}\Omega)(2 \times 10^{-9})(4)$$

$$= 2.02 \text{ mV} + 0.8 \text{ mV} = 2.82 \text{ mV}$$

5-9 OTHER TEMPERATURE- AND SUPPLY VOLTAGE-SENSITIVE PARAMETERS

The error voltage in the op-amp's output voltage is not the only possible result of temperature changes. A collection of other temperature-sensitive characteristics for the μA741C op-amp is shown in Figure 5-29. The curve in Figure 5-29(a) shows that the input resistance R_i of the 741C op-amp increases with an increasing temperature when supply voltage equals ± 15 V. Recall that the input resistance of an op-amp is defined as the equivalent resistance that would be measured at either an inverting or noninverting input terminal with the other terminal grounded. A changing input resistance, which is the load resistance on the signal source, can cause a drifting input signal amplitude. In fact, for a given signal source, the amount of drift in input signal amplitude depends on the relative values of the external components and also on the operating mode of the op-amp amplifier (inverting or noninverting). Thus, for a given signal source and op-amp circuit, we can minimize the drift in input signal amplitude without affecting the desired performance of the circuit if we select relatively small values for external components.

The power consumption as a function of ambient temperature curve for the 741C op-amp is shown in Figure 5-29(b). As can be seen from this curve, the power consumption, which is the dc power required to operate the op-amp under no-load conditions, decreases with increasing ambient temperature.

Finally, as shown in Figure 5-29(c), the output short-circuit current of the 741C op-amp decreases with an increasing temperature. In other words, the amount of output current obtainable with the output terminal shorted to ground or to either $+V_{CC}$ or $-V_{EE}$ supply decreases with increasing temperature.

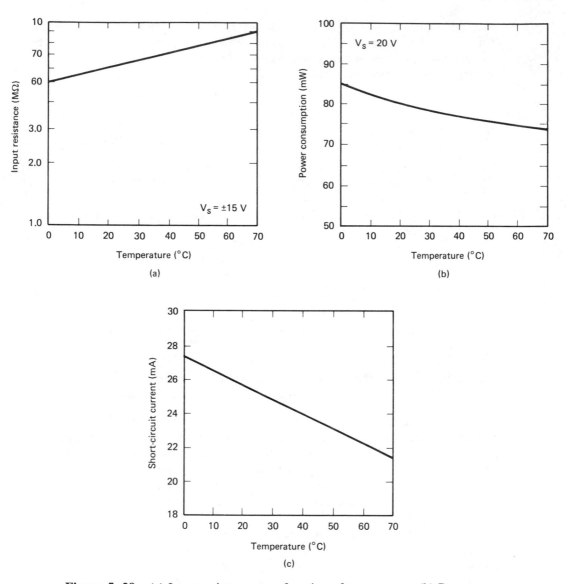

Figure 5–29 (a) Input resistance as a function of temperature. (b) Power consumption as a function of temperature. (c) Output short-circuit current as a function of temperature. (Courtesy of Fairchild Semiconductor Corporation)

Figure 5-30 shows a typical collection of supply voltage-sensitive parameters of the 741C op-amp. The open-loop voltage gain in decibels increases as the supply voltage values are increased, as shown in Figure 5-30(a). The open-loop voltage gain varies between 82 and 100 dB as the supply voltages are changed from ± 2 V to ± 18 V, respectively, when the ambient temperature is 25°C. Note that the voltage gain in decibels is equal to $20 \log(v_o/v_{in})$.

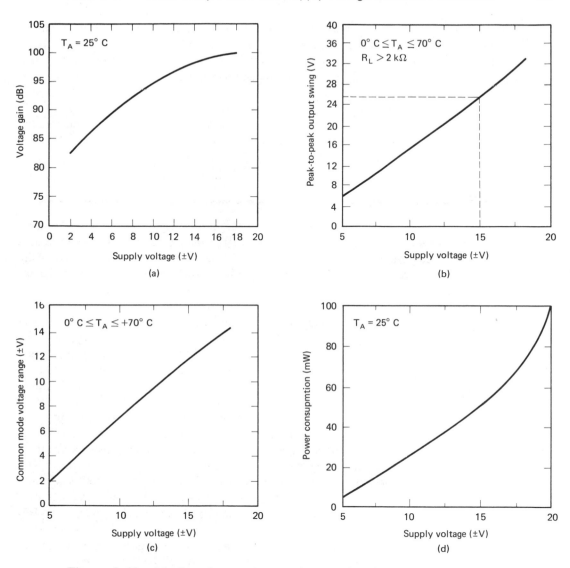

Figure 5–30 (a) Open-loop voltage gain as a function of supply voltage. (b) Output voltage swing as a function of supply voltage. (c) Input common-mode voltage range as a function of supply voltage. (d) Power consumption as a function of supply voltage. (Courtesy of Fairchild Semiconductor Corporation.)

The curve in Figure 5-30(b) shows the output voltage swing as a function of supply voltages $+V_{CC}$ and $-V_{EE}$. The peak-to-peak output swing depends on the value of the supply voltages and is always less than that value. For example, when the supply voltage is ±15 V, the peak-to-peak output swing is roughly 26 V, as shown in Figure 5-30(b).

Figure 5-30(c) shows that the input common-mode voltage range (CMVR) increases with increasing supply voltage values. That is, the range of common-mode input voltage over which the op-amp operates within specifications increases when the supply voltage values are increased. The CMVR varies between ± 2 V and ± 14 V as the supply voltages are changed from ± 5 V to ± 17.5 V, respectively [see Figure 5-30(c)]. For instance, when the supply voltage is ± 5 V, the common-mode voltage could be as high as either $+2$ V or as low as -2 V.

The power consumption as a function of supply voltage curve is shown in Figure 5-30(d). It can be seen here that the amount of dc power required to operate the op-amp under no-load conditions increases with an increase in supply voltages. Note that the curve is drawn at 25°C.

5-10 NOISE

Noise is a major source of interference with the desired signal in electronic systems composed of either discrete (separate) components or integrated circuits. Any unwanted signal associated with the desired signal is noise. Because it is the combined effect of many sources, noise is random in nature and hard to predict or analyze. In any electronic system, noise can come from many external sources as well as be self-induced, a result of the circuitry itself. Examples of external noise sources are switching of rotating machinery, ignition systems, and various control circuits. Natural phenomena such as lighting may also be external noise sources.

The self-induced or internal noise may be caused by the ac random voltages and currents generated within conductors and semiconductors of one circuit as a result of the switching of another circuit. The rate of change of current and voltage per unit of time, the speed of operation of the circuit, and the type of coupling between two circuits are some of the factors that determine the amount of noise induced in a given circuit.

Different types of noise phenomena are associated with op-amps: Schottky noise, thermal noise, and $1/f$ noise are the most important. The thermal noise increases with an increase in temperature. Like thermal noise, the amount of Schottky noise is greater with wider bandwidths and larger resistances; on the other hand, $1/f$ noise increases with a decrease in frequency f. Sometimes, manufacturers provide the data on noise, as shown in Figure 5-31. Figure 5-31(a) and (b) show the input mean square noise voltage (V^2/Hz) and mean-square noise current (A^2/Hz) at various frequencies for the $\mu A741$ op-amp. These noise characteristic curves indicate that the noise level is higher at lower frequencies. Broadband noise versus source resistance curves for the 741 op-amp are shown in Figure 5-31(c). A glance at these curves reveals that the wider bandwidths for a given value of source resistance generate more noise; also, larger source resistances for a given bandwidth appreciably increase the noise level.

To reduce the effect of electrical noise on ICs, several schemes have been commonly used. Physical shielding of the ICs and associated wiring helps to prevent external electromagnetic radiation from inducing noise into the internal circuitry. Special buffering and filtering circuits can be used between the electronic circuits and signal leads. To provide a path for any radio frequency (RF), all linear IC power supply terminals should generally be bypassed to ground. The breadboard layout should be such that the bypass capacitors are as near the IC terminals as possible. Internal noise generation can be reduced by

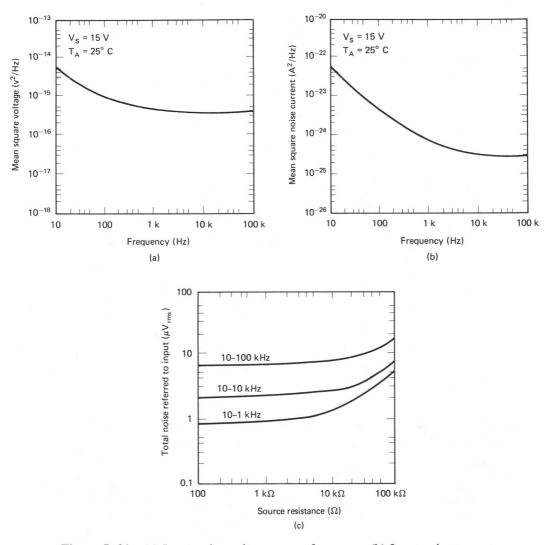

Figure 5–31   (a) Input noise voltage versus frequency. (b) Input noise current versus frequency. (c) Broadband noise for various bandwidths. (Courtesy of Fairchild Semiconductor Corporation.)

keeping input and output lead lengths as short as practically possible. Use one common tie point near the IC for all grounds. In a high electrical noise environment an IC with a high degree of noise immunity will minimize the amount of special care needed for proper circuit operation.

Among inverting, noninverting, and differential amplifiers, the latter type offers the best immunity to induced noise. The differential amplifier circuit is shown in Figure 5-32. In this circuit, v_d is the desired differential input signal to be amplified, and v_{ni} is the noise

Figure 5–32   Noise in a differential amplifier.

voltage induced into each input terminal with respect to ground. Since $R_1 = R_2$ and $R_F = R_3$, the voltage at the inverting and noninverting input terminals will be amplified by the same factor, R_F/R_1. Thus, if the noise voltage to ground at the inverting terminal is the same as the noise voltage to ground at the noninverting terminal, the noise voltage at the output should be negligibly small if not zero. Thus the ratio of the output noise voltage to the input noise voltage in practice will be much smaller than 1.

With proper component selection such that both inverting and noninverting input impedances are equal, both of the induced noise voltages v_{ni} are equal in amplitude and phase. Therefore, in differential amplifier circuits the noise voltages at the output are greatly reduced.

5-11 COMMON-MODE CONFIGURATION AND COMMON-MODE REJECTION RATIO

When the same input voltage is applied to both input terminals of an op-amp, the op-amp is said to be operating in an common-mode configuration. Since the input voltage applied is common to both the inputs, it is referred to as a common-mode voltage v_{cm}, as shown in Figure 5-33. A common-mode voltage v_{cm} can be ac, dc, or a combination of ac and dc.

Because ideally an op-amp amplifies only differential input voltages, no common-mode output voltage v_{ocm} should appear at the output. However, due to imperfections within an actual op-amp, some common-mode voltage v_{ocm} will appear at the output. The amplitude of this v_{ocm} is very small and often insignificant compared to v_{cm}. Therefore, in practice the ratio of the output common-mode voltage v_{ocm} to the input common-mode voltage v_{cm}, which is called the common-mode voltage gain A_{cm}, is generally much smaller than 1. In equation form,

$$A_{cm} = \frac{v_{ocm}}{v_{cm}} \tag{5-37}$$

Ideally, the common-mode voltage gain A_{cm} is zero.

Figure 5–33 Op-amp connected in common-mode configuration. (a) Without feedback. (b) With feedback.

Although A_{cm} is usually not specified on the op-amp data sheets, A_{cm} can be calculated for a given op-amp by applying a known value of common-mode input voltage v_{cm} and measuring the resultant output common-mode voltage v_{ocm}. Op-amp manufacturers usually list a common-mode rejection ratio CMRR. The CMRR is defined in several essentially equivalent ways by the various manufacturers. Generally, it can be defined as the ratio of the differential gain A_D to the common-mode gain A_{cm}, that is,

$$\text{CMRR} = \frac{A_D}{A_{cm}} \qquad (5\text{-}38)$$

Note that, in Figure 5-33(a), A_D is equal to the internal gain A of the op-amp.

The CMRR can also be expressed as the ratio of the change in input offset voltage to the total change in common-mode voltage. Thus

$$\text{CMRR} = \frac{V_{io}}{v_{cm}} \qquad (5\text{-}39)$$

From Equations (5-37) and (5-38), we can then establish the relationship between the v_{ocm} and CMRR:

$$\text{CMRR} = \frac{A_D}{A_{cm}} = \frac{A_D}{v_{ocm}/v_{cm}}$$

$$= \frac{A_D v_{cm}}{v_{ocm}}$$

$$v_{ocm} = \frac{A_D v_{cm}}{\text{CMRR}} \tag{5-40}$$

Equation (5-40) indicates that the higher the value of CMRR, the smaller will be the amplitude of the output common-mode voltage v_{ocm}. Generally, the CMRR value is very large and is therefore usually specified in decibels (dB), where

$$\text{CMRR (dB)} = 20 \log\left(\frac{A_D}{A_{cm}}\right) \tag{5-41a}$$

or, from Equation (5-39),

$$\text{CMRR (dB)} = 20 \log\left(\frac{V_{io}}{v_{cm}}\right) \tag{5-41b}$$

The value of the CMRR listed on the data sheets is specified for the open-loop common-mode op-amp configuration as shown in Figure 5-33(a). For example, for μA741C op-amp CMRR (dB) = 90 dB. However, the CMRR value listed on the data sheets can be used as an approximate value for the closed-loop common-mode configuration shown in Figure 5-33(b). Being the ratio of differential gain A_D to the common-mode gain A_{cm}, the CMRR should be the same for open-loop as well as closed-loop common-mode configurations [see Equation (5-38)].

Whether the CMRR is defined as in Equation (5-38) or (5-39), it is a measure of the degree of matching between two input terminals; that is, the larger the value of CMRR (dB), the better is the matching between the two input terminals and the smaller is the output common-mode voltage v_{ocm}. On the other hand, a large voltage v_{ocm} for a given common-mode input voltage v_{cm} is an indication of a large degree of imbalance between the two input terminals or of poor common-mode rejection. Thus, in practice, it is advantageous to use op-amps with higher CMRRs since these op-amps have better ability to reject common-mode voltage, such as 60-Hz induced noise voltages. The CMRR is a function of frequency and decreases as the frequency is increased as shown in Figure 5-34.

EXAMPLE 5-14

In the circuit of Figure 5-32, if $R_1 = R_2 = 1$ kΩ, $R_F = R_3 = 10$ kΩ, $v_d = 5$-mV sine wave at 1 kHz, and $v_{ni} = 2$ mV at 60 Hz, calculate (a) the output voltage at 1 kHz and (b) the amplitude of the induced 60-Hz noise at the output. The op-amp is the μA741 with CMRR (dB) = 90 dB.

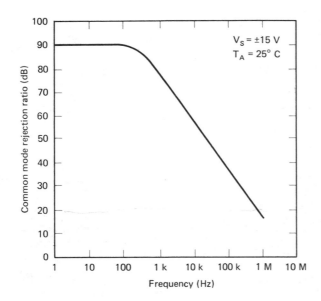

Figure 5–34 Common-mode rejection ratio as a function of frequency. (Courtesy of Fairchild Camera and Instrument Corporation.)

SOLUTION   (a) The closed-loop differential gain of the circuit is

$$A_D = \frac{R_F}{R_1} = \frac{10 \text{ k}\Omega}{1 \text{ k}\Omega} = 10$$

Since the input signal is applied in a differential mode, it is amplified by the differential gain A_D. Therefore, at 1 kHz the output signal is

$$v_o = A_D v_d = (10)(5 \text{ mV}) = 50 \text{ mV}$$

(b) The 60-Hz noise voltage v_{ni} appears in common mode at the input terminals of the op-amp; therefore, $v_{ni} = v_{cm}$. Thus, using Equation (5-40), we can calculate the value of the output common-mode voltage v_{ocm}. To implement Equation (5-40), we first have to convert the CMRR (dB) value into its equivalent numerical value. The CMRR (dB) = 90 dB implies that

$$20 \log \text{CMRR} = 90 \text{ dB}$$

$$\log \text{CMRR} = \frac{90}{20}$$

$$\text{CMRR} = 10^{4.5}$$

$$\text{CMRR} = 31,622.78$$

Therefore,

$$v_{ocm} = \frac{A_D v_{cm}}{\text{CMRR}} = \frac{(10)(2 \text{ mV})}{31,622.78}$$

$$= 0.63 \text{ } \mu\text{V at 60 Hz}$$

Thus the op-amp used in the differential mode is an effective circuit in reducing noise problems since the 60-Hz output is much smaller than the input-induced 60-Hz noise. The differential-mode op-amp circuit is particularly helpful in reducing induced input noise voltages when the small-amplitude differential input signal is being amplified.

SUMMARY

1. All operational amplfiers have finite values of input offset voltage, input offset current, and input bias current. These values result in a dc output offset voltage.

2. The op-amp can be nulled by using an offset voltage-compensating network. The amount of output offset voltage due to input bias currents can be significantly reduced if we use offset-minimizing resistors.

3. The input offset voltage and current tend to drift with change in temperature; the drift results in a finite dc error voltage in the desired output voltage at temperatures other than 25°C.

4. The amount of error voltage affects the accuracy of the dc amplifiers and limits the amplitude of the ac signal that can be amplified without distortion.

5. The input offset voltage of an op-amp changes with the change in supply voltages; the change in input offset voltage in turn causes a change in output offset voltage. To keep the change in output offset voltage to a minimum, use a regulated power supply or use an op-amp with a relatively high supply voltage rejection ratio.

6. The op-amp's input offset voltage and input offset current change with time. Therefore, for better long-term stability, use an op-amp with smaller drifts in input offset voltage and current per unit of time, and also calibrate the op-amp circuit periodically.

7. The input resistance, power consumption, and output short-circuit current of an op-amp are temperature-sensitive parameters. On the other hand, the open-loop voltage gain, output voltage swing, power consumption, and common-mode voltage range are all supply voltage-dependent parameters.

8. Any unwanted signal associated with the op-amp's desired output signal is noise. Schottky noise, thermal noise, and $1/f$ noise are the most important types of noise associated with op-amps.

9. The differential op-amp offers the best immunity to induced noise, since input noise voltages appear as common-mode voltage.

10. The common-mode rejection ratio of an op-amp is the ratio of differential gain to the common-mode gain. The larger the CMRR value, the smaller is the common-mode output voltage.

QUESTIONS

5-1. Define input offset voltage and explain why it exists in all op-amps.

5-2. Why is it necessary to use an external offset voltage-compensating network with practical op-amp circuits?

5-3. What is the offset-minimizing resistor R_{OM}?

5-4. Why is a resistor R_{OM} not needed in differential op-amp circuits?

5-5. Why is the output offset voltage generated by the input bias current always larger than that generated by the input offset current?

5-6. What are the factors that affect the input offset voltage, input bias, and input offset currents?

5-7. What is thermal drift? How does it affect the performance of an op-amp circuit?

5-8. What is error voltage? How can it be reduced?

5-9. Define supply voltage sensitivity. What is meant by a poorly regulated power supply?

5-10. What is an electrical noise? What precautions can be taken to minimize the effect of noise on an op-amp circuit?

5-11. Define the common-mode rejection ratio (CMRR) and explain the significance of a relatively large value of CMRR.

PROBLEMS

5-1. For the inverting amplifier shown in Figure 5-7(b), if $R_1 = 100\ \Omega$ and $R_F = 4.7\ \text{k}\Omega$, what is the maximum possible output offset voltage V_{oo}? The op-amp is an LM307 with $V_{io} = 10$ mV and supply voltages ±15 V.

5-2. Repeat Problem 5-1 if the LM307 op-amp is replaced by a μA715 for which $V_{io} = 5$ mV maximum.

5-3. Design an input offset voltage-compensating network for the μA715 op-amp of Problem 5-2. Draw the complete circuit diagram.

5-4. The μA715 op-amp is to be used as a voltage follower with the offset voltage-compensating network designed in Problem 5-3 (see Figure 5-12). Compute the output voltage v_o if the input is 1 V pp sine wave.

5-5. **(a)** For the noninverting amplifier of Figure 5-7(a), $R_1 = 100\ \Omega$ and $R_F = 10\ \text{k}\Omega$. Determine the maximum possible output offset voltage due to (1) the input offset voltage V_{io} and (2) the input bias current I_B. The op-amp is an LM307 with $V_{io} = 10$ mV and $I_B = 300$ nA.
(b) What value of R_{OM} is needed to reduce the effect of input bias current I_B?

5-6. For the inverting amplifier in Figure 5-21, determine the maximum output offset voltage $V_{ol_{io}}$ if the op-amp is an LM307 with $I_{io} = 70$ nA.

5-7. Compute the maximum possible total output offset voltage V_{ooT} in the amplifier circuits shown in Figure 5-22. Assume that the op-amp is an LM307 with supply voltages ±15 V.

5-8. Repeat Problem 5-7 if the op-amp is a μA741.

5-9. Refer to the inverting amplifier in Figure 5-24. The op-amp is a type 307 with the following specifications:

$$\frac{\Delta V_{io}}{\Delta T} = 30 \ \mu V/°C \ \text{maximum}$$

$$\frac{\Delta I_{io}}{\Delta T} = 0.3 \ \text{nA}/°C \ \text{maximum}$$

$$V_S = \pm 15 \ V$$

$$R_1 = 1 \ k\Omega, \quad R_F = 100 \ k\Omega, \quad \text{and} \quad R_L = 10 \ k\Omega$$

Assume that the amplifier is nulled at 25°C. Calculate the value of the error voltage and the output voltage V_o at 35°C if:

(a) $V_{in} = 1$ mV dc

(b) $V_{in} = 10$ mV dc

5-10. Repeat Problem 5-9 with the type 307 op-amp configured as a noninverting amplifier. Refer to Figure 5-26.

5-11. The LM312 op-amp is used as an inverting amplifier, as shown in Figure 5-24, with the following specifications:

$$\frac{\Delta V_{io}}{\Delta T} = 30 \ \mu V/°C, \qquad \frac{\Delta I_{io}}{\Delta T} = 10 \ \text{nA}/°C, \qquad V_S = \pm 15 \ V$$

$$R_1 = 100 \ \Omega, \qquad R_F = 8.2 \ k\Omega, \qquad R_L = 10 \ k\Omega$$

Assume that the amplifier is nulled at 25°C. If v_{in} is 20 mV peak sine wave at 100 Hz:

(a) Calculate E_V and v_o values at 45°C.

(b) Draw the output voltage waveform at 25°C and at 45°C.

5-12. Repeat Problem 5-11 with the same op-amp reconnected as a noninverting amplifier.

5-13. Referring to Figure 5-24, suppose that the circuit is nulled when the voltage across terminals $+V_{CC}$ and $-V_{EE}$ measures 24 V dc. Because of poor regulation, the voltage across terminals $+V_{CC}$ and $-V_{EE}$ varies with time from 20 V to 28 V. Determine:

(a) The change in the output offset voltage caused by the change in dc supply.

(b) The total output voltage V_O if $V_{in} = 5$ mV dc. The op-amp is the MC1741, and $R_1 = 100 \ \Omega$ and $R_F = 4.7 \ k\Omega$.

4-14. Repeat Problem 5-13 with $R_F = 10$ kΩ and $v_{in} = 10$ mV peak sine wave at 100 Hz. Also draw the output waveform.

5-15. Repeat Problem 5-13 if the same op-amp is reconnected as the noninverting amplifier shown in Figure 5-26. Use the same values for R_1 and R_F.

5-16. Referring to Figure 5-26, suppose that the circuit is nulled when the voltage across terminals $+V_{CC}$ and $-V_{EE}$ measures 30 V dc. Also suppose that, because of poor filtering, 10-mV rms ac ripple is measured across the terminals $+V_{CC}$ and $-V_{EE}$. While the input signal $v_{in} = 0$ V, how much ripple voltage can we expect at the output if the op-amp is an LM312 with SVRR = 96 dB typical? Use $R_1 = 1$ kΩ and $R_F = 47$ kΩ.

5-17. Determine the maximum possible change in output offset voltage in the circuit of Figure 5-24 after 6 months if the op-amp is the LH0041C with the following specifications:

$$\frac{\Delta V_{io}}{\Delta t} = 5 \ \mu\text{V/week} \quad \text{and} \quad \frac{\Delta I_{io}}{\Delta t} = 2 \ \text{nA/week}$$

$$R_1 = 82 \ \Omega \quad \text{and} \quad R_F = 8.2 \ \text{k}\Omega$$

Assume that the circuit is initially nulled and that room temperature and the voltage across terminals $+V_{CC}$ and $-V_{EE}$ remain constant.

5-18. Repeat Problem 5-17 with the op-amp LH0041C configured as shown in Figure 5-26. Use the same specifications as given in Problem 5-17.

5-19. In the circuit of Figure 5-32, if $R_1 = R_2 = 1 \ \text{k}\Omega$, $R_F = R_3 = 47 \ \text{k}\Omega$, $v_d = 10 \ \text{mV}$ sine wave at 1 kHz, calculate:
   (a) The output voltage at 1 kHz.
   (b) The amplitude of the induced 60-Hz noise at the output.
   The op-amp is the μA741 with CMRR (dB) = 90 dB.

5-20. Repeat Problem 5-19 if the μA741 op-amp is replaced by a LM312 for which CMRR (dB) = 100 dB typical.

5-21. In the circuit of Figure 5-33(b), $R_1 = 100 \ \Omega$, $R_F = 4.7 \ \text{k}\Omega$, and the op-amp is the μA741 with CMRR (dB) = 90 dB. If the amplitude of the induced 60-Hz noise at the output is 5 mV (rms), calculate the amplitude of the common-mode input voltage v_{cm}.

5-22. Repeat Problem 5-21 with the op-amp LM312. Use CMRR (dB) = 100 dB.

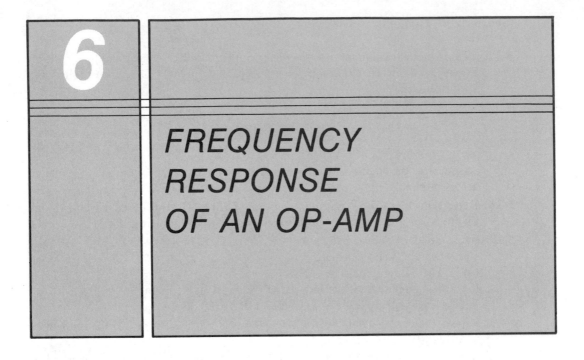

6

FREQUENCY RESPONSE OF AN OP-AMP

6-1 INTRODUCTION

As we saw in Chapter 3, some of the characteristics of the op-amp are frequency dependent. One of the most important of these characteristics is the open-loop voltage gain A. The gain A decreases as the operating frequency increases. This variation in gain as a function of frequency imposes a limitation not only on the performance of the op-amp but also on its use in ac applications. This chapter investigates the factors responsible for variations in open-loop gain as a function of frequency. It will also discuss the differences between internally and externally compensated op-amps and the effect of negative feedback on variations in open-loop gain. The chapter concludes with another important frequency-dependent parameter: slew rate and its effect on op-amp applications.

6-2 FREQUENCY RESPONSE

Up to now we have treated the gain of the op-amp as a constant. However, it is a complex number that is a function of frequency. Therefore, at a given frequency the gain will have a specific magnitude as well as a phase angle. This means that the variation in operating frequency will cause the variation in gain magnitude and its phase angle. The manner in which the gain of the op-amp *responds* to different frequencies is called the *frequency response*. A graph of the magnitude of the gain versus frequency is called a *frequency response plot*. This plot is included on the data sheets of most op-amps. Although gain magnitude may be expressed either in decibels (dB) or as a numerical value, the frequency is always plotted on a logarithmic scale. Remember that to accommodate large frequency ranges the frequency is assigned a logarithmic scale, just as the gain magnitude is assigned a linear scale and is expressed in decibels to accommodate very high gain, on the order of 10^5 or higher. The frequency response for the amplifiers is obtained from the experimental results by measuring its input and output voltages at different frequencies.

Another technique used in the ac analysis of networks is the *Bode plot,* composed of magnitude versus frequency and phase angle versus frequency plots. In magnitude versus frequency plots the magnitude is always taken in decibels. Although both frequency response and Bode plots indicate the effect of frequency variation on gain, the Bode plot is generally used for stability determination and network design.

6-3 COMPENSATING NETWORKS

Generally, for an amplifier, as the operating frequency increases, two effects become more evident: (1) The gain (magnitude) of the amplifier decreases, and (2) the phase shift between the output and input signals increases. In the case of an op-amp the change in gain and phase shift as a function of frequency is attributed to the internally integrated capacitor(s), as well as stray capacitors. These capacitors are due to the physical characteristics of semiconductor devices (BJTs and FETs) and the internal construction of the op-amp.

The manner in which the gain of the op-amp changes with variation in frequency is known as the *magnitude plot,* and the manner in which the phase shift changes with variation in frequency is known as the *phase angle plot.* Generally, manufacturers supply magnitude plots for the op-amps they produce. However, phase angle plots are not generally provided, especially for later-generation op-amps such as the 741, 351, and 771. This is because phase shifts of these op-amps are less than 90° even at cross-over frequencies. Recall that the cross-over frequency, also referred to as the unity gain bandwidth (UGB), is the maximum usable frequency for a given op-amp. For the 741 op-amp, UGB = 1 MHz (refer to Figure 6-1).

The rate of change of gain as well as the phase shift can be changed by using specific components with the op-amp. The most commonly used components are resistors and capacitors. The network formed by such components and used for modifying the rate of change of gain and the phase shift is called a *compensating network.* The *phase lag* and *phase lead* are the most commonly used compensating networks in op-amps. These two networks are indicative of their functions, since phase lag contributes a negative phase angle and phase lead a positive phase angle. Thus the main purpose of the compensating networks is to modify the performance of an op-amp circuit over the desired frequency range by controlling its gain and phase shift.

First-generation op-amps such as the 709 require *external* compensating networks, whereas later-generation op-amps such as the 741 and 351 have *internal* compensating networks. Therefore, there are two types of op-amps: internally compensated and externally compensated.

In internally compensated op-amps, the compensating network is designed into the circuit to control the gain and the phase shift of the op-amp. On the other hand, external (discrete) compensating components, that is, resistors and/or capacitors, are added at designated terminals in noncompensated op-amps. For proper operation, the manufacturer recommends appropriate compensating components for the uncompensated op-amps. The

Figure 6–1 Frequency response of (a) internally compensated op-amp 741C and (b) noncompensated op-amp μA709 for various values of C_1, R_1, and C_2. (c) Frequency compensation circuit of the μA709. Use $R_2 = 50\Omega$ when the amplifier is operated with capacitive loading.

741C is an internally compensated op-amp, while the 709C is a noncompensated op-amp. The open-loop frequency response curves of these op-amps, as well as the connection diagram of the 709C for the external compensating components, are shown in Figure 6-1. From Figure 6-1 (b) and (c) it is obvious that the 709C requires three compensating components: a resistor and two capacitors. Also, the nature of the frequency response curve for the 709C op-amp depends on the values of the compensating components used. On the other hand, the 741C op-amp has an internal compensating capacitor C_1 (30 pF), as shown in its equivalent circuit in Figure 2-3.

6-4 FREQUENCY RESPONSE OF INTERNALLY COMPENSATED OP-AMPS

Let us examine the open-loop frequency response of the internally compensated op-amp. In the plot of Figure 6-1(a), to accommodate the entire operating range, the gain has a linear scale and the frequency is assigned a logarithmic scale. In addition, the voltage gain is expressed in decibels. The unity gain–bandwidth of the 741C is approximately 1 MHz and the 741C has a single break frequency f_o before the unity gain–bandwidth. Recall that at the break frequency f_o the open-loop gain is down 3 dB or 0.707 of its value at 0 Hz (dc). The gain of the op-amp remains essentially constant from 0 Hz to the break frequency f_o, and thereafter rolls off at a constant rate, that is, 20 dB per decade (tenfold increase in frequency) [see Figure 6-1(a)]. Thus the open-loop bandwidth is the frequency band extending from 0 Hz to f_o, or simply f_o. Hence the open-loop bandwidth of the 741C is approximately 5 Hz.

In the 741 op-amp a 30-pF capacitor is the internal compensating component (Figure 2-3), which helps to control the open-loop gain to allow it to roll off at a rate of 20 dB/decade. In fact, even if the 741 is configured as a closed-loop amplifier, inverting or noninverting, using only resistive components, the gain will always roll off at a rate of 20 dB/decade, regardless of the value of its closed-loop gain. The internally compensated op-amps are sometimes simply called *compensated* op-amps and generally have very small open-loop bandwidths. For example, for the 741 op-amp, an open-loop bandwidth or the break frequency f_o is approximately 5 Hz.

6-5 FREQUENCY RESPONSE OF NONCOMPENSATED OP-AMPS ———

Op-amps requiring external compensating components are called *noncompensated* op-amps. They are sometimes called *tailored frequency response* op-amps because the user has to provide the compensation if it is needed to tailor the response. The open-loop gain characteristics of the noncompensated op-amp are altered by the compensating components as shown in Figure 6-1(b). The roll-off rate with various compensating components that are specified along the gain versus frequency curves is about 20 dB/decade. Note that the open-loop bandwidth of a 709C decreases from the outermost compensated curve to the innermost. That is, if $C_1 = 10$ pF, $R_1 = 0$ Ω, and $C_2 = 3$ pF, the bandwidth is approximately 5 kHz, while if $C_1 = 5000$ pF, $R_1 = 1.5$ Ω, and $C_2 = 200$ pF, the bandwidth is 100 Hz, as shown in Figure 6-1(b). The frequency compensation circuit of the

μA709 is shown in Figure 6-1(c). The uncompensated op-amps offer relatively broader open-loop bandwidths, as can be seen from Figure 6-1(a) and (b).

Next we obtain the general mathematical expression for the gain, primarily to confirm that it is a complex quantity and is a function of frequency. Then, to prove the validity of this equation, we construct the gain versus frequency plot from it. However, to derive the gain equation, a high-frequency op-amp model must first be developed.

6-6 HIGH-FREQUENCY OP-AMP EQUIVALENT CIRCUIT

What causes the gain of the op-amp to roll off after a certain frequency is reached? Obviously, there must be a capacitive component in the equivalent circuit of the op-amp since its reactance decreases as the frequency increases. This means that we have to include a capacitor in the high-frequency model of the op-amp at the output terminal. But how do we account for this capacitor? Two major sources are responsible for capacitive effects:

1. *Physical characteristics of semiconductor devices.* Recall that op-amps are composed of BJTs and FETs which contain junction capacitors. These junction capacitors are very small (on the order of picofarads) and act as open circuits at low frequencies but take finite values at higher frequencies. In fact, as frequency increases, the reactances of these capacitors decrease.

2. The *internal construction* of the op-amp is a second source of capacitive effects. In op-amps a number of transistors as well as resistors and sometimes a capacitor are integrated on the same material, called a *substrate*. In fact, the substrate acts as an insulator and helps to separate these components. The various components are connected by conducting paths, and the paths are separated by insulators. However, whenever two conducting paths are separated by an insulator, it acts as a capacitor. This means that because of its construction the op-amp may contain a number of such stray capacitors.

The cumulative effect of these capacitors due to the characteristics of semiconductor devices and the internal construction of the op-amp causes the gain to decrease as the frequency increases.

For an op-amp with only one break frequency, we will represent all the capacitive effects by a single capacitor, as shown in Figure 6-2. This figure represents the high-frequency model of the op-amp with a single break frequency. Similarly, op-amps with more than one break frequency may be represented by using as many capacitors as the break frequencies they have.

Note that the high-frequency model of Figure 6-2 is a modified version of the equivalent circuit of the op-amp presented in Chapter 2. Here we have simply added a capacitor C at the output.

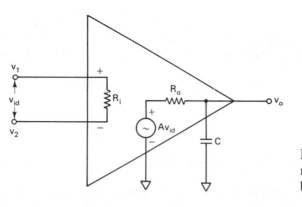

Figure 6–2 High-frequency model of an op-amp with single break frequency.

6-7 OPEN-LOOP VOLTAGE GAIN AS A FUNCTION OF FREQUENCY

Let us now obtain an expression for the gain as a function of frequency. From Figure 6-2, using a voltage-divider rule, we get

$$v_o = \frac{-jX_C}{R_o - jX_C} (Av_{id})$$

Since $-j = 1/j$ and $X_C = 1/2\pi fC$,

$$v_o = \frac{1/j2\pi fC}{R_o + 1/j2\pi fC} (Av_{id})$$

$$= \frac{Av_{id}}{1 + j2\pi fR_o C}$$

Hence the open-loop voltage gain is

$$A_{OL}(f) = \frac{(v_o)}{(v_{id})}$$

$$A_{OL}(f) = \frac{A}{1 + j2\pi fR_o C}$$

Let $f_o = 1/2\pi R_o C$; then

$$A_{OL}(f) = \frac{A}{1 + j(f/f_o)} \tag{6-1}$$

where $A_{OL}(f)$ = open-loop voltage gain as a function of frequency
   A = gain of the op-amp at 0 Hz (dc)
   f = operating frequency (Hz)
   f_o = break frequency of the op-amp (Hz)
Note that the break frequency f_o depends on the value of C and on output resistance R_o. Therefore, f_o is fixed for a given op-amp.

Equation (6-1) is important because it indicates that the open-loop gain of the op-amp $A_{OL}(f)$ is a complex quantity and is a function of operating frequency. Being complex in nature, the gain can be expressed in a polar form as follows. The open-loop gain magnitude is

$$|A_{OL}(f)| = \frac{A}{\sqrt{1 + (f/f_o)^2}} \qquad (6\text{-}2a)$$

and phase angle

$$\phi(f) = -\tan^{-1}\left(\frac{f}{f_o}\right) \qquad (6\text{-}2b)$$

where f is used to indicate that the magnitude of gain and phase angle of gain are functions of frequency. Using Equation (6-2a), we can obtain a magnitude versus frequency plot, while the phase angle versus frequency plot can be obtained by using Equation (6-2b).

Before proceeding with the graphical representation of Equation (6-2), we will present a qualitative analysis of it. Examine first the magnitude Equation (6-2a). At 0 Hz the denominator is 1, and the open-loop gain $A_{OL}(f)$ is equal to A. In fact, for any frequency less than f_o the gain is approximately constant and is equal to A. However, at frequencies above f_o, the denominator value increases, causing the gain $A_{OL}(f)$ to decrease. Thus, as the frequency increases, the gain $A_{OL}(f)$ continues to drop.

As for the phase shift Equation (6-2b), at 0 Hz the phase shift between input and output voltages is zero. In fact, for any frequency below f_o the absolute value of the phase shift is less than 45°. Above f_o the absolute value of the phase shift increases toward 90° with increases in frequency.

Now let us verify the validity of Equation (6-2) by using it to plot the Bode diagrams for the 741C op-amp. Let us begin by expressing the magnitude equation, (6-2a), in decibels:

$$20 \log |A_{OL}(f)| = 20 \log\left[\frac{A}{\sqrt{1 + (f/f_o)^2}}\right]$$

$$A_{OL}(f)\,\mathrm{dB} = 20 \log A - 20 \log \sqrt{1 + \left(\frac{f}{f_o}\right)^2} \qquad (6\text{-}3)$$

If $f_o \cong 5$ Hz and $A = 200{,}000$ for the 741C, let us substitute these values in Equation (6-3):

$$A_{OL}(f)\,\mathrm{dB} = 20 \log(0.2 \times 10^6) - 20 \log \sqrt{1 + \left(\frac{f}{5}\right)^2}$$

We now determine the gain in decibels at different frequencies, including at f_o:

At $f = 0$ Hz, $A_{OL}(f)\,\mathrm{dB} = 106.02 - 20 \log \sqrt{1 + \left(\frac{0}{5}\right)^2} \cong 106$

At $f = 5$ Hz $= f_o$, $A_{OL}(f)\,\mathrm{dB} = 106.02 - 3.01 \cong 103$

At $f = 50$ Hz, $A_{OL}(f)\,\mathrm{dB} = 106.02 - 20.04 \cong 86$

At $f = 500$ Hz, $A_{OL}(f)\,\mathrm{dB} = 106.02 - 40 \cong 66$

At $f = 5$ kHz, $A_{OL}(f)$ dB $= 106.02 - 60 \cong 46$

At $f = 50$ kHz, $A_{OL}(f)$ dB $= 106.02 - 80 \cong 26$

At $f = 100$ kHz, $A_{OL}(f)$ dB $= 106.02 - 86.02 = 20$

At $f = 1$ MHz, $A_{OL}(f)$ dB $= 106.02 - 106.02 = 0$

From these gain calculations the following observations can be made:

1. The open-loop gain $A_{OL}(f)$ dB is approximately constant from 0 Hz to the break frequency f_o.

2. When the input signal frequency f is equal to the break frequency f_o, the gain $A_{OL}(f)$ dB is 3 dB down from its value at 0 Hz. For this reason the break frequency is sometimes called the -3 *dB frequency*. It is also known as the *corner frequency*.

3. The open-loop gain $A_{OL}(f)$ dB is approximately constant up to the break frequency f_o, but thereafter it decreases 20 dB each time there is a tenfold increase (one decade) in frequency. Therefore, it may be said that the gain rolls off at the rate of 20 dB/decade. Stated another way, the gain rolls off at the rate of 6 dB/octave, where *octave* represents a twofold increase in frequency. This can be seen from the difference in gains at 50 kHz and 100 kHz. Thus we can describe the roll-off rate as either -6 dB/octave or -20 dB/decade.

4. Finally, at some specific value of the input signal frequency, the open-loop gain $A_{OL}(f)$ dB is zero. This specific frequency is called the *unity gain–bandwidth*. Other equivalent terms for it are *gain–bandwidth product, closed-loop bandwidth, small-signal bandwidth,* and *unity gain crossover frequency*. Remember that the unity gain–bandwidth is equal to the gain–bandwidth product only if an op-amp has a single break frequency before $A_{OL}(f)$dB is zero. This is precisely the case for op-amps like 741. The unity gain–bandwidth of the 741C is approximately 1 MHz and is verified by the decibel gain calculation at $f = 1$ MHz. Note that a gain of 0 dB is the same as a numerical gain of 1.

Using the values calculated previously, the magnitude of the open-loop gain in decibels versus frequency plot is shown in Figure 6-3(a). Note that this plot is almost a duplicate of the frequency response in Figure 6-1(a), which proves the validity of the gain equation, (6-2a). Thus, if we know the break frequency and the gain of the op-amp, it is possible to plot the frequency response using Equation (6-2a).

Next, we obtain the data for the phase angle versus frequency curve using Equation (6-2b). The phase angles may be computed at the same frequencies as were used to calculate the open-loop voltage gain $A_{OL}(f)$ in decibels.

$$\phi(f) = -\tan^{-1}\left(\frac{f}{5}\right) \qquad (6\text{-}2b)$$

At $f = 0$ Hz, $\phi(f) = -\tan^{-1}\left(\frac{0}{5}\right) = 0°$

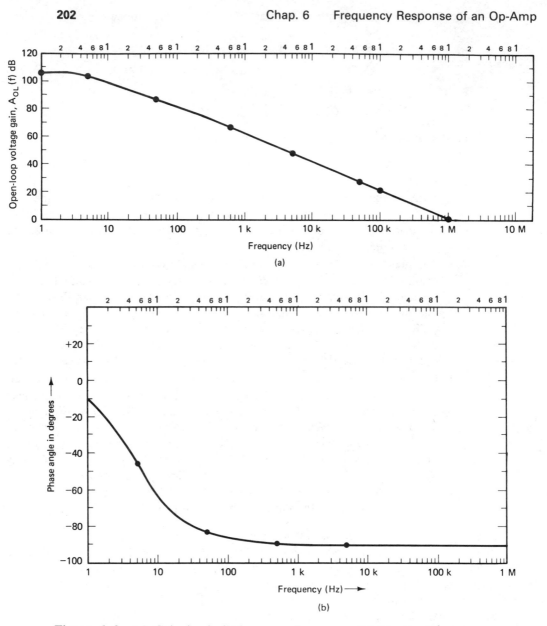

Figure 6–3 (a) Gain in decibels versus frequency. (b) Phase angle versus frequency.

$$\text{At } f = f_o = 5 \text{ Hz,} \quad \phi(f) = -\tan^{-1}\left(\frac{5}{5}\right) = -45°$$

$$\text{At } f = 50 \text{ Hz,} \qquad \phi(f) = -84.29°$$

$$\text{At } f = 500 \text{ Hz,} \qquad \phi(f) = -89.43°$$

At $f = 5$ kHz, $\phi(f) = -89.94°$

At $f = 50$ kHz, $\phi(f) = -89.99°$

At $f = 100$ kHz, $\phi(f) = -90°$

At $f = 1$ MHz, $\phi(f) = -90°$

The phase angle versus frequency plot of Figure 6-3(b) is constructed using these calculated values. From this plot it is obvious that the phase angle of the op-amp with a single break frequency varies between 0° and −90°. In other words, the phase shift between the input and output voltages is 90° maximum; that is, the output voltage lags the input voltage by 90°.

Remember that to implement the gain equation, (6-1), we must know the dc gain A and break frequency f_o. However, on a typical op-amp data sheet, the value of unity gain–bandwidth (UGB) is given instead of the value of the break frequency f_o. Therefore, f_o can be calculated by using Equation (4-10a), which is repeated here for convenience.

$$f_o = \frac{\text{UGB}}{A} \qquad (4\text{-}10a)$$

EXAMPLE 6-1

The 741C is connected as a noninverting amplifier. What maximum gain can be used that will still keep the amplifier's response flat to 10 kHz?

SOLUTION   The gain that will produce a flat response up to approximately 10 kHz can be obtained from the frequency response curve of the 741C. To do this, project a 10-kHz frequency vertically on the curve as shown in Figure 6-4 and then read the corresponding value of the gain on the vertical axis. This value of gain is equal to 40 dB, that is, 100.

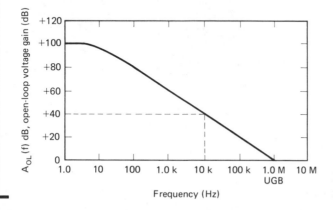

Figure 6–4   The 741C frequency response plot for Example 6–1.

Up to this point we have been studying op-amps with only one break frequency. Actually, op-amps have more than one break frequency because there are quite a few capacitors present, as pointed out in Section 6-6. Often, the upper break frequencies are well above the unity gain bandwidth and are, therefore, not shown in the frequency response plot. Since each break frequency contributes at most a 90° phase shift, the presence of multiple break frequencies should, therefore, be evident in the phase response plot. One such plot is shown in Figure 6-5(b). However, our major concern is the number of break frequencies before the unity gain–bandwidth because (1) we seldom use an op-amp above its unity gain–bandwidth, and (2) a stable circuit always requires a phase angle less than 180°. For more information on circuit stability refer to Section 6-9.

Using the concepts developed for a single break frequency op-amp, we can write the gain equation for the multiple break frequency op-amp, as shown in Example 6-2.

EXAMPLE 6-2

Using the frequency response and phase response curves shown in Figure 6-5, obtain the gain equation for the MC1556 op-amp. Also determine the approximate values of the break frequencies.

(a)

(b)

Figure 6–5   (a) Frequency response and (b) phase response curves of the MC1556. (Courtesy of Motorola Semiconductor.)

SOLUTION   Figure 6-5(b) shows that the phase shift is almost $-180°$ at 10 MHz. This means that two break frequencies must be present, since each break frequency contributes at the most $-90°$. Therefore, the gain equation can be written as

$$A_{OL}(f) = \frac{A}{[1 + j(f/f_{o1})][1 + j(f/f_{o2})]} \tag{6-4a}$$

which is simply an extension of Equation (6-1), where

$$f_{o1} = \text{first break frequency}$$
$$f_{o2} = \text{second break frequency}$$

However, from the graph of Figure 6-5(a) the first break frequency is approximately at 6 Hz. The value of the second break frequency can be computed by substituting a specific known value of the phase shift in a phase angle equation. The equation is

$$\phi(f) = -\tan^{-1}\left(\frac{f}{f_{o1}}\right) - \tan^{-1}\left(\frac{f}{f_{o2}}\right) \tag{6-4b}$$

The phase shift is $-157.5°$ at about 3 MHz [see Figure 6-5(b)]. Substituting these values in Equation (6-4b), we get

$$-157.5° = -\tan^{-1}\left(\frac{3\text{ MHz}}{6\text{ Hz}}\right) - \tan^{-1}\left(\frac{3\text{ MHz}}{f_{o2}}\right)$$

$$\tan^{-1}\left(\frac{3\text{ MHz}}{f_{o2}}\right) = 67.50°$$

$$\frac{3\text{ MHz}}{f_{o2}} = \tan(67.50°)$$

$$f_{o2} = \frac{[3\text{ MHz}]}{\tan(67.50)} = 1.24\text{ MHz}$$

Thus the gain equation with break frequency values is as follows:

$$A_{OL}(f) = \frac{140,000}{[1 + j(f/6)][1 + j(f/1.24\text{ MHz})]}$$

where $A \cong 140,000$ or 103 dB, $f_{o1} = 6$ Hz, and $f_{o2} = 1.24$ MHz.

6-8 CLOSED-LOOP FREQUENCY RESPONSE

So far the discussion has been centered around the nature of the open-loop frequency response. Since the op-amp is generally used in a closed-loop configuration, a study of the closed-loop performance of the amplifier is in order.

We saw that the open-loop gain $A_{OL}(f)$ is constant only up to the first break frequency. Therefore, the bandwidth of the op-amp is simply the first break frequency f_{o1}; that is, f_{o1} is the maximum useful frequency of the open-loop op-amp. But this bandwidth f_{o1} is very small, and therefore the open-loop configuration is of little use practically. Ideally, of course, we expect the bandwidth to be infinite so that the gain is the same at

Figure 6–6 Frequency response of the μA709C for various closed-loop gains. (Courtesy of Fairchild Semiconductor Corporation.)

all frequencies. To increase the bandwidth of an op-amp and thus make it a versatile device, a negative feedback must be used.

As shown in Chapter 4, the maximum bandwidth of a single break frequency op-amp is equal to its unity gain–bandwidth. Recall that the 741C has a *unity gain–bandwidth* of approximately 1 MHz. This means that for the 741C the product of the coordinates (gain and frequency) of any point beyond the break frequency (5 Hz) on the open-loop frequency response curve is about 1 MHz. That is, given a voltage gain, the closed-loop bandwidth can be determined using a frequency response curve. For instance, if the 741C is wired for a gain of 100 or 40 dB, its bandwidth will be about 10 kHz. This bandwidth is obtained by projecting to the right on the open-loop frequency response curve from 40 dB and reading the frequency on the horizontal axis corresponding to the point of intersection (see Figure 6-4).

Often, manufacturers' data sheets give the frequency response for various closed-loop gains. For example, Figure 6-6 shows the frequency response of the 709C for various closed-loop gains using recommended compensation networks. Obviously, if high gain and relatively wide bandwidth are required, the compensating components for curve 1 should be used. Note that the closed-loop frequency response curves of Figure 6-6 can be obtained from the open-loop frequency response of Figure 6-1(b) by using the same procedure as stated for the 741C (see Figure 6-4).

Remember that besides using the graphical method, depending on the amplifier configuration, Equation (4-10d) or (4-21b) can also be used to compute closed-loop bandwidth.

6-9 CIRCUIT STABILITY

A circuit or a group of circuits connected together as a system is said to be stable if its output reaches a fixed value in a finite time. On the other hand, a circuit/system is said to be unstable if its output increases with time instead of achieving a fixed value. In fact, the output keeps on increasing until the system breaks down. Therefore, unstable systems are impractical and need to be made stable. The criterion given for stability is used when the system is to be tested *practically.* However, theoretically, analytical and/or graphical methods are almost always used to test systems for stability before they are built. The most common example of an analytical method is *Routh–Hurwitz criteria,* whereas a graphical method is the *Bode plots.* Bode plots are composed of magnitude versus frequency and phase angle versus frequency graphs.

Any system whose stability is to be determined can be represented by the block diagram of Figure 4-3. In fact, in control system analysis the block diagram of Figure 4-3 is the *standard* form of representing a system. The standard block diagram is composed of two blocks, as shown in Figure 6-7. The block between the output and the input is referred to as the *forward block* (a block in the forward path), and the block between the output signal and the feedback signal is referred to as the *feedback block* (a block in the feedback path). The content of each block commonly referred to as the *transfer function* (in control system theory) depends on the complexity of a system. In-depth analysis of a system using the control system theory and concepts is beyond the scope of this book. However, we will use a specific case (Figure 6-7) to illustrate some basic concepts so that the performance evaluation of a given op-amp circuit can be made.

In Section 6-7 we saw that the *open-loop gain* of an op-amp is a complex quantity and is a function of frequency. Therefore, in Figure 6-7 we have represented it by $A_{OL}(f)$, which is given by

$$A_{OL}(f) = \frac{v_o}{v_{in}} \quad \text{if } v_f = 0 \tag{6-5a}$$

where $A_{OL}(f)$ = open-loop voltage gain

Similarly the closed-loop gain A_F is given by

$$A_F = \frac{v_o}{v_{in}}$$

$$A_F = \frac{A_{OL}}{1 + (A_{OL})(B)} \tag{6-5b}$$

where B = gain of a feedback circuit.

B is a constant if the feedback circuit uses only resistive components such as in a noninverting amplifier of Chapter 4. In Equation (6-5b), f is dropped for simplicity.

Bode plots always use the loop-gain $(A_{OL})(B)$ to determine system stability. To obtain $(A_{OL})(B)$, the feedback is broken at the input summing junction as shown in Figure 6-7. The loop-gain $(A_{OL})(B)$ is then used to obtain the data as shown in Section 6-7

Figure 6–7   A typical closed-loop system (noninverting amplifier).

for the magnitude and the phase angle plots over a desired frequency range. Once the magnitude versus frequency and the phase angle versus frequency plots are drawn, system stability may be determined as follows:

1. Determine the phase angle when the magnitude of (A_{OL}) (B) is 0 dB or 1. If the phase angle is $> -180°$, the system is stable. However, for some systems the magnitude may never be 0 dB; in that case, method 2 must be used to determine the system stability.
2. Determine the magnitude of (A_{OL}) (B) when the phase angle is $-180°$. If the magnitude is negative decibels, then the system is stable. However, sometimes the phase angle of a system may never reach $-180°$; under such conditions, method 1 must be used to determine the system stability.

EXAMPLE 6-3

Determine the stability of the voltage follower shown in Figure 4-7. Assume that the op-amp is a 741C.

SOLUTION   For the voltage follower, $B = 1$. Also, for the 741C op-amp we know

$$A_{OL}(f) = \frac{A}{1 + j\left(\dfrac{f}{f_o}\right)}$$

where $A = 20,000$
$f_o \cong 5$ Hz

Therefore,

$$(A_{OL})(B) = \frac{(2 \times 10^5)(1)}{1 + j\left(\dfrac{f}{5}\right)}$$

The data for the magnitude and the phase angle plots for this function were obtained in Section 6-7 and for convenience are repeated in Table 6-1. Using these data, the magnitude and phase angle plots (Bode plots) of Figure 6-8 are drawn.

TABLE 6-1

Frequency f in Hz	Magnitude $(A_{OL})(B)$ in dB	Phase angle $\phi(f)$ in degrees
0	106	0
10	99.03	−63.43
50	86	−84.29
500	66	−89.43
5K	46	−89.94
50K	26	−89.99
100K	20	−90
1M	0	−90

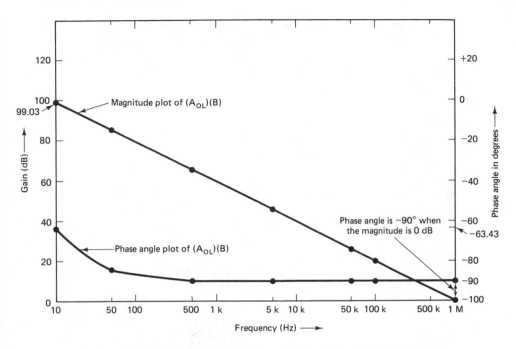

Figure 6–8   Magnitude and phase angle plots for Example 6–3.

Using these plots we see that the phase angle is $-90°$ when the magnitude is 0 dB. However, since phase angle never reaches $-180°$, the magnitude corresponding to $-180°$ cannot be determined. In other words, since the phase angle is $> -180°$ when the magnitude of $(A_{OL})(B)$ is 0 dB, the voltage follower of Figure 4-7 is a stable circuit.

From the preceding discussion, it is obvious that the op-amp with single break frequency is inherently stable because the total phase shift never exceeds $-90°$. However, when the op-amp is configured with nonresistive components, the total phase shift for $(A_{OL})(B)$ is equal to the phase shift due to feedback network plus the phase shift due to the internal circuitry of the op-amp. If at any frequency total phase shift (feedback plus internal) becomes $< -180°$ when the magnitude of $(A_{OL})(B)$ is 0 dB, the circuit may become unstable. In other words, the composition of a feedback circuit is important for the stability of an op-amp circuit with feedback.

6-10 SLEW RATE

Another important frequency-related parameter of an op-amp is the slew rate. The slew rate, as mentioned in Chapter 3, is the maximum rate of change of output voltage with respect to time, usually specified in V/μs. For example, a 1 V/μs slew rate means that the output rises or falls no faster than 1 V every microsecond. Ideally, we would like an infinity slew rate so that the op-amp's output voltage would change simultaneously with the input. Practical op-amps are available with slew rates from 0.1 V/μs to well above 1000 V/μs. The National Semiconductor LH0063C has a slew rate of 6000 V/μs.

Generally, the slew rate is specified for unity gain and is measured by applying a step input (dc) voltage. Slew rate is sometimes given indirectly in data sheets as *output voltage swing* as a *function of frequency* or as *voltage follower large-signal pulse response*, as shown in Figure 6-9. Often the slew rate improves with higher closed-loop gains and dc supply voltages. This is true for the μA715 op-amp, as shown in Figure 6-10(a) and (b), page 212. Also, for the 715C, the slew rate is a function of temperature and generally decreases with an increase in temperature, as shown in Figure 6-10(c). The slew rate of some op-amps is improved when they are wired with *feed-forward compensation*. The LM101 and LM108 are such op-amps, the compensation connections for which are given in the data sheets.

6-10.1 Causes of Slew Rate

The unity gain–bandwidth is the *small-signal* high-frequency limitation on the use of the op-amp. Slew rate, on the other hand, is a *large-signal* phenomenon. A large signal is one whose amplitude is comparable with the power supply voltages. Typically, the large signal is on the order of volts, while a small signal is in the range of milli- or even microvolts. Slew rate is caused by current limiting and the saturation of internal stages of an op-amp when a high-frequency, large-amplitude signal is applied. The resulting current is the maximum current available to charge the compensation capacitance network. We know that the capacitor requires a finite amount of time to charge and discharge. This means that internal capacitors prevent the output voltage from responding immediately to a fast-changing input. The rate at which the voltage across the capacitor rises is

$$\frac{dV_C}{dt} = \frac{I}{C} \tag{6-6}$$

Thus slew rate limiting is caused by this capacitor charging rate, in which the voltage across the capacitor is the output voltage.

6-10.2 Slew Rate Equation

Since the slew rate on a data sheet is generally listed for a unity gain, let us consider the voltage follower shown in Figure 6-11, page 213. Furthermore, let us assume that the input is a large-amplitude and high-frequency sine wave. The equation for the sine wave is

$$v_{in} = V_P \sin \omega t$$

or

$$v_o = V_P \sin \omega t \tag{6-7}$$

The rate of change of the output is

$$\frac{dv_o}{dt} = V_P \omega \cos \omega t$$

and the maximum rate of change of the output occurs when $\cos \omega t = 1$. That is,

$$\left. \frac{dv_o}{dt} \right|_{max} = V_P \omega$$

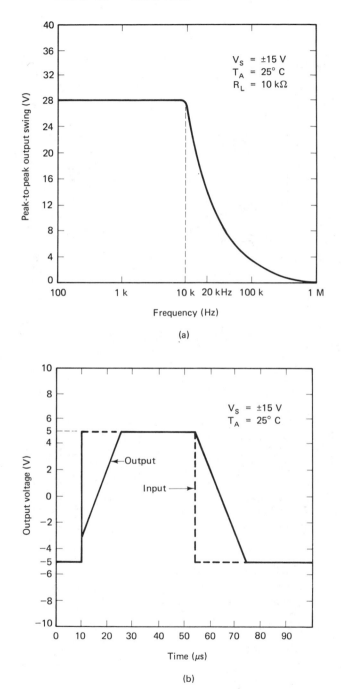

Figure 6–9   Specifying slew rate indirectly for the 741C as (a) output voltage swing versus frequency and (b) voltage follower large-signal pulse response. (Courtesy of Fairchild Semiconductor Corporation.)

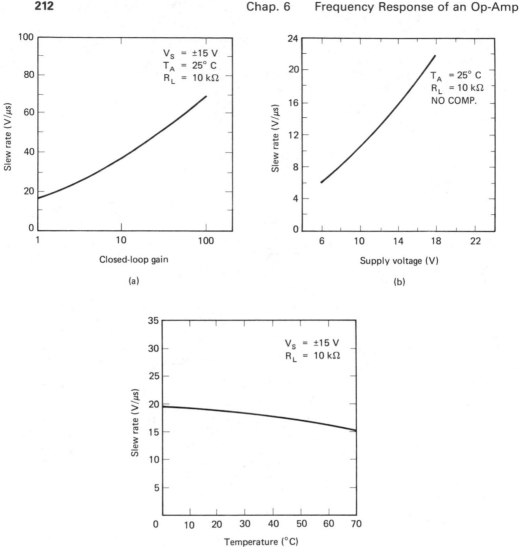

Figure 6–10 Slew rate of the μA715C. (a) Increases with closed-loop gain. (b) Increases with supply voltage. (c) Decreases with increase in temperature. (Courtesy of Fairchild Semiconductor Corporation.)

or

$$SR = 2\pi f V_P \quad \text{V/s}$$

$$= \frac{2\pi f V_P}{10^6} \quad \text{V/}\mu\text{s} \qquad (6\text{-}8)$$

where SR = slew rate (V/μs)

     f = input frequency (Hz)

    V_P = peak value of the output sine wave (volts)

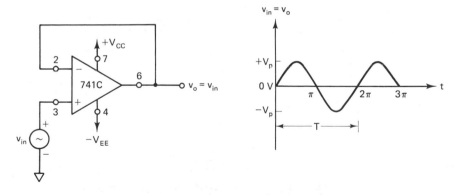

Figure 6–11   Deriving the slew rate equation.

For output free of distortion, the slew rate determines the maximum frequency of operation f_{max} for a desired output swing. Stated another way, with a lower frequency the slew rate determines the maximum undistorted output voltage swing. Thus, as long as the value of the right-hand side of Equation (6-8) is less than the slew rate of the op-amp, the output waveform will always be undistorted. If either the frequency or the amplitude of the input signal is increased to exceed the slew rate of the op-amp, the output will distort.

6-10.3 Effect of Slew Rate in Applications

The slew rate has important effects on both open-loop and closed-loop op-amp circuits. Figure 6-12(a) shows the open-loop configuration using the 741C. Since the open-loop voltage gain is very large, the output will go to about $+14$ V and then to -14 V each time the input sine wave crosses zero volts, as shown in Figure 6-12(b). The time taken by the output to go from $+14$ V to -14 V can be determined by using the slew rate of the 741C

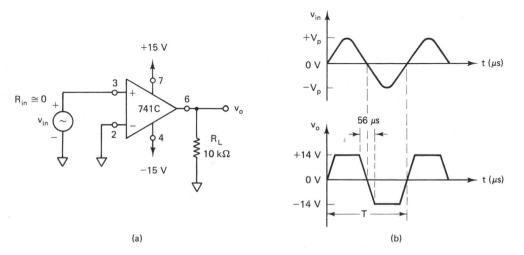

(a)   (b)

Figure 6–12   (a) Open-loop configuration using the 741C. (b) Input and output waveforms.

listed in the data sheet. The slew rate can also be calculated from the slope of the voltage-follower large-signal pulse response curve of Figure 6-9(b).

The 741C has a typical slew rate of 0.5 V/μs; therefore,

$$\frac{28 \text{ V}}{0.5 \text{ V}/\mu s} = 56 \ \mu s$$

must be the minimum time between the zero crossings. Hence the maximum input frequency f_{max} at which the output will be distorted is given by

$$f_{max} = \frac{1}{(2) (56 \ \mu s)} = 8.93 \text{ kHz}$$

Therefore, at f_{max} the output may be a triangular instead of square wave. Thus, to have a more square-wave output either we keep the input frequency below f_{max} or choose an op-amp with a faster slew rate. However, faster slew rate op-amps are sometimes characterized by overshoot and ringing, which cause the output to take longer to reach a steady state than with slower slew rate op-amps. For this reason, settling time is another important parameter that should be considered, especially in applications such as digital-to-analog (D/A) or analog-to-digital (A/D) converters.

Next, consider the effect of slew rate on the inverting amplifier with feedback. Figure 6-13 shows the 741C op-amp used as an inverting amplifier with a gain of 50. Obviously, the amplifier will operate with a gain of 50 up to about 20 kHz (Refer to Equation (4-21b)). Using Equation (6-8), the maximum output voltage at 20 kHz is

$$0.5 = \frac{(2\pi) (20 \text{ k}) (V_P)}{10^6}$$

$$V_P = \frac{(0.5) (10^6)}{(2\pi) (20 \text{ k})} = 3.98 \text{ V peak} \quad \text{(undistorted)}$$

or

$$v_o = 7.96 \text{ V peak-to-peak sine wave}$$

Figure 6–13   Inverting amplifier with a gain of 50.

Hence, for the output to be a sine wave, the maximum input signal $v_{in\,max}$ should be less than

$$\frac{7.96}{50} = 159 \text{ mV peak to peak}$$

6-10.4 Difference between Bandwidth, Transient Response, and Slew Rate

Table 6-2 is a summary of three important ac parameters of the op-amp: bandwidth, transient response, and slew rate. A clear idea of the differences between these parameters is important in ac applications.

TABLE 6-2 SUMMARY OF AC PARAMETERS

Bandwidth	Transient response	Slew rate
A small-signal phenomenon	A small-signal phenomenon	A large-signal phenomenon
Band of frequencies for which the gain remains constant	That part of the total response before the response reaches a steady state	The maximum time rate of change of the output voltage
Depends on compensating components and closed-loop gain	Composed of overshoot and rise time; rise time is related to bandwidth and overshoot is a measure of stability	Slew rate limiting depends on both frequency and amplitude; often increases with closed-loop gain and power supply voltages
If exceeded, results in a reduction of output voltage	Affects settling time	If exceeded, results in distortion

SUMMARY ━━━━━━━━━━━━━━━━━━━━━━━━

1. Frequency response is the manner in which the gain magnitude and the phase angle between the input and output respond to different frequencies.
2. Compensating networks are used to control the phase shift and thus improve the stability of the op-amps. These networks are typically composed of resistors and capacitors. The compensating network is either designed into the circuit or is added at designated terminals.
3. In internally compensated op-amps, the compensating network is designed into the circuit. On the other hand, a compensating network is added externally in non-compensated op-amps. Generally, open-loop noncompensated op-amps have wider bandwidths than those of compensated op-amps.

4. Because of capacitances within the op-amp, the gain decreases and the phase shift between input and output voltages increases as frequency increases.

5. The open-loop gain of the op-amp is relatively constant at frequencies below the break frequency but successively decreases at a rate of -20 dB/decade after each break frequency.

6. The very small bandwidth of an open-loop op-amp limits its use. The closed-loop op-amp has the advantage of a much larger bandwidth.

7. To achieve a stable circuit, the phase angle of the loop gain must be greater than $-180°$ when its magnitude reaches unity.

8. Slew-rate limiting occurs with all large, fast-changing signals. If slew rate is exceeded, distortion of the output waveform results.

QUESTIONS ━━━━━━━━━━━━━━━━━━━━━━━━━━━

6-1. What is a frequency response?

6-2. Briefly explain the need for compensating networks in op-amps.

6-3. What is the difference between compensated and noncompensated op-amps?

6-4. How does the high-frequency model differ from the equivalent circuit of an op-amp? Explain.

6-5. Define the *break frequency* and *bandwidth*.

6-6. Explain the effect of negative feedback on the frequency response.

6-7. What is a slew rate? List causes of the slew rate and explain its significance in applications.

6-8. Explain the difference between the slew rate and the transient response.

PROBLEMS ━━━━━━━━━━━━━━━━━━━━━━━━━━━

6-1. The 741C is connected as a noninverting amplifier and is required to have a closed-loop gain of 20. What is its bandwidth? Is the circuit stable?

6-2. The 709C is used as a noninverting amplifier with a gain of 50 dB. What is its bandwidth if $C_1 = 500$ pF, $R_1 = 1.5$ kΩ, and $C_2 = 20$ pF?

6-3. Repeat Problem 6-2 with a gain of 70 dB.

6-4. Repeat Problem 6-2 with the compensating components as follows: $C_1 = 5000$ pF, $R_1 = 1.5$ kΩ, and $C_2 = 200$ pF.

6-5. What maximum gain can be used while keeping the amplifier's response flat to 1 kHz? Assume that the 741C is wired as an inverting amplifier.

6-6. The frequency response of a certain op-amp is shown in Figure 6-14. Write the open-loop gain equation for the op-amp as a function of break frequencies and dc gain A.

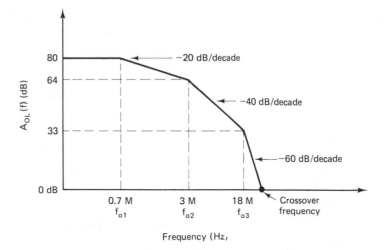

Figure 6–14   Frequency response for Problem 6–6.

6-7. Determine the magnitude of the open-loop gain and phase shift between the input and output for the op-amp in Problem 6-6 if the frequency of operation is 10 MHz. Is the circuit stable? Explain.

6-8. Compute the cross-over frequency of the op-amp whose frequency response is given in Figure 6-14.

6-9. Noncompensated op-amp MC1539 has a dc gain $A = 120,000$ and the following break frequencies: $f_{o1} = 5$ kHz, $f_{o2} = 320$ kHz, $f_{o3} = 1$ MHz, and $f_{o4} = 2$ MHz. Write the open-loop gain equation for the op-amp as a function of break frequencies and dc gain A. Then draw the Bode diagrams.

6-10. For the op-amp of Problem 6-9, compute the phase shift between input and output signals at 50 kHz and 1.5 MHz.

6-11. The μA709C is connected as shown in Figure 6-15. What is its bandwidth when the 20-kΩ pot is adjusted to (a) 1 kΩ and (b) 10 kΩ? (*Hint*: Use the frequency response characteristic curves of the μA709C.)

6-12. The μA709C is wired up as a noninverting amplifier for a gain of 1000 and is required to give a flat response to about 300 kHz. What compensating components would you recommend for the op-amp? (*Hint:* Refer to the frequency response characteristic curves of μA709C.)

6-13. The 741C is connected as a noninverting amplifier for a gain of 100. Determine the stability of the amplifier at this gain.

6-14. The 741C is used as an inverting amplifier with a gain of 50. The sinusoidal input signal has a variable frequency and maximum amplitude of 20 mV peak. What is the maximum frequency of the input at which the output will be undistorted? Assume that the amplifier is initially nulled.

6-15. Determine the slew rate of the 741C using the voltage-follower large-signal pulse response of Figure 6-9(b).

Figure 6–15 Amplifier for Problem 6–11.

6-16. An inverting amplifier using the 741C must have a flat response up to 40 kHz. The gain of the amplifier is 10. What maximum peak-to-peak input signal can be applied without distorting the output?

6-17. Determine the slew rate of the 715C if it is used as a noninverting amplifier with a gain of (a) 5 and (b) 50. [*Note:* Refer to Figure 6-10(a).]

6-18. When an 8-V peak-to-peak square wave of 3.6-MHz frequency is the input to a voltage follower, the output is a triangular wave, as shown in Figure 6-16. What is the slew rate of the op-amp?

6-19. In Problem 6-18, what must the minimum slew rate of the op-amp be in order to get the square-wave output?

6-20. The 715C is connected as a noninverting amplifier with a gain of 10. If the input is a sine wave at a frequency of 2 MHz, what is the largest undistorted output voltage? [Refer to the op-amp characteristic in Figure 6-10(a).]

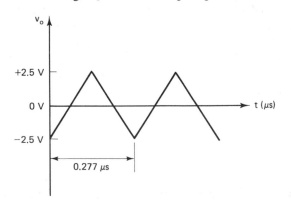

Figure 6–16 Waveform for Problem 6–18.

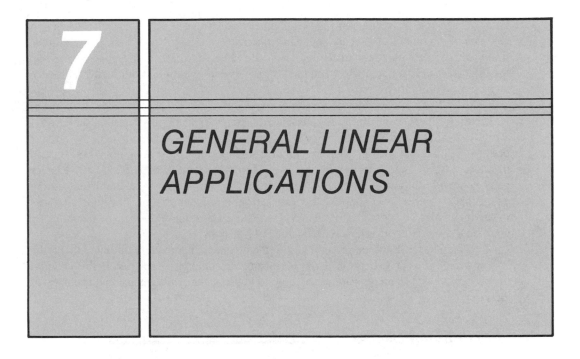

GENERAL LINEAR APPLICATIONS

7-1 INTRODUCTION

All op-amps are basically alike in being direct-coupled high-gain differential amplifiers. However, many different specific op-amps are available because no one circuit design can possibly optimize all the dc and ac parameters. For particular applications, op-amps are designed to optimize a parameter such as slew rate, bandwidth, gain, or power consumption. Therefore, to achieve the best performance in a desired application, it is essential to select a suitable op-amp. Often, a choice has to be made between performance and cost since an op-amp optimized for a particular parameter is more expensive than a general-purpose op-amp.

The op-amp parameters were defined and discussed in Chapter 3. These parameters are very useful in circuit design as well as in selecting an op-amp for a specific application. Also, Chapter 6 showed how data sheet information can be used to determine circuit stability. The data books contain a wealth of information, and we should consult them to select the most suitable device for a desired application.

Having considered the op-amp's basic construction, characteristics, parameter limitations, and various configurations, our next logical step is to discuss op-amp applications. Since there are literally countless applications of op-amps, they will be arranged into the following categories: general linear applications, filter and oscillator applications, comparator and detector applications, special integrated circuit applications, and selected system applications. Each category will include the most important applications in that area. Thus the remainder of the book is devoted entirely to op-amp applications. We begin with *general linear applications* in this chapter. Recall that in linear circuits the output signal is of the same nature as the input and varies in accordance with the input within the limits set by the saturation levels and slew rate. The first part of the chapter presents the important types of amplifiers and the last part discusses simple modifications of basic amplifier configurations.

In all these applications a general-purpose op-amp such as the 741 will be used. However, to improve the circuit performance, in some applications the use of the LF351, a low-cost high-speed JFET op-amp, is recommended. For this reason in some circuits both the 741 and the 351 op-amps are included. In fact, the LF351 is an improved version of and is also pin compatible with the 741. More importantly, the LF351 offers larger gain–bandwidth product, faster slew rate, and higher input impedance compared to the 741. It also requires lower supply current, has lower noise and offset voltage drift, and has lower input bias and offset currents. In short, the LF351 may be used to immediately upgrade the overall performance of existing 741 designs.

Remember that no one op-amp has all the dc and ac parameters optimized, and the 741 or even 351 is no exception. However, using the concepts developed in this chapter, the reader may select an appropriate op-amp to improve the performance and accuracy of the circuit.

7-2 DC AND AC AMPLIFIERS

As pointed out in Chapter 4, the properties of closed-loop amplifiers depend primarily on characteristics of the feedback network rather than those of the op-amp itself. Since a typical feedback network is composed of resistors and capacitors, the accuracy and stability of circuits using them can be improved significantly because these components are available in conditions of high precision and low drift.

7-2.1 DC Amplifier

Basically, an op-amp can amplify two types of signals: dc and ac. In a *dc amplifier* the output signal changes in response to changes in its dc input levels. A dc amplifier can be inverting, noninverting, or differential, as shown in Figure 7-1. To reduce the output offset voltage to zero, that is, to improve the accuracy of the dc amplifier, the offset null circuitry of the op-amp should be used. For op-amps without offset null capability, the external offset voltage compensating network should be used, as shown in Figure 7-2. Otherwise, a high-precision op-amp such as the μA714, which has smaller offsets and drifts, must be used. Recall that all the circuits in Figures 7-1 and 7-2 have been discussed previously and their output voltage equations derived.

7-2.2 AC Amplifier

The circuits of Figures 7-1 and 7-2 respond to ac input signals as well. However, if the designer needs the ac response characteristics of the op-amp, that is, low- and high-frequency limits, or if the ac input is riding on some dc level, it is necessary to use an *ac amplifier* with a coupling capacitor. For example, in an *audio receiver system* that consists

(a)

$$V_o = -\frac{R_F}{R_1}V_{in}$$

(b)

$$V_o = \left(1 + \frac{R_F}{R_1}\right)V_{in}$$

(c)

$$V_o = -\frac{R_F}{R_1}(V_x - V_y)$$

Figure 7–1 (a) Inverting amplifier, (b) noninverting amplifier, and (c) differential amplifier with offset null circuitry.

(a)

(b)

(c)

Figure 7–2   (a) Inverting amplifier, (b) noninverting amplifier, and (c) differential amplifier with external offset voltage-compensating networks.

of a number of stages, because of thermal drift, component tolerances, and variations, the dc level is produced. To prevent the amplification of such dc levels, the coupling capacitors must be used between the stages. Figure 7-3 shows the ac inverting and noninverting amplifiers with coupling capacitors.

The coupling capacitor not only blocks the dc voltage but also sets the low-frequency cutoff limit, which is given by

$$f_L = \frac{1}{2\pi C_i(R_{iF} + R_o)} \tag{7-1}$$

(a)

(b)

Figure 7–3   (a) AC inverting amplifier. (b) AC noninverting amplifier. (Pin numbers are for the 14-pin DIP.)

where f_L = low-frequency cutoff or low end of the bandwidth

C_i = capacitance between two stages being coupled or dc blocking capacitor

R_{iF} = ac input resistance of the second stage

R_o = ac output resistance of the first stage or the source resistance, R_{in}

For the derivation of Equation (7-1), refer to Appendix C. Remember that the high-frequency cutoff f_H or high end of the bandwidth depends on the closed-loop gain of the amplifier.

Thus the bandwidth of the amplifier,

$$BW = f_H - f_L \qquad (7\text{-}2)$$

depends on the desired value of f_L and the closed-loop gain of the amplifier. The required value of capacitor C_i can be calculated using Equation (7-1). The coupling capacitor C_i, besides providing a low-frequency cutoff limit, also helps to eliminate dc level amplification from stage to stage. In Figure 7-3, for example, the input voltage v_{in} could be a preceding amplifier output with a dc level; the coupling capacitor C_i will block this dc level. Thus the coupling capacitor prevents the amplification of the dc level. However, C_i has no effect on the output offset voltage produced by the input offset voltage and input offset current of an op-amp. Although not critical, the output offset voltage may cause the output of the amplifiers in Figure 7-3 to distort based on the amplitude of the input signals and the gain of the amplifiers. Therefore, to minimize the effect of output offset voltage, use either offset minimizing resistor R_{OM}, or offset voltage compensating network, or output coupling capacitor C_o. The output capacitor C_o may be used between the output terminal of an amplifier and the following stage. However, the use of C_o will affect the bandwidth of the amplifier. In Figure 7-3, offset minimizing resistors and offset voltage-compensating networks are used to reduce the effect of output offset voltages on the outputs of amplifiers.

Figure 7-3(a) shows an ac inverting amplifier, the closed-loop gain of which is still

$$A_F \cong -\frac{R_F}{R_1} \qquad (7\text{-}3)$$

since the reactance of C_i is negligible within the bandwidth. Similarly, the closed-loop gain of the noninverting amplifier of Figure 7-3(b) is

$$A_F \cong 1 + \frac{R_F}{R_1} \qquad (7\text{-}4)$$

EXAMPLE 7-1

In the circuit of Figure 7-3(a), R_{in} = 50 Ω, C_i = 0.1 μF, R_1 = 100 Ω, R_F = 1 kΩ, R_L = 10 kΩ, and supply voltages = $\pm$15 V. Determine the bandwidth of the amplifier.

SOLUTION   Recall that the input resistance of the inverting amplifier with feedback is

$$R_{iF} = R_1 \cong 100 \ \Omega \quad \text{(see Equation 4-19)}$$

The source resistance R_{in} = R_o = 50 Ω. Therefore, from Equation (7-1),

$$f_L = \frac{1}{(2\pi)(10^{-7})(150)} = 10.6 \text{ kHz}$$

Since $A_F = -R_F/R_1 = -10$, the high-frequency cutoff f_H using Equation (4-21b) is

$$f_H = \frac{(\text{UGB})(K)}{A_F} = \frac{(10^6)(0.909)}{(10)} \cong 90.9 \text{ kHz}$$

Thus, from Equation (7-2), BW $= 90.9 \text{ kHz} - 10.6 \text{ kHz} = 80.3 \text{ kHz}$.

7-3 AC AMPLIFIERS WITH A SINGLE SUPPLY VOLTAGE

An ac amplifier can be powered by a single supply voltage if an additional coupling capacitor is used on the output of an amplifier, as shown in Figure 7-4. This capacitor blocks dc; therefore, an output offset voltage as well as a dc level in the output signal has little effect on the operation of the amplifier. Besides that, in Figure 7-4(a) positive dc level is intentionally inserted using a voltage-divider network at the noninverting input terminal so that output can swing both positively as well as negatively. Furthermore, the dc level inserted is generally $(+V_{CC}/2)$, so the positive output swing equals the negative. This is accomplished by selecting $R_2 = R_3$. In the inverting amplifier of Figure 7-4(a), since the reactance of input capacitor C_i at 0 Hz (dc) is infinity, the resistor R_1 is an open component; hence the circuit effectively operates as a voltage follower, producing an output of $(+V_{CC}/2)$ V dc. However, with ac input single v_{in}, the reactance of the input coupling capacitor C_i is negligible, and the ac output voltage v_o at frequencies within the bandwidth is $v_o = (-R_F/R_1)v_{\text{in}}$. Thus, as shown in Figure 7-4(b) the ac output is riding on a dc level of $(+V_{CC}/2)$ volts. The output coupling capacitor C_o blocks this dc voltage, and the resultant waveform is therefore ac. Note that the maximum peak-to-peak output voltage swing is approximately equal to $+V_{CC}$ volts. The noninverting ac amplifier using a single supply is shown in Figure 7-4(c). Again, at 0 Hz or dc the input capacitors C_i and C_1 act as open components; therefore, the circuit effectively operates as a voltage follower, producing the dc output voltage of $(+V_{CC}/2)$ volts. Note that the noninverting amplifier requires an extra capacitor C_1 to produce an output dc level of $(+V_{CC}/2)$ volts. At frequencies within the bandwidth the input capacitors C_i and C_1 have negligible reactances; as a result, the ac output voltage v_o of the amplifier is

$$v_o = \left(1 + \frac{R_F}{R_1}\right)v_{\text{in}}$$

The total output voltage (dc plus ac) v_o' is as shown in Figure 7-4(d). The output capacitor C_o blocks dc voltage of $(+V_{CC}/2)$ volts, and hence the ac output voltage v_o results.

Although the 741C was used as an ac amplifier with a single supply voltage, some op-amps, such as the LM124 series, are designed specifically to operate on a single supply. These op-amps have a wide power supply range and low input offsets. For example, for the LM324,

$$\text{power supply range} = 3 \text{ to } 30 \text{ V dc}$$

$$I_B = 45 \text{ nA}$$

$$I_{io} = 5 \text{ nA}$$

$$V_{io} = 2 \text{ mV}$$

$$\text{supply current drain} = 800 \text{ } \mu\text{A independent of supply voltage}$$

(a)

(b)

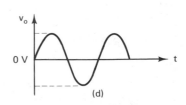

(c)

(d)

Figure 7–4 (a) AC inverting amplifier with single supply. (b) Its input and output waveforms. (c) AC noninverting amplifier with single supply. (d) Its input and output waveforms.

EXAMPLE 7-2
For the noninverting amplifier of Figure 7-4(c), $R_{in} = 50\ \Omega$, $C_i = C_1 = 0.1\ \mu F$, $R_1 = R_2 = R_3 = 100\ k\Omega$, $R_F = 1\ M\Omega$, and $+V_{CC} = +15$ V. Determine (a) the bandwidth of the amplifier and (b) maximum output voltage swing.

SOLUTION   (a) The ac input resistance of the amplifier is

$$R_{iF} = (R_2)\|(R_3)\|[R_i(1 + AB)]$$

$$R_{iF} \cong (R_2)\|(R_3)\quad \text{since } [R_i(1 + AB)] >> R_2 \quad \text{or} \quad R_3$$

$$R_{iF} \cong 100\ k\Omega\|100\ k\Omega = 50\ k\Omega$$

Therefore, from Equation (7-1),

$$f_L = \frac{1}{(2\pi)(10^{-7})(50\ k + 50)} = 31.8\ \text{Hz}$$

The gain of the amplifier is $(1 + R_F/R_1) = 11$. Hence from Equation (4-10b)

$$f_H = \frac{\text{UGB}}{A_F} = \frac{1\ \text{MHz}}{11} = 90.91\ \text{kHz}$$

and

$$\text{BW} = f_H - f_L = 90.91\ \text{kHz} - 0.0318\ \text{kHz} = 90.88\ \text{kHz}$$

(b) The ideal maximum output voltage swing $= +V_{CC} = +15$ V pp.

7-4 THE PEAKING AMPLIFIER

The *peaking response*, that is, the frequency response that peaks at a certain frequency, can be obtained by using a parallel LC network with the op-amp. Figure 7-5(a) shows a peaking amplifier that uses a parallel LC network in the feedback path. The frequency response of the amplifier is shown in Figure 7-5(b). The resonant frequency or peak frequency at which the peaking occurs is determined by the combination of L and C.

$$f_p = \frac{1}{2\pi\sqrt{LC}} \quad \text{if } Q_{coil} \ge 10 \tag{7-5a}$$

where Q_{coil} = figure of merit of the coil. The impedance of the parallel LC network is very large at the resonant frequency. Therefore, the gain of the amplifier at resonance is maximum and given by

$$A_F = -\frac{R_F\|R_P}{R_1} \tag{7-5b}$$

where R_P = equivalent parallel resistance of the tank circuit $= (Q_{coil}^2 R)$
$\quad\quad R$ = internal resistance of the coil

The impedance of the parallel LC network below and above the peak frequency is less than R_P; therefore, the gain of the amplifier is less than $(R_F\|R_P/R_1)$ at any other frequency than f_P. Note that, using Equation (7-5b), the impedance of the tank circuit R_P at resonance can easily be computed from the measured value of gain and known values of R_1 and R_F. The

Figure 7–5 The peaking amplifier. (a) Schematic diagram. (b) Frequency response.

bandwidth of the peaking amplifier of Figure (7-5a) can be determined by using the following equation:

$$BW = \frac{f_p}{Q_p} \qquad (7\text{-}5c)$$

where f_p = peaking frequency
Q_p = figure of merit of the parallel resonant circuit = $(R_F \| R_P / X_L)$

EXAMPLE 7-3

The circuit of Figure 7-5(a) is to provide a gain of 10 at a peak frequency of 16 kHz. Determine the values of all components.

SOLUTION The gain times peak frequency = (10)(16 kHz) = 160 kHz. Therefore, the 741C can be used in the circuit of Figure 7-5(a) since its 1-MHz unity gain–bandwidth is larger than 160 kHz.

Although any combination of L and C that provides the peak frequency of 16 kHz can be used, the value of the capacitor should be less than 1 μF to avoid leakage problems. Similary, to reduce the large size of the inductors, it should be less than 1 H in value. Let $C = 0.01$ μF. Then, using Equation (7-5a), the value of the inductor is

$$L = \frac{1}{(2\pi)^2 (f_p)^2 (C)}$$

$$L = \frac{1}{(4\pi^2)(16 \text{ k})^2 (10^{-8})} = 9.9 \text{ mH}$$

Let L be 10 mH. Also, let the internal resistance R of the inductor be 30 Ω. Then

$$Q_{coil} = \frac{X_L}{R} = \frac{(2\pi)(16\text{ k})(10)(10^{-3})}{30} = 33.5$$

and

$$R_p = (Q_{coil})^2(R) = (33.5)^2(30) = 33.67\text{ k}\Omega$$

Since a gain of 10 is needed at 16 kHz, let R_1 be equal to 100 Ω; therefore, using Equation (7-5b), the value of R_F is

$$10 = \frac{(R_F)(33.67\text{ k})}{100(R_F + 33.67\text{ k})}$$

that is,

$$R_F = 1.03\text{ k}\Omega \quad (\text{use } R_F = 1\text{ k}\Omega)$$

To minimize the effect of input bias currents, use R_{OM} approximately equal to $R_1\|R_F = 100\ \Omega$. Thus, the component values are as follows:

Op-amp: 741C
$$R_1 = 100\ \Omega = R_{OM}$$
$$R_F = 1\text{ k}\Omega$$
$$L = 10\text{ mH}, \quad \text{having } R = 30\ \Omega$$
$$C = 0.01\ \mu\text{F}$$

7-5 SUMMING, SCALING, AND AVERAGING AMPLIFIERS

This section shows how the inverting, noninverting, and differential configurations are useful in such applications as summing, scaling, and averaging amplifiers.

7-5.1 Inverting Configuration

Figure 7-6 shows the inverting configuration with three inputs V_a, V_b, and V_c. Depending on the relationship between the feedback resistor R_F and the input resistors R_a, R_b, and R_c, the circuit can be used as either a summing amplifier, scaling amplifier, or averaging amplifier. The circuit's function can be verified by examining the expression for the output voltage V_o, which is obtained from Kirchhoff's current equation written at node V_2. Referring to Figure 7-6,

$$I_a + I_b + I_c = I_B + I_F \tag{7-6}$$

Since R_i and A of the op-amp are ideally infinity, $I_B = 0$ A and $V_1 = V_2 \cong 0$ V. Therefore,

$$\frac{V_a}{R_a} + \frac{V_b}{R_b} + \frac{V_c}{R_c} = -\frac{V_o}{R_F}$$

or

$$V_o = -\left(\frac{R_F}{R_a}V_a + \frac{R_F}{R_b}V_b + \frac{R_F}{R_c}V_c\right) \tag{7-7}$$

Figure 7–6 Inverting configuration with three inputs can be used as a summing amplifier, scaling amplifier, or averaging amplifier.

7-5.1(a) Summing amplifier.   If in the circuit of Figure 7-6, $R_a = R_b = R_c = R$, for example, then Equation (7-7) can be rewritten as

$$V_o = -\frac{R_F}{R}(V_a + V_b + V_c) \tag{7-8a}$$

This means that the output voltage is equal to the *negative* sum of all the inputs times the gain of the circuit R_F/R; hence the circuit is called a *summing amplifier*. Obviously, when the gain of the circuit is 1, that is, $R_a = R_b = R_c = R_F$, the output voltage is equal to the *negative* sum of all input voltages. Thus

$$V_o = -(V_a + V_b + V_c) \tag{7-8b}$$

7-5.1(b) Scaling or weighted amplifier.   If each input voltage is amplified by a different factor, in other words, weighted differently at the output, the circuit in Figure 7-6 is then called a *scaling* or *weighted amplifier*. This condition can be accomplished if R_a, R_b, and R_c are different in value. Thus the output voltage of the scaling amplifier is

$$V_o = -\left(\frac{R_F}{R_a}V_a + \frac{R_F}{R_b}V_b + \frac{R_F}{R_c}V_c\right) \tag{7-8c}$$

where

$$\frac{R_F}{R_a} \neq \frac{R_F}{R_b} \neq \frac{R_F}{R_c}$$

7-5.1(c) Averaging circuit.   The circuit of Figure 7-6 can be used as an *averaging circuit,* in which the output voltage is equal to the average of all the input voltages. This is accomplished by using all input resistors of equal value, $R_a = R_b = R_c = R$ say. In addition, the gain by which each input is amplified must be equal to 1 over the number of inputs; that is,

$$\frac{R_F}{R} = \frac{1}{n}$$

where n is the number of inputs.

Thus, if there are three inputs (as shown in Figure 7-6), we want $R_F/R = 1/3$. Consequently, from Equation (7-7),

$$V_o = -\left(\frac{V_a + V_b + V_c}{3}\right) \qquad (7\text{-}8d)$$

Remember that in the preceding applications the inputs V_a, V_b, and V_c could be either ac or dc. These circuits are commonly used in analog computers and audio mixers, in which a number of inputs is added up (mixed) to produce a desired output.

In Figure 7-6 the offset minimizing resistor R_{OM} is used to minimize the effect of input bias currents on the output offset voltage. However, to reduce the output offset voltage to zero, the offset voltage-compensating network must be used, especially when the inputs are dc voltages.

EXAMPLE 7-4

In the circuit of Figure 7-6, $V_a = +1$ V, $V_b = +2$ V, $V_c = +3$ V, $R_a = R_b = R_c = 3$ kΩ, $R_F = 1$ kΩ, $R_{OM} = 270$ Ω, and supply voltages $= \pm 15$ V. Assuming that the op-amp is initially nulled, determine the output voltage V_o.

SOLUTION   Using Equation (7-8a), we obtain

$$V_o = -\frac{1\ \mathrm{k}}{3\ \mathrm{k}}(1 + 2 + 3) = -2\ \mathrm{V}$$

This value is equal to the average of three inputs with a negative sign.

7-5.2 Noninverting Configuration

If input voltage sources and resistors are connected to the noninverting terminal as shown in Figure 7-7, the circuit can be used either as a summing or averaging amplifier through selection of appropriate values of resistors, that is, R_1 and R_F.

Again, to verify the functions of the circuit, the expression for the output voltage must be obtained. Recall that the input resistance R_{iF} of the noninverting amplifier is very large (see Figure 7-7). Therefore, using the superposition theorem, the voltage V_1 at the noninverting terminal is

$$V_1 = \frac{R/2}{R + R/2}V_a + \frac{R/2}{R + R/2}V_b + \frac{R/2}{R + R/2}V_c$$

or

$$V_1 = \frac{V_a}{3} + \frac{V_b}{3} + \frac{V_c}{3} = \frac{V_a + V_b + V_c}{3} \qquad (7\text{-}9)$$

Hence the output voltage V_o is

$$V_o = \left(1 + \frac{R_F}{R_1}\right)\left(\frac{V_a + V_b + V_c}{3}\right)$$

$$R_{iF} \cong R_i \frac{AR_1}{R_1 + R_F}$$

Figure 7–7   Noninverting configuration with three inputs can be used as an averaging amplifier or a summing amplifier.

$$V_o = \left(1 + \frac{R_F}{R_1}\right) V_1$$

$$= \left(1 + \frac{R_F}{R_1}\right)\frac{V_a + V_b + V_c}{3} \tag{7-10a}$$

7-5.2(a) Averaging amplifier.   Equation (7-10a) shows that the output voltage is equal to the average of all input voltages times the gain of the circuit $(1 + R_F/R_1)$, hence the name *averaging amplifier*. Depending on the application requirement, the gain $(1 + R_F/R_1)$ can be set to a specific value. Obviously, if the gain is 1, the output voltage will be equal to the average of all input voltages.

Note that there are two basic differences between this averaging amplifier and that using the inverting configuration [see Equations (7-8d) and (7-10a)]:

1. No sign change or phase reversal occurs between the average of the inputs and output.
2. The noninverting input voltage V_1 is the average of all inputs, whereas in the inverting averaging amplifier the output is the average of all inputs, with a negative sign.

7-5.2(b) Summing amplifier.   A close examination of Equation (7-10a) reveals that if the gain $(1 + R_F/R_1)$ is equal to the number of inputs, the output voltage becomes equal to the *sum* of all input voltages. That is, if $(1 + R_F/R_1) = 3$ [from Equation (7-10a)],

$$V_o = V_a + V_b + V_c \tag{7-10b}$$

Hence the circuit is called a *noninverting summing amplifier*.

Again, when the circuit of Figure 7-7 is used as either an averaging or summing amplifier, offset null circuitry or an offset null compensating network must be used to improve its accuracy.

EXAMPLE 7-5

In the circuit of Figure 7-7, supply voltages $= \pm 15$ V, $V_a = +2$ V, $V_b = -3$ V, $V_c = +4$ V, $R = R_1 = 1$ kΩ, and $R_F = 2$ kΩ. Determine the voltage V_1 at the noninverting terminal and the output voltage V_o. Assume that the op-amp is initially nulled.

SOLUTION   Using Equation (7-9),

$$V_1 = \frac{2 - 3 + 4}{3} = 1 \text{ V}$$

which is the average of three inputs: $+2$ V, -3 V, and $+4$ V. From Equation (7-10a),

$$V_o = \left(1 + \frac{2 \text{ k}}{1 \text{ k}}\right)(1) = 3 \text{ V}$$

which is the sum of the three inputs.

7-5.3 Differential Configuration

Using a basic differential op-amp configuration, a subtractor and a summing amplifier may be constructed as described below.

7-5.3(a) A subtractor.   A basic differential amplifier can be used as a *subtractor* as shown in Figure 7-8. In this figure, input signals can be scaled to the desired values by selecting appropriate values for the external resistors; when this is done, the circuit is referred to as *scaling amplifier*. However, in Figure 7-8, all external resistors are equal in value, so the gain of the amplifier is equal to 1.

From this figure, the output voltage of the differential amplifier with a gain of 1 is

$$V_o = -\frac{R}{R}(V_a - V_b) \quad \text{(see Section 4-5.1)}$$

That is,

$$V_o = V_b - V_a \tag{7-11}$$

Figure 7-8   Basic differential amplifier used as a subtractor.

Thus the output voltage V_o is equal to the voltage V_b applied to the noninverting terminal *minus* the voltage V_a applied to the inverting terminal; hence the circuit is called a *subtractor*.

7-5.3(b) Summing amplifier.

A four-input summing amplifier may be constructed using the basic differential amplifier of Figure 7-8 if two additional input sources are connected, one each to the inverting and noninverting input terminals through resistor R (see Figure 7-9).

The output voltage equation for this circuit can be obtained by using the superposition theorem. For instance, to find the output voltage due to V_a alone, reduce all other input voltages V_b, V_c, and V_d to zero as shown in Figure 7-10. In fact, this circuit is an inverting amplifier in which the inverting input is at *virtual ground* ($V_2 = 0$ V). Therefore, the output voltage is

$$V_{oa} = -\frac{R}{R}V_a = -V_a$$

This result can also be obtained by Thévenizing the input circuit looking back from node V_2.

Similarly, the output voltage due to V_b alone is

$$V_{ob} = -V_b$$

Now if input voltages V_a, V_b, and V_d are set to zero, the circuit in Figure 7-9 becomes a noninverting amplifier in which the voltage V_1 at the noninverting input is

$$V_1 = \frac{R/2}{R + R/2}V_c = \frac{V_c}{3}$$

This means that the output voltage due to V_c alone is

$$V_{oc} = \left(1 + \frac{R}{R/2}\right)V_1 = (3)\left(\frac{V_c}{3}\right) = V_c$$

Figure 7–9 Summing amplifier using differential configuration.

Figure 7–10 Deriving the output voltage equation for the summing amplifier of Figure 7–9.

Similarly, the output voltage due to input voltage V_d alone is

$$V_{od} = V_d$$

Thus by using the superposition theorem the output voltage due to all four input voltages is given by

$$V_o = V_{oa} + V_{ob} + V_{oc} + V_{od}$$
$$= -V_a - V_b + V_c + V_d \tag{7-12}$$

Notice that the output voltage is equal to the *sum* of the input voltages applied to the noninverting terminal plus the negative sum of the input voltages applied to the inverting terminal. Even though in Figure 7-9 the gain of the summing amplifier is 1, any scale factor can be used for the inputs by selecting proper external resistors.

EXAMPLE 7-6

In the circuit of Figure 7-9, $R = 1$ kΩ, $V_a = +2$ V, $V_b = +3$ V, $V_c = +4$ V, $V_d = +5$ V, and supply voltages $= \pm 15$ V. Determine the output voltage V_o. Assume that the op-amp is initially nulled.

SOLUTION   From Equation (7-12),

$$V_o = -2 - 3 + 4 + 5 = +4 \text{ V}$$

7-6 INSTRUMENTATION AMPLIFIER

In many industrial and consumer applications the measurement and control of physical conditions are very important. For example, measurements of temperature and humidity inside a dairy or meat plant permit the operator to make necessary adjustments to maintain product quality. Similarly, precise temperature control of a plastic furnace is needed to produce a particular type of plastic.

Generally, a transducer is used at the measuring site to obtain the required information easily and safely. The *transducer* is a device that converts one form of energy into another. For example, a strain gage when subjected to *pressure* or force (physical energy) undergoes a change in its *resistance* (electrical energy). An instrumentation system is used to measure the output signal produced by a transducer and often to control the physical signal producing it. Figure 7-11 shows a simplified form of such a system. The input stage is composed of a preamplifier and some sort of transducer, depending on the physical quantity to be measured. The output stage may use devices such as meters, oscilloscopes, charts, or magnetic recoders.

In Figure 7-11 the connecting lines between the blocks represent *transmission lines,* used especially when the transducer is at a remote test site monitoring hazardous conditions such as high temperatures or the liquid levels of flammable chemicals. These transmission lines permit signal transfer from unit to unit. The length of the transmission lines depends primarily on the physical quantities to be monitored and on system requirements.

The signal source of the instrumentation amplifier is the output of the transducer. Although some transducers produce outputs with sufficient strength to permit their use directly, many do not. To amplify the low-level output signal of the transducer so that it can drive the indicator or display is the major function of the *instrumentation amplifier*. In short, the instrumentation amplifier is intended for precise, low-level signal amplification where low noise, low thermal and time drifts, high input resistance, and accurate closed-loop gain are required. Besides, low power consumption, high common-mode rejection ratio, and high slew rate are desirable for superior performance.

There are many instrumentation operational amplifiers, such as the μA725, ICL7605, and LH0036, that make a circuit extremely stable and accurate. These ICs are, however, relatively expensive; they are very precise special-purpose circuits in which most of the electrical parameters, such as offsets, drifts, and power consumption, are minimized, whereas input resistance, CMRR, and supply range are optimized. Some instrumentation amplifiers are even available in modular form to suit special installation requirements.

Obviously, the requirements for instrumentation op-amps are more rigid than those for general-purpose applications. However, where the requirements are not too strict, the general-purpose op-amp can be employed in the differential mode. We will call such

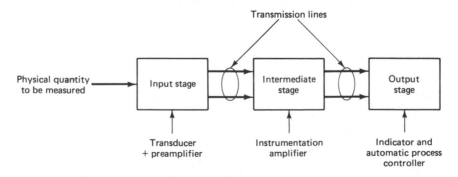

Figure 7–11   Block diagram of an instrumentation system.

amplifiers *differential instrumentation amplifiers*. Since most of the instrumentation systems use a transducer in a bridge circuit, we will consider a simplified differential instrumentation system arrangement using a transducer bridge circuit.

7-6.1 Instrumentation Amplifier Using Transducer Bridge

Figure 7-12 shows a simplified differential instrumentation amplifier using a transducer bridge. A *resistive transducer* whose resistance changes as a function of some physical energy is connected in one arm of the bridge with a small circle around it and is denoted by $(R_T \pm \Delta R)$, where R_T is the resistance of the transducer and ΔR the change in resistance R_T.

The bridge in the circuit of Figure 7-12 is dc excited but could be ac excited as well. For the balanced bridge at some reference condition,

$$V_b = V_a$$

or

$$\frac{R_B(V_{dc})}{R_B + R_C} = \frac{R_A(V_{dc})}{R_A + R_T}$$

Figure 7–12   Differential instrumentation amplifier using a transducer bridge.

That is,

$$\frac{R_C}{R_B} = \frac{R_T}{R_A} \qquad (7\text{-}13)$$

Generally, resistors R_A, R_B, and R_C are selected so that they are equal in value to the transducer resistance R_T at some *reference condition*. The reference condition is the specific value of the physical quantity under measurement at which the bridge is balanced. This value is normally established by the designer and depends on the transducer's characteristics, the type of physical quantity to be measured, and the desired application.

The bridge is balanced initially at a desired reference condition. However, as the physical quantity to be measured changes, the resistance of the transducer also changes, which causes the bridge to unbalance ($V_a \neq V_b$). The output voltage of the bridge can be expressed as a function of the change in resistance of the transducer, as described next.

Let the change in resistance of the transducer be ΔR. Since R_B and R_C are fixed resistors, the voltage V_b is constant. However, voltage V_a varies as a function of the change in transducer resistance. Therefore, according to the voltage-divider rule,

$$V_a = \frac{R_A(V_{dc})}{R_A + (R_T + \Delta R)}$$

$$V_b = \frac{R_B(V_{dc})}{R_B + R_C}$$

Consequently, the voltage V_{ab} across the output terminals of the bridge is

$$V_{ab} = V_a - V_b$$

$$= \frac{R_A V_{dc}}{R_A + R_T + \Delta R} - \frac{R_B V_{dc}}{R_B + R_C}$$

However, if $R_A = R_B = R_C = R_T = R$, say, then

$$V_{ab} = -\frac{\Delta R(V_{dc})}{2(2R + \Delta R)} \qquad (7\text{-}14)$$

The negative $(-)$ sign in this equation indicates that $V_a < V_b$ because of the increase in the value of R_T.

The output voltage V_{ab} of the bridge is then applied to the differential instrumentation amplifier composed of three op-amps (see Figure 7-12). The voltage followers preceding the basic differential amplifier help to eliminate loading of the bridge circuit. The gain of the basic differential amplifier is $(-R_F/R_1)$; therefore, the output V_o of the circuit is

$$V_o = V_{ab}\left(-\frac{R_F}{R_1}\right) = \frac{(\Delta R)V_{dc}}{2(2R + \Delta R)}\frac{R_F}{R_1} \qquad (7\text{-}15a)$$

Generally, the change in resistance of the transducer ΔR is very small. Therefore, we can approximate $(2R + \Delta R) \cong 2R$. Thus, the output voltage

$$V_o = \frac{R_F}{R_1}\frac{\Delta R}{4R}V_{dc} \qquad (7\text{-}15b)$$

This equation indicates that V_o is directly proportional to the change in resistance ΔR of the transducer. Since the change in resistance is caused by a change in physical energy, a meter connected at the output can be calibrated in terms of the units of that physical energy.

Before proceeding with specific bridge applications, let us briefly consider the important characteristics of some resistive types of transducers. In these resistive types of transducers the resistance of the transducer changes as a function of some physical quantity. The thermistors, photoconductive cells, and strain gages are some of the most commonly used resistive transducers; hence they will be further discussed here.

Thermistors are essentially semiconductors that behave as resistors, usually with a *negative temperature coefficient of resistance*. That is, as the temperature of a thermistor increases, its resistance decreases. The temperature coefficient of resistance is expressed in ohms per unit change in degrees Celsius (°C). Thermistors with a high temperature coefficient of resistance are more sensitive to temperature change and are therefore well suited to temperature measurement and control. Thermistors are available in a wide variety of shapes and sizes. However, thermistor beads sealed in the tips of glass rods are most commonly used because they are relatively easy to mount.

The photoconductive cell belongs to the family of photodetectors (photosensitive devices) whose resistance varies with an incident radiant energy or with light. As the intensity of incident light increases, the resistance of the cell decreases. The resistance of the photoconductive cell in darkness is typically on the order of 100 kΩ. Generally, the resistance of the cell in darkness and at particular light intensities is listed on the data sheet. The intensity of light is expressed in meter candles (lux).

Materials such as cadmium sulfide and silicon, whose conductivity is a function of incident radiant energy, are used for photoconductive cells. Some cells are extremely sensitive to light and hence can be used into the ultraviolet and infrared regions. The photoconductive cell is typically composed of a ceramic base, a layer of photoconductive material, a moisture-proof enclosure, and metallic leads. Photoconductive cells are also known as photocells or light-dependent resistors (LDRs).

Another important resistive transducer is the strain gage, whose resistance changes due to elongation or compression when an external stress is applied. The stress is defined as force per unit area [newtons/(meter)^2] and can be related to pressure, torque, and displacement. Therefore, a strain gage may be used to monitor change in applied pressure, torque, and displacement by measuring the corresponding change in the gage's resistance.

Two basic types of strain gages are the *wire* and *semiconductor*. Semiconductor strain gages are much more sensitive than the wire type and therefore provide better accuracy and resolution. The *sensitivity* of a strain gage is defined as unit change in resistance per unit change in length and is a dimensionless quantity.

The thermistor, photocell, and strain gage are all passive transducers, meaning that they require external voltage (ac or dc) for their operation.

7-6.1(a) Temperature indicator.   The circuit of Figure 7-12 can be used as a temperature indicator if the transducer in the bridge circuit is a thermistor and the output meter is calibrated in degrees Celsius or Fahrenheit. The bridge can be balanced at a desired reference condition, say 25°C. As the temperature varies from its reference value, the resistance of the thermistor changes and the bridge becomes unbalanced. This unbalanced bridge in turn produces the meter movement. The meter can be calibrated to read a desired temperature range by selecting an appropriate gain for the differential instrumentation amplifier. In Figure 7-12 the meter movement is dependent on the amount of imbalance in the bridge, that is, the change ΔR in the value of the thermistor resistance. The ΔR for the thermistor, however, can be determined as follows: ΔR = (temperature coefficient of resistance) (final temperature — reference temperature.) See Example 7-7.

EXAMPLE 7-7

In the circuit of Figure 7-12, R_1 = 1 kΩ, R_F = 4.7 kΩ, R_A = R_B = R_C = 100 kΩ, V_{dc} = +5 V, and op-amp supply voltages = ±15 V. The transducer is a thermistor with the following specifications: R_T = 100 kΩ at a reference temperature of 25°C; temperature coefficient of resistance = −1 kΩ/°C or 1%/°C. Determine the output voltage at 0°C and at 100°C.

SOLUTION   At 25°C, R_A = R_B = R_c = R_T = 100 kΩ. Therefore, the bridge is balanced (V_a = V_b) and V_o = 0 V. However, at 0°C the change ΔR in the resistance of the thermistor is

$$\Delta R = \frac{-1 \text{ k}\Omega}{°C}(0°C - 25°C) = 25 \text{ k}\Omega$$

Therefore, using Equation (7-15), at 0°C,

$$V_o = \frac{4.7 \text{ k}}{1 \text{ k}} \frac{25 \text{ k}}{400 \text{ k}}(5)$$

$$V_o = 1.47 \text{ V}$$

Similarly, at 100°C,

$$\Delta R = \frac{-1 \text{ k}}{°C}(100°C - 25°C) = -75 \text{ k}\Omega$$

$$V_o = \frac{4.7 \text{ k}}{1 \text{ k}} \frac{(-75 \text{ k})}{400 \text{ k}}(5)$$

$$V_o = -4.41 \text{ V}$$

Thus, when V_o = 1.47 V, the meter face can be marked as 0°C, and when V_o = −4.41 V, it can be marked as 100°C. Note that, at 25°C, V_o = 0 V; therefore, a center-zero meter is required. Thus, using the resistance–temperature characteristic of the thermistor, the meter can be calibrated from 0° to 100°C.

Thermistors with relatively higher resistance ($R_T \geq$ 1 MΩ) and sensitivity (temperature coefficient of resistance $\geq$3%/°C) are best suited for *remote measurements* because the effect of transmission-line resistance is negligible.

7-6.1(b) Temperature controller.   A simple and inexpensive temperature control circuit may be constructed by using a thermistor in the bridge circuit and by replacing a meter with a relay in the circuit of Figure 7-12. The output of the differential instrumentation amplifier drives a relay that controls the current in the heat-generating circuit. A properly designed circuit should energize a relay when the temperature of the thermistor drops below a desired value, causing the heat unit to turn on.

7-6.1(c) Light-intensity meter.

The circuit in Figure 7-12 can be used as a light-intensity meter if a transducer is a photocell. The bridge can be balanced for darkness conditions. Therefore, when exposed to light, the bridge will be unbalanced and cause the meter to deflect. The meter can be calibrated in terms of *lux* to measure the change in light intensity.

The light-intensity meter using an instrumentation bridge amplifier is more accurate and stable than single-input inverting or noninverting configurations because the common-mode (noise) voltages are effectively rejected by the differential configuration.

7-6.1(d) Measurement of flow and thermal conductivity.

A *flow meter* or a *thermal conductivity meter* may be constructed using the circuit of Figure 7-12, provided that two thermistors are used adjacent to each other in the bridge. For instance, assume that R_C and R_T represent two identical thermistors. For the flow measurement, one thermistor is sealed in a small-cavity copper cylinder and the other installed in a small copper pipe. When no air flows through the pipe, the bridge can be balanced so that the output voltage is zero. When air flows over the thermistor, its temperature decreases and in turn the resistance increases, causing the bridge to be unbalanced. This unbalanced voltage is then amplified by the differential instrumentation amplifier and applied to the meter. Thus the amount of meter deflection is proportional to the flow rate of the air in the pipe. The meter can be calibrated in (meter)^3/second to accommodate a desired flow rate range.

For the thermal conductivity measurement, two thermistors are mounted in separate small copper cylinders. With air in both cylinders, the bridge can be balanced, and hence the output will be zero. When air in one cylinder is replaced by a medium being tested, which has a different thermal conductivity than air, the bridge becomes unbalanced. This happens because the temperature of the thermistor changes, thus changing its resistance. Suppose that the medium being tested is carbon dioxide (CO_2). Because of its lower thermal conductivity, CO_2 will increase the temperature of the thermistor, which will cause a decrease in the thermistor's resistance. This results in unbalancing the bridge, which in turn produces a meter deflection. The meter can be calibrated in terms of relative thermal conductivity (cal/s-cm-°C).

7-6.1(e) Analog weight scale.

By connecting a strain gage in the bridge, the circuit of Figure 7-12 can be converted into a simple and inexpensive *analog weight scale*.

In the analog weight scale, strain gage elements are connected in all four arms of the bridge. The elements are mounted on the base of the weight platform so that, when an external force or weight is applied to the platform, one pair of elements in the opposite arms elongates, whereas the other pair of elements in the opposite arms compresses. In other words, when the weight is placed on the platform, R_{T1} and R_{T3} both decrease in resistance, and R_{T2} and R_{T4} both increase in resistance, or vice versa (see Figure 7-13).

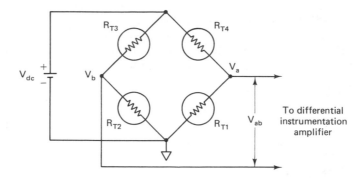

Figure 7–13   Strain-gage bridge circuit for analog weight scale, $R_{T1} = R_{T2} = R_{T3} = R_{T4}$.

When no weight is placed on the platform, the bridge is balanced, $R_{T1} = R_{T2} = R_{T3} = R_{T4} = R$, say, and the output voltage of the weight scale can be zero. When a weight is placed on the scale platform, the bridge becomes unbalanced. Assuming that R_{T1} and R_{T3} decrease in resistance and R_{T2} and R_{T4} increase in resistance by the same number of ohms ΔR, the unbalanced voltage V_{ab} is given by

$$V_{ab} = -V_{dc}\left(\frac{\Delta R}{R}\right) \tag{7-16}$$

where V_{dc} = dc excitation voltage of the bridge
$\quad R = R_{T1} = R_{T2} = R_{T3} = R_{T4}$ = unstrained gage resistance
$\quad \Delta R$ = change in gage resistance
Remember that, if the decrease in gage resistance R_{T1} and R_{T3} is ΔR, the increase in resistance of R_{T2} and R_{T4} is also ΔR. Therefore, with this assumption, the voltage $V_a < V_b$ and the output voltage V_{ab} is negative, as indicated by Equation (7-16).

The voltage V_{ab} is then amplified by the differential instrumentation amplifier, which drives the meter. Since the gain of the amplifier is $(-R_F/R_1)$, the output voltage V_o is

$$V_o = V_{dc}\left(\frac{\Delta R}{R}\right)\frac{R_F}{R_1} \tag{7-17}$$

The gain of the amplifier can be selected according to the sensitivity of the strain gage and the full-scale deflection requirements of the meter. The meter can be calibrated in terms of kilograms.

For better accuracy and resolution, a microprocessor-based digital weight scale may be constructed. However, such a scale is much more complex and expensive than the analog scale.

EXAMPLE 7-8

The circuit of Figure 7-12 is used as an analog weight scale with the following specifications. The gain of the differential instrumentation amplifier $= -100$. Assume that $V_{dc} = +10$ V and that the op-amp supply voltages $= \pm 10$ V. The unstrained resistance of each of the four elements of the strain gage is 100 Ω. When a certain weight is placed on the scale platform, the output voltage $V_o = 1$ V. Assuming that the output is initially zero, determine the change in the resistance of each strain-gage element.

SOLUTION   Using Equation (7-17),

$$1 = (10)\frac{\Delta R}{100}(100)$$

$$\Delta R = 0.1 \ \Omega$$

This means that R_{T1} and R_{T3} will decrease by 0.1 Ω if R_{T2} and R_{T4} increase by 0.1 Ω when a certain weight is placed on the scale platform.

7-7 DIFFERENTIAL INPUT AND DIFFERENTIAL OUTPUT AMPLIFIER

In all the applications discussed so far, the op-amp is used with a single-ended or unbalanced output. However, in certain applications a differential output is required. The differential input and differential output amplifier is most commonly used as a preamplifier and in driving push-pull arrangements.

Figure 7-14(a) shows one possible arrangement of a differential input and differential output amplifier using two identical op-amps, that is, a dual op-amp. For example, the μAF772 dual op-amp incorporates two well-matched op-amps. The connection diagrams of the 8-pin mini DIP and 14-pin DIP are shown in Figure 7-14(b). The analysis of the circuit in Figure 7-14(a) can be accomplished by determining the output of each op-amp due to the differential input. Using the superposition theorem, the output V_{ox} due to inputs V_x and V_y is

$$V_{ox} = \left(1 + \frac{R_F}{R_1}\right)V_x - \left(\frac{R_F}{R_1}\right)V_y \tag{7-18a}$$

Similarly, the output V_{oy} is

$$V_{oy} = \left(1 + \frac{R_F}{R_1}\right)V_y - \left(\frac{R_F}{R_1}\right)V_x \tag{7-18b}$$

However, the differential output V_o is

$$V_o = V_{ox} - V_{oy}$$

Therefore, from Equations (7-18a) and (7-18b),

$$V_o = \left(1 + \frac{R_F}{R_1}\right)V_x - \left(\frac{R_F}{R_1}\right)V_y - \left(1 + \frac{R_F}{R_1}\right)V_y + \left(\frac{R_F}{R_1}\right)V_x$$

$$= \left(1 + \frac{2R_F}{R_1}\right)(V_x - V_y)$$

or

$$V_o = \left(1 + \frac{2R_F}{R_1}\right)V_{in} \tag{7-18c}$$

(a)

(b)

Figure 7–14   (a) Differential input and differential output amplifier using a dual op-amp. (b) The connection diagram of a typical dual op-amp. (Note especially the difference between the supply voltage and offset-null connections.) (Courtesy of Fairchild Camera and Instrument Corporation.)

This means that the differential input and output are in phase or of the same polarity provided that $V_{in} = V_x - V_y$ and $V_o = V_{ox} - V_{oy}$.

The differential input and output amplifier of Figure 7-14(a) is very useful in noisy environments, especially if the input signal is relatively smaller, because it rejects the common-mode noise voltages.

EXAMPLE 7-9

The differential input and output amplifier of Figure 7-14(a) is used as a preamplifier and requires a differential output of at least 3.7 V. Determine the gain of the circuit if the differential input $V_{in} = 100$ mV.

SOLUTION   Using Equation (7-18c), we get

$$3.7 = \left(1 + \frac{2R_F}{R_1}\right)(100)(10^{-3})$$

$$1 + \frac{2R_F}{R_1} = 37 \quad \text{or} \quad R_F = 18 R_1$$

If $R_1 = 100\ \Omega$, then $R_F = 1.8$ kΩ.

7-8 VOLTAGE-TO-CURRENT CONVERTER WITH FLOATING LOAD

Figure 7-15 shows a voltage-to-current converter in which load resistor R_L is *floating* (not connected to ground). The input voltage is applied to the noninverting input terminal, and the feedback voltage across R_1 drives the inverting input terminal. This circuit is also called a *current-series negative feedback* amplifier because the feedback voltage across R_1 (applied to the inverting terminal) depends on the output current i_o and is in series with the input difference voltage v_{id}.

Writing Kirchhoff's voltage equation for the input loop,

$$v_{in} = v_{id} + v_f$$

Figure 7–15   Voltage-to-current converter with floating load.

But $v_{id} \cong 0$ V, since A is very large; therefore,

$$v_{in} = v_f$$
$$v_{in} = R_1 i_o$$

or

$$i_o = \frac{v_{in}}{R_1} \tag{7-19}$$

This means that in the circuit of Figure 7-15 an input voltage v_{in} is converted into an output current of v_{in}/R_1. In other words, input voltage v_{in} appears across R_1. If R_1 is a precision resistor, the output current $(i_o = v_{in}/R_1)$ will be precisely fixed.

The voltage-to-current converter can be used in such applications as low-voltage dc and ac voltmeters, diode match finders, light-emitting diodes (LEDs), and zener diode testers.

7-8.1 Low-Voltage DC Voltmeter

If in the converter circuit of Figure 7-15 we replace the load resistor R_L by an ammeter (D'Arsonval meter movement) with a full-scale deflection of 1 mA, the resulting circuit is the dc voltmeter of Figure 7-16. This figure shows an external offset voltage compensating network, which is used because it makes the nulling of the op-amp relatively easy. Using the *offset null circuitry* of the 741 (10-kΩ potentiometer), the op-amp sometimes cannot be nulled because the output is very sensitive to even slight variations in wiper position.

Figure 7–16   DC voltmeter with 1- to 13-V full-scale range.

The Thévenin equivalent resistance of the compensating network is approximately equal to 10 Ω (16.25 k$\Omega \parallel$ 10 Ω). Therefore, when the switch is in the $\times 1$ position,

$$R_1 = 10 + 1000 \cong 1 \text{ k}\Omega$$

If $V_{in} = 1$ V, then

$$I_o = \frac{V_{in}}{R_1} \cong \frac{1 \text{ V}}{1 \text{ k}\Omega} = 1 \text{ mA}$$

This means that 1 V causes the full-scale deflection of the ammeter.

If the range switch is changed to the $\times 10$ position (10 Ω + 10 k$\Omega \cong$ 10 kΩ), it will require a 10-V input to get full-scale deflection. Thus successively higher resistance values are required to measure relatively higher input voltages. However, the input voltage range for the 741C op-amp is ± 14 V. Therefore, with ± 15-V supply voltages, the maximum input voltage has to be $\leq \pm 14$ V. In Figure 7-16 the maximum full-scale input voltage of 13 V can be applied when the range switch is in the $\times 13$ position.

Thus, by calibrating the face of the ammeter in volts, a dc voltmeter with a full-scale voltage range of 1 to 13 V can be constructed. Note that meter resistance R_m does not affect I_o. Only V_{in} and R_1 determine the I_o value.

A centered-zero ammeter can be used to measure positive as well as negative input voltages. To improve the accuracy of voltage measurements, the ammeter should be nulled each time before input voltage is applied.

7-8.2 Low-Voltage AC Voltmeter

It is also possible to modify the meter movement so that it will indicate values of ac voltage or current. A combination of an ammeter and a full-wave rectifier can be employed in the feedback loop to form an ac voltmeter, as shown in Figure 7-17. In this circuit an alternating current is converted into a direct current. During the positive half-cycle of the ac input, diodes D_1 and D_3 conduct, whereas diodes D_2 and D_4 conduct during the negative half-cycle of v_{in}. Thus the current through the ammeter flows only in one direction (A to B) for the entire cycle of the ac input. In other words, the ammeter registers the average (dc) value of the rectified current. The amount of deflection of the pointer must be considered for proper scale calibration. For full-wave rectification, meter current can be expressed as

$$i_o = \frac{0.9 v_{in}}{R_1} \tag{7-20a}$$

or

$$v_{in} = (1.1R_1)i_o \tag{7-20b}$$

where i_o = dc meter current (amperes)
   v_{in} = ac rms input voltage (volts)
   R_1 = resistance in series with the meter (ohms)
Using Equation (7-20b), the ammeter face can be calibrated to read an rms value of the input voltage.

Figure 7–17 AC voltmeter.

Rectifier-type meters are designed primarily for ac frequencies in the audio range. However, because of the low slew rate of the 741C op-amp, the frequency of the input signal should be less than 4 kHz for proper operation of the voltmeter. Also, the *peak inverse voltage* (PIV) of the diodes is recommended to be larger than the saturation voltages of the op-amp.

EXAMPLE 7-10

In the circuit of Figure 7-17, for the indicated values of resistors, determine the full-scale range for the input voltage.

SOLUTION   The full-scale meter movement is 1 mA. Therefore, substituting minimum and maximum values of resistor R_1 in Equation (7-20b), we can determine the range for the input voltage.

$$v_{in}(rms)\,|_{minimum} = (1.1)\,(1\ k\Omega)\,(1\ mA) = 1.1\ V$$

$$v_{in}(rms)\,|_{maximum} = (1.1)\,(6.8\ k\Omega)\,(1\ mA) = 7.48\ V$$

Thus the full-scale input voltage ranges from 1.1 to 7.48 V rms or 3.1 to 21.16 V peak to peak. Note that since the input voltage range for the 741C is ±14 V, the input voltage swing must be less than 28 V peak to peak for safe operation of the circuit.

7-8.3 Diode Match Finder

In some circuits, such as the *ring modulator* and *Foster–Sealy discriminator,* it is necessary to have matched diodes with equal voltage drops at a particular value of diode current.
The circuit shown in Figure 7-18 can be used in finding matched diodes and is obtained
from Figure 7-15 by replacing R_L with a diode.

When the switch is in position 1, the rectifier diode 1N4001 is placed in the feedback
loop; the current through this loop is set by input voltage V_{in} and resistor R_1. For $V_{in} = 1$ V
and $R_1 = 100\ \Omega$, the current through the diode is

$$I_o = \frac{V_{in}}{R_1} = \frac{1}{100} = 10\ \text{mA}$$

As long as V_{in} and R_1 are constant, the current I_o will be constant. The voltage drop across
the diode can be found either by measuring the voltage across it or the output voltage. The
output voltage is equal to $(V_{in} + V_D)$. To avoid an error in output voltage, the op-amp
should be *initially* nulled.

Thus the matched diodes can be found by connecting diodes one after another in the
feedback path and measuring the voltage across them. A desired testing current I_o can be
established by selecting an appropriate combination of V_{in} and R_1.

EXAMPLE 7-11

In the circuit of Figure 7-18, when the switch is in position 1, $V_{in} = 0.5$ V and
$V_o = 1.2$ V. Determine the current through the diode and the voltage drop across it.
Assume that the op-amp is initially nulled.

Figure 7–18   Diode match finder.

SOLUTION

$$I_o = \frac{V_{in}}{R_1} = \frac{0.5}{100} = 5 \text{ mA}$$

$$V_D = V_o - V_{in} = 1.2 - 0.5 = 0.7 \text{ V}$$

7-8.4 Zener Diode Tester

The circuit of Figure 7-18 becomes a zener diode tester when the switch is placed in position 2. This circuit can be used to find the breakdown voltage of zener diodes. V_{in} and R_1 set the zener current at a constant value. If this current is larger than the knee current (I_{zk}) of the zener, the zener blocks V_z volts. For example, the 1N4735 zener in the circuit of Figure 7-18 has $I_{zk} = 1$ mA and $V_z = 6.2$ V. Since the current through the zener is

$$I_o = \frac{V_{in}}{R_1} = \frac{1}{100} = 10 \text{ mA} > I_{zk}$$

the voltage across the zener will be approximately equal to 6.2 V.

Matched zeners are useful in obtaining split ($\pm$) supply voltages from a single supply.

7-8.5 Light-Emitting Diode Tester

The circuit of Figure 7-18 can be converted into a light-emitting diode (LED) tester when the switch is in position 3. Again the LED current is set at a constant value by V_{in} and R_1. The LEDs can be tested for brightness one after another at this current. Matched LEDs with equal brightness at a specific value of current are useful as indicators and display devices in digital applications.

In all the applications described, it should be remembered that the maximum current through the load (ammeter, rectifier and zener diodes, or LED) cannot exceed the *short-circuit current* of the op-amp. For the 741C this current is 25 mA.

7-9 VOLTAGE-TO-CURRENT CONVERTER WITH GROUNDED LOAD ⎯⎯⎯⎯

Another version of the voltage-to-current converter is shown in Figure 7-19. In this circuit, one terminal of the load is grounded, and load current is controlled by an input voltage. The analysis of the circuit is accomplished by first determining the voltage V_1 at the noninverting input terminal and then establishing the relationship between V_1 and the load current.

Writing Kirchhoff's current equation at node V_1,

$$I_1 + I_2 = I_L$$

$$\frac{V_{in} - V_1}{R} + \frac{V_o - V_1}{R} = I_L$$

$$V_{in} + V_o - 2V_1 = I_L R$$

Therefore,

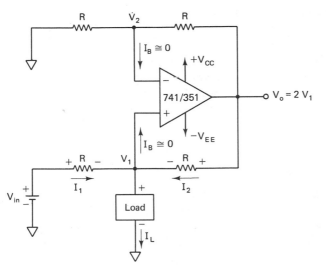

Figure 7–19 Voltage-to-current converter with grounded load.

$$V_1 = \frac{V_{in} + V_o - I_L R}{2} \tag{7-21a}$$

Since the op-amp is connected in the noninverting mode, the gain of the circuit in Figure 7-19 is $1 + R/R = 2$. Then the output voltage is

$$V_o = 2V_1 = V_{in} + V_o - I_L R$$

That is,

$$V_{in} = I_L R$$

or

$$I_L = \frac{V_{in}}{R} \tag{7-21b}$$

This means that the load current depends on the input voltage V_{in} and resistor R. Notice that all resistors must be equal in value.

The voltage-to-current converter of Figure 7-19 may also be used in testing such devices as zeners and LEDs forming a ground load. However, the circuit will perform satisfactorily provided that load size $\leq R$ value.

EXAMPLE 7-12

In the circuit of Figure 7-19, $V_{in} = 5$ V, $R = 10$ kΩ, and $V_1 = 1$ V. Find (a) the load current and (b) the output voltage V_o. Assume that the op-amp is initially nulled.

SOLUTION   (a) Using Equation (7-21b),

$$I_L = \frac{V_{in}}{R} = \frac{5}{10 \text{ k}\Omega} = 0.50 \text{ mA}$$

(b) Since $V_{in} = I_L R$, from Equation (7-21a), $V_o = 2V_1 = 2$ V.

7-10 CURRENT-TO-VOLTAGE CONVERTER

In Chapter 4 the current-to-voltage (I-to-V) converter was presented as a special case of the inverting amplifier in which an input current is converted into a proportional output voltage. One of the most common uses of the current-to-voltage converter is in digital-to-analog circuits (DACs) and in sensing current through photodetectors such as photocells, photodiodes, and photovoltaic cells. Photosensitive devices produce a current that is proportional to an incident radiant energy or light and therefore can be used to detect the light.

7-10.1 DAC Using Current-to-Voltage Converter

Figure 7-20 shows a combination of a DAC and current-to-voltage converter. The eight-digit binary signal is the input to the MC1408 DAC, and V_o is the corresponding analog output of the current-to-voltage converter. The output of the MC1408 is current I_o, the value of which depends on the logic state (0 or 1) of the binary inputs as indicated by the following equation:

$$I_o = \frac{V_{\text{ref}}}{R_1}\left(\frac{D_7}{2} + \frac{D_6}{4} + \frac{D_5}{8} + \frac{D_4}{16} + \frac{D_3}{32} + \frac{D_2}{64} + \frac{D_1}{128} + \frac{D_0}{256}\right) \qquad (7\text{-}22a)$$

where I_o = output current of the DAC (mA)
               R_1 = resistance (kΩ)

$$I_o = \frac{V_{\text{ref}}}{R_1}\left(\frac{D7}{2} + \frac{D6}{4} + \frac{D5}{8} + \frac{D4}{16} + \frac{D3}{32} + \frac{D2}{64} + \frac{D1}{128} + \frac{D0}{256}\right)$$

Figure 7–20   DAC using current-to-voltage converter.

V_{ref} = reference voltage (volts)

D_0 through D_7 = eight binary inputs

This means that I_o is zero when all the inputs are logic 0, and I_o is maximum when all the inputs are logic 1. Thus the value of I_o is a function of the state of eight binary inputs. The variations in I_o can be converted into a desired output voltage range by selecting a proper value for R_F, since

$$V_o = I_o R_F \qquad (7\text{-}22b)$$

where I_o is given by Equation (7-22a). It is common to parallel R_F with a capacitance C to minimize overshoot and ringing. Note that in Figure 7-20 the output voltage V_o of the current-to-voltage converter is positive because the direction of input current I_o is opposite to that in Figure 4-12.

EXAMPLE 7-13

In the circuit of Figure 7-20, $V_{ref} = 2$ V, $R_1 = 1$ kΩ, and $R_F = 2.7$ kΩ. Assuming that the op-amp is initially nulled, determine the range for the output voltage V_o.

SOLUTION   When all the binary inputs D_o through D_7 are logic 0, the current $I_o = 0$; therefore, the minimum value of $V_o = 0$ V. However, when all the inputs are logic 1, the I_o [using Equation (7-22a)] is

$$I_o = \frac{2}{1\,\text{k}} \left(\frac{1}{2} + \frac{1}{4} + \frac{1}{8} + \frac{1}{16} + \frac{1}{32} + \frac{1}{64} + \frac{1}{128} + \frac{1}{256} \right)$$

$$= 1.992 \text{ mA}$$

Hence the maximum value of output voltage is

$$V_o = I_o R_F = (1.992 \text{ mA})(2.7 \text{ k}\Omega) = 5.38 \text{ V}$$

Thus the output voltage range is 0 to 5.38 V.

7-10.2 Detecting Current through Photosensitive Devices

Photocells, photodiodes, and photovoltaic cells give an output current that depends on the intensity of an incident radiant energy or light and is independent of the load. The current through these devices can be converted to voltage by using an *I*-to-*V* converter and can be used as a measure of the amount of light or radiant energy incident in the device.

Figure 7-21 shows a photocell, the Clairex CL505L, connected to the *I*-to-*V* converter. Note that, since the photocell is a *passive* transducer, it requires an external voltage V_{dc}. The CL505L photocell has the following specifications:

resistance when illuminated (at 0.61 lux) = 1.5 kΩ

minimum dark resistance = 100 kΩ

measurement voltage = 10 V

temperature range = $-50°$ to $75°$C

Figure 7–21 *I*-to-*V* converter used to measure the current through the photocell.

EXAMPLE 7-14

In the circuit of Figure 7-21, $V_{dc} = 5$ V and $R_F = 3$ kΩ. Determine the change in the output voltage if the photocell is exposed to a light of 0.61 lux from a dark condition. Assume that the op-amp is initially nulled.

SOLUTION   The resistance R_T of the CL505L in darkness is 100 kΩ. The minimum output voltage in darkness is

$$V_{o\,min} = -\frac{V_{dc}}{R_T}R_F = -\frac{5}{100\text{ k}}(3\text{ k}) = -0.15 \text{ V}$$

When the CL505L is illuminated, its resistance $R_T = 1.5$ kΩ. Therefore, the maximum output voltage is

$$V_{o\,max} = -\frac{V_{dc}}{R_T}R_F = -\frac{5}{1.5\text{ k}}(3\text{ k}) = -10 \text{ V}$$

Thus V_o varies from -0.15 to -10 V as the photocell is exposed to light from a dark condition. The capacitor C is used to reduce high-frequency noise.

The circuit in Figure 7-21 can be used as a *light-intensity meter* by connecting at the output a meter that is calibrated for light intensity. The dc voltage V_{dc} in Figure 7-21 can be eliminated if a photovoltaic cell is used instead of a photocell. The photovoltaic cell is a semiconductor junction device that converts radiation energy into electrical power. It is a *self-generating* transducer because it does not require external voltage. The most common example of a photovoltaic cell is the *solar cell* used in space applications and watches.

The lower limit on current measurement with an *I*-to-*V* converter is set by the bias current I_B of the op-amp. This means that op-amps with smaller I_B values, such as the μA714 ($I_B = 3$ nA), can be used to detect lower currents.

7-11 VERY HIGH INPUT IMPEDANCE CIRCUIT

Recall that the voltage follower has the highest input resistance of any op-amp circuit [see Section (4-3.8)]. For this reason it is used to reduce voltage error caused by source loading and to isolate high-impedance sources from following circuitry. Figure 7-22 (a) shows a direct-coupled dc voltage follower in which the input voltage is directly applied to the noninverting input terminal. Bias current I_B for the op-amp is supplied through the input source resistances R_{in}.

If the input source resistance is high (or if the voltage follower is driven from high-input source resistance) because of the voltage drop across it, the voltage at the noninverting input will be in error. In other words, the output and input voltages will not be equal. To remedy this problem, an op-amp with low input bias current should be chosen as a voltage follower when working from high-input source resistances. Obviously, the best choices are op-amps specially designed as buffers or voltage followers. These op-amps have low bias currents and generally higher slew rates. For example, the LM310 has $I_B = 10$ nA maximum and SR $= 30$ V/μs.

When an ac input voltage rides on a dc level, in order to block the dc level a coupling capacitor must be used in series with the input, as shown in Figure 7-22(b). But whenever a voltage follower is ac-coupled, it is necessary to connect a bias resistor to provide a ground path for I_B. However, this bias resistor drastically reduces the input resistance of

(a)

(b)

Figure 7–22   (a) DC-coupled voltage follower. (b) AC-coupled voltage follower with input resistance bootstrapped.

the follower circuit. In fact, the input resistance is equal to the bias resistance. Therefore, to get higher input resistance, the bias resistance is *bootstrapped* as shown in Figure 7-22(b). In this circuit the input is applied through a coupling capacitor C_1 to the top of R_1, and simultaneously the output voltage is coupled through capacitor C_2 to the bottom of R_1. Since the gain of the circuit is 1, the voltage drop across R_1 is $(v_{in} - v_o)$, which is almost zero. Therefore, the current through R_1 is almost zero, and the input resistance, in turn, is incredibly high.

Voltage followers are useful in such applications as active filters, sample-and-hold circuits, and bridge circuits using transducers.

7-12 THE INTEGRATOR

A circuit in which the output voltage waveform is the integral of the input voltage waveform is the *integrator* or the *integration amplifier*. Such a circuit is obtained by using a basic inverting amplifier configuration if the feedback resistor R_F is replaced by a capacitor C_F [see Figure 7-23(a)].

The expression for the output voltage v_o can be obtained by writing Kirchhoff's current equation at node v_2:

$$i_1 = I_B + i_F$$

Since I_B is negligibly small,

$$i_1 \cong i_F$$

Recall that the relationship between current through and voltage across the capacitor is

$$i_c = C\frac{dv_c}{dt}$$

Therefore,

$$\frac{v_{in} - v_2}{R_1} = C_F\left(\frac{d}{dt}\right)(v_2 - v_o)$$

However, $v_1 = v_2 \cong 0$ because A is very large. Therefore,

$$\frac{v_{in}}{R_1} = C_F\frac{d}{dt}(-v_o)$$

The output voltage can be obtained by integrating both sides with respect to time:

$$\int_0^t \frac{v_{in}}{R_1}\,dt = \int_0^t C_F\frac{d}{dt}(-v_o)\,dt$$

$$= C_F(-v_o) + v_o|_{t=0}$$

Therefore,

$$v_o = -\frac{1}{R_1C_F}\int_0^t v_{in}\,dt + C \qquad (7\text{-}23)$$

where C is the integration constant and is proportional to the value of the output voltage v_o at time $t = 0$ seconds.

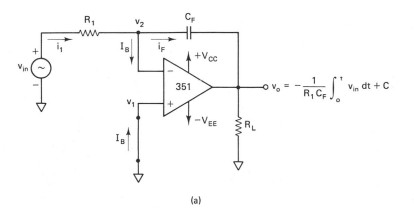

$$v_o = -\frac{1}{R_1 C_F} \int_o^t v_{in}\, dt + C$$

(a)

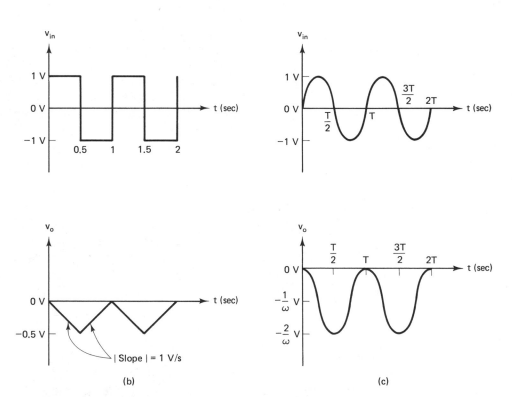

(b)   (c)

Figure 7–23 (a) The integrator circuit. (b) and (c) Input and ideal output waveforms using a sine wave and square wave, respectively. $R_1 C_F = 1$ second and $V_{ooT} = 0$ V assumed.

Equation (7-23) indicates that the output voltage is directly proportional to the negative integral of the input voltage and inversely proportional to the time constant $R_1 C_F$. For example, if the input is a sine wave, the output will be a cosine wave; or if the input is a square wave, the output will be a triangular wave, as shown in Figure 7-23(b) and (c), respectively. Note that these waveforms are drawn with the assumption that $R_1C_F = 1$ second and $V_{ooT} = 0$ V, that is, $C = 0$.

When $v_{in} = 0$, the integrator of Figure 7-23(a) works as an open-loop amplifier. This is because the capacitor C_F acts as an open circuit ($X_{CF} = \infty$) to the input offset voltage V_{io}. In other words, the input offset voltage V_{io} and the part of the input current charging capacitor C_F produce the error voltage at the output of the integrator. Therefore, in the practical integrator shown in Figure 7-25, to reduce the error voltage at the output, a resistor R_F is connected across the feedback capacitor C_F. Thus, R_F limits the low-frequency gain and hence minimizes the variations in the output voltage.

The frequency response of the basic integrator is shown in Figure 7-24. In this figure, f_b is the frequency at which the gain is 0 dB and is given by

$$f_b = \frac{1}{2\pi R_1 C_F} \tag{7-24}$$

For the derivation of Equation (7-24), refer to Appendix C.

Both the stability and the low-frequency roll-off problems can be corrected by the addition of a resistor R_F as shown in the *practical integrator* of Figure 7-25. The frequency response of the practical integrator is shown in Figure 7-24 by a dashed line. In this figure, f is some relative operating frequency, and for frequencies f to f_a the gain R_F/R_1 is constant. However, after f_a the gain decreases at a rate of 20 dB/decade. In other words, between f_a and f_b the circuit of Figure 7-25 acts as an integrator. The gain-limiting frequency f_a is given by

$$f_a = \frac{1}{2\pi R_F C_F} \tag{7-25}$$

Refer to Appendix C for derivation.

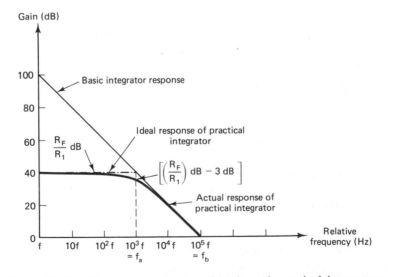

Figure 7–24 Frequency response of basic and practical integrators. $f_a = 1/(2\pi R_F C_F)$ and $f_b = 1/(2\pi R_1 C_F)$.

Figure 7–25   Practical integrator.

Generally, the value of f_a and in turn $R_1 C_F$ and $R_F C_F$ values should be selected such that $f_a < f_b$. For example, if $f_a = f_b/10$, then $R_F = 10R_1$. In fact, the input signal will be integrated properly if the time period T of the signal is larger than or equal to $R_F C_F$. That is,

$$T \geq R_F C_F \tag{7-26}$$

where

$$R_F C_F = \frac{1}{2\pi f_a}$$

The integrator is most commonly used in analog computers and analog-to-digital (ADC) and signal-waveshaping circuits.

EXAMPLE 7-15

In the circuit of Figure 7-23, $R_1 C_F = 1$ second, and the input is a step (dc) voltage, as shown in Figure 7-26(a). Determine the output voltage and sketch it. Assume that the op-amp is initially nulled.

SOLUTION   The input function is constant beginning at $t = 0$ seconds. That is, $V_{in} = 2$ V for $0 \leq t \leq 4$. Therefore, using Equation (7-23),

$$V_o = -\int_0^{t=4} 2\,dt$$

$$= -\left[\int_0^1 2\,dt + \int_1^2 2\,dt + \int_2^3 2\,dt + \int_3^4 2\,dt \right]$$

$$= -(2 + 2 + 2 + 2) = -8 \text{ V}$$

The output voltage waveform is drawn in Figure 7-26(b); the waveform is called a *ramp function*. The slope of the ramp is -2 V/s. Thus, with a constant voltage applied at the input, the integrator gives a ramp at the output.

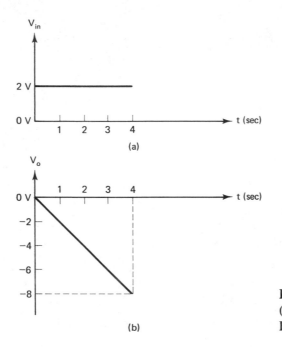

Figure 7–26   (a) Input and (b) output waveforms for Example 7–15.

7-13 THE DIFFERENTIATOR

Figure 7-27(a) shows the *differentiator* or *differentiation amplifier*. As its name implies, the circuit performs the mathematical operation of differentiation; that is, the output waveform is the derivative of the input waveform. The differentiator may be constructed from a basic inverting amplifier if an input resistor R_1 is replaced by a capacitor C_1.

The expression for the output voltage can be obtained from Kirchhoff's current equation written at node v_2 as follows:

$$i_C = I_B + i_F$$

Since $I_B \cong 0$,

$$i_C = i_F$$

$$C_1 \frac{d}{dt}(v_{in} - v_2) = \frac{v_2 - v_o}{R_F}$$

But $v_1 = v_2 \cong 0$ V, because A is very large. Therefore,

$$C_1 \frac{dv_{in}}{dt} = -\frac{v_o}{R_F}$$

or

$$v_o = -R_F C_1 \frac{dv_{in}}{dt} \qquad (7\text{-}27)$$

Thus the output v_o is equal to the $R_F C_1$ times the negative instantaneous rate of change of the input voltage v_{in} with time. Since the differentiator performs the reverse of the

Figure 7–27   Basic differentiator. (a) Circuit. (b) Frequency response.

integrator's function, a cosine wave input will produce a sine wave output, or a triangular input will produce a square wave output. However, the differentiator of Figure 7-27(a) will not do this because it has some practical problems. The gain of the circuit (R_F/X_{C1}) increases with increase in frequency at a rate of 20 dB/decade. This makes the circuit unstable. Also, the input impedance X_{C1} decreases with increase in frequency, which makes the circuit very susceptible to high-frequency noise. When amplified, this noise can completely override the differentiated output signal. The frequency response of the basic differentiator is shown in Figure 7-27(b). In this figure, f_a is the frequency at which the gain is 0 dB and is given by

$$f_a = \frac{1}{2\pi R_F C_1} \tag{7-28}$$

Also, f_c is the unity gain–bandwidth of the op-amp, and f is some relative operating frequency. For the derivation of Equation (7-28), refer to Appendix C.

Both the stability and the high-frequency noise problems can be corrected by the addition of two components: R_1 and C_F, as shown in Figure 7-28(a). This circuit is a *practical differentiator*, the frequency response of which is shown in Figure 7-27(b) by a dashed line. From frequency f to f_b, the gain increases at 20 dB/decade. However, after f_b the gain decreases at 20 dB/decade. This 40-dB/decade change in gain is caused by the R_1C_1 and R_FC_F combinations. The gain-limiting frequency f_b is given by

$$f_b = \frac{1}{2\pi R_1 C_1} \tag{7-29}$$

where $R_1C_1 = R_F C_F$. For the derivation of Equation (7-29), refer to Appendix C. Thus $R_1 C_1$ and $R_F C_F$ help to reduce significantly the effect of high-frequency input, amplifier noise, and offsets. Above all, it makes the circuit more stable by preventing the increase in gain with frequency. Generally, the value of f_b and in turn $R_1 C_1$ and $R_F C_F$ values should be selected such that

$$f_a < f_b < f_c \tag{7-30}$$

where

$$f_a = \frac{1}{2\pi R_F C_1}$$

$$f_b = \frac{1}{2\pi R_1 C_1} = \frac{1}{2\pi R_F C_F}$$

$$f_c = \text{unity gain–bandwidth}$$

The input signal will be differentiated properly if the time period T of the input signal is larger than or equal to $R_F C_1$. That is,

$$T \geq R_F C_1$$

Figure 7-28(b) and (c) show the sine wave and square wave inputs and resulting differentiated outputs, respectively, for the practical differentiator. A workable differentiator can be designed by implementing the following steps (see Example 7-16):

1. Select f_a equal to the highest frequency of the input signal to be differentiated. Then, assuming a value of $C_1 < 1\ \mu\text{F}$, calculate the value of R_F.
2. Choose $f_b = 20f_a$ and calculate the values of R_1 and C_F so that $R_1 C_1 = R_F C_F$.

The differentiator is most commonly used in waveshaping circuits to detect high-frequency components in an input signal and also as a rate-of-change detector in FM modulators.

EXAMPLE 7-16

(a) Design a differentiator to differentiate an input signal that varies in frequency from 10 Hz to about 1 kHz.
(b) If a sine wave of 1 V peak at 1000 Hz is applied to the differentiator of part (a), draw its output waveform.

$$v_c = -R_F C_1 \frac{dv_{in}}{dt}$$

if $R_F C_1 > R_1 C_1$ or $R_F C_F$

(a)

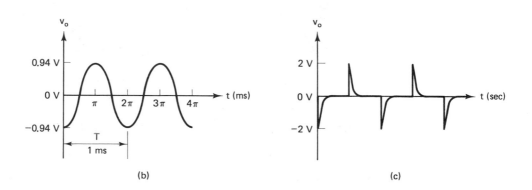

(b)

(c)

Figure 7–28   Practical differentiator. (a) Circuit. (b) Sine wave input and resulting cosine wave output. (c) Square wave input and resulting spike output.

SOLUTION   (a) To design a differentiator, we simply follow the steps outlined previously:

1. $f_a = 1$ kHz

$$= \frac{1}{2\pi R_F C_1} \quad \text{[from Equation (7-28)]}$$

Let $C_1 = 0.1 \; \mu\text{F}$; then

$$R_F = \frac{1}{(2\pi)(10^3)(10^{-7})} = 1.59 \; \text{k}\Omega$$

Let R_F be 1.5 kΩ.

2. $f_b = 20$ kHz

$$= \frac{1}{2\pi R_1 C_1} \quad \text{[from Equation (7-29)]}$$

Hence

$$R_1 = \frac{1}{(2\pi)(2)(10^4)(10^{-7})} = 79.5 \; \Omega$$

Let R_1 be 82 Ω. Since $R_1 C_1 = R_F C_F$,

$$C_F = \frac{(82)(10^{-7})}{1.5 \; \text{k}\Omega} \cong 0.0055 \; \mu\text{F}$$

Let C_F be 0.005 μF. Finally, $R_{OM} = R_1 \| R_F \cong 78 \; \Omega (\text{use } 82 \; \Omega)$.

The complete circuit with component values is shown in Figure 7-28(a).
(b) Since $V_P = 1$ V and $f = 1000$ Hz, the input voltage is

$$v_{\text{in}} = V_p \sin \omega t$$
$$= \sin(2\pi)(10^3)t$$

Hence, from Equation (7-27),

$$v_o = -R_F C_1 \frac{dv_{\text{in}}}{dt}$$

$$= -(1.5 \; \text{k}\Omega)(0.1 \; \mu\text{F}) \frac{d}{dt}[\sin(2\pi)(10^3)t]$$

$$= -(1.5 \; \text{k}\Omega)(0.1 \; \mu\text{F})(2\pi)(10^3)\cos[(2\pi)(10^3)t]$$

$$= -0.94 \cos[(2\pi)(10^3)t]$$

The input and differentiated output waveforms are as shown in Figure 7-28(b).

SUMMARY

1. So many different op-amps are available because no one circuit design can possibly optimize all the dc and ac parameters. To select the most suitable device for a desired application, it is therefore necessary to refer to the op-amp data book.

2. Op-amps are designed to optimize a parameter such as slew rate, bandwidth, or low power consumption. However, the use of a general-purpose op-amp such as the 741 will give satisfactory results in most applications.

3. Although a direct-coupled op-amp can amplify both ac and dc inputs, a capacitive coupling is used when ac response characteristics are needed or when it is essential to block the dc associated with the input signal. An ac amplifier can be powered by a single supply, provided that a dc level is injected at the input and removed at the output by using a voltage divider and coupling capacitor, respectively. Besides that, the output voltage swing is limited to a single supply voltage.

4. A peaking or narrow band-pass response can be obtained by using a parallel LC circuit in the feedback path of an inverting amplifier.

5. Summing, scaling, and averaging amplifiers can be constructed by using inverting, noninverting, and differential configurations.

6. The instrumentation amplifier is intended for precise, low-level signal amplification where low noise, low thermal and time drifts, and high input resistance are required. Various differential amplifier configurations can be used as instrumentation amplifiers, depending on the application requirements. When employed in conjunction with the transducer bridge and indicating meter, the instrumentation amplifier can be used for a variety of applications, such as temperature and light indicators, temperature controller, and analog weight scale.

7. The differential input and output amplifier can be used as a preamplifier and also in driving push-pull arrangements.

8. The V-to-I converter can be used in such applications as dc and ac voltmeters, diode, zener, and LED testers.

9. The I-to-V converter, on the other hand, is used with digital-to-analog circuits, as well as in testing photosensitive devices.

10. Using a bootstrap technique, the input impedance of an ac voltage follower can be increased significantly.

11. The integrator and differentiator are most commonly used in signal waveshaping applications. In addition, the integrator is used in analog computers and the differentiator is used as a rate-of-change detector in FM modulators.

QUESTIONS

7-1. Briefly explain the difference between the dc and ac amplifiers.

7-2. What are the major advantage and disadvantage of a single-supply ac amplifier?

7-3. What determines the peak frequency f_p in the peaking amplifier?

7-4. Explain the difference between **(a)** inverting and differential summing amplifiers and **(b)** inverting and noninverting averaging amplifiers.

7-5. What is an instrumentation amplifier? List three applications of the instrumentation amplifier besides those discussed in this chapter.

7-6. Explain briefly the characteristics of a thermistor, photocell, and strain gage.

7-7. What is the major difference between the photocell and the photovoltaic cell?

7-8. Explain briefly the advantages of the differential input and output amplifier.

7-9. What modifications are needed in the dc voltmeter circuit of Figure 7-16 so that it may be used to measure higher dc voltage ($> \pm 13$ V)?

7-10. The maximum ideal value of current I_o in the circuit of Figure 7-18 is equal to the output short circuit current of the op-amp. True or false?

7-11. Is there any limitation on the size of the load in the V-to-I converter with grounded load of Figure 7-19? Explain.

7-12. If a 741C is configured as an I-to-V converter, what is the lowest value of current that may be measured?

7-13. Explain the difference between the dc and ac voltage followers.

7-14. Explain the difference between the integrator and differentiator and give one application of each.

PROBLEMS

In all the following problems, assume that the op-amp is initially nulled and the supply voltages $= \pm 15$ V unless otherwise specified.

7-1. In the circuit of Figure 7-1(a), $R_F = 4.7$ kΩ, $R_1 = 100$ Ω, $R_L = 10$ kΩ, and $R_{OM} = 100$ kΩ. Determine the output voltage V_o if $V_{in} = 100$ mV.

7-2. Repeat Problem 7-1 for the circuit in Figure 7-1(b).

7-3. For the noninverting amplifier of Figure 7-3(b), $R_{in} = 50$ Ω, $C_i = 0.01$ μF, $R_1 = 1$ kΩ, $R_{OM} = 820$ Ω, $R_F = 5.6$ kΩ, and $R_L = 10$ kΩ. Determine:
(a) The gain
(b) The bandwidth of the amplifier

7-4. For the inverting amplifier with a single supply shown in Figure 7-4(a), $R_{in} = 50$ Ω, $R_1 = 10$ kΩ, $R_2 = R_3 = R_F = 100$ kΩ, and $C_i = C_o = 0.1$ μF.
(a) Determine the bandwidth of the amplifier.
(b) Determine the maximum ideal output voltage swing.
(c) Sketch the output voltage waveforms v_o' and v_o if $v_{in} = 200$ mV peak sine wave at 1 kHz.

7-5. The peaking amplifier of Figure 7-5(a) has the following values: $R_1 = 1$ kΩ, $L = 100$ μH with 3-Ω internal resistance, $C = 0.01$ μF, $R_F = 6.8$ kΩ, $R_{OM} = 820$ Ω, and $R_L = 10$ kΩ. Determine:
(a) The peak frequency f_p
(b) The gain of the amplifier at f_p
(c) The bandwidth of the amplifier

7-6. In the circuit of Figure 7-6, $V_a = 100$ mV, $V_b = -200$ mV, $V_c = 300$ mV, $R_a = 3$ kΩ, $R_b = 2.2$ kΩ, $R_c = 1$ kΩ, $R_F = 4.7$ kΩ, $R_{OM} = 470$ Ω, and $R_L = 10$ kΩ.

(a) Determine the output voltage V_o.

(b) Identify the circuit from its operation.

7-7. Repeat Problem 7-6 with $R_a = R_b = R_c = 1$ kΩ.

7-8. The circuit of Figure 7-7 is to be used as an averaging amplifier with the following specifications: $V_a = V_b = 1.5$ V, $V_c = 3$ V, $R_1 = R = 1.5$ kΩ, and $V_o = 5.2$ V. Determine the required value of R_F.

7-9. In the circuit of Figure 7-9, $V_a = 1$ V, $V_b = 2$ V, $V_c = -3$ V, $V_d = -4$ V, $R = 1$ kΩ, and $R_L = 10$ kΩ. Determine the output voltage V_o.

7-10. In the circuit of Figure 7-12, $R_1 = 1.8$ kΩ, $R_F = 4.7$ kΩ, $R_A = R_B = R_C = 500$ kΩ, $V_{dc} = 10$ V, and supply voltages $= \pm 15$ V. The transducer is a thermistor with the following specifications: $R_T = 500$ kΩ at 25°C. The temperature coefficient of resistance $= -1$ kΩ/°C. If the temperature changes from 0° to 70°C, find the variations in (a) the input signal V_{ab} and (b) the output signal V_o.

7-11. The transducer in the circuit of Figure 7-12 is a CL605L photocell with the following specifications: minimum dark resistance $= 500$ kΩ, and resistance at 0.61 meter-candle (lux) $= 7.5$ kΩ. Using the circuit specifications of Problem 7-10, determine the input voltage V_{ab} and output voltage V_o (a) at a dark condition and (b) at 0.61 lux.

7-12. The circuit of Figure 7-12 uses a strain gage and has the following specifications. The unstrained resistance of each of the four elements of the strain gage is 120 Ω, $V_{dc} = 5$ V, and the op-amp supply voltages $= \pm 10$ V. $R_1 = 100$ Ω and $R_F = 47$ kΩ. Determine the output voltage V_o if the change in the resistance of each strain-gage element is 0.1 Ω.

7-13. For the strain-gage bridge circuit shown in Figure 7-13, show that

$$V_{ab} = -V_{dc}\left(\frac{\Delta R}{R}\right)$$

Assume that under the strained condition the resistance of R_{T1} and R_{T3} decreases and that of R_{T2} and R_{T4} increases by the same amount $\Delta R\,\Omega$. Also, $R_{T1} = R_{T2} = R_{T3} = R_{T4} = R$ under the unstrained condition.

7-14. In the differential amplifier of Figure 7-14(a), the desired output voltage $V_o = 5$ V. Determine the required value of differential input voltage V_{in} if $R_1 = 100$ Ω and $R_F = 22$ kΩ. The op-amp supply voltages $= \pm 10$ V.

7-15. In the circuit of Figure 7-16, $V_{in} = 2$ V and the switch is in position 1. Determine the current through the meter.

7-16. A 100-μA full-scale meter movement is used in the ac voltmeter of Figure 7-17. Determine the full-scale input voltage v_{in} if $R_1 = 4.7$ kΩ.

7-17. In the circuit of Figure 7-18, if $R_1 = 1.5$ kΩ and $V_{in} = 5$ V, determine (a) the current through the diode and (b) the output voltage V_o.

7-18. Repeat Problem 7-17 if a diode is replaced by a zener with $V_z = 5.1$ V and $I_{zk} = 1$ mA.

7-19. Determine the load current I_L in the circuit of Figure 7-19 if $R = 10$ kΩ and $V_{in} = 2$ V. For the proper operation of this circuit, the value of the load has to be less than R. Why?

7-20. Referring to the circuit of Figure 7-20, determine the output voltage V_o if D_0 through D_3 are connected to ground and D_4 through D_7 are connected to +5 V (logic 1.)

7-21. In the circuit of Figure 7-21, $V_{dc} = 10$ V, the dark resistance of the CL505L is 100 kΩ, and $R_F = 10$ kΩ. Determine the output voltage for the dark condition. What is the function of capacitor C?

7-22. In the integrator circuit of Figure 7-25, the input is a sine wave with a peak-to-peak amplitude of 5 V at 1 kHz. Draw the output voltage waveform if $R_1 C_F = 0.1$ ms and $R_F = 10R_1$. Assume that the voltage across C_F is initially zero.

7-23. In the differentiator circuit of Figure 7-28(a), the input is a sine wave with a peak-to-peak amplitude of 3 V at 200 Hz. Sketch the output voltage waveform.

7-24. Design a differentiator that will differentiate an input signal with $f_{max} = 100$ Hz.

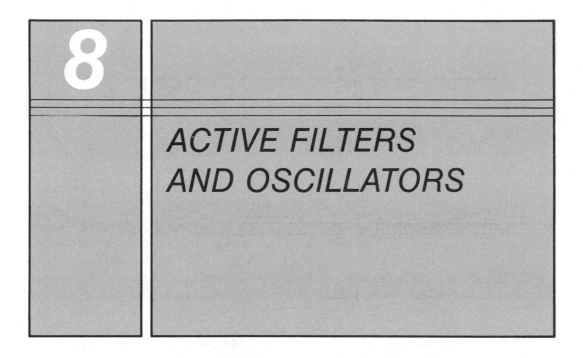

8

ACTIVE FILTERS
AND OSCILLATORS

8-1 INTRODUCTION

In Chapter 7 you saw how op-amp circuits are used to provide ac/dc amplification, perform such mathematical operations as summing, averaging, differentiation, and integration, convert *I*-to-*V* and *V*-to-*I* signals, and provide very high input impedance. This chapter presents another important field of application using op-amps: filters and oscillators. The chapter begins with the analysis and design of basic and inexpensive filter types and then discusses the various oscillator circuits. At the end of the chapter, a voltage-controlled oscillator (VCO) using the NE/SE566 integrated circuit is presented.

8-2 ACTIVE FILTERS

An electric filter is often a *frequency-selective* circuit that passes a specified band of frequencies and blocks or attenuates signals of frequencies outside this band. Filters may be classified in a number of ways:

1. Analog or digital
2. Passive or active
3. Audio (AF) or radio frequency (RF)

Analog filters are designed to process analog signals, while *digital* filters process analog signals using digital techniques. Depending on the type of elements used in their construction, filters may be classified as passive or active. Elements used in *passive* filters are resistors, capacitors, and inductors. *Active* filters, on the other hand, employ transistors or op-amps in addition to the resistors and capacitors. The type of element used dictates the operating frequency range of the filter. For example, *RC* filters are commonly used for audio or low-frequency operation, whereas *LC* or crystal filters are employed at RF or high frequencies. Especially because of their high *Q* value (figure of merit), the crystals provide more stable operation at higher frequencies.

First, this chapter presents the analysis and design of analog active-*RC* (audio-frequency) filters using op-amps. In the audio frequencies, inductors are often not used because they are very large, costly, and may dissipate more power. Inductors also emit magnetic fields.

An active filter offers the following advantages over a passive filter:

1. *Gain and frequency adjustment flexibility.* Since the op-amp is capable of providing a gain, the input signal is not attenuated as it is in a passive filter. In addition, the active filter is easier to tune or adjust.

2. *No loading problem.* Because of the high input resistance and low output resistance of the op-amp, the active filter does not cause loading of the source or load.

3. *Cost.* Typically, active filters are more economical than passive filters. This is because of the variety of cheaper op-amps and the absence of inductors.

Although active filters are most extensively used in the field of communications and signal processing, they are employed in one form or another in almost all sophisticated electronic systems. Radio, television, telephone, radar, space satellites, and biomedical equipment are but a few systems that employ active filters.

The most commonly used filters are these:

1. Low-pass filter
2. High-pass filter
3. Band-pass filter
4. Band-reject filter
5. All-pass filter

Each of these filters uses an op-amp as the active element and resistors and capacitors as the passive elements. Although the 741 type op-amp works satisfactorily in these filter circuits, high-speed op-amps such as the LM318 or ICL8017 improve the filter's performance through their increased slew rates and higher unity gain–bandwidths.

Figure 8-1 shows the frequency response characteristics of the five types of filters. The ideal response is shown by dashed curves, while the solid lines indicate the practical filter response. A low-pass filter has a constant gain from 0 Hz to a high cutoff frequency f_H. Therefore, the bandwidth is also f_H. At f_H the gain is down by 3 dB; after that ($f > f_H$) it decreases with the increase in input frequency. The frequencies between 0 Hz and f_H are known as the *passband* frequencies, whereas the range of frequencies, those beyond f_H, that are attenuated includes the *stopband* frequencies.

Figure 8-1(a) shows the frequency response of the low-pass filter. As indicated by the dashed line, an *ideal* filter has a zero loss in its passband and infinite loss in its stopband. Unfortunately, ideal filter response is not practical because linear networks cannot produce the discontinuities. However, it is possible to obtain a practical response

Figure 8–1 Frequency response of the major active filters. (a) Low pass. (b) High pass. (c) Band pass. (d) Band reject. (e) Phase shift between input and output voltages of an all-pass filter.

that approximates the ideal response by using special design techniques, as well as precision component values and high-speed op-amps.

Butterworth, Chebyshev, and Cauer filters are some of the most commonly used practical filters that approximate the ideal response. The key characteristic of the Butterworth filter is that it has a flat passband as well as stopband. For this reason, it is sometimes called a *flat-flat* filter. The Chebyshev filter has a ripple passband and flat stopband, while the Cauer filter has a ripple passband and a ripple stopband. Generally, the Cauer filter gives the best stopband response among the three. Because of their simplicity of design, the low-pass and high-pass Butterworth filters are discussed here.

Figure 8-1(b) shows a high-pass filter with a stopband $0 < f < f_L$ and a passband $f > f_L \cdot f_L$ is the low cutoff frequency, and f is the operating frequency. A band-pass filter has a passband between two cutoff frequencies f_H and f_L, where $f_H > f_L$, and two stopbands: $0 < f < f_L$ and $f > f_H$. The bandwidth of the band-pass filter, therefore, is equal to $f_H - f_L$. The band-reject filter performs exactly opposite to the band-pass; that is, it has a bandstop between two cutoff frequencies f_H and f_L and two passbands: $0 < f < f_L$ and $f > f_H$. The band-reject is also called a *band-stop* or *band-elimination* filter. The frequency responses of band-pass and band-reject filters are shown in Figure 8-1(c) and (d), respectively. In these figures, f_C is called the center frequency since it is approximately at the center of the passband or stopband.

Figure 8-1(e) shows the phase shift between input and output voltages of an all-pass filter. This filter passes all frequencies equally well; that is, output and input voltages are equal in amplitude for all frequencies, with the phase shift between the two a function of frequency. The highest frequency up to which the input and output amplitudes remain equal is dependent on the unity gain–bandwidth of the op-amp. At this frequency, however, the phase shift between the input and output is maximum.

Before proceeding with specific filter types, let us reexamine the filter characteristics, especially in the stopband region. As shown in Figure 8-1(a)–(d), the actual response curves of the filters in the stopband either steadily decrease or increase or both with increase in frequency. The rate at which the gain of the filter changes in the stopband is determined by the order of the filter. For example, for the first-order low-pass filter the gain rolls off at the rate of 20 dB/decade in the stopband, that is, for $f > f_H$; on the other hand, for the second-order low-pass filter the roll-off rate is 40 dB/decade; and so on. By contrast, for the first-order high-pass filter the gain increases at the rate of 20 dB/decade in the stopband, that is, until $f = f_L$; the increase is 40 dB/decade for the second-order high-pass filter; and so on.

8-3 FIRST-ORDER LOW-PASS BUTTERWORTH FILTER

Figure 8-2 shows a first-order low-pass Butterworth filter that uses an RC network for filtering. Note that the op-amp is used in the noninverting configuration; hence it does not load down the RC network. Resistors R_1 and R_F determine the gain of the filter.

According to the voltage-divider rule, the voltage at the noninverting terminal (across capacitor C) is

$$v_1 = \frac{-jX_C}{R - jX_C} v_{\text{in}} \tag{8-1a}$$

where

$$j = \sqrt{-1} \quad \text{and} \quad -jX_C = \frac{1}{j2\pi fC}$$

Figure 8–2   First-order low-pass Butterworth filter. (a) Circuit. (b) Frequency response.

Simplifying Equation (8-1a), we get

$$v_1 = \frac{v_{in}}{1 + j2\pi fRC}$$

and the output voltage

$$v_o = \left(1 + \frac{R_F}{R_1}\right)v_1$$

That is,

$$v_o = \left(1 + \frac{R_F}{R_1}\right)\frac{v_{in}}{1 + j2\pi fRC}$$

or

$$\frac{v_o}{v_{in}} = \frac{A_F}{1 + j(f/f_H)} \tag{8-1b}$$

where $\dfrac{v_o}{v_{in}}$ = gain of the filter as a function of frequency

$A_F = 1 + \dfrac{R_F}{R_1}$ = passband gain of the filter

f = frequency of the input signal

$f_H = \dfrac{1}{2\pi RC}$ = high cutoff frequency of the filter

The gain magnitude and phase angle equations of the low-pass filter can be obtained by converting Equation (8-1b) into its equivalent polar form, as follows:

$$\left|\frac{v_o}{v_{in}}\right| = \frac{A_F}{\sqrt{1 + (f/f_H)^2}} \tag{8-2a}$$

$$\phi = -\tan^{-1}\left(\frac{f}{f_H}\right) \tag{8-2b}$$

where ϕ is the phase angle in degrees.

The operation of the low-pass filter can be verified from the gain magnitude equation, (8-2a):

1. At very low frequencies, that is, $f < f_H$,

$$\left|\frac{v_o}{v_{in}}\right| \cong A_F$$

2. At $f = f_H$,

$$\left|\frac{v_o}{v_{in}}\right| = \frac{A_F}{\sqrt{2}} = 0.707A_F$$

3. At $f > f_H$,

$$\left|\frac{v_o}{v_{in}}\right| < A_F$$

Thus the low-pass filter has a constant gain A_F from 0 Hz to the high cutoff frequency f_H. At f_H the gain is $0.707A_F$, and after f_H it decreases at a constant rate with an increase in frequency [see Figure 8-2(b)]. That is, when the frequency is increased tenfold (one decade), the voltage gain is divided by 10. In other words, the gain decreases 20 dB (= 20 log 10) each time the frequency is increased by 10. Hence the rate at which the gain rolls off after f_H is 20 dB/decade or 6 dB/octave, where octave signifies a twofold increase in frequency. The frequency $f = f_H$ is called the *cutoff frequency* because the gain of the filter at this frequency is down by 3 dB (= 20 log 0.707) from 0 Hz. Other equivalent terms for cutoff frequency are *−3 dB frequency, break frequency,* or *corner frequency.*

8-3.1 Filter Design

A low-pass filter can be designed by implementing the following steps:

1. Choose a value of high cutoff frequency f_H.
2. Select a value of C less than or equal to 1 μF. Mylar or tantalum capacitors are recommended for better performance.
3. Calculate the value of R using

$$R = \frac{1}{2\pi f_H C}$$

4. Finally, select values of R_1 and R_F dependent on the desired passband gain A_F using

$$A_F = 1 + \frac{R_F}{R_1}$$

8-3.2 Frequency Scaling

Once a filter is designed, there may sometimes be a need to change its cutoff frequency. The procedure used to convert an original cutoff frequency f_H to a new cutoff frequency

f_H' is called *frequency scaling*. Frequency scaling is accomplished as follows. To change a high cutoff frequency, multiply R or C, but not both, by the ratio of the original cutoff frequency to the new cutoff frequency. In filter design the needed values of R and C are often not standard. Besides, a variable capacitor C is not commonly used. Therefore, choose a standard value of capacitor, and then calculate the value of resistor for a desired cutoff frequency. This is because for a nonstandard value of resistor a potentiometer can be used (see Examples 8-1 and 8-2).

EXAMPLE 8-1

Design a low-pass filter at a cutoff frequency of 1 kHz with a passband gain of 2.

SOLUTION   Follow the preceding design steps.
1. $f_H = 1$ kHz.
2. Let $C = 0.01 \ \mu F$.
3. Then $R = 1/(2\pi)(10^3)(10^{-8}) = 15.9 \ k\Omega$. (Use a 20-k$\Omega$ potentiometer.)
4. Since the passband gain is 2, R_1 and R_F must be equal. Therefore, let $R_1 = R_F = 10 \ k\Omega$. The complete circuit with component values is shown in Figure 8-2(a).

EXAMPLE 8-2

Using the frequency scaling technique, convert the 1-kHz cutoff frequency of the low-pass filter of Example 8-1 to a cutoff frequency of 1.6 kHz.

SOLUTION   To change a cutoff frequency from 1 kHz to 1.6 kHz, we multiply the 15.9-kΩ resistor by

$$\frac{\text{original cutoff frequency}}{\text{new cutoff frequency}} = \frac{1 \text{ kHz}}{1.6 \text{ kHz}} = 0.625$$

Therefore, new resistor $R = (15.9 \ k\Omega)(0.625) = 9.94 \ k\Omega$. However, 9.94 k$\Omega$ is not a standard value. Therefore, use $R = 10 \ k\Omega$ potentiometer and adjust it to 9.94 kΩ. Thus the new cutoff frequency is

$$f_H = \frac{1}{(2\pi)(0.01 \ \mu F)(9.94 \ k\Omega)}$$

$$= 1.6 \text{ kHz}$$

EXAMPLE 8-3

Plot the frequency response of the low-pass filter of Example 8-1.

SOLUTION   To plot the frequency response, we have to use Equation (8-2a). The data of Table 8-1 are, therefore, obtained by substituting various values for f in this equation. Equation (8-2a) will be repeated here for convenience:

$$\left| \frac{v_o}{v_{in}} \right| = \frac{A_F}{\sqrt{1 + (f/f_H)^2}}$$

where $A_F = 2$ and $f_H = 1$ kHz. The data of Table 8-1 are plotted as shown in Figure 8-3.

TABLE 8-1   FREQUENCY RESPONSE DATA FOR EXAMPLE 8-3

Input frequency, f (Hz)	Gain magnitude, $\|v_0/v_{in}\|$	Magnitude (dB) = $20 \log\|v_0/v_{in}\|$
10	2	6.02
100	1.99	5.98
200	1.96	5.85
700	1.64	4.29
1,000	1.41	3.01
3,000	0.63	−3.98
7,000	0.28	−10.97
10,000	0.20	−14.02
30,000	0.07	−23.53
100,000	0.02	−33.98

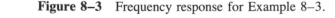

Figure 8–3   Frequency response for Example 8–3.

8-4 SECOND-ORDER LOW-PASS BUTTERWORTH FILTER

A stop-band response having a 40-dB/decade roll-off is obtained with the second-order low-pass filter. A first-order low-pass filter can be converted into a second-order type simply by using an additional RC network, as shown in Figure 8-4.

(a)   (b)

Figure 8–4   Second-order low-pass Butterworth filter. (a) Circuit. (b) Frequency response.

Second-order filters are important because higher-order filters can be designed using them. The gain of the second-order filter is set by R_1 and R_F, while the high cutoff frequency f_H is determined by R_2, C_2, R_3, and C_3, as follows:

$$f_H = \frac{1}{2\pi\sqrt{R_2 R_3 C_2 C_3}} \tag{8-3}$$

For the derivation of f_H, refer to Appendix C.

Furthermore, for a second-order low-pass Butterworth response, the voltage gain magnitude equation is

$$\left|\frac{v_o}{v_{in}}\right| = \frac{A_F}{\sqrt{1 + (f/f_H)^4}} \tag{8-4}$$

where $A_F = 1 + \dfrac{R_F}{R_1}$ = passband gain of the filter

f = frequency of the input signal (Hz)

$f_H = \dfrac{1}{2\pi\sqrt{R_2 R_3 C_2 C_3}}$ = high cutoff frequency (Hz)

8-4.1 Filter Design

Except for having twice the roll-off rate in the stopband, the frequency response of the second-order low-pass filter is identical to that of the first-order type. Therefore, the design steps of the second-order filter are identical to those of the first-order filter, as follows:

1. Choose a value for the high cutoff frequency f_H.
2. To simplify the design calculations, set $R_2 = R_3 = R$ and $C_2 = C_3 = C$. Then choose a value of $C \le 1\ \mu F$.

3. Calculate the value of R using Equation (8-3):

$$R = \frac{1}{2\pi f_H C}$$

4. Finally, because of the equal resistor ($R_2 = R_3$) and capacitor ($C_2 = C_3$) values, the passband voltage gain $A_F = (1 + R_F/R_1)$ of the second-order low-pass filter has to be equal to 1.586. That is, $R_F = 0.586R_1$. This gain is necessary to guarantee Butterworth response. Hence choose a value of $R_1 \leq 100$ kΩ and calculate the value of R_F.

As outlined in Section 8-3.2, the frequency scaling method of the first-order filter is also applicable to the second-order low-pass filter.

EXAMPLE 8-4

(a) Design a second-order low-pass filter at a high cutoff frequency of 1 kHz.
(b) Draw the frequency response of the network in part (a).

SOLUTION   (a) To design the second-order low-pass filter, simply follow the steps just presented:

1. $f_H = 1$ kHz.
2. Let $C_2 = C_3 = 0.0047$ μF.
3. Then

$$R_2 = R_3 = \frac{1}{(2\pi)(10^3)(47)(10^{-10})} = 33.86 \text{ k}\Omega$$

(Use $R_2 = R_3 = 33$ kΩ.)

4. Since R_F must be equal to $0.586R_1$, let R_1 equal 27 kΩ. Therefore,

$$R_F = (0.586)(27 \text{ k}\Omega) = 15.82 \text{ k}\Omega$$

(Use $R_F = 20$ kΩ pot.) Thus the required components are

$$R_2 = R_3 = 33 \text{ k}\Omega$$

$$C_2 = C_3 = 0.0047 \text{ μF}$$

$$R_1 = 27 \text{ k}\Omega \quad \text{and} \quad R_F = 15.8 \text{ k}\Omega \quad (20\text{k} - \Omega \text{ pot})$$

Another method to design the second-order low-pass filter is to use the same values of resistor and capacitor obtained for the first-order filter in Example 8-1. This is because the cutoff frequency of both the seond-order and first-order filters is 1 kHz. Therefore, we may use $R_2 = R_3 = 15.9$ kΩ and $C_2 = C_3 = 0.01$ μF. However, the values of R_1 and R_F must be chosen such that $R_F = 0.586R_1$. Therefore, use $R_1 = 27$ kΩ and $R_F = 15.8$ kΩ.

(b) The frequency response data shown in Table 8-2 are obtained from the magnitude equation, (8-4), by substituting various values from 10 Hz to 100 kHz for f. Equation (8-4) is repeated here for convenience:

TABLE 8-2 FREQUENCY RESPONSE DATA FOR EXAMPLE 8-4

Frequency, f (Hz)	Gain magnitude, $\lvert v_o/v_{in} \rvert$	Magnitude (dB) = $20 \log \lvert v_o/v_{in} \rvert$
10	1.59	4.01
100	1.59	4.01
200	1.58	4.00
700	1.42	3.07
1,000	1.12	1.00
3,000	0.18	−15.13
7,000	0.03	−29.80
10,000	0.02	−35.99
30,000	1.76×10^{-3}	−55.08
100,000	1.59×10^{-4}	−75.99

$$\left| \frac{v_o}{v_{in}} \right| = \frac{A_F}{\sqrt{1 + (f/f_H)^4}}$$

where $A_F = 1.586$ and $f_H = 1$ kHz. The frequency response of the second-order low-pass filter of Example 8-4 is shown in Figure 8-5.

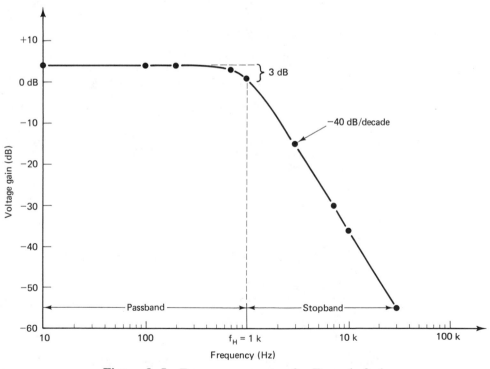

Figure 8–5 Frequency response for Example 8–4.

8-5 FIRST-ORDER HIGH-PASS BUTTERWORTH FILTER —————

High-pass filters are often formed simply by interchanging frequency-determining resistors and capacitors in low-pass filters. That is, a first-order high-pass filter is formed from a first-order low-pass type by interchanging components R and C. Similarly, a second-order high-pass filter is obtained from a second-order low-pass filter if R and C are interchanged, and so on. Figure 8-6 shows a first-order high-pass Butterworth filter with a low cutoff frequency of f_L. This is the frequency at which the magnitude of the gain is 0.707 times its passband value. Obviously, all frequencies higher than f_L are passband frequencies, with the highest frequency determined by the closed-loop bandwidth of the op-amp.

Note that the high-pass filter of Figure 8-6(a) and the low-pass filter of Figure 8-2(a) are the same circuits, except that the frequency-determining components (R and C) are interchanged.

For the first-order high-pass filter of Figure 8-6(a), the output voltage is

$$v_o = \left(1 + \frac{R_F}{R_1}\right) \frac{j2\pi fRC}{1 + j2\pi fRC} v_{in}$$

or

$$\frac{v_o}{v_{in}} = A_F\left[\frac{j(f/f_L)}{1 + j(f/f_L)}\right] \tag{8-5}$$

where $A_F = 1 + \dfrac{R_F}{R_1}$ = passband gain of the filter

f = frequency of the input signal (Hz)

$f_L = \dfrac{1}{2\pi RC}$ = low cutoff frequency (Hz)

Hence the magnitude of the voltage gain is

(a)   (b)

Figure 8–6   (a) First-order high-pass Butterworth filter. (b) Its frequency response.

$$\left|\frac{v_o}{v_{in}}\right| = \frac{A_F(f/f_L)}{\sqrt{1 + (f/f_L)^2}} \tag{8-6}$$

Since high-pass filters are formed from low-pass filters simply by interchanging R's and C's, the design and frequency scaling procedures of the low-pass filters are also applicable to the high-pass filters (see Sections 8-3.1 and 8-3.2).

EXAMPLE 8-5

(a) Design a high-pass filter at a cutoff frequency of 1 kHz with a passband gain of 2.
(b) Plot the frequency response of the filter in part (a).

SOLUTION   (a) Use the same values of R and C that were determined for the low-pass filter of Example 8-1, since $f_L = f_H = 1$ kHz. That is, $C = 0.01$ μF and $R = 15.9$ kΩ. Similarly, use $R_1 = R_F = 10$ kΩ, since $A_F = 2$.

(b) The data for the frequency response plot can be obtained by substituting for the input frequency f values from 100 Hz to 100 kHz in Equation (8-6). These data are included in Table 8-3. Equation (8-6) is repeated here for convenience:

$$\left|\frac{v_o}{v_{in}}\right| = \frac{A_F(f/f_L)}{\sqrt{1 + (f/f_L)^2}}$$

where $A_F = 2$ and $f_L = 1$ kHz. The frequency response data of Table 8-3 are plotted in Figure 8-7. In the stopband (from 100 Hz to 1 kHz) the gain increases at the rate of 20 dB/decade. However, in the passband (after $f = f_L = 1$ kHz) the gain remains constant at 6.02 dB. Moreover, the upper-frequency limit of the passband is set by the closed-loop bandwidth of the op-amp.

TABLE 8-3 FREQUENCY RESPONSE DATA FOR THE FIRST-ORDER
HIGH-PASS FILTER OF EXAMPLE 8-5

Frequency, f (Hz)	Gain magnitude, $\|v_o/v_{in}\|$	Magnitude (dB) = $20\log\|v_o/v_{in}\|$
100	0.20	−14.02
200	0.39	−8.13
400	0.74	−2.58
700	1.15	1.19
1,000	1.41	3.01
3,000	1.90	5.56
7,000	1.98	5.93
10,000	1.99	5.98
30,000	2	6.02
100,000	2	6.02

Figure 8–7   Frequency response for Example 8–5.

8-6 SECOND-ORDER HIGH-PASS BUTTERWORTH FILTER

As in the case of the first-order filter, a second-order high-pass filter can be formed from a second-order low-pass filter simply by interchanging the frequency-determining resistors and capacitors. Figure 8-8(a) shows the second-order high-pass filter.

The voltage gain magnitude equation of the second-order high-pass filter is as follows:

Figure 8–8   (a) Second-order high-pass Butterworth filter. (b) Its frequency response.

$$\left|\frac{v_o}{v_{\text{in}}}\right| = \frac{A_F}{\sqrt{1 + (f_L/f)^4}} \tag{8-7}$$

where $A_F = 1.586 =$ passband gain for the second-order Butterworth response

$\quad f =$ frequency of the input signal (Hz)

$\quad f_L =$ low cutoff frequency (Hz)

Since second-order low-pass and high-pass filters are the same circuits except that the positions of resistors and capacitors are interchanged, the design and frequency scaling procedures for the high-pass filter are the same as those for the low-pass filter.

EXAMPLE 8-6

(a) Determine the low cutoff frequency f_L of the filter shown in Figure 8-8(a).

(b) Draw the frequency response plot of the filter.

SOLUTION   (a)

$$f_L = \frac{1}{2\pi\sqrt{R_2 R_3 C_2 C_3}}$$

$$= \frac{1}{2\pi\sqrt{(33\,\text{k}\Omega)^2 (0.0047\,\mu\text{F})^2}} \cong 1\,\text{kHz}$$

(b) The frequency response data in Table 8-4 are obtained from the voltage gain magnitude equation, (8-7), which is repeated here for convenience:

$$\left|\frac{v_o}{v_{\text{in}}}\right| = \frac{A_F}{\sqrt{1 + (f_L/f)^4}}$$

TABLE 8-4   FREQUENCY RESPONSE DATA FOR SECOND-ORDER HIGH-PASS FILTER OF EXAMPLE 8-6

Input frequency, f(Hz)	Gain magnitude, $\lvert v_o/v_{\text{in}}\rvert$	Magnitude (dB) = $20\log\lvert v_o/v_{\text{in}}\rvert$
100	0.01586	−35.99
200	0.0634	−23.96
700	0.6979	−3.124
1,000	1.1215	0.9960
3,000	1.5763	3.953
7,000	1.5857	4.004
10,000	1.5859	4.006
30,000	1.5860	4.006
100,000	1.5860	4.006

where $A_F = 1.586$ and $f_L = 1$ kHz. The resulting frequency response plot is shown in Figure 8-9.

Figure 8–9   Frequency response for Example 8–6.

8-7 HIGHER-ORDER FILTERS

From the preceding discussions of filters we can conclude that in the stopband the gain of the filter changes at the rate of 20 dB/decade for first-order filters and at 40 dB/decade for second-order filters. This means that, as the order of the filter is increased, the actual stopband response of the filter approaches its ideal stopband characteristic.

Higher-order filters, such as third, fourth, fifth, and so on, are formed simply by using the first- and second-order filters. For example, a third-order low-pass filter is formed by connecting in series or cascading first- and second-order low-pass filters; a fourth-order low-pass filter is composed of two cascaded second-order low-pass sections, and so on. Although there is no limit to the order of the filter that can be formed, as the order of the filter increases, so does its size. Also, its accuracy declines, in that the difference between the actual stopband response and the theoretical stopband response increases with an increase in the order of the filter. Figure 8-10 shows third- and fourth-order low-pass Butterworth filters. Note that in the third-order filter the voltage gain of the first-order section is *one*, and that of the second-order section is *two*. On the other hand, in the fourth-order filter the gain of the first section is *1.152*, while that of the second section is *2.235*. These gain values are necessary to guarantee Butterworth response and have to remain the same regardless of the filter's cutoff frequency. Furthermore, the overall gain of the filter is equal to the product of the individual voltage gains of the filter sections. Thus the overall gain of the third-order filter is 2.0, and that of the fourth order is $(1.152)(2.235) = 2.57$.

Since the frequency-determining resistors are equal and the frequency-determining capacitors are also equal, the high cutoff frequencies of the third- and fourth-order low-pass filters in Figure 8-10(a) and (b) must also be equal. That is,

(a)

(b)

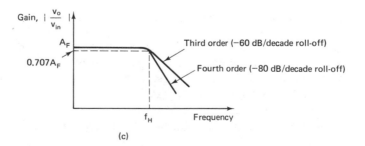

(c)

Figure 8–10   (a) Third-order and (b) fourth-order low-pass Butterworth filters.
(c) Their frequency responses. A_1 and A_2 dual op-amp: 1458/353.

$$f_H = \frac{1}{2\pi RC} \tag{8-8}$$

As with the first- and second-order filters, the third- and fourth-order high-pass filters are formed by simply interchanging the positions of the frequency-determining resistors and capacitors in the corresponding low-pass filters. The high-order filters can be designed by following the procedures outlined for the first- and second-order filters. However, note that the overall gain of the higher-order filters is *fixed* because all the frequency-determining resistors and capacitors are equal.

Generally, the minimum-order filter required depends on the application specifications. Although a higher-order filter than necessary gives a better stopband response, the higher-order type is more complex, occupies more space, and is more expensive.

8-8 BAND-PASS FILTERS

A band-pass filter has a passband between two cutoff frequencies f_H and f_L such that $f_H > f_L$. Any input frequency outside this passband is attenuated.

Basically, there are two types of band-pass filters: (1) wide band pass, and (2) narrow band pass. Unfortunately, there is no set dividing line between the two. However, we will define a filter as wide band pass if its *figure of merit* or *quality factor* $Q < 10$. On the other hand, if $Q > 10$, we will call the filter a narrow band-pass filter. Thus Q is a measure of selectivity, meaning the higher the value of Q, the more selective is the filter or the narrower its bandwidth (BW). The relationship between Q, the 3-dB bandwidth, and the center frequency f_C is given by

$$Q = \frac{f_C}{BW} = \frac{f_C}{f_H - f_L} \tag{8-9a}$$

For the wide band-pass filter the center frequency f_C can be defined as

$$f_C = \sqrt{f_H f_L} \tag{8-9b}$$

where f_H = high cutoff frequency (Hz)
   f_L = low cutoff frequency of the wide band-pass filter (Hz)
In a narrow band-pass filter, the output voltage peaks at the center frequency.

8-8.1 Wide Band-Pass Filter

A wide band-pass filter can be formed by simply cascading high-pass and low-pass sections and is generally the choice for simplicity of design and performance. To obtain a ± 20 dB/decade band-pass, first-order high-pass and first-order low-pass sections are cascaded; for a ± 40-dB/decade band-pass filter, second-order high-pass and second-order low-pass sections are connected in series, and so on. In other words, the order of the band-pass filter depends on the order of the high-pass and low-pass filter sections.

Figure 8-11 shows the ± 20-dB/decade wide band-pass filter, which is composed of first-order high-pass and first-order low-pass filters. To realize a band-pass response, however, f_H must be larger than f_L, as illustrated in Example 8-7.

(a)

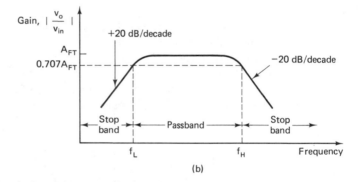

(b)

Figure 8–11   (a) ±20 dB/decade-wide band-pass filter. (b) Its frequency response. A_1 and A_2 dual op-amp: 1458/353.

EXAMPLE 8-7

(a) Design a wide band-pass filter with $f_L = 200$ Hz, $f_H = 1$ kHz, and a passband gain = 4.

(b) Draw the frequency response plot of this filter.

(c) Calculate the value of Q for the filter.

SOLUTION   (a) A low-pass filter with $f_H = 1$ kHz was designed in Example 8-1; therefore, the same values of resistors and capacitors can be used here, that is, $R' = 15.9$ kΩ and $C' = 0.01$ μF. As in the case of the high-pass filter, it can be designed by following the steps of Section 8-3.1:

1. $f_L = 200$ Hz.
2. Let $C = 0.05$ μF.
3. Then

$$R = \frac{1}{2\pi f_L C} = \frac{1}{(2\pi)(200)(5)(10^{-8})}$$
$$= 15.9 \text{ k}\Omega$$

Since the band-pass gain is 4, the gain of the high-pass as well as low-pass sections could be set equal to 2. That is, input and feedback resistors must be equal in value, say 10 kΩ each. The complete band-pass filter is shown in Figure 8-11(a).

(b) The voltage gain magnitude of the band-pass filter is equal to the product of the voltage gain magnitudes of the high-pass and low-pass filters. Therefore, from Equations (8-2a) and (8-6),

$$\left|\frac{v_o}{v_{in}}\right| = \frac{A_{FT}(f/f_L)}{\sqrt{[1 + (f/f_L)^2][1 + (f/f_H)^2]}} \tag{8-10}$$

where A_{FT} = total passband gain
f = frequency of the input signal (Hz)
f_L = low cutoff frequency (Hz)
f_H = high cutoff frequency (Hz)

Here $A_{FT} = 4$, $f_L = 200$ Hz, and $f_H = 1$ kHz. The frequency response data in Table 8-5 are obtained by substituting into Equation (8-10) the values of f from 10 Hz to 10 kHz. The frequency response plot is shown in Figure 8-12.

(c) From Equation (8-9b),

$$f_C = \sqrt{(1000)(200)} = 447.2 \text{ Hz}$$

TABLE 8-5 FREQUENCY RESPONSE DATA FOR THE BAND-PASS FILTER OF EXAMPLE 8-7

Input frequency, f(Hz)	Gain magnitude, $\|v_o/v_{in}\|$	Magnitude (dB) = $20 \log\|v_o/v_{in}\|$
10	0.1997	-13.99
30	0.5931	-4.54
100	1.780	5.01
200	2.774	8.861
447.2	3.33	10.46
700	3.151	9.969
1,000	2.774	8.861
2,000	1.780	5.001
7,000	0.5655	-4.95
10,000	0.3979	-8.004

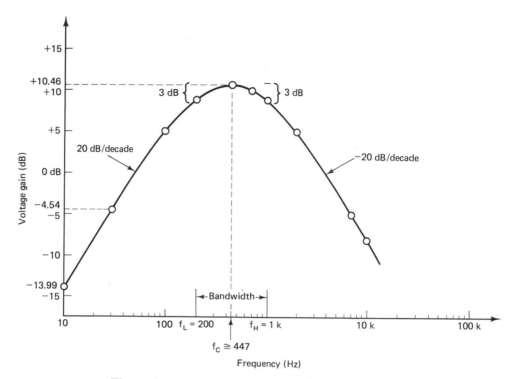

Figure 8–12   Frequency response for Example 8–7.

Substituting this value in Equation (8-9a),

$$Q = \frac{447.2}{1000 - 200} = 0.56$$

Thus Q is less than 10, as expected for the wide band-pass filter.

8-8.2 Narrow Band-Pass Filter

The narrow band-pass filter using multiple feedback is shown in Figure 8-13. As shown in this figure, the filter uses only one op-amp. Compared to all the filters discussed so far, this filter is unique in the following respects:

1. It has two feedback paths, hence the name *multiple-feedback filter*.
2. The op-amp is used in the *inverting* mode.

    Generally, the narrow band-pass filter is designed for specific values of center frequency f_C and Q or f_C and bandwidth [see Equation (8-9a)]. The circuit components are determined from the following relationships.

(a)

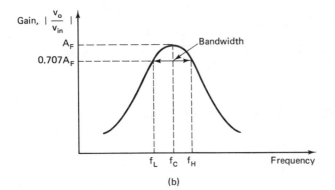

(b)

Figure 8–13 (a) Multiple-feedback narrow band-pass filter. (b) Its frequency response.

To simplify the design calculations, choose $C_1 = C_2 = C$.

$$R_1 = \frac{Q}{2\pi f_C C A_F} \qquad (8\text{-}11)$$

$$R_2 = \frac{Q}{2\pi f_C C(2Q^2 - A_F)} \qquad \simeq \frac{R_1}{A(2Q^2 - A)} \qquad (8\text{-}12)$$

$$R_3 = \frac{Q}{\pi f_C C} \qquad (8\text{-}13)$$

where A_F is the gain at f_C, given by

$$A_F = \frac{R_3}{2R_1} \qquad (8\text{-}14a)$$

The gain A_F, however, must satisfy the condition

$$A_F < 2Q^2 \qquad (8\text{-}14b)$$

Another advantage of the multiple feedback filter of Figure 8-13 is that its center frequency f_C can be changed to a new frequency f'_C without changing the gain or bandwidth. This is accomplished simply by changing R_2 to R'_2 so that

$$R'_2 = R_2\left(\frac{f_C}{f'_C}\right)^2 \tag{8-15}$$

(see Example 8-8).

EXAMPLE 8-8

(a) Design the bandpass filter shown in Figure 8-13(a) so that f_C = 1 kHz, Q = 3, and A_F = 10.
(b) Change the center frequency to 1.5 kHz, keeping A_F and the bandwidth constant.

SOLUTION

(a) Choose the values of C_1 and C_2 first and then calculate the values of R_1, R_2, and R_3 from Equations (8-11) through (8-13). Let $C_1 = C_2 = C = 0.01$ μF.

$$R_1 = \frac{3}{(2\pi)(10^3)(10^{-8})(10)} = 4.77 \text{ k}\Omega$$

$$R_2 = \frac{3}{(2\pi)(10^3)(10^{-8})[2(3)^2 - 10]} = 5.97 \text{ k}\Omega$$

$$R_3 = \frac{3}{(\pi)(10^3)(10^{-8})} = 95.5 \text{ k}\Omega$$

Use $R_1 = 4.7$ kΩ, $R_2 = 6.2$ kΩ, and $R_3 = 100$ kΩ.

(b) Using Equation (8-15), the value of R'_2 required to change the center frequency from 1 kHz to 1.5 kHz is

$$R'_2 = (5.97 \text{ k}\Omega)\left(\frac{1 \text{ k}}{1.5 \text{ k}}\right)^2 = 2.65 \text{ k}\Omega$$

(Use $R'_2 = 2.7$ kΩ.)

8-9 BAND-REJECT FILTERS

The band-reject filter is also called a *band-stop* or *band-elimination* filter. In this filter, frequencies are attenuated in the stopband while they are passed outside this band, as shown in Figure 8-1(d). As with band-pass filters, the band-reject filters can also be classified as (1) wide band-reject or (2) narrow band-reject. The narrow band-reject filter is commonly called the *notch filter*. Because of its higher Q (>10), the bandwidth of the narrow band-reject filter is much smaller than that of the wide band-reject filter.

8-9.1 Wide Band-Reject Filter

Figure 8-14(a) shows a wide band-reject filter using a low-pass filter, a high-pass filter, and a summing amplifier. To realize a band-reject response, the low cutoff frequency f_L of the high-pass filter must be larger than the high cutoff frequency f_H of the low-pass filter. In addition, the passband gain of both the high-pass and low-pass sections must be equal (see Example 8-9). The frequency response of the wide band-reject filter is shown in Figure 8-14(b).

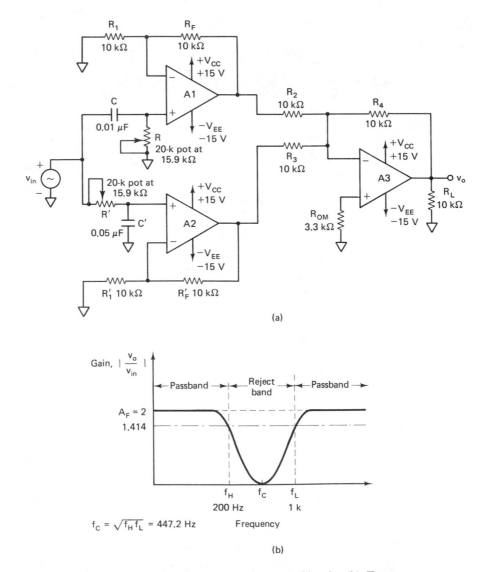

(a)

(b)

Figure 8–14 Wide band-reject filter. (a) Circuit. (b) Frequency response. For A_1, A_2, and A_3 use quad op-amp $\mu AF774/MC34004$.

EXAMPLE 8-9

Design a wide band-reject filter having $f_H = 200$ Hz and $f_L = 1$ kHz.

SOLUTION   In Example 8-7, a wide band-pass filter was designed with $f_L = 200$ Hz and $f_H = 1$ kHz. In this example these band frequencies are interchanged, that is, $f_L = 1$ kHz and $f_H = 200$ Hz. This means that we can use the same components as in Example 8-7, but interchanged between high-pass and low-pass sections. Therefore, for the low-pass section, $R' = 15.9$ kΩ and $C' = 0.05$ μF, while for the high-pass section

$$R = 15.9 \text{ k}\Omega \quad \text{and} \quad C = 0.01 \ \mu\text{F}$$

Since there is no restriction on the passband gain, use a gain of 2 for each section. Hence let

$$R_1 = R_F = R'_1 = R'_F = 10 \text{ k}\Omega$$

Furthermore, the gain of the summing amplifier is set at 1; therefore,

$$R_2 = R_3 = R_4 = 10 \text{ k}\Omega$$

Finally, the value of $R_{OM} = R_2 \| R_3 \| R_4 \cong 3.3$ kΩ.

The complete circuit is shown in Figure 8-14(a), and its response is shown in Figure 8-14(b). The voltage gain changes at the rate of 20 dB/decade above f_H and below f_L, with a maximum attenuation occurring at f_C.

8-9.2 Narrow Band-Reject Filter

The narrow band-reject filter, often called the *notch filter,* is commonly used for the rejection of a single frequency such as the 60-Hz power line frequency hum. The most commonly used notch filter is the *twin-T* network shown in Figure 8-15(a). This is a *passive filter* composed of two T-shaped networks. One T network is made up of two resistors and a capacitor, while the other uses two capacitors and a resistor. The *notch-out* frequency is the frequency at which maximum attenuation occurs; it is given by

$$f_N = \frac{1}{2\pi RC} \tag{8-16}$$

Unfortunately, the passive twin-T network has a relatively low figure of merit Q. However, the Q of the network can be increased significantly if it is used with the voltage follower as shown in Figure 8-15(b). Here the output of the voltage follower is fed back to the junction of $R/2$ and $2C$. The frequency response of the active notch filter of Figure 8-15(b) is shown in Figure 8-15(c). The most common use of notch filters is in communications and biomedical instruments for eliminating undesired frequencies. To design an active notch filter for a specific notch-out frequency f_N, choose the value of $C \leq 1$ μF and then calculate the required value of R from Equation (8-16). For the best response, the circuit components should be very close to their indicated values.

EXAMPLE 8-10

Design a 60-Hz active notch filter.

(a)

(b)

(c)

Figure 8–15 (a) Twin-T notch filter. (b) Active notch filter. (c) Frequency response of the active notch filter.

SOLUTION Let $C = 0.068\ \mu F$. Then, from Equation (8-16), the value of R is

$$R = \frac{1}{2\pi f_N C} = \frac{1}{(2\pi)\,(60)\,(68)\,(10^{-9})} = 39.01\ k\Omega$$

(Use 39 kΩ.) For $R/2$, parallel two 39-kΩ resistors; for the $2C$ component, parallel two 0.068-μF capacitors.

8-10 ALL-PASS FILTER

As the name suggests, an all-pass filter passes all frequency components of the input signal without attenuation, while providing predictable phase shifts for different frequencies of the input signal. When signals are transmitted over transmission lines, such as telephone wires, they undergo change in phase. To compensate for these phase changes, all-pass filters are required. The all-pass filters are also called *delay equalizers* or *phase cor-*

(a)

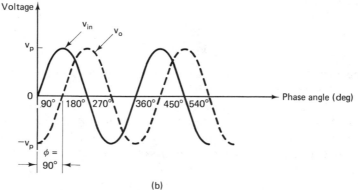

(b)

Figure 8–16   All-pass filter. (a) Circuit. (b) Phase shift between input and output voltages.

rectors. Figure 8-16(a) shows an all-pass filter wherein $R_F = R_1$. The output voltage v_o of the filter can be obtained by using the superposition theorem:

$$v_o = -v_{in} + \frac{-jX_C}{R - jX_C} v_{in}(2) \qquad (8\text{-}17)$$

But $-j = 1/j$ and $X_C = 1/2\pi fC$. Therefore, substituting for X_C and simplifying, we get

$$v_o = v_{in}\left(-1 + \frac{2}{j2\pi fRC + 1}\right)$$

or

$$\frac{v_o}{v_{in}} = \frac{1 - j2\pi fRC}{1 + j2\pi fRC} \qquad (8\text{-}18)$$

where f is the frequency of the input signal in hertz.

Equation (8-18) indicates that the amplitude of v_o/v_{in} is unity; that is, $|v_o| = |v_{in}|$ throughout the useful frequency range, and the phase shift between v_o and v_{in} is a function of input frequency f. The phase angle ϕ is given by

$$\phi = -2 \tan^{-1}\left(\frac{2\pi fRC}{1}\right) \qquad (8\text{-}19)$$

where ϕ is in degrees, f in hertz, R in ohms, and C in farads. Equation (8-19) is used to find the phase angle ϕ if f, R, and C are known. Figure 8-16(b) shows a phase shift of 90° between the input v_{in} and output v_o. That is, v_o lags v_{in} by 90°. For fixed values of R and C, the phase angle ϕ changes from 0 to $-180°$ as the frequency f is varied from 0 to ∞. In Figure 8-16(a), if the positions of R and C are interchanged, the phase shift between input and output becomes positive. That is, output v_o leads input v_{in}.

EXAMPLE 8-11

For the all-pass filter of Figure 8-16(a), find the phase angle ϕ if the frequency of v_{in} is 1 kHz.

SOLUTION   From Equation (8-19),

$$\phi = -2 \tan^{-1}\left[\frac{(2\pi)(10^3)(15.9)(10^3)(10^{-8})}{1}\right]$$

$$= -90°$$

This means that the output voltage v_o has the same frequency and amplitude but lags v_{in} by 90°, as shown in Figure 8-16(b).

With the advance of integrated-circuit technology, a number of manufacturers now offer ready-to-use *universal filters* having simultaneous low-pass, high-pass, and band-pass output responses. Notch and all-pass functions are also available by combining these output responses in the uncommitted op-amp. Because of its versatility, this filter is called the *universal filter*. It provides the user with easy control of the gain and Q factor. The universal filter, sometimes called a *state-variable filter*, is presented in Chapter 10.

8-11 OSCILLATORS

Thus far we have examined op-amps wired as amplifiers or filters. This section will introduce the use of op-amps as oscillators capable of generating a variety of output waveforms. Basically, the function of an oscillator is to generate alternating current or voltage waveforms. More precisely, an oscillator is a circuit that generates a repetitive waveform of fixed amplitude and frequency without any external input signal. Oscillators are used in radio, television, computers, and communications. Although there are different types of oscillators, they all work on the same basic principle.

8-11.1 Oscillator Principles

An oscillator is a type of feedback amplifier in which part of the output is fed back to the input via a feedback circuit. If the signal fed back is of proper magnitude and phase, the circuit produces alternating currents or voltages. To visualize the requirements of an oscillator, consider the block diagram of Figure 8-17. This diagram looks identical to that of the feedback amplifiers of Chapter 4 (see Figures 4-3 and 4-9). However, here the input voltage is zero ($v_{in} = 0$). Also, the feedback is positive because most oscillators use positive feedback. Finally, the closed-loop gain of the amplifier is denoted by A_v rather than A_F.

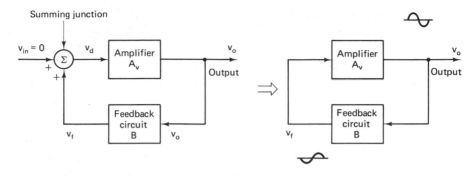

Figure 8–17   Oscillator block diagram.

In the block diagram of Figure 8-17,

$$v_d = v_f + v_{in}$$

$$v_o = A_v v_d$$

$$v_f = B v_o$$

Using these relationships, the following equation is obtained:

$$\frac{v_o}{v_{in}} = \frac{A_v}{1 - A_v B}$$

However, $v_{in} = 0$ and $v_o \neq 0$ implies that

$$A_v B = 1 \tag{8-20}$$

Expressed in polar form,

$$A_v B = 1 \underline{/0°} \text{ or } 360° \tag{8-21}$$

Equation (8-21) gives the two requirements for oscillation: (1) the magnitude of the loop gain $A_v B$ must be at least 1, and (2) the total phase shift of the loop gain $A_v B$ must be equal to 0° or 360°. For instance, as indicated in Figure 8-17, if the amplifier causes a phase shift of 180°, the feedback circuit must provide an additional phase shift of 180° so that the total phase shift around the loop is 360°. The waveforms shown in Figure 8-17 are sinusoidal and are used to illustrate the circuit's action. The type of waveform generated by an oscillator depends on the components in the circuit and hence may be sinusoidal, square, or triangular. In addition, the frequency of oscillation is determined by the components in the feedback circuit.

8-11.2 Oscillator Types

Because of their widespread use, many different types of oscillators are available. These oscillator types are summarized in Table 8-6.

8-11.3 Frequency Stability

The ability of the oscillator circuit to oscillate at one exact frequency is called *frequency stability*. Although a number of factors may cause changes in oscillaator frequency, the primary factors are temperature changes and changes in the dc power supply. Temperature

TABLE 8-6 OSCILLATOR TYPES

Types of components used	Frequency of oscillation	Types of waveform generated
RC oscillator LC oscillator Crystal oscillator	Audio frequency (AF) Radio frequency (RF)	Sinusoidal Square wave Triangular wave Sawtooth wave, etc.

and power supply changes cause variations in the op-amp's gain, in junction capacitances and resistances of the transistors in an op-amp, and in external circuit components. In most cases these variations can be kept small by careful design, by using regulated power supplies, and by temperature control.

Another important factor that determines frequency stability is the *figure of merit Q* of the circuit. The higher the Q, the greater the stability. For this reason, crystal oscillators are far more stable than RC or LC oscillators, especially at higher frequencies. LC circuits and crystals are generally used for the generation of high-frequency signals, while RC components are most suitable for audio-frequency applications. Here we discuss audio-frequency RC oscillators only. We begin with the sinusoidal oscillators.

8-12 PHASE SHIFT OSCILLATOR

Figure 8-18 shows a phase shift oscillator, which consists of an op-amp as the amplifying stage and three RC cascaded networks as the feedback circuit. The feedback circuit provides feedback voltage from the output back to the input of the amplifier. The op-amp is used in the inverting mode; therefore, any signal that appears at the inverting terminal is shifted by 180° at the output. An additional 180° phase shift required for oscillation is provided by the cascaded RC networks. Thus the total phase shift around the loop is 360° (or 0°). At some specific frequency when the phase shift of the cascaded RC networks is exactly 180° and the gain of the amplifier is sufficiently large, the circuit will oscillate at that frequency. This frequency is called the frequency of oscillation f_o and is given by

$$f_o = \frac{1}{2\pi\sqrt{6}\,RC} = \frac{0.065}{RC} \tag{8-22a}$$

At this frequency, the gain A_v must be *at least* 29. That is,

$$\left|\frac{R_F}{R_1}\right| = 29$$

or

$$R_F = 29R_1 \tag{8-22b}$$

For the derivation of Equation (8-22a) and (8-22b), refer to Appendix C. Thus the circuit will produce a sinusoidal waveform of frequency f_o if the gain is 29 and the total phase

Figure 8–18   Phase shift oscillator and its output waveform.

shift around the circuit is exactly 360°. For a desired frequency of oscillation, choose a capacitor C, and then calculate the value of R from Equation (8-22a). A desired output amplitude, however, can be obtained with back-to-back zeners connected at the output terminal.

EXAMPLE 8-12
Design the phase shift oscillator of Figure 8-18 so that $f_o = 200$ Hz.

SOLUTION   Let $C = 0.1\ \mu F$. Then, from Equation (8-22a),

$$R = \frac{0.065}{(200)\,(10^{-7})} = 3.25\ k\Omega$$

(Use $R = 3.3\ k\Omega$.)

To prevent the loading of the amplifier because of RC networks, it is necessary that $R_1 \geq 10\ R$. Therefore, let $R_1 = 10R = 33\ k\Omega$. Then, from Equation (8-22b),

$$R_F = 29(33\ k\Omega) = 957\ k\Omega$$

(Use $R_F = 1\text{-}M\Omega$ potentiometer.)

When choosing an op-amp, type 741 can be used at lower frequencies (<1 kHz); however, at higher frequencies, an op-amp such as the LM318 or LF351 is recommended because of its increased slew rate.

8-13 WIEN BRIDGE OSCILLATOR

Because of its simplicity and stability, one of the most commonly used audio-frequency oscillators is the Wien bridge. Figure 8-19 shows the Wien bridge oscillator in which the Wien bridge circuit is connected between the amplifier input terminals and the output terminal. The bridge has a series RC network in one arm and a parallel RC network in the adjoining arm. In the remaining two arms of the bridge, resistors R_1 and R_F are connected (see Figure 8-19).

Figure 8–19   Wien bridge oscillator.

The phase angle criterion for oscillation is that the total phase shift around the circuit must be 0°. This condition occurs only when the bridge is balanced, that is, *at resonance*. The frequency of oscillation f_o is exactly the resonant frequency of the balanced Wien bridge and is given by

$$f_o = \frac{1}{2\pi RC} = \frac{0.159}{RC} \qquad (8\text{-}23a)$$

assuming that the resistors are equal in value, and capacitors are equal in value in the reactive leg of the Wien bridge. At this frequency the gain required for sustained oscillation is given by

$$A_v = \frac{1}{B} = 3$$

That is,

$$1 + \frac{R_F}{R_1} = 3$$

or

$$R_F = 2R_1 \qquad (8\text{-}23b)$$

For the derivation of Equations (8-23a) and (8-23b), refer to Appendix C. The Wien bridge oscillator is designed using Equations (8-23a) and (8-23b), as illustrated in Example 8-13.

EXAMPLE 8-13
   Design the Wien bridge oscillator of Figure 8-19 so that $f_o = 965$ Hz.

SOLUTION   Let $C = 0.05\ \mu$F. Therefore, from Equation (8-23a),

$$R = \frac{0.159}{(5)(10^{-8})(965)} = 3.3\ \text{k}\Omega$$

Now, let $R_1 = 12\ \text{k}\Omega$. Then, from Equation (8-23b),

$$R_F = (2)(12\ \text{k}\Omega) = 24\ \text{k}\Omega$$

(Use $R_F = 50\ \text{k}\Omega$ potentiometer.)

8-14 QUADRATURE OSCILLATOR

As its name implies, the quadrature oscillator generates two signals (sine and cosine) that are in quadrature, that is, out of phase by 90°. Although the actual location of the sine and cosine is arbitrary, in the quadrature oscillator of Figure 8-20 the output of A_1 is labeled a sine and the output of A_2 is a cosine. This oscillator requires a dual op-amp and three RC combinations. The first op-amp A_1 is operating in the noninverting mode and appears

Figure 8–20   Quadrature oscillator. A_1 and A_2 dual op-amp: 1458/353.

as a noninverting integrator. The second op-amp A_2 is working as a pure integrator. Furthermore, A_2 is followed by a voltage divider consisting of R_3 and C_3. The divider network forms a feedback circuit, whereas A_1 and A_2 form the amplifier stage.

The total phase shift of 360° around the loop required for oscillation is obtained in the following way. The op-amp A_2 is a pure integrator and inverter. Hence it contributes $-270°$ or (90°) of phase shift. The remaining $-90°$ (or 270°) of phase shift needed are obtained at the voltage divider R_3C_3 and the op-amp A_1. The total phase shift of 360°, however, is obtained at only one frequency f_o, called the *frequency of oscillation*. This frequency is given by

$$f_o = \frac{1}{2\pi RC} \tag{8-24a}$$

where $R_1 C_1 = R_2 C_2 = R_3 C_3 = RC$. At this frequency,

$$A_v = \frac{1}{B} = 1.414 \tag{8-24b}$$

which is the second condition for oscillation.

Thus, to design a quadrature oscillator for a desired frequency f_o, choose a value of C; then, from Equation (8-24a), calculate the value of R. To simplify design calculations, choose $C_1 = C_2 = C_3$ and $R_1 = R_2 = R_3$. In addition, R_1 may be a potentiometer in order to eliminate any possible distortion in the output waveforms.

EXAMPLE 8-14

Design the quadrature oscillator of Figure 8-20 so that $f_o = 159$ Hz. The op-amp is the 1458/772.

SOLUTION   Let $C = 0.01$ μF. Then, from Equation (8-24a),

$$R = \frac{0.159}{(159)(10^{-8})} = 100 \text{ k}\Omega$$

Thus $C_1 = C_2 = C_3 = 0.01$ μF and $R_1 = R_2 = R_3 = 100$ kΩ. However, R_1 may be a 200-kΩ potentiometer, which can be adjusted for undistorted output waveforms.

Using the oscillator principles of Section 8-11.1, a number of other sine wave RC oscillators, such as the twin-T and biquad (biquadratic), can also be designed.

Before proceeding with the other types of oscillators, consider the difference between the signal generator and the function generator. The function generator has more than one function, such as sine wave, square wave, and triangular wave; the signal generator often generates only one type of waveform, such as a sine wave. However, both types of generators have amplitude and frequency modulation capabilities, unlike the oscillator. Since the generator and the oscillator are equivalent terms, they can be used interchangeably. Usually, circuits producing sine waves are called oscillators, while those generating a square wave, triangular wave, or sawtooth wave are called generators.

8-15 SQUARE WAVE GENERATOR

In contrast to sine wave oscillators, square wave outputs are generated when the op-amp is forced to operate in the saturated region. That is, the output of the op-amp is forced to swing repetitively between positive saturation $+V_{\text{sat}}$ ($\cong +V_{CC}$) and negative saturation $-V_{\text{sat}}$($\cong -V_{EE}$), resulting in the square-wave output. One such circuit is shown in Figure 8-21(a). This square wave generator is also called a *free-running* or *astable* multivibrator. The output of the op-amp in this circuit will be in positive or negative saturation, depending on whether the differential voltage v_{id} is negative or positive, respectively.

Assume that the voltage across capacitor C is zero volts at the instant the dc supply voltages $+V_{CC}$ and $-V_{EE}$ are applied. [See Figure 8-21(b).] This means that the voltage at the inverting terminal is zero initially. At the same instant, however, the voltage v_1 at the noninverting terminal is a very small finite value that is a function of the output offset voltage V_{ooT} and the values of R_1 and R_2 resistors. Thus the differential input voltage v_{id} is equal to the voltage v_1 at the noninverting terminal. Although very small, voltage v_1 will

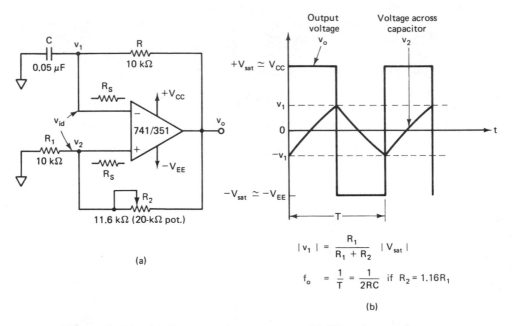

Figure 8–21 (a) Square wave generator. (b) Waveforms of output voltage v_o and capacitor voltage v_2 of the square-wave generator.

start to drive the op-amp into saturation. For example, suppose that the output offset voltage V_{ooT} is positive and that, therefore, voltage v_1 is also positive. Since initially the capacitor C acts as a short circuit, the gain of the op-amp is very large (A); hence v_1 drives the output of the op-amp to its positive saturation $+V_{sat}$. With the output voltage of the op-amp at $+V_{sat}$, the capacitor C starts charging toward $+V_{sat}$ through resistor R. However, as soon as the voltage v_2 across capacitor C is slightly more positive than v_1, the output of the op-amp is forced to switch to a negative saturation, $-V_{sat}$. With the op-amp's output voltage at negative saturation, $-V_{sat}$, the voltage v_1 across R_1 is also negative, since

$$v_1 = \frac{R_1}{R_1 + R_2}(-V_{sat}) \tag{8-25a}$$

Thus the net differential voltage $v_{id} = v_1 - v_2$ is negative, which holds the output of the op-amp in negative saturation. The output remains in negative saturation until the capacitor C discharges and then recharges to a negative voltage slightly higher than $-v_1$. [See Figure 8-21(b).] Now, as soon as the capacitor's voltage v_2 becomes more negative than $-v_1$, the net differential voltage v_{id} becomes positive and hence drives the output of the op-amp back to its positive saturation $+V_{sat}$. This completes one cycle. With output at $+V_{sat}$, voltage v_1 at the noninverting input is

$$v_1 = \frac{R_1}{R_1 + R_2}(+V_{sat}) \tag{8-25b}$$

The time period T of the output waveform is given by

$$T = 2RC \ln\left(\frac{2R_1 + R_2}{R_2}\right) \tag{8-26a}$$

or

$$f_o = \frac{1}{2RC \ln[(2R_1 + R_2)/R_2]} \tag{8-26b}$$

Equation (8-26b) indicates that the frequency of the output f_o is not only a function of the RC time constant but also of the relationship between R_1 and R_2. For example, if $R_2 = 1.16R_1$, Equation (8-26b) becomes

$$f_o = \frac{1}{2RC} \tag{8-27}$$

Equation (8-27) shows that the smaller the RC time constant, the higher the output frequency f_o, and vice versa. As with sine wave oscillators, the highest frequency generated by the square wave generator is also set by the slew rate of the op-amp. An attempt to operate the circuit at relatively higher frequencies causes the oscillator's output to become triangular. In practice, each inverting and noninverting terminal needs a series resistance R_S to prevent excessive differential current flow because the inputs of the op-amp are subjected to large differential voltages. The resistance R_S used should be 100 kΩ or higher. A reduced peak-to-peak output voltage swing can be obtained in the square-wave generator of Figure 8-21(a) by using back-to-back zeners at the output terminal.

EXAMPLE 8-15

Design the square-wave oscillator of Figure 8-21(a) so that $f_o = 1$ kHz. The op-amp is a 741 with dc supply voltages = ±15 V.

SOLUTION   Use $R_2 = 1.16 R_1$ so that the simplified frequency Equation (8-27) can be applied. Let $R_1 = 10$ kΩ. Then

$$R_2 = (1.16)(10 \text{ k}\Omega) = 11.6 \text{ k}\Omega$$

(Use $R_2 = 20$-kΩ potentiometer.)

Next, choose a value of C and calculate the value of R from Equation (8-27). Hence let $C = 0.05$ μF. By Equation (8-27),

$$R = \frac{1}{(10)(10^{-8})(10^3)} = 10 \text{ k}\Omega$$

Thus

$$R_1 = 10 \text{ k}\Omega$$
$$R_2 = 11.6 \text{ k}\Omega \quad [20\text{-k}\Omega \text{ potentiometer}]$$
$$R = 10 \text{ k}\Omega$$
$$C = 0.05 \text{ μF}$$

8-16 *TRIANGULAR WAVE GENERATOR*

Recall that the output waveform of the integrator is triangular if its input is a square wave. (Refer to Section 7-12.) This means that a triangular wave generator can be formed by simply connecting an integrator to the square wave generator of Figure 8-21(a). The resultant circuit is shown in Figure 8-22(a). This circuit requires a dual op-amp, two capacitors, and at least five resistors. The frequencies of the square wave and triangular wave are the same. For fixed R_1, R_2, and C values, the frequency of the square wave as well as the triangular wave depends on the resistance R. [See Equation (8-26b).] As R is increased or decreased, the frequency of the triangular wave will decrease or increase, respectively. Although the amplitude of the square wave is constant ($\pm V_{sat}$), the amplitude of the triangular wave decreases with an increase in its frequency, and vice versa. [See Figure 8-22(b).]

(a)

(b)

Figure 8–22 Triangular wave generator. (a) Circuit. (b) Its output waveform. A_1 and A_2 dual op-amp: 1458/353.

The input of integrator A_2 is a square wave, while its output is a triangular wave. However, for the output of A_2 to be a triangular wave requires that $5R_3 C_2 > T/2$, where T is the period of the square wave input. As a general rule, $R_3 C_2$ should be equal to T. To obtain a stable triangular wave, it may also be necessary to shunt the capacitor C_2 with resistance $R_4 = 10R_3$ and connect an offset voltage-compensating network at the non-inverting terminal of A_2. As with any other oscillator, the frequency of the triangular wave generator is limited by the slew rate of the op-amp. Therefore, high-slew-rate op-amps should be used for the generation of relatively higher frequencies.

Another triangular wave generator, which requires fewer components, is shown in Figure 8-23(a). The generator consists of a *comparator* A_1 and an *integrator* A_2. The comparator A_1 compares the voltage at point P continuously with the inverting input that

(a)

(b)

Figure 8–23   Triangular wave generator. (a) Circuit. (b) Its wave-forms. A_1 and A_2 dual op-amp: 1458/353.

is at 0 V. When the voltage at P goes slightly below or above 0 V, the output of A_1 is at the negative or positive saturation level, respectively.

To illustrate the circuit's operation, let us set the output of A_1 at positive saturation $+V_{sat}$ ($\cong +V_{CC}$). This $+V_{sat}$ is an input of the integrator A_2. The output of A_2, therefore, will be a negative-going ramp. Thus one end of the voltage divider R_2–R_3 is the positive saturation voltage $+V_{sat}$ of A_1 and the other is the negative-going ramp of A_2. When the negative-going ramp attains a certain value $-V_{Ramp}$, point P is slightly below 0 V; hence the output of A_1 will switch from positive saturation to negative saturation $-V_{sat}(\cong -V_{EE})$. This means that the output of A_2 will now stop going negatively and will begin to go positively. The output of A_2 will continue to increase until it reaches $+V_{Ramp}$. At this time the point P is slightly above 0 V; therefore, the output of A_1 is switched back to the positive saturation level $+V_{sat}$. The sequence then repeats. The output waveform is as shown in Figure 8-23(b).

The frequencies of the square wave and the triangular wave are the same. The amplitude of the square wave is a function of the dc supply voltages. However, a desired amplitude can be obtained by using appropriate zeners at the output of A_1. [See Figure 8-23(a).]

The amplitude and the frequency of the triangular wave can be determined as follows: From Figure 8-23(b), when the output of the comparator A_1 is $+V_{sat}$, the output of the integrator A_2 steadily decreases until it reaches $-V_{Ramp}$. At this time the output of A_1 switches from $+V_{sat}$ to $-V_{sat}$. Just before this switching occurs, the voltage at point P (+input) is 0 V. This means that the $-V_{Ramp}$ must be developed across R_2, and $+V_{sat}$ must be developed across R_3. That is,

$$\frac{-V_{Ramp}}{R_2} = -\frac{+V_{sat}}{R_3}$$

or

$$-V_{Ramp} = -\frac{R_2}{R_3}(+V_{sat}) \qquad (8\text{-}28a)$$

Similarly, $+V_{Ramp}$, the output voltage of A_2 at which the output of A_1 switches from $-V_{sat}$ to $+V_{sat}$, is given by

$$+V_{Ramp} = -\frac{R_2}{R_3}(-V_{sat}) \qquad (8\text{-}28b)$$

Thus, from Equations (8-28a) and (8-28b), the peak-to-peak (pp) output amplitude of the triangular wave is

$$v_o(pp) = +V_{Ramp} - (-V_{Ramp})$$

$$v_o(pp) = (2)\frac{R_2}{R_3}(V_{sat}) \qquad (8\text{-}29)$$

where $V_{sat} = |+V_{sat}| = |-V_{sat}|$. Equation (8-29) indicates that the amplitude of the triangular wave decreases with an increase in R_3.

The time it takes for the output waveform to swing from $-V_{Ramp}$ to $+V_{Ramp}$ (or from $+V_{Ramp}$ to $-V_{Ramp}$) is equal to half the time period $T/2$. [See Figure 8-23(b).] This time can be calculated from the integrator output equation, (7-23), by substituting $v_i = -V_{sat}$,

$v_o = v_o(\text{pp})$, and $C = 0$.

$$v_o(\text{pp}) = -\frac{1}{R_1 C_1} \int_0^{T/2} (-V_{sat})\, dt \qquad (7\text{-}23)$$

$$= \left(\frac{V_{sat}}{R_1 C_1}\right)\left(\frac{T}{2}\right)$$

Hence

$$\frac{T}{2} = \frac{v_o(\text{pp})}{V_{sat}}(R_1 C_1)$$

or

$$T = (2R_1 C_1)\frac{v_o(\text{pp})}{V_{sat}} \qquad (8\text{-}30\text{a})$$

where $V_{sat} = |+V_{sat}| = |-V_{sat}|$. Substituting the value of $v_o(\text{pp})$ from Equation (8-29), the time period of the triangular wave is

$$T = \frac{4R_1 C_1 R_2}{R_3} \qquad (8\text{-}30\text{b})$$

The frequency of oscillation then is

$$f_o = \frac{R_3}{4R_1 C_1 R_2} \qquad (8\text{-}30\text{c})$$

Equation (8-30c) shows that the frequency of oscillation f_o increases with an increase in R_3.

The triangular wave generator is designed for a desired amplitude and frequency f_o by using Equations (8-29) and (8-30c).

EXAMPLE 8-16

Design the triangular wave generator of Figure 8-23 so that $f_o = 2$ kHz and $v_o(\text{pp}) = 7$ V. The op-amp is a 1458/772 and supply voltages $= \pm 15$ V.

SOLUTION   For the 1458, $V_{sat} = 14$ V. Therefore, from Equation (8-29),

$$\frac{R_2}{R_3} = \frac{7}{(2)(14)}$$

$$R_2 = \frac{R_3}{4}$$

Let $R_2 = 10$ kΩ; then $R_3 = 40$ kΩ. (Use a 50-kΩ potentiometer.)
Now, from Equation (8-30c),

$$2 \text{ kHz} = \frac{40 \text{ k}\Omega}{(4)(R_1 C_1)(10 \text{ k}\Omega)}$$

Therefore, $R_1 C_1 = 0.5$ ms. Let $C_1 = 0.05$ μF; then $R_1 = 10$ kΩ. *Thus* $R_1 = R_2 = 10$ kΩ, $C_1 = 0.05$ μF, and $R_3 = 40$ kΩ (50-kΩ potentiometer). [See Figure 8-23(a).]

8-17 SAWTOOTH WAVE GENERATOR

The difference between the triangular and sawtooth waveforms is that the rise time of the triangular wave is always equal to its fall time. That is, the same amount of time is required for the triangular wave to swing from $-V_{Ramp}$ to $+V_{Ramp}$ as from $+V_{Ramp}$ to $-V_{Ramp}$. [See Figure 8-23(b).] On the other hand, the sawtooth waveform has unequal rise and fall times. That is, it may rise positively many times faster than it falls negatively, or vice versa. The triangular wave generator of Figure 8-23(a) can be converted into a sawtooth wave generator by injecting a variable dc voltage into the noninverting terminal of the integrator A_2. This can be accomplished by using the potentiometer and connecting it to the $+V_{CC}$ and $-V_{EE}$ as shown in Figure 8-24(a). Depending on the R_4 setting, a certain

(a)

(b)

Figure 8–24 Sawtooth wave generator. (a) Circuit. A_1 and A_2 dual op-amp: 1458/353. D_1 and D_2: 1N4735 with $V_z = 6.2$ V. (b) Output waveform when noninverting input of A_2 is at some negative dc level.

dc level is inserted in the output of A_2. Now, suppose that the output of A_1 is a square wave and the potentiometer R_4 is adjusted for a certain dc level. This means that the output of A_2 will be a triangular wave, riding on some dc level that is a function of the R_4 setting. The duty cycle of the square wave will be determined by the polarity and amplitude of this dc level. A duty cycle less than 50% will then cause the output of A_2 to be a sawtooth. [See Figure 8-24(b).] With the wiper at the center of R_4, the output of A_2 is a triangular wave. For any other position of R_4 wiper, the output is a sawtooth waveform. Specifically as the R_4 wiper is moved toward $-V_{EE}$, the rise time of the sawtooth wave becomes longer than the fall time, as shown in Figure 8-24(b). On the other hand, as the wiper is moved toward $+V_{CC}$, the fall time becomes longer than the rise time. Also, the frequency of the sawtooth wave decreases as R_4 is adjusted toward $+V_{CC}$ or $-V_{EE}$. However, the amplitude of the sawtooth wave is independent of the R_4 setting.

8-18 *VOLTAGE-CONTROLLED OSCILLATOR*

In all the preceding oscillators, the frequency is determined by the *RC* time constant. However, there are applications, such as *frequency modulation* (FM), *tone generators,* and *frequency shift keying* (FSK), where the frequency needs to be controlled by means of an input voltage called *control voltage*. This function is achieved in the *voltage-controlled oscillator* (VCO), also called a *voltage-to-frequency converter*. A typical example is the Signetics NE/SE 566 VCO, which provides simultaneous square wave and triangular wave outputs as a function of input voltage. Figure 8-25(b) is a block diagram of the 566. The frequency of oscillation is determined by an external resistor R_1, capacitor C_1, and the voltage V_C applied to the control terminal 5. [See Figure 8.25(c).] The triangular wave is generated by alternately charging the external capacitor C_1 by one current source and then linearly discharging it by another. [See Figure 8-25(b).] The charge–discharge levels are determined by Schmitt trigger action. The Schmitt trigger also provides the square wave output. Both the output waveforms are buffered so that the output impedance of each is 50 Ω. The typical amplitude of the triangular wave is 2.4 V pp and that of the square wave is 5.4 V pp.

Figure 8-25(c) is a typical connection diagram. In this arrangement, the $R_1 C_1$ combination determines the free-running frequency, and the control voltage V_C at terminal 5 is set by the voltage divider formed with R_2 and R_3. The initial voltage V_C at terminal 5 must be in the range

$$\frac{3}{4}(+V) \leq V_C \leq +V \tag{8-31a}$$

where $+V$ is the total supply voltage. The modulating signal is ac coupled with the capacitor C and must be $< 3\ V$ pp. The frequency of the output waveforms is approximated by

$$f_o \cong \frac{2(+V - V_C)}{R_1 C_1(+V)} \tag{8-31b}$$

where R_1 should be in the range 2 kΩ $< R_1 <$ 20 kΩ. For a fixed V_C and constant C_1, the frequency f_o can be varied over a 10 : 1 frequency range by the choice of R_1 between 2 kΩ

Figure 8-25 Voltage-controlled oscillator 566. (a) Pin configuration. (b) Block diagram. (c) Typical connection diagram. (d) Output waveforms. (Courtesy of Signetics Corporation.)

and 20 kΩ. Similarly, for a constant R_1C_1 product, the frequency f_o can be modulated over a 10:1 range by the control voltage V_C. In either case the maximum output frequency is 1 MHz. A small capacitor of 0.001 μF should be connected between pins 5 and 6 to eliminate possible oscillations in the control current source.

  If the VCO is to be used to drive standard logic circuitry, a dual supply of ±5 V is recommended so that the square wave output has the proper dc levels for logic circuitry.

  The VCO is commonly used in converting low-frequency signals such as electro-encephalograms (EEG) or electrocardiograms (EKG) into an audio-frequency range. These audio signals can then be transmitted over telephone lines or two-way radio communication for diagnostic purposes or can be recorded on a magnetic tape for documentation or further reference. For more information on the VCO applications, refer to Section 10-5.

EXAMPLE 8-17

  In the circuit of Figure 8-25(c), $+V = 12$ V, $R_2 = 1.5$ kΩ, $R_1 = R_3 = 10$ kΩ, and $C_1 = 0.001$ μF.
(a) Determine the nominal frequency of the output waveforms.
(b) Compute the modulation in the output frequencies if V_C is varied between 9.5 V and 11.5 V.
(c) Draw the square wave output waveform if the modulating input is a sine wave, as shown in Figure 8-26.

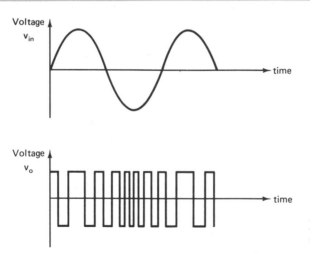

Figure 8–26 Input and output waveforms for Example 8–17.

SOLUTION (a) Using the voltage-divider rule, the initial control voltage V_C at terminal 5 is

$$V_C = \frac{(10\text{ k})(12)}{11.5\text{ k}} = 10.43\text{ V}$$

From Equation (8-31b), the approximate nominal frequency f_o is

$$f_o \cong \frac{(2)(12 - 10.43)}{(10^4)(10^{-9})(12)} = 26.17\text{ kHz}$$

(b) The modulation in the output frequencies can be calculated from Equation (8-31b) by substituting for V_C, first 9.5 V and then 11.5 V, as follows:

$$f_o \cong \frac{(2)(12 - 9.5)}{(10^4)(10^{-9})(12)} = 41.67 \text{ kHz}$$

$$f_o \cong \frac{(2)(12 - 11.5)}{(10^4)(10^{-9})(12)} = 8.33 \text{ kHz}$$

Thus the change in the output frequency is

$$41.67 \text{ kHz} - 8.33 \text{ kHz} = 33.34 \text{ kHz}$$

(c) During the positive half-cycle of the sine wave input, the control voltage V_C will increase. Therefore, according to Equation (8-31b), the frequency of the output waveform will decrease and the time period will increase. Exactly the opposite action will take place during the negative half-cycle of the input, as shown in Figure 8-26.

SUMMARY

1. A filter is often a frequency-selective circuit that passes a specified band of frequencies and attenuates signals of frequencies outside this band.
2. Filters may be classified in a number of ways: analog or digital, passive or active, audio or radio frequency.
3. The most commonly used filters are the following:
   (a) Low-pass filter
   (b) High-pass filter
   (c) Band-pass filter
   (d) Band-reject filter
   (e) All-pass filter
   A low-pass filter has a constant gain from 0 Hz to a certain frequency called the cutoff frequency f_H, at which the gain is down by 3 dB. Above f_H, the gain decreases with an increase in frequency. On the other hand, a high-pass filter passes all the frequencies above a certain frequency called the low cutoff frequency f_L. The upper cutoff frequency of the high-pass filter, however, depends on the bandwidth of the op-amp. A band-pass filter has a passband between two cutoff frequencies f_H and f_L such that $f_H > f_L$. Any input frequency outside this passband is attenuated. The band-reject filter performs exactly opposite to the band-pass, in that it has a stopband between two cutoff frequencies f_H and f_L. Finally, the all-pass filter has input and output amplitudes equal at all frequencies; however, the phase shift between the two is a function of frequency.
4. The order of the filter indicates the rate at which the gain changes while the input frequency is approaching or exceeding the cutoff frequency of the filter. For example, in the first-order low-pass filter, for $f > f_H$, the gain decreases at the rate of 20 dB/decade, whereas in the first-order high-pass filter the gain increases at the rate of 20 dB/decade until $f = f_L$. Similarly, in the second-order filter the change in the rate of the gain is 40 dB/decade, in the third-order 60 dB/decade, and so on.

5. To convert a low-pass filter into a high-pass filter, or vice versa, simply interchange frequency-determining components, that is, resistors and capacitors.

6. Higher-order filters can be formed by simply using first- and second-order filters. For example, a third-order filter is constructed by cascading first- and second-order filters; a fourth-order is formed by cascading two second-order filters; and so on.

7. There are two types of band-pass filters: wide band pass and narrow band pass. A wide band-pass filter can be formed simply by cascading high-pass and low-pass sections. On the other hand, the multiple-feedback filter is one of the narrow band-pass filters.

8. As with band-pass filters, band-reject filters can be either a wide band-reject or narrow band-reject. The wide band-reject filter is formed using a high-pass filter, a low-pass filter, and a summing amplifier. The outputs of low-pass and high-pass filters are added together through a summing amplifier. The narrow band-reject filter, often called the notch filter, uses the twin-T network. It is usually used for the rejection of a single frequency, such as 60-Hz power line frequency hum.

9. The all-pass filter does just what its name suggests. It provides unity gain with predictable phase shifts for different input frequencies.

10. Basically, the function of an oscillator is to generate alternating current or voltage waveforms. Oscillators are classified according to the type of components used, the frequency of oscillation, and the type of waveform generated. There are two requirements for oscillation:
   (a) The magnitude of the loop gain $A_v B$ must be greater than or equal to 1.
   (b) The total phase shift of the loop gain must be $0°$.
   The Wien bridge, phase shift, and quadrature oscillators are the most commonly used sinusoidal oscillators. In all these oscillators the frequency of oscillation is a function of the RC time constant.

11. A square wave output waveform is generated if the output of the op-amp is forced to swing repetitively between positive and negative saturation. One way to obtain the triangular wave is to integrate the square wave. Therefore, the triangular wave generator can be formed by using a comparator and an integrator.

12. The sawtooth waveform has a rise time many times longer than the fall time, or vice versa. The triangular wave generator using a comparator and an integrator can be converted to obtain a sawtooth waveform.

13. The oscillator whose output frequency depends on the amplitude of the input voltage is called a voltage-controlled oscillator (VCO) or voltage-to-frequency (V/F) converter. The VCO is used in phase-locked-loop circuits and for frequency modulation (FM), among other things.

QUESTIONS ────────────────────────────────────

8-1. Define a filter. How are filters classified?

8-2. List the most commonly used filters.

8-3. What is a passband and a stopband for a filter?

8-4. What are the advantages of active filters over passive ones?

8-5. What is the Butterworth response?

8-6. What is an all-pass filter? Where and why is it needed?

8-7. Define an oscillator.

8-8. What are the two requirements for oscillation?

8-9. How are oscillators classified?

8-10. What is frequency stability? Explain its significance.

8-11. What is the difference between the sawtooth wave and the triangular wave?

8-12. In the sawtooth generator of Figure 8-24(a), how does the potentiometer R_4 affect the frequency and amplitude of the output waveform?

8-13. What is a VCO? Give two applications that require a VCO.

PROBLEMS

8-1. Design a first-order low-pass filter so that it has a cutoff frequency of 2 kHz and a pass-band gain of 1.

8-2. Convert the 2-kHz low-pass filter of Problem 8-1 to a cutoff frequency of 3 kHz.

8-3. Obtain the frequency response data similar to that in Table 8-1 for the filter of Problem 8-1. Construct the frequency response plot from this data.

8-4. Design a second-order low-pass filter at a cutoff frequency of 1.2 kHz.

8-5. Design a first-order high-pass filter at a cutoff frequency of 400 Hz and a passband gain of 1.

8-6. Obtain the frequency response data similar to that in Table 8-3 for the filter of Problem 8-5. Construct the frequency response plot from the data.

8-7. In the circuit of Figure 8-8, $C_2 = C_3 = 0.047~\mu F$, $R_2 = R_3 = 3.3~k\Omega$, $R_1 = 27~k\Omega$, and $R_F = 15.8~k\Omega$.
(a) Determine the low cutoff frequency f_L of the filter.
(b) Obtain the frequency response data and draw the frequency response plot for the filter.

8-8. Design a wide band-pass filter with $f_L = 400$ Hz, $f_H = 2$ kHz, and passband gain = 4. Also draw an approximate frequency response plot for the filter.

8-9. (a) Design a narrow band-pass filter so that $f_C = 2$ kHz, $Q = 20$, and $A_F = 10$.
(b) Change the center frequency f_C to 1 kHz, keeping gain and bandwidth constant.

8-10. Design a wide band-reject filter using first-order high-pass and low-pass filters having $f_L = 2$ kHz and $f_H = 400$ Hz, respectively.

8-11. Design a 400-Hz active notch filter.

8-12. For the all-pass filter of Figure 8-16(a), determine the phase shift ϕ between the input and output at $f = 2$ kHz. To obtain a positive phase shift ϕ, what modifications are necessary in the circuit?

8-13. Design a phase shift oscillator so that $f_o = 1$ kHz.

8-14. Design a Wien bridge oscillator that will oscillate at 2 kHz.

8-15. In the circuit of Figure 8-20, $R_1 = R_2 = R_3 = 82$ kΩ and $C_1 = C_2 = C_3 = 0.05$ μF. Determine the frequency of oscillation.

8-16. In the square-wave generator of Figure 8-21(a), if $R_1 = 12$ kΩ, $R_2 = 13.92$ kΩ, $R = 100$ kΩ, and $C = 0.01$ μF, what is the frequency of oscillation?

8-17. Design an integrator that can be used with the signal generator of Problem 8-16 so that the combination can be used as a triangular-wave generator.

8-18. In the triangular-wave generator of Figure 8-23(a), $R_2 = 1.2$ kΩ, $R_3 = 6.8$ kΩ, $R_1 = 120$ kΩ, and $C_1 = 0.01$ μF. Determine (a) the peak-to-peak output amplitude of the triangular wave and (b) the frequency of the triangular wave.

8-19. For the VCO of Figure 8-25(c), determine the change in output frequency if V_C is varied between 9 V and 11 V. Assume that $+V = 12$ V, $R_2 = 15$ kΩ, $R_3 = 100$ kΩ, $R_1 = 6.8$ kΩ, and $C_1 = 75$ pF.

8-20. In the VCO of Figure 8-25(c), if $+V = 15$ V, $V_C = 13$ V, and $C_1 = 0.0068$ μF, determine the approximate change in output frequency if R_1 is varied from 4 kΩ to 18 kΩ.

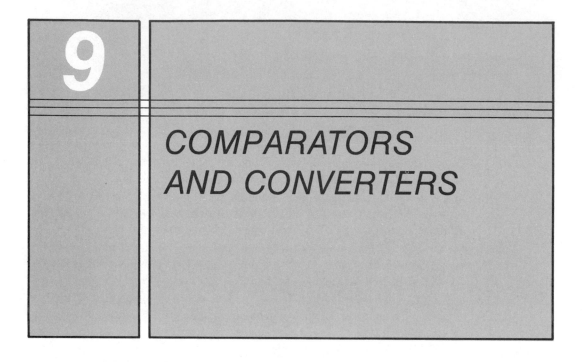

9

COMPARATORS AND CONVERTERS

9-1 INTRODUCTION

The amplifier, filter, and oscillator applications presented so far illustrate a fair sampling of typical op-amp uses. However, op-amps are used in many other circuits, where they are employed under specific names. Such functions include comparators, detectors, limiters, and digital interface devices, that is, converters. In this chapter we discuss comparators, limiters, detectors, and converters using a general-purpose op-amp. To obtain far better performance, we shall also look at integrated circuits designed specifically as comparators and converters.

A comparator, as its name implies, compares a signal voltage on one input of an op-amp with a known voltage called the *reference voltage* on the other input. In its simplest form, it is nothing more than an open-loop op-amp, with two analog inputs and a digital output; the output may be (+) or (−) saturation voltage, depending on which input is the larger. Comparators are used in circuits such as digital interfacing, Schmitt triggers, discriminators, voltage-level detectors, and oscillators.

9-2 BASIC COMPARATOR

Figure 9-1(a) shows an op-amp used as a comparator. A fixed reference voltage V_{ref} of 1 V is applied to the (−) input, and the other time-varying signal voltage v_{in} is applied to the (+) input. Because of this arrangement, the circuit is called the *noninverting comparator*. When v_{in} is less than V_{ref}, the output voltage v_o is at $-V_{sat}$ ($\cong -V_{EE}$) because the voltage at the (−) input is higher than that at the (+) input. On the other hand, when v_{in} is greater than V_{ref}, the (+) input becomes positive with respect to the (−) input, and v_o goes to $+V_{sat}$

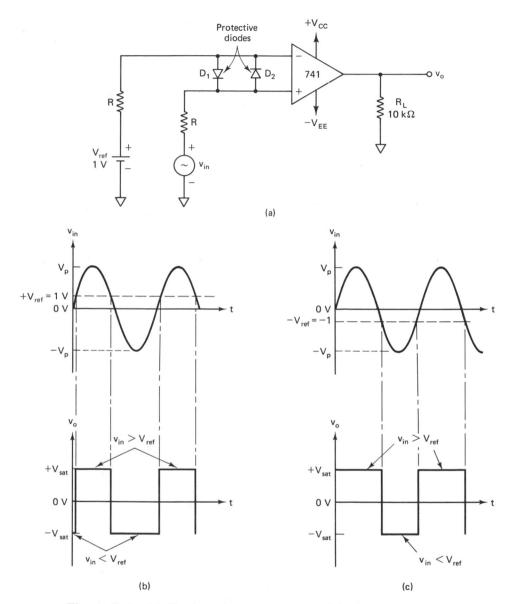

Figure 9–1   (a) Noninverting comparator and its input and output waveforms. (b) If V_{ref} is positive. (c) If V_{ref} is negative.

($\cong +V_{CC}$). Thus v_o changes from one saturation level to another whenever $v_{in} \cong V_{ref}$, as shown in Figure 9-1(b). In short, the comparator is a type of analog-to-digital converter. At any given time the v_o waveform shows whether v_{in} is greater or less than V_{ref}. The comparator is sometimes also called a *voltage-level detector* because, for a desired value of V_{ref}, the voltage level of the input v_{in} can be detected.

In Figure 9-1(a) the diodes $D1$ and $D2$ protect the op-amp from damage due to excessive input voltage v_{in}. Because of these diodes, the difference input voltage v_{id} of the op-amp is clamped to either 0.7 V or −0.7 V; hence the diodes are called *clamp diodes*. There are some op-amps with built-in input protection; in such op-amps the input diodes $D1$ and $D2$ are unnecessary. The resistance R in series with v_{in} is used to limit the current through $D1$ and $D2$. To reduce offset problems, a resistance $R_{OM} \cong R$ is connected between (−) input and V_{ref}. [See Figure 9-1(a).]

If the reference voltage V_{ref} is negative with respect to ground, with sinusoidal signal applied to the (+) input, the output waveform will be as shown in Figure 9-1(c). When $v_{in} > V_{ref}$, v_o is at $+V_{sat}$; on the other hand, when $v_{in} < V_{ref}$, v_o is at $−V_{sat}$. Obviously, the amplitude of v_{in} must be large enough to pass through V_{ref} if the switching action is to take place.

Figure 9-2(a) shows an inverting comparator in which the reference voltage V_{ref} is applied to the (+) input and v_{in} is applied to the (−) input. In this circuit, V_{ref} is obtained by using a 10-kΩ potentiometer that forms a voltage divider with the dc supply voltages $+V_{CC}$ and $−V_{EE}$ and the wiper connected to the (+) input. As the wiper is moved toward $−V_{EE}$, V_{ref} becomes more negative, while if it is moved toward $+V_{CC}$, V_{ref} becomes more positive. Thus a V_{ref} of a desired amplitude and polarity can be obtained by simply adjusting the 10-kΩ potentiometer. With the sinusoidal input waveform, the output v_o has the waveform shown in Figure 9-2(b) or (c), depending on whether V_{ref} is positive or negative, respectively.

9-3 ZERO-CROSSING DETECTOR

An immediate application of the comparator is the *zero-crossing detector* or *sine wave-to-square wave converter*. The basic comparator of Figure 9-1(a) or Figure 9-2(a) can be used as the zero-crossing detector provided that V_{ref} is set to zero ($V_{ref} = 0$ V). Figure 9-3(a) shows the inverting comparator used as a zero-crossing detector. The output voltage v_o waveform in Figure 9-3(b) shows when and in what direction an input signal v_{in} crosses zero volts. That is, the output v_o is driven into negative saturation when the input signal v_{in} passes through zero in the positive direction. Conversely, when v_{in} passes through zero in the negative direction, the output v_o switches and saturates positively.

In some applications, the input v_{in} may be a slowly changing waveform, that is, a low-frequency signal. Therefore, it will take v_{in} more time to cross 0 V; therefore, v_o may not switch quickly from one saturation voltage to the other. On the other hand, because of the noise at the op-amp's input terminals, the output v_o may fluctuate between two saturation voltages $+V_{sat}$ and $−V_{sat}$, detecting zero reference crossings for noise voltages as well as v_{in}. Both of these problems can be cured with the use of regenerative or *positive feedback* that causes the output v_o to change faster and eliminate any false output transitions due to noise signals at the input.

9-4 SCHMITT TRIGGER

Figure 9-4(a) shows an inverting comparator with *positive feedback*. This circuit converts an irregular-shaped waveform to a square wave or pulse. The circuit is known as the

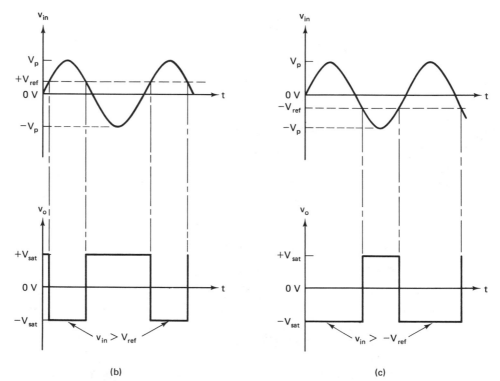

Figure 9–2 (a) Inverting comparator with input and output wave-forms. (b) If V_{ref} is positive. (c) If V_{ref} is negative.

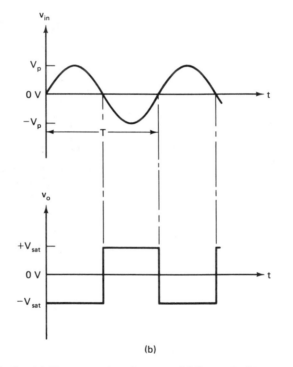

Figure 9–3 (a) Zero-crossing detector. (b) Its typical input and output waveforms.

Schmitt trigger or squaring circuit. The input voltage v_{in} triggers (changes the state of) the output v_o every time it exceeds certain voltage levels called the upper threshold voltage V_{ut} and lower threshold voltage V_{lt}, as shown in Figure 9-4(b).

In Figure 9-4(a), these threshold voltages are obtained by using the voltage divider R_1–R_2, where the voltage across R_1 is fed back to the (+) input. The voltage across R_1 is a variable reference threshold voltage that depends on the value and polarity of the output voltage v_o. When $v_o = +V_{sat}$, the voltage across R_1 is called the *upper threshold voltage*, V_{ut}. The input voltage v_{in} must be slightly more positive than V_{ut} in order to cause the output

(a)

(b)

(c)

Figure 9–4   (a) Inverting comparator as Schmitt trigger. (b) Input and output waveforms of Schmitt trigger. (c) v_o versus v_{in} plot of the hysteresis voltage.

v_o to switch from $+V_{sat}$ to $-V_{sat}$. As long as $v_{in} < V_{ut}$, v_o is at $+V_{sat}$. Using the voltage-divider rule,

$$V_{ut} = \frac{R_1}{R_1 + R_2}(+V_{sat}) \qquad (9\text{-}1a)$$

On the other hand, when $v_o = -V_{sat}$, the voltage across R_1 is referred to as *lower threshold voltage*, V_{lt}. v_{in} must be slightly more negative than V_{lt} in order to cause v_o to switch from $-V_{sat}$ to $+V_{sat}$. In other words, for v_{in} values greater than V_{lt}, v_o is at $-V_{sat}$. V_{lt} is given by the following equation:

$$V_{lt} = \frac{R_1}{R_1 + R_2}(-V_{sat}) \qquad (9\text{-}1b)$$

Thus, if the threshold voltages V_{ut} and V_{lt} are made larger than the input noise voltages, the positive feedback will eliminate the false output transitions. Also, the positive feedback, because of its regenerative action, will make v_o switch faster between $+V_{sat}$ and $-V_{sat}$. In Figure 9-4(a), resistance $R_{OM} \cong R_1 \| R_2$ is used to minimize the offset problems.

Figure 9-4(b) shows that the output of the Schmitt trigger is a square wave when the input is a sine wave. Recall that a slightly different version of the Schmitt trigger is used in the triangular wave and sawtooth wave generators of Figures 8-23 and 8-24, respectively. In these generators a noninverting comparator is used as a Schmitt trigger. When the input is a triangular wave, the output of the Schmitt trigger is a square wave, whereas if the input is a sawtooth wave, the output is a pulse waveform.

The comparator with positive feedback is said to exhibit *hysteresis*, a deadband condition. That is, when the input of the comparator exceeds V_{ut}, its output switches from $+V_{sat}$ to $-V_{sat}$ and reverts back to its original state, $+V_{sat}$, when the input goes below V_{lt} [see Figure 9-4(c)]. The hysteresis voltage is, of course, equal to the difference between V_{ut} and V_{lt}. Therefore,

$$V_{hy} = V_{ut} - V_{lt}$$

$$= \frac{R_1}{R_1 + R_2}[+V_{sat} - (-V_{at})] \qquad (9\text{-}2)$$

EXAMPLE 9-1

In the circuit of Figure 9-4(a), $R_1 = 100\ \Omega$, $R_2 = 56\ k\Omega$, $v_{in} = 1$ V pp sine wave, and the op-amp is type 741 with supply voltages $= \pm 15$ V. Determine the threshold voltages V_{ut} and V_{lt} and draw the output waveform.

SOLUTION For 741 the maximum output voltage swing is ± 14 V, that is, $+V_{sat} = 14$ V and $-V_{sat} = -14$ V. From Equations (9-1a) and (9-1b),

$$V_{ut} = \frac{100}{56,100}(14) = 25\ mV$$

$$V_{lt} = \frac{100}{56,100}(-14) = -25\ mV$$

The output v_o waveform is shown in Figure 9-4(b). From Equation (9-2), the hysteresis voltage $V_{hy} = 50$ mV.

9-5 COMPARATOR CHARACTERISTICS

The important characteristics of a comparator are these:

1. Speed of operation
2. Accuracy
3. Compatibility of output

The output of the comparator must switch rapidly between saturation levels and also respond instantly to any change of conditions at its inputs. This implies that the bandwidth of the op-amp comparator must be rather wide; in fact, the wider the bandwidth, the higher is the speed of operation. As discussed in Section 9-4, the speed of operation of the comparator is improved with positive feedback (hysteresis).

The accuracy of the comparator depends on its voltage gain, common-mode rejection ratio, input offsets, and thermal drifts. High voltage gain requires a smaller difference voltage (hysteresis voltage) to cause the comparator's output voltage to switch between saturation levels. On the other hand, a high CMRR helps to reject the common-mode input voltages, such as noise, at the input terminals. Finally, to minimize the offset problems, the input offset current and input offset voltage must be negligible; also, the *changes* in these offsets due to temperature variations should be very slight.

Since the comparator is a form of analog-to-digital converter, its output must swing between two logic levels suitable for a certain logic family such as transistor–transistor logic (TTL).

9-6 LIMITATIONS OF OP-AMPS AS COMPARATORS

A general-purpose op-amp such as the 741 can be used in relatively less critical comparator applications in which speed and accuracy are not major factors. As illustrated in Section 9-4, with positive feedback (hysteresis), the switching speed of the op-amp comparator can be improved and false transition due to noise can be eliminated. In addition, an offset voltage-compensating network and offset minimizing resistor can be used to minimize offset problems. However, the output voltage swing of an op-amp is relatively large because it is designed primarily as an amplifier. In other words, the output of an op-amp comparator is generally not compatible with a particular logic family such as the TTL, which requires input voltages of either approximately +5 V or 0 V. Therefore, to keep the output voltage swing within specific limits, op-amps are used with externally wired components such as zeners or diodes. The resulting circuits, in which the outputs are limited to predetermined values, are called *limiters*. The next section presents an analysis of typical limiters in order to suggest other such possibilities.

9-7 VOLTAGE LIMITERS

Several op-amp comparator circuits with output voltage limiting are shown in Figures 9-5 through 9-7. In the circuit of Figure 9-5(a), the zener diodes D_1 and D_2 are connected in the feedback path; this arrangement limits the positive and negative values of the output

(a)

(b)

Figure 9–5 (a) Basic comparator with positive and negative output voltage limiting. (b) Its input and output waveforms.

voltage v_o. When the input voltage v_{in} crosses 0 V and increases in the positive direction, the output voltage v_o increases in the negative direction until diode D_1 is forward biased and D_2 goes into avalanche conduction. Therefore, the maximum negative value of v_o is equal to $(V_Z + V_{D1})$, where V_Z is the zener voltage and V_{D1} is the voltage drop across the *forward*-biased zener D_1 ($= 0.7$ V typically). On the other hand, when v_{in} crosses 0 V and starts increasing in the negative direction, v_o starts increasing positively until diode D_2 is forward biased and D_1 goes into avalanche conduction. Thus the maximum positive v_o is equal to $(V_Z + V_{D2})$, where V_Z is the zener voltage and V_{D2} the voltage drop across the forward-biased zener $D_2(= 0.7 \text{ V}$ typically). Thus the output voltage swing is limited to $+(V_Z + 0.7)$ and $-(V_Z + 0.7)$ [see Figure 9-5(b)].

Note that, in the circuit of Figure 9-5(a), since the input terminals of the op-amp are at virtual ground $(v_1 = v_2 \cong 0 \text{ V})$, the input voltage v_{in} appears across R, and v_o appears across D_1 and D_2. The resistance R_{OM} is used to minimize offset problems.

If there is a need to limit the swing of the output voltage v_o to a positive direction only, a combination of zener and rectifier diodes is used as shown in the circuit of Figure 9-6(a). When the input voltage v_{in} crosses 0 V and increases in the positive direction, the output voltage v_o is at $-V_{\text{sat}}$. This happens because diode D_2 is reverse biased, causing the op-amp to operate in the open-loop configuration. However, when v_{in} crosses 0 V and starts increasing in the negative direction, v_o starts increasing positively until D_2 is forward biased and D_1 goes into avalanche conduction. Therefore, the maximum positive $v_o = V_Z + V_{D2}$ [see Figure 9-6(b)].

In the circuit of Figure 9-6(a), if the position of D_1 and D_2 diodes is interchanged, the output voltage v_o will then be limited in a negative direction to $-(V_Z + V_{D2})$. However, the maximum positive $v_o = +V_{\text{sat}}$.

Finally, if only a single zener is used in the feedback path of an op-amp as shown in Figure 9-7(a), the output voltage v_o is limited to $+V_Z$ and $-V_D$, where V_Z is the zener voltage and V_D the voltage drop across the forward-biased zener ($= +0.7 \text{ V}$ typically)

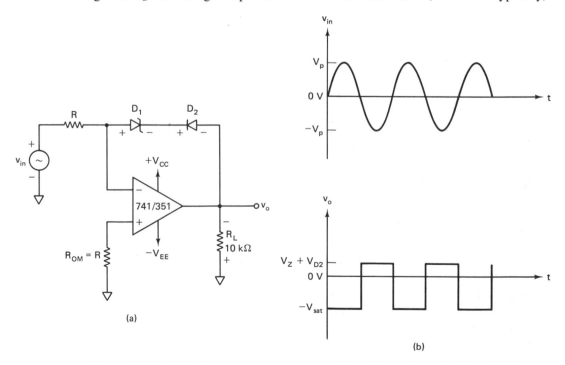

Figure 9–6 (a) Basic op-amp comparator with positive voltage limiting. (b) Its input and output waveforms.

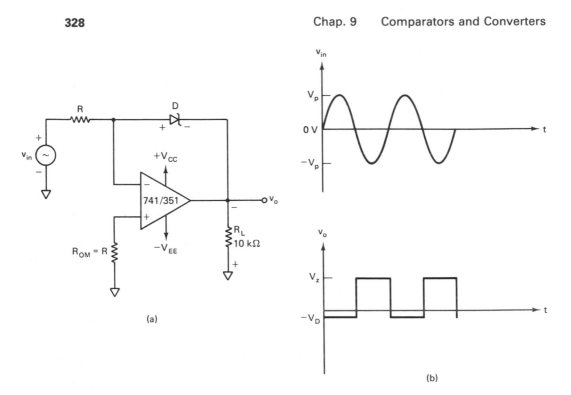

Figure 9–7   (a) Basic op-amp comparator with positive and negative limiting. (b) Its input and output waveforms.

[see Figure 9-7(b)]. The exact opposite result can be obtained by reversing the direction of the zener diode. That is, v_o will be limited to $-V_Z$ and $+V_D$.

   Voltage limiters are commonly used in communications devices such as TV and FM receivers.

EXAMPLE 9-2

In the circuit of Figure 9-7(a), v_{in} = 500 mV peak 60-Hz sine wave, R = 100 Ω, 1N3826 zener with V_Z = 5.1 V, and the supply voltages = ±15 V. Determine the output voltage swing. Assume that the voltage drop across the forward-biased zener = 0.7 V.

SOLUTION   During the positive half-cycle of the input waveform, the output voltage would be at $-V_D$ = -0.7 V because the zener will be forward biased. However, during the negative half-cycle of v_{in}, v_o would be at $+V_Z$ = $+5.1$ V since the zener will be reverse biased into zener breakdown. Thus, the use of a zener diode in the feedback path limits v_o to $+5.1$ V and -0.7 V. This output voltage swing will, therefore, make the op-amp comparator of Figure 9-7(a) compatible with TTL.

9-8 HIGH-SPEED AND PRECISION-TYPE COMPARATORS ──────────

Although op-amps can be used as comparators, specially designed comparators outperform the op-amp types. The specially designed devices are optimized for the most desirable parameters, such as speed and accuracy. Their output is compatible with RTL, DTL, TTL, and MOS logic. A variety of integrated comparators are available, including FET-input, low-power, low-offset voltage, high-speed, and precision voltage comparators. The information included on the comparator data sheets is identical to that on op-amp data sheets in most cases. A typical comparator data sheet includes a general description of the device, a schematic diagram, a connection diagram, absolute maximum ratings, electrical characteristics, typical performance curves, and applications. Electrical parameters such as response time, positive output level, negative output level, strobe current, strobe release time, and saturation voltage are typical of comparators. These parameters are defined as follows. The *response time* is the interval between the application of an input step function and the time when the output crosses the logic threshold voltage. The *positive output level* or *output high voltage* is the high output voltage level with a given load and the input drive equal to or greater than a specified value. The *negative output level* or *output low voltage* is the negative dc output voltage with the comparator saturated by a differential input equal to or greater than a specified voltage. The *strobe current* is the current out of the strobe terminal when it is at the zero logic level. The *strobe release time* is the time required for the output to rise to the logic threshold voltage after the strobe terminal has been driven from *zero* to the *one* logic level. The *saturation voltage* is the low-output voltage level with the input drive equal to or greater than a specified value. Furthermore, parameters such as strobe current and strobe release time are only listed for comparators with strobe capability.

The μA311 precision comparator is designed for low-level signal detection and high-level output drive capability. It can operate from ± 15-V op-amp supplies down to the single $+5$-V supply used for IC logic. In addition, its output can drive RTL, DTL, TTL, and MOS logic, as well as lamps or relays. Outputs can also be wire ORed. The input slew rate can be increased by increasing the input stage current. The μA311 has input offset voltage balancing and TTL strobe capability [see Figure 9-8(a)]. The strobe capability allows the comparator's output to either respond to input signals or be independent of input signals. The response time is typically 200 ns for a 100-mV input step with 5-mV overdrive.

The μAF311 is a FET input comparator that has input currents (I_{io} and I_B) more than 1000 times lower than for the μA311. Except for this difference, the μAF311 has the same characteristics and features as the μA311.

As shown in Figure 9-8(b), the LM1414 has two totally separate comparators with independent strobe capability. Compared to the 311s, it has lower response time (30 ns typical) at the expense of higher input currents.

The features of the ICL8001 include low input currents, low power consumption (30 mW), and 250-ns response time. An output stage enables the designer to control the output voltage swing. That is, the positive output level can be adjusted to suit a desired logic family by changing V+ [see Figure 9-8(c)].

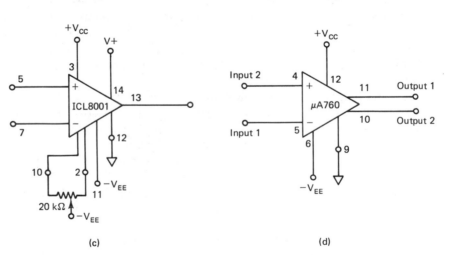

Figure 9–8 Connection diagrams of comparators. (a) μA311: when not used, strobe terminal 6 should either be open or connected to positive supply. (b) LM1414: pin 4 NC. (c) ICL8001: pin connections shown are for 14-pin DIP package. (d) μA760: pin connections shown are for 14-pin DIP package.

The μA760 is a high-speed differential voltage comparator with *complementary* TTL outputs. It has a typical response time of 16 ns and operates from supplies of ±4.5 to ±6.5 V [see Figure 9-8(d)]. Its typical applications are high-speed analog-to-digital converters and zero-crossing detectors.

9-9 WINDOW DETECTOR

Sometimes there is a need to determine when an unknown input is between two precise reference thresholds V_{ut} and V_{lt}. This determination can be made by a circuit called the window detector. Figure 9-9(a) shows such a circuit using a dual comparator LM1414. An unknown voltage v_{in} is applied to the $(-)$ input of comparator C_1 (½ LM1414) and to the $(+)$ input of comparator C_2 (remaining ½ LM1414). An upper threshold voltage V_{ut} is applied to the $(+)$ input of C_1, whereas the lower threshold voltage V_{lt} is connected to the $(-)$ input of C_2. In addition, the outputs of C_1 and C_2 are connected to form a single output v_o. When v_{in} is between V_{ut} and V_{lt}, the output v_o is *high*. However, when v_{in} goes above V_{ut} or drops below V_{lt}, the output v_o switches *low* or to 0 V. As a visual aid an appropriate lamp circuit may be connected at the output of the comparator. The window detector can also be "strobed" when needed. Window detectors are usually used in industrial alarms, level detectors and controls, digital computers, and production-line testing.

The Burr-Brown 4115/04 is a hybrid IC window detector that is available in a 14-pin double-width DIP. Figure 9-9(b) and (c) show the pin diagram and block diagram of the 4115/04. As shown in Figure 9-9(c), the inputs are diode protected and the value of the input voltages can be as high as ± 15 V. The device has three mutually exclusive outputs, HIGH, LOW, and GO, and sinks up to 200 mA of current when an output is *on*. The required power supply voltages are ± 15 V, while the maximum load supply voltage $+V_L$ can be up to $+30$ V. For proper operation of the device, separate connections should be made from each power supply common ($+V_{CC}$, $-V_{EE}$, and $+V_L$) to the device common, that is, pin 8. The 4115/04 is designed to drive TTL and DTL logic devices as well as lamps and relays.

9-10 VOLTAGE-TO-FREQUENCY AND FREQUENCY-TO-VOLTAGE CONVERTERS

The SE/NE 566 voltage-to-frequency converter is discussed in Section 8-18. This section investigates the Teledyne 9400 series, which can be used as voltage-to-frequency (V/F) or frequency-to-voltage (F/V) converters. The series includes the 9400, 9401, and 9402 converters. These converters have the same internal circuitry and connection diagrams, but they differ slightly in electrical characteristics. A complete V/F or F/V system can be formed simply by using two external capacitors, three resistors, and a reference voltage. In addition, the 9400 series consists of CMOS and bipolar devices that can operate on dual or single supply voltages.

9-10.1 V/F Converter

The 9400 is designed for pulse and square wave outputs having a frequency range of 1 Hz to 100 kHz. Furthermore, the input can be either current or voltage, and the output can interface with most forms of logic. When it is used as a V/F converter, the equivalent circuit (inside the dashed box) and connection diagram of the 9400 is as shown in Figure 9-10. The equivalent circuit consists of an integrator, comparators, a delay network, a divide-by-2 network, and open-collector output transistors. The input current $I_{in} = V_{in}/R_{in}$ is converted to a charge by the integrating capacitor C_{int} and shows up as a

Figure 9–9 (a) Window detector using LM1414. The BB4415/04 window detector. (b) Pin diagram. (c) Block diagram. (d) Transfer characteristics. (Courtesy of Burr-Brown Research Corporation.)

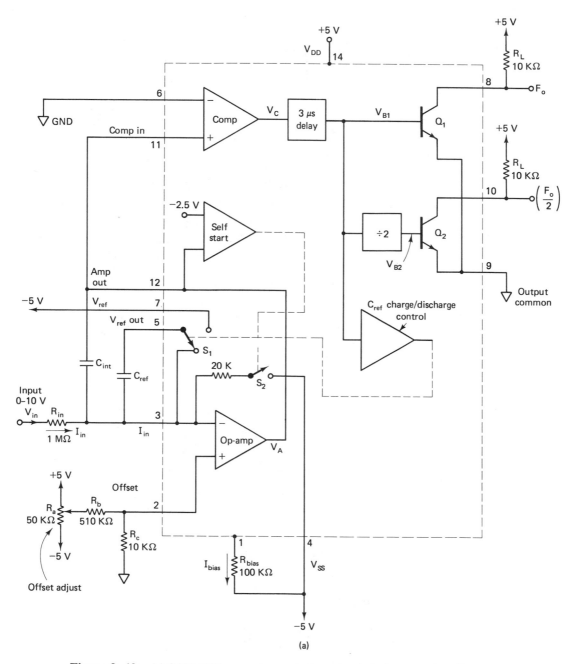

Figure 9–10   (a) 9400 V/F converter equivalent circuit and connection diagram. (b) Its waveforms. (Courtesy of Teledyne Semiconductor.)

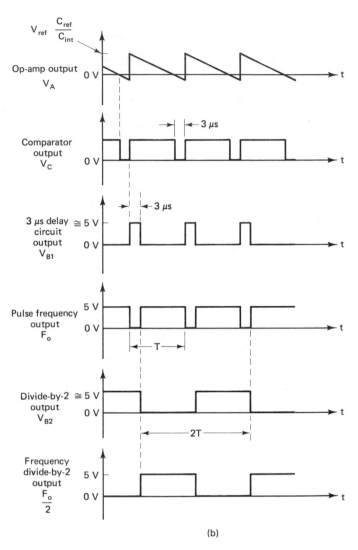

(b)

Figure 9–10 (Continued)

linearly decreasing voltage V_A at the output of the op-amp integrator, as shown in Figure 9-10(b). In equation form,

$$V_A = -\left(\frac{I_{in}}{C_{int}}\right)t \qquad (9\text{-}3)$$

where I_{in} = input current (amperes)
     C_{int} = integrating capacitor (farads)
       t = time (seconds)

The output V_A of the integrator is sensed by the comparator. The output of the comparator is then applied via the 3-μs delay network to the output transistor Q_1, the divide-by-2 network, and the C_{ref} charge/discharge control circuit. The output of the divide-by-2 network drives the output transistor Q_2. When the output V_A of the op-amp is positive, the output V_C of the noninverting comparator is *high,* or $\cong +5$ V, and the output V_{B1} of the 3-μs delay network is *low* or 0 V. Since V_{B1} is *low,* the transistor Q_1 is *off,* and the output F_o is *high,* or $\cong +5$ V [see Figure 9-10(b)]. The divide-by-two network is a negative-edge-triggered flip-flop whose output voltage V_{B2} is a complement (inversion) of V_{B1}; hence the transistor Q_2 will be on, and output $F_o/2$ will be *low.* Finally, with V_{B1} *low,* the C_{ref} charge/discharge control circuit is disabled, and capacitor C_{ref} remains discharged (is shorted out). However, from Equation (9-3), the output V_A of the op-amp is linearly decreasing. Therefore, when V_A goes slightly below 0 V, the output V_C of the comparator switches "low" to 0 V. Three microseconds after V_C goes low, V_{B1} switches from low to high. This forces output F_o *low* and enables the C_{ref} charge/discharge control circuit, which connects C_{ref} to the reference voltage V_{ref} of -5 V. At this instant the negative voltage at the $(-)$ input causes the output of the op-amp to step up to a finite amount, as shown in Figure 9-10(b). The reference capacitor C_{ref} remains connected to V_{ref} for a time period long enough to virtually charge it to V_{ref}. The charging path for C_{ref} is through the output terminal of the op-amp, through C_{int}, through C_{ref}, and finally through the negative reference voltage. Now, since the output of the op-amp V_A is positive, the output of the comparator switches from low to high, and 3 μs later V_{B1} goes low. This disables the C_{ref} charge/discharge control circuit, and C_{ref} is shorted out, dissipating the stored reference charge. When V_{B1} goes low, the output F_o goes high, the output of the divide-by-2 circuit goes low, and output $F_o/2$ therefore goes high [see Figure 9-10(b)]. The integrating capacitor C_{int} again begins converting the input current I_{in} to a charge, and in turn the output of the op-amp starts decreasing linearly [see Equation (9-3)]. When the output of the op-amp goes slightly below 0 V, the output of the comparator goes from high to low, and hence the cycle repeats. In short, the continued discharging of C_{int} by the input current I_{in} is balanced out by fixed charge from V_{ref}. As the input voltage V_{in} is increased, the rate of decrease of output voltage V_A also increases [see Equation (9-3)], causing output frequency F_o to also increase. Since each positive voltage increment in V_A is fixed, the increase in frequency with voltage is linear.

As shown in Figure 9-10(a), the 9400 contains a "self-start" circuit that assures proper V/F operation when power is first applied. If the op-amp output V_A is below the comparator threshold of 0 V and C_{ref} is already charged, then, when the power is first applied, a positive voltage step will not occur. In this situation the output V_A of the op-amp continues to decrease until it crosses the -2.5-V threshold of the "self-start" comparator. When this happens, the self-start comparator connects through the 20-kΩ resistor a negative supply voltage V_{SS} to the $(-)$ input of the op-amp, which in turn causes the output of the op-amp to quickly go positive. As soon as V_A goes positive, the self-start comparator is disabled, and the 9400 resumes its normal operating mode.

The 9400 is a 14-pin DIP with plastic and ceramic options. The pin diagram is shown in Figure 9-11. The functions of the pins are as follows:

Figure 9–11 Pin diagram for the 9400.

Pin 11: Comparator input. This pin is the (+) input of the comparator and is connected to the output of the op-amp in the V/F mode. In the F/V mode, the input frequency is applied to the comparator input.

Pin 8: Pulse frequency output. At this pin a pulse waveform output is available whose frequency is proportional to the input voltage. Since this is an open-collector output, it requires a pull-up resistor and interfaces directly with all forms of logic.

Pin 10: Frequency divide-by-2 output. This pin provides a square-wave output that is one-half the frequency of the pulse output at pin 8. This output is an open collector and hence requires a pull-up resistor. It is also compatible with all logic families.

Pin 9: Output common. The emitters of both the output transistors are connected to this pin. A unipolar or bipolar output voltage swing can be obtained by connecting this pin either to ground or to the negative supply V_{SS}, respectively. For high-performance applications, the output ground (pin 9) should be separated from the input ground (pin 6).

Pin 1: I_{bias}. A resistance R_{bias} is connected between pins 1 and 4. Specifications for the 9400 are based on $R_{bias} = 100$ k$\Omega \pm 10\%$.

Pin 12: Amplifier out. This is the output of the op-amp, which is a negative-going ramp signal in the V/F mode. In the F/V mode, a voltage proportional to the frequency input is generated at this pin.

Pin 2: Zero adjust. The circuit that can be used to reduce the output initially to zero is connected to this pin. The low-frequency set point is determined by adjusting the voltage at this pin.

Pin 3: I_{in}. This pin is the (−) input of the op-amp, and the input current I_{in} is applied to it. I_{in} is 10 μA for nominal full scale with an overrange current up to 50 μA.

Pin 7: V_{ref}. A reference voltage from either a separate negative supply or the V_{SS} may be applied to this pin.

Pin 5: V_{ref} out. When C_{ref} is to be charged, the C_{ref} charge/discharge control circuit connects this pin to the reference voltage V_{ref} (pin 7).

Pin 13: NC. No connection.

Pin 6: GND. Input ground.

9-10.1(a) V/F design procedure.   The output frequency F_o of the 9400 V/F converter is related to the analog input voltage V_{in} by the equation

$$F_o = \frac{V_{in}}{R_{in}} \frac{1}{(V_{ref})(C_{ref})} \tag{9-4}$$

The V/F converter shown in Figure 9-12 can be designed by using the following steps:

1. Choose V_{DD} and V_{SS} such that

$$4 \text{ V} \le V_{DD} \le 7.5 \text{ V}$$

$$-7.5 \text{ V} \le V_{SS} \le -4 \text{ V}$$

Generally, $V_{DD} = +5$ V and $V_{SS} = -5$ V. For high accuracy, 0.1-μF disk decoupling capacitors located near pins 4 and 14 are recommended.

2. The reference voltage V_{ref} can be a separate precision source or can be the negative supply voltage V_{SS}. Usually, $V_{ref} = V_{SS} = -5$ V.

3. R_{in} is calculated from the equation

$$R_{in} = \frac{V_{in} \text{ full scale}}{I_{in} \text{ full scale}}$$

For the 9400 V/F, the specified values of V_{in} full scale = 10 V and I_{in} full scale = 10 μA. Hence $R_{in} = 1$ MΩ. Metal film resistors with $\le 1\%$ tolerance are recommended for high accuracy.

Figure 9–12   A 10-Hz to 10-kHz V/F converter. To increase $F_{o\,max}$ to 100 kHz, change C_{ref} to 27 pF and C_{int} to 75 pF. (Courtesy of Teledyne Semiconductor.)

4. For the 9400 V/F, the specified value of $R_{bias} = 100$ kΩ.
5. The specified value of pull-up resistors R_L is 10 kΩ each (see Figure 9-12).
6. Although the exact value of C_{ref} is not critical, it should be <500 pF. This capacitor should be located as close as possible to pins 3 and 5. Glass film capacitors are recommended for high accuracy.
7. Choose the value of C_{int} such that

$$3C_{ref} \leq C_{int} \leq 10C_{ref}$$

However, for improved stability and linearity, C_{int} must be $\geq 4C_{ref}$. This capacitor should be located as close as possible to pins 3 and 12.
8. The exact values of offset adjust resistors R_a, R_b, and R_c are not critical; however, they are related by the relationship

$$R_c < R_a < R_b$$

(see Figure 9-12).

9-10.1(b) Adjustment/calibration procedure.   For a 10-kHz full-scale output frequency F_o, the adjustment/calibration procedure is as follows:

1. Set V_{in} to 10 mV and adjust the R_a potentiometer to obtain a 10-Hz output frequency (see Figure 9-12).
2. Set V_{in} to 10 V and adjust R_{in}, V_{ref}, or C_{ref} to obtain a 10-kHz output frequency.

9-10.1(c) Single-supply operation.   The 9400 V/F converter can also be operated using a single supply, as shown in Figure 9-13.

EXAMPLE 9-3
The V/F converter of Figure 9-12 is initially adjusted for a 10-kHz full-scale output frequency. Determine the output frequencies F_o and $F_o/2$ if the input signal $V_{in} = 2$ V.

SOLUTION   The output frequency F_o of the 9400 is directly proportional to the input voltage V_{in}. In addition, it has a guaranteed linearity of 0.05%. Since the circuit is initially adjusted for a 10-kHz full-scale output frequency (at $V_{in} = 10$ V), when the input $V_{in} = 2$ V, the output frequency F_o will be 2kHz± 0.05% and $F_o/2$ will be 1 kHz± 0.05%. The nature of these output waveforms will be as shown in Figure 9-10(b).

The V/F converter is used in instrumentation and control, digital, and communication systems. Typical applications that use V/F converters are in temperature sensing and control, transducer encoding, analog-to-digital (A/D) converters, microprocessor data acquisition, digital panel meters, phase-locked loops, and analog data transmission and recording. For specific application of the V/F converter, refer to Chapter 11.

9-10.2 F/V Converter

When used as an F/V converter the 9400 generates an output voltage that is linearly proportional to the input frequency waveform. The features of the 9400 F/V converter

Figure 9–13   The 9400 V/F converter using single-variable supply voltage with offset and gain adjust. (Courtesy of Teledyne Semiconductor.)

include dc to 100-kHz operation, op-amp output, programmable scale factor, high input impedance (>10 MΩ), and, above all, its capability to accept any voltage wave shape. Figure 9-14(a) shows an equivalent circuit (inside the dashed box) and connection diagram of the 9400 when it is used as a F/V converter. The input frequency is applied to the $(+)$ input of the comparator (pin 11). Since the comparator hysteresis voltage is ±200 mV, the input signal amplitude must be greater than ±200 mV in order to trip the comparator. If only a unipolar input signal is available, it can be converted into a bipolar waveform by using the offset circuit of Figure 9-15.

Each time the input signal crosses zero in the negative direction, the output of the comparator goes low. Three microseconds later the C_{ref} *charge/discharge control circuit* is enabled, which instantaneously connects the reference capacitor C_{ref} to the reference voltage V_{ref} [see Figure 9-14(a)]. This action charges C_{int} each time with a precise amount of voltage until the voltage across it can no longer increase. The charging path is through the output terminal of the op-amp, through C_{int}, through C_{ref}, and finally through V_{ref}. On the other hand, each time the input waveform crosses zero in the positive direction, the output of the comparator switches high. This disables the C_{ref} *charge/discharge control*

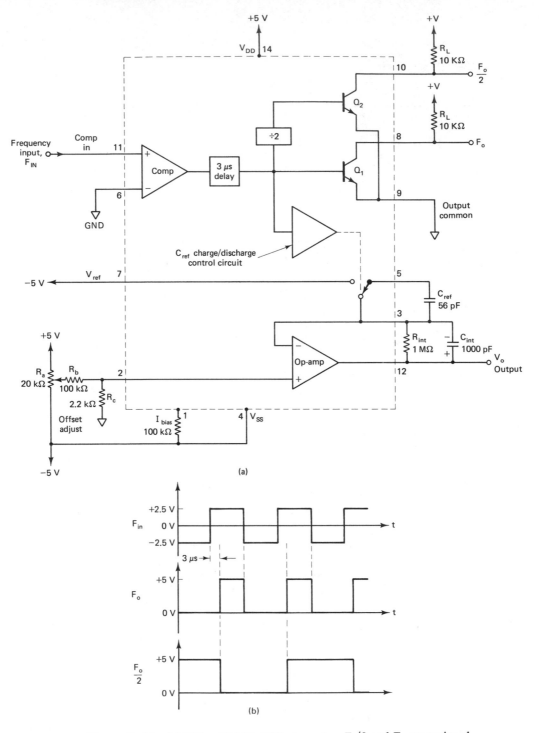

Figure 9–14 (a) DC to 10-kHz F/V converter. $F_o/2$ and F_o are optional if a buffer is needed. When not used, pins 8, 9, and 10 may be connected to ground. (b) F/V digital outputs. (Courtesy of Teledyne Semiconductor.)

Figure 9–15   Offset circuit to convert unipolar waveform into bipolar waveform.

circuit, and C_{ref} is shorted out. However, the voltage across C_{int} is retained because the only discharge path for C_{int} is through R_{int}, which is very large (1 MΩ). The voltage across C_{int} is the output voltage V_o.

The amount of ripple on V_o is inversely proportional to C_{int} and the frequency of the input F_{in}. This means that for low frequencies C_{int} can be increased in the range from 1 to 100 μF to reduce the ripple. To eliminate the ripple on V_o, an extra op-amp that is operating in the common-mode configuration shown in Figure 9-16 can be connected to the output of the F/V converter of Figure 9-14(a). Because of the common-mode configuration, the ac ripple is canceled at the output of an op-amp provided that both (+) and (−) inputs have the same gain. The 10-kΩ potentiometer is used to make the gain of (+) and (−) inputs equal. Besides that, the circuit in Figure 9-16 has a *dc* gain of unity, so output voltage is equal to input.

In the F/V converter of Figure 9-14(a), outputs F_o and $F_o/2$ are optional because these outputs are useful in only some applications. The output F_o follows the input frequency waveform with a 3-μS delay on the rising edge [see Figure 9-14(b)]. $F_o/2$ is a square wave with a frequency one-half of F_o. If F_o and $F_o/2$ outputs are not used, pins 8 and 10 may be connected to ground.

Figure 9–16   F/V ripple eliminator.

Although the F//V converter accepts any input wave shape,, the circuit will work only if the positive pulse width of the input waveform is at least 5 μs and the negative pulse width is ≥ 0.5 μs. When $F_{in\,max}$ is less than 1 kHz, the duty cycle [= (pulse width)(100)/time period] should be greater than 20% to ensure that C_{ref} is fully charged and discharged.

9-10.2(a) F/V design procedure.   The output voltage V_o of the F/V converter is related to the input frequency F_{in} by the equation

$$V_o = (V_{ref}\,C_{ref}\,R_{int})F_{in} \tag{9-5}$$

The response time to a change in F_{in} is equal to $R_{int}\,C_{int}$. The F/V converter can be designed by following these steps:

1. Choose V_{DD} and V_{SS} such that

$$4\ V \leq V_{DD} \leq 7.5\ V$$

$$-7.5\ V \leq V_{SS} \leq -4\ V$$

   (Usually, $V_{DD} = +5$ V and $V_{SS} = -5$ V.)
2. Choose $V_{ref} = V_{SS} = -5$ V.
3. Choose $R_{int} = 1$ MΩ for $F_{in\,max}$ of 10 kHz. However, for $F_{in\,max}$ of 100 kHz, R_{int} should be decreased to 100 kΩ.
4. $R_{bias} = 100$ kΩ is the specified value for the 9400 F/V converter.
5. Choose pull-up resistors $R_L = 10$ kΩ.
6. Choose $C_{ref} \cong 56$ pF for a $F_{in\,max}$ of 10 kHz. However, C_{ref} should be increased for a lower $F_{in\,max}$.
7. Choose $C_{int} = 1000$ pF for $F_{in\,max}$ of 10 kHz. However, for a lower $F_{in\,max}$, C_{int} can be increased in the range from 1 to 100 μF.
8. Choose offset voltage-compensating network resistors R_a, R_b, and R_c such that $R_c < R_a < R_b$ [see Figure 9-14(a)].

9-10.2(b) Adjustment/calibration procedure.   For a 10-kHz maximum input frequency $F_{in\,max}$, the adjustment/calibration procedure is as follows:

1. When no input signal is applied ($F_{in} = 0$), adjust the offset null circuit to obtain a 0-V dc output voltage.
2. Set $F_{in} = 10$ kHz and adjust C_{ref} so that V_o is approximately 2.5 to 3 V.

9-10.2(c) Single-supply operation.   The 9400 F/V converter can also be operated using a single supply as shown in Figure 9-17. In this figure, V_{ref} is defined as the voltage difference between pins 2 and 7. The comparator's $(-)$ input is referenced to 6.2 V; therefore, the input signal must satisfy the requirement

$$4\ V < \text{input voltage} < V_{DD}$$

The output voltage in turn is also referenced to 6.2 V. Additionally, if the input signal is ac-coupled, a 100-kΩ to 10-MΩ resistor must be connected between $(+)$ and $(-)$ inputs (pins 11 and 6) of the comparator.

Figure 9–17   The 9400 F/V converter using single supply. (Courtesy of Teledyne Semiconductor.)

The 9400 F/V converter can be used in applications such as frequency meters and tachometers, speedometers, rpm (revolutions per minute) indicators, FM demodulation frequency multipliers and dividers, and motor control.

EXAMPLE 9-4

The F/V converter of Figure 9-14(a) is initially adjusted for $V_o = 2.8$ V at $F_{in\,max}$ of 10 kHz. Determine the output voltage V_o if $F_{in} = 1$ kHz.

SOLUTION   The 9400 F/V converter generates an output voltage that is linearly proportional to the input frequency F_{in}. Therefore, at $F_{in} = 1$ kHz, $V_o = 2.8$ V/10 $= 0.28$ V.

9-11 ANALOG-TO-DIGITAL AND DIGITAL-TO-ANALOG CONVERTERS

Digital systems are used in ever more applications because of their increasingly efficient, reliable, and economical operation. With the development of the microprocessor, data processing has become an integral part of various systems. Data processing involves transfer of data to and from the microcomputer via input/output devices. Since digital systems such as microcomputers use a *binary system* of ones and zeros, the data to be put into the microcomputer have to be converted from analog form to digital form. The circuit that performs this conversion is called an *analog-to-digital (A/D) converter.* On the other hand, a *digital-to-analog (D/A) converter* is used when a binary output from a digital system must be converted to some equivalent analog voltage or current. The binary output (a sequence of 1's and 0's) from a digital system is difficult to interpret; however, a D/A converter makes the interpretation easier. The function of a D/A converter is exactly opposite that of an A/D converter. This section presents both types of converters. The D/A converter is presented first because (1) it is simpler than the A/D converter and (2) it can be used to form the A/D converter.

9-11.1 D/A Converters

A D/A converter in its simplest form uses an op-amp and either *binary-weighted resistors* or R and $2R$ resistors (see Figures 9-18 and 9-19).

9-11.1(a) D/A converter with binary-weighted resistors.   Figure 9-18 shows a D/A converter using an op-amp and binary-weighted resistors. Although in this figure the op-amp is connected in the inverting mode, it can also be connected in the noninverting mode. Since the number of binary inputs is four, the converter is called a 4-*bit* (*bi*nary dig*it*) converter. Because there are 16 (2^4) combinations of binary inputs for b0 through b3, an analog output should have 16 possible corresponding values. In Figure 9-18, four switches (b0 to b3) are used to simulate the binary inputs; in practice, a 4-bit binary counter such as the 7493 may be used instead. When switch b0 is closed (connected to +5 V), the voltage across R is 5 V because $V_2 = V_1 = 0$ V. Therefore, the current through R is 5 V/10 kΩ = 0.5 mA. However, the input bias current I_B is negligible; hence the current through feedback resistor R_F is also 0.5 mA, which, in turn produces an output voltage of $-(1 \text{ k}\Omega)(0.5 \text{ mA}) = -0.5$ V. Note that the op-amp is working as a current-to-voltage converter. Now suppose that switch b1 is closed and b0 is opened. This action connects $R/2$ to the positive supply of +5 V, causing twice as much current (1 mA) to flow through R_F, which in turn doubles the output voltage. Thus the output voltage V_o is -1 V when switch b1 is closed. Similarly, if both switches b0 and b1 are closed, the current through R_F will be 1.5 mA, which will be converted to an output voltage of $-(1 \text{ k}\Omega)(1.5 \text{ mA}) = -1.5$ V.

Thus, depending on whether switches b0 to b3 are open or closed, the binary-weighted currents will be set up in input resistors. The sum of these currents is equal to the current through R_F, which in turn is converted to a proportional output voltage. When all the switches are closed, obviously the output will be maximum. The output voltage equation is given by

$$V_o = -R_F\left(\frac{b0}{R} + \frac{b1}{R/2} + \frac{b2}{R/4} + \frac{b3}{R/8}\right) \qquad (9\text{-}6)$$

where each of the inputs b3, b2, b1, and b0 may either be high (+5 V) or low (0 V).

Figure 9-18(b) shows analog outputs versus possible combinations of inputs. The output is a negative-going staircase waveform with 15 steps of -0.5 V each. In practice, however, the steps may not all be the same size because of the variations in *logic high*

(a)

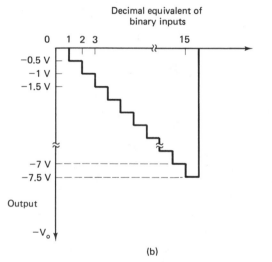

(b)

Figure 9–18   (a) D/A converter with binary-weighted resistors. (b) Graph of output versus inputs.

voltage levels. Notice that the size of the steps depends on the value of R_F. Therefore, a desired step can be obtained by selecting a proper value of R_F, provided that the maximum output voltage does not exceed the saturation levels of an op-amp. For accurate operation of the D/A converter precision metal film resistors are recommended.

The problem with the D/A converter of Figure 9-18 is that it requires binary-weighted resistors, which may not be readily available, especially if the number of inputs is more than four. An attractive alternative is to use R and $2R$ resistors for the D/A converter since it requires only two sets of precision resistance values.

9-11.1(b) D/A converter with *R* and *2R* resistors.   Figure 9-19(a) shows a D/A converter with R and $2R$ resistors. As before, the binary inputs are simulated by

(b)

Decimal equivalent of binary inputs	Input (V)				Output voltage (V)
	b3	b2	b1	b0	
0	0	0	0	0	0
1	0	0	0	5	−0.625
2	0	0	5	0	−1.25
3	0	0	5	5	−1.875
4	0	5	0	0	−2.50
5	0	5	0	5	−3.125
6	0	5	5	0	−3.750
7	0	5	5	5	−4.375
8	5	0	0	0	−5.0
9	5	0	0	5	−5.625
10	5	0	5	0	−6.25
11	5	0	5	5	−6.875
12	5	5	0	0	−7.50
13	5	5	0	5	−8.125
14	5	5	5	0	−8.875
15	5	5	5	5	−9.375

(c)

Figure 9–19 (a) D/A converter with $R/2R$ resistors. (b) Equivalent circuit when b3 is high and b0, b1, and b2 are low. (c) Output versus inputs.

switches b0 through b3, and the output is proportional to the binary inputs. Binary inputs can either be in the *high* (+5 V) or *low* (0 V) state. Assume that the most significant bit (MSB) switch b3 is connected to +5 V and other switches are connected to ground, as in Figure 9-19(a). Thévenizing the circuit to the left of switch b3, Thévenin's equivalent resistance R_{TH} is

$$R_{TH} = [\{[(2R \parallel 2R + R) \parallel 2R] + R\} \parallel 2R] + R$$
$$= 2R = 20 \text{ k}\Omega \tag{9-7a}$$

The resultant circuit is shown in Figure 9-19(b). In this figure the (−) input is at virtual ground ($V_2 \cong 0$ V); therefore, the current through $R_{TH}(= 2R)$ is zero. However, the current through $2R$ connected to +5 V is 5 V/20 kΩ = 0.25 mA. The same current flows through R_F and in turn produces the output voltage

$$V_o = -(20 \text{ k}\Omega)(0.25 \text{ mA}) = -5 \text{ V} \tag{9-7b}$$

Using the same analysis, the output voltage corresponding to all possible combinations of binary inputs can be calculated as shown in Figure 9-19(c). The maximum or full-scale output of −9.375 V is obtained when all the inputs are high. The output voltage equation can be written as

$$V_o = -R_F\left(\frac{b3}{2R} + \frac{b2}{4R} + \frac{b1}{8R} + \frac{b0}{16R}\right) \tag{9-8}$$

where each of the inputs b3, b2, b1, and b0 may be either high (+5 V) or low (0 V).

The great advantage of the D/A converter of Figure 9-19(a) is that it requires only two sets of precision resistance values; nevertheless, it requires more resistors and is also more difficult to analyze than the binary-weighted resistor type. As the number of binary inputs is increased beyond four, both D/A converter circuits get complex and their accuracy degenerates. Therefore, in critical applications an integrated circuit specially designed as a D/A converter should be used.

9-11.1(c) Monolithic/hybrid D/A converters.   At present, 8-, 10-, 12-, 14-, and 16-bit D/A converters are available with either a current output, a voltage output, or both current and voltage outputs. The MC1408L is a common 8-bit monolithic D/A converter with a current output that can be converted to a voltage type by using an I-to-V converter op-amp [see Figure 9-20(a)]. On the other hand, the SE/NE 5018 is a typical 8-bit D/A converter with voltage output. Figure 9-20(b) shows the SE/NE 5018 configured for unipolar output (0 to 10 V). For 12 bits of resolution as well as current and voltage outputs, hybrid D/A converters such as the DATEL DAC-HZ series are an excellent choice. To make an intelligent selection from a variety of D/A converters to suit a given application, it is necessary to know the key specifications of D/A converters, such as resolution, nonlinearity, gain error, and settling time.

Resolution. Resolution is determined by the number of input bits of the D/A converter. For example, an 8-bit converter has 2^8 (= 256) possible output levels; therefore, its resolution is 1 part in 256. In short, resolution is a value of the LSB.

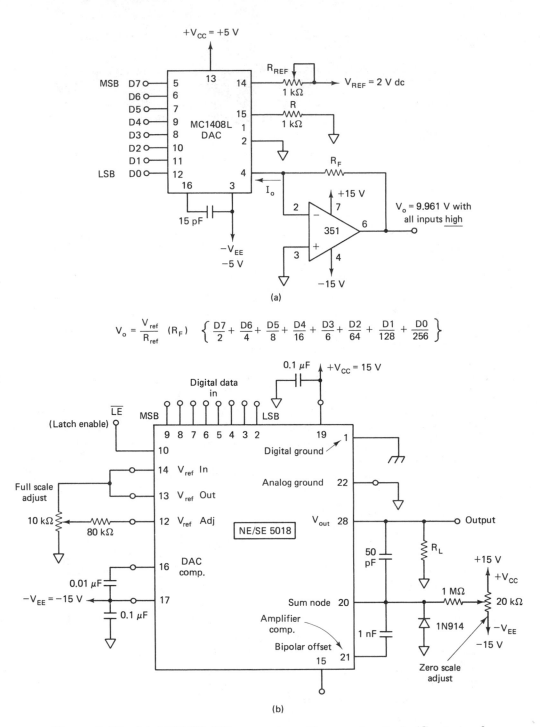

$$V_o = \frac{V_{ref}}{R_{ref}} (R_F) \left\{ \frac{D7}{2} + \frac{D6}{4} + \frac{D5}{8} + \frac{D4}{16} + \frac{D3}{6} + \frac{D2}{64} + \frac{D1}{128} + \frac{D0}{256} \right\}$$

Figure 9–20 (a) MC1408 D/A converter with current output. (Courtesy of Motorola Semiconductor.) (b) NE/SE 5018 D/A converter with voltage output. (Courtesy of Signetics Corporation.)

Nonlinearity or linearity error. This is the difference between the actual output of the DAC and its ideal straight-line output. The error is normally expressed as a percentage of the full-scale range.

Gain error and offset error. Gain error is usually caused by the deviations in the feedback resistor on the *I*-to-*V* converter. On the other hand, offset error implies that the output of the DAC is not zero when the binary inputs are all zero. This error stems from the input offsets (*V* and *I*) of the op-amp as well as the DAC.

Settling time. This is the time required for the output of the DAC to settle to within $\pm 1/2$ LSB of the final value for a given digital input, that is, zero to full scale.

Typical applications for D/A converters include microcomputer interfacing, CRT graphics generation, programmable power supplies, digitally controlled gain circuits, digital filters, and others.

9-11.2 A/D Converters

A/D converters convert an analog voltage to the digital output that best represents the input. As in the case of D/A converters, analog converters are also specified as 8, 10, 12, or 16 bit. There are many types of A/D converters: single-ramp integrating, dual-ramp integrating, single counter, tracking, and successive approximation. This section discusses only the successive-approximation type, which uses a comparator, a successive-approximation register, output latches, and a D/A converter.

9-11.2(a) Successive-approximation A/D converter.   Figure 9-21 shows a successive-approximation type of A/D converter. The heart of the circuit is an 8-bit successive-approximation register (SAR), whose output is applied to an 8-bit D/A converter. The analog output (V_a) of the D/A converter is then compared to an analog input signal V_{in} by the comparator. The output of the comparator is a serial data input to the SAR. The SAR then adjusts its digital output (8 bits) until it is equivalent to analog input V_{in}. The 8-bit latch at the end of conversion holds onto the resultant digital data output. The circuit works as follows. At the start of a conversion cycle, the SAR is reset by holding the start (S) signal high. On the first clock pulse LOW-to-HIGH transition, the most significant output bit Q_7 of the SAR is set. The D/A converter then generates an analog equivalent to the Q_7 bit, which is compared with the analog input V_{in}. If the comparator output is *low,* the D/A output $> V_{in}$ and the SAR will clear its MSB Q_7. On the other hand, if the comparator output is *high,* the D/A output $< V_{in}$ and the SAR will keep the MSB Q_7 set. In any case, on the next clock pulse LOW-to-HIGH transition, the SAR will set the next MSB Q_6. Depending on the output of the comparator, the SAR will then either keep or reset the bit Q_6. This process is continued until the SAR tries all the bits. As soon as the LSB Q_0 is tried, the SAR forces the conversion complete (CC) signal HIGH to indicate that the parallel output lines contain valid data. The CC signal in turn enables the latch, and digital data appear at the output of the latch. Digital data are also available serially as the SAR determines each bit. To cycle the converter continuously, the CC signal may be connected to the start conversion input (see Figure 9-21). The advantage of the successive-approximation A/D converter is its high speed and excellent resolution. For

Figure 9–21 Successive-approximation-type A/D converter.

example, the 8-bit successive-approximation A/D converter of Figure 9-21 requires only eight clock pulses.

9-11.2(b) Monolithic/hybrid A/D converters.

There are many monolithic A/D converters, such as the integrating A/D, integrating A/D with three-state outputs, and the tracking A/D with latched outputs. In addition, the outputs of A/D are coded in straight binary, binary-coded decimal (BCD), complementary binary (1's or 2's), or sign-magnitude binary. Figure 9-22 shows the Teledyne 8703, an 8-bit monolithic CMOS A/D converter with three-state output. The converter is microprocessor compatible, exhibits high stability over a full temperature range, contains all required active elements, and has latched parallel binary outputs with strobed or free-running conversion. It has an infinite input range since any positive voltage can be applied via a scaling register R_{in}.

As in the case of monolithics, there are many hybrid A/D converters, such as the successive-approximation A/D with input buffer amplifier, low-power CMOS A/D, the fast A/D with sample-and-hold, and the ultrafast A/D with input buffer amplifier. Datel-Intersil's hybrid ADC-815MC is a very high speed 8-bit successive-approximation A/D converter. It is capable of 8-bit resolution in only 600 ns. It has a six analog-input voltage range with parallel or serial outputs and requires no calibration. Datel's ADC-MC8B is an 8-bit monolithic multifunction A/D–D/A converter that operates on a single +5-V supply. It is a complete D/A converter that can be configured as an A/D converter by using an external comparator and a quad two-input Schmitt trigger NAND gate.

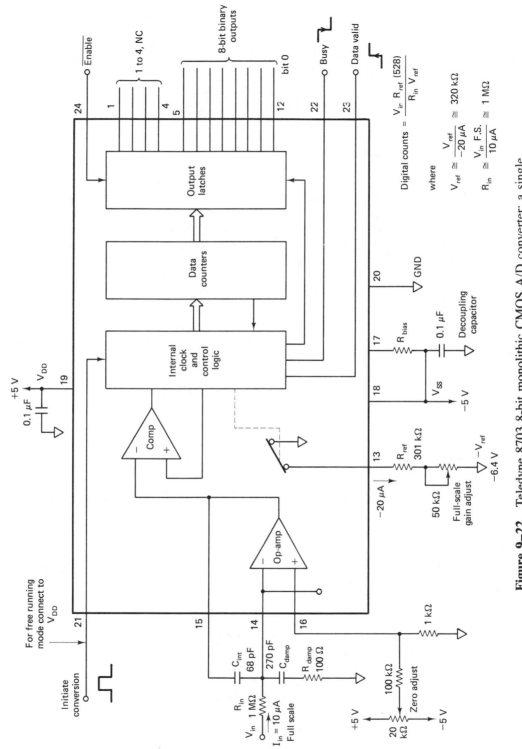

Figure 9–22 Teledyne 8703 8-bit monolithic CMOS A/D converter: a single 24-pin DIP package. (Courtesy of Teledyne Semiconductor.)

351

As with D/A converters, specifications such as resolution or nonlinearity are also used for A/D converters. Another important parameter for A/D is *conversion time*, the time required to convert an analog input into valid digital outputs.

Typical applications of A/D converters include microprocessor interfacing, data printing and recording, digital voltmeters, and control of LED or LCD displays.

9-12 CLIPPERS AND CLAMPERS

Waveshaping circuits are commonly used in digital computers and communications such as TV and FM receivers. Waveshaping techniques include limiting, clipping, and clamping. Voltage limiting was discussed in Section 9-7. This section presents typical circuits that use clipping and clamping techniques. In op-amp clipper circuits a rectifier diode may be used to clip off a certain portion of the input signal to obtain a desired output waveshape. An op-amp half-wave rectifier and an absolute-value output circuit use diodes to remove or modify certain portions of the input waveform. In the applications described previously, the diode works as an *ideal diode* (switch) because, when *on,* the voltage drop across the diode is divided by the open-loop gain of the op-amp. On the other hand, when reverse biased (off), the diode is an open circuit. In an op-amp clamper circuit, however, a predetermined dc level is deliberately inserted in the output voltage. For this reason, the clamper is sometimes called a *dc inserter.*

9-12.1 Positive and Negative Clippers

A *positive clipper,* a circuit that removes positive parts of the input signal, can be formed by using an op-amp with a rectifier diode as shown in Figure 9-23(a). In this circuit the op-amp is basically used as a voltage follower with a diode in the feedback path. The clipping level is determined by the reference voltage V_{ref}, which should be less than the input voltage range of the op-amp. Additionally, since V_{ref} is derived from the positive supply voltage ($+V_{CC}$), the dc supply voltages must be well regulated. As shown in Figure 9-23(b), the output voltage has portions of the positive half-cycles above V_{ref} clipped off.

The circuit works as follows. During the positive half-cycle of the input, the diode D_1 conducts only until $v_{\text{in}} = V_{\text{ref}}$. This happens because, when $v_{\text{in}} < V_{\text{ref}}$, the voltage ($V_{\text{ref}}$) at the ($-$) input is higher than that at the ($+$) input; hence the output voltage v'_o of the op-amp becomes sufficiently negative to drive D_1 into conduction. When D_1 conducts, it closes the feedback loop and the op-amp operates as a voltage follower; that is, output v_o follows input v_{in} until $v_{\text{in}} = V_{\text{ref}}$. However, when v_{in} is slightly higher than V_{ref}, the output v'_o of the op-amp becomes sufficiently positive to drive D_1 into cutoff. This opens the feedback loop and the op-amp operates open loop; therefore, it further drives its output v'_o toward positive saturation ($\cong +V_{CC}$). With D_1 reverse biased, the output voltage $v_o = V_{\text{ref}}$. Thus, when $v_{\text{in}} > V_{\text{ref}}$, $v'_o \cong +V_{CC}$ and $v_o = V_{\text{ref}}$ [see Figure 9-23(b)].

When v_{in} drops below V_{ref}, the output of the op-amp v'_o again becomes sufficiently negative to drive D_1 into conduction. This closes the feedback loop; hence the output follows the input. Thus diode D_1 is *on* for $v_{\text{in}} < V_{\text{ref}}$ and *off* for $v_{\text{in}} > V_{\text{ref}}$. The output follows the input only when the diode is *on*. The op-amp alternates between open-loop and

Figure 9–23 (a) Positive clipper circuit. (b) Input and output waveforms with $+V_{\text{ref}} = 1$ V. (c) Input and output waveforms with $-V_{\text{ref}} = -1$ V.

closed-loop operations as the diode D_1 is turned *off* and *on*, respectively. For this reason the op-amp used must be high speed and preferably compensated for unity gain. HA2500, LM310, and μA318 are examples of high-speed op-amps. In addition, the difference input voltage v_{id} is high during the time when the feedback loop is open (D_1 off); hence an op-amp with high difference input voltage is necessary to prevent input breakdown.

In Figure 9-23(a), if pot R_P is connected to the negative supply $-V_{EE}$ instead of $+V_{CC}$, the reference voltage V_{ref} will be negative. This will cause the entire output waveform above $-V_{ref}$ to be clipped off, as shown in Figure 9-23(c). The output follows the input only when $v_{in} < -V_{ref}$.

The positive clipper of Figure 9-23(a) is converted into a negative clipper by simply reversing diode D_1 and changing the polarity of reference voltage V_{ref}. The resultant circuit is shown in Figure 9-24(a). The negative clipper clips off the negative parts of the input signal below the reference voltage [see Figure 9-24(b)]. Diode D_1 conducts when $v_{in} > -V_{ref}$, and therefore during this period output v_o follows the input v_{in}. However, the

Figure 9–24 (a) Negative clipper circuit. (b) Input and output waveforms with $-V_{ref} = -1$ V. (c) Input and output waveforms with $+V_{ref} = +1$ V.

negative portion of the output voltage below $-V_{ref}$ is clipped off because D_1 is off for $v_{in} < -V_{ref}$. In the circuit of Figure 9-24(a), if $-V_{ref}$ is changed to $+V_{ref}$ by connecting the potentiometer R_P to the $+V_{CC}$, the output voltage below $+V_{ref}$ will be clipped off, as shown in Figure 9-24(c). The diode D_1 must be *on* for $v_{in} > +V_{ref}$ and *off* for $v_{in} < +V_{ref}$.

9-12.2 Small-Signal Half-Wave Rectifiers

The circuit of Figure 9-24(a) can be used as a positive small-signal half-wave rectifier provided that $-V_{ref} = 0$ V. Shown in Figure 9-25(a), the resultant circuit can rectify signals with peak values down to a few millivolts, unlike conventional diodes. This is possible because the high open-loop gain of the op-amp automatically adjusts the voltage drive to the diode D_1 so that the rectified output peak is the same as the input [see Figure 9-25(b)]. In fact, the diode acts as an *ideal diode* (switch), since the voltage drop across the *on* diode is divided by the open-loop gain of the op-amp. As v_{in} starts increasing in the positive direction, the v_o' also starts increasing positively until diode D_1 is forward biased. When D_1 is forward biased, it closes a feedback loop and the op-amp works as a voltage follower. Therefore, the output voltage v_o follows the input voltage v_{in} during the positive half-cycle, as shown in Figure 9-25(b). However, when v_{in} starts increasing in the negative direction, v_o' also increases negatively until it is equal to the negative saturation voltage ($\cong -V_{EE}$). This reverse biases diode D_1 and opens the feedback loop. Therefore, during the negative half-cycle of the input signal, v_o is 0 V.

The op-amp in the circuit of Figure 9-25(a) must be a high-speed op-amp since it alternates between open-loop and closed-loop operations. μA318, HA2500, and LM310 are typical examples of high-speed op-amps.

Figure 9-26(a) shows a negative small-signal half-wave rectifier. This circuit, in fact, can be obtained from the circuit of Figure 9-23(a) by setting $V_{ref} = 0$ V. During the

(a)

(b)

Figure 9–25   (a) Positive small-signal half-wave rectifier circuit. (b) Its input and output waveforms.

(a)   (b)

Figure 9–26 (a) Negative small-signal half-wave rectifier circuit.
(b) Its input and output waveforms.

positive alternation of v_{in}, D_1 is reverse biased; therefore, $v_o = 0$ V. On the other hand, during the negative alternation, D_1 is forward biased; hence v_o follows v_{in}.

Yet another negative half-wave rectifier is shown in Figure 9-27(a). In this circuit two diodes are used so that the output v_o' of the op-amp does not saturate. This minimizes the response time and increases the operating frequency range of the op-amp. However, notice that the op-amp is used in the inverting configuration, and the output is measured

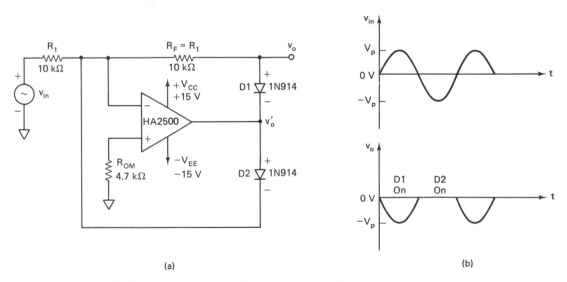

(a)   (b)

Figure 9–27 (a) Negative half-wave rectifier. (b) Its input and output wave-forms.

at the anode of diode D_1 with respect to ground. Also, the output resistance is nonuniform since it depends on the *state* of diode D_1. In other words, the output impedance is low when D_1 is on and high ($\cong R_F$) when D_1 is off. This problem, however, can be cured by connecting a voltage follower stage at the output. During the positive half-cycle of v_{in}, output v_o' is negative, which forward biases diode D_1 and closes the feedback loop through R_F. Since $R_1 = R_F$, $v_o = v_{in}$. However, on the negative alternation of v_{in}, output v_o' is positive; hence diode D_2 is forward biased. In fact, it is this diode that prevents the op-amp from going into positive saturation. Since diode D_1 is *off*, output $v_o = 0$ V. To obtain positive half-wave rectified outputs, diodes D_1 and D_2 must be reversed.

EXAMPLE 9-5

In the circuit of Figure 9-25(a), $v_{in} = 200$ mV peak-to-peak sine wave at 100 Hz. Briefly describe the operation of the circuit and draw the output waveform.

SOLUTION   During the positive half-cycle of the input, diode D_1 conducts; therefore, output follows the input; that is, $v_o = 100$ mV peak. During the negative half-cycle of v_{in}, diode D_1 is reverse biased; hence $v_o = 0$ V. The output waveform is shown in Figure 9-25(b).

9-12.3 Positive and Negative Clampers

In clamper circuits a predetermined dc level is added to the output voltage. In other words, the output is clamped to a desired dc level. If the clamped dc level is positive, the clamper is a *positive clamper*. On the other hand, if the clamped dc level is negative, the clamper is called a *negative clamper*. Other equivalent terms used for clamper are *dc inserter* or *restorer*.

Recall that op-amps designed to operate from a single power supply, such as the 124 series, employ a dc insertion technique. The inverting and noninverting amplifiers that use this technique are shown in Figure 9-28(a) and (b). The value of the capacitors in these circuits depends on different input rates and pulse widths. Notice that in both amplifier circuits the dc level added to the output voltage is approximately equal to $V_{CC}/2$. This fixed positive dc level is needed to obtain a maximum undistorted symmetrical sine wave. A clamper with a variable positive dc level is shown in Figure 9-29(a). However, this circuit functions differently from those of Figure 9-28 in that the input waveform peak is clamped at V_{ref} [see Figure 9-29(b)]. For this reason the circuit is called the *peak clamper*.

The output voltage of the clamper is a net result of ac and dc input voltages applied to $(-)$ and $(+)$ input terminals, respectively. Therefore, to understand the circuit operation, each input must be considered separately. First consider the input voltage V_{ref} at the $(+)$ input. Since this voltage is positive, v_o' is also positive, which forward biases diode D_1. This closes the feedback loop and the op-amp operates as a voltage follower. This is possible because C_i is an open circuit for dc voltage. Therefore, $v_o = +V_{ref}$. As far as voltage v_{in} at the $(-)$ input is concerned, during its negative half-cycle, diode D_1 conducts, charging C_i to the negative peak value of V_P. However, during the positive half-cycle of v_{in}, diode D_1 is reverse biased; hence, the voltage V_P across the capacitor acquired during the negative half-cycle is retained. Since this voltage V_P is in series with the positive peak

$$v_o = \left[V_{ref} - \frac{R_F}{R_1} v_{in}\right]$$

where $V_{ref} \cong \dfrac{V_{CC}}{2}$

(a)

$$v_o = \left[V_{ref} + \left(1 + \frac{R_F}{R_1}\right)v_{in}\right]$$

where $V_{ref} \cong \dfrac{V_{CC}}{2}$

(b)

Figure 9–28 (a) Inverting amplifier with dc level. (b) Noninverting amplifier with dc level.

voltage V_P, the output voltage $v_o = 2V_P$. Thus the net output is V_{ref} plus $2V_P$, so the negative peak of $2V_P$ is at V_{ref} [see Figure 9-29(b)]. For precision clamping,

$$C_i R_d << \frac{T}{2}$$

where R_d = resistance of diode D_1 when it is forward biased
            = 100 Ω typically
        T = time period of the input waveform

(a)

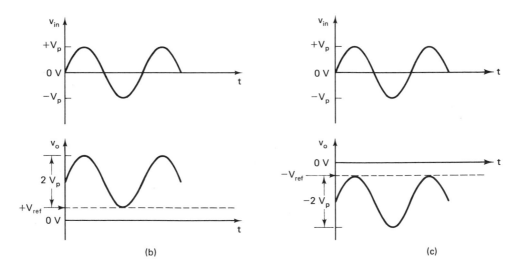

(b)   (c)

Figure 9–29   (a) Peak clamper circuit. (b) Input and output waveforms with $+V_{\text{ref}}$. (c) Input and output waveforms with $-V_{\text{ref}}$.

Resistor R is used to protect the op-amp against excessive discharge currents from capacitor C_i, especially when the dc supply voltages are switched off. A positive-peak clamping is accomplished by reversing D_1 and using negative reference voltage ($-V_{\text{ref}}$) [see the waveform in Figure 9-29(c)].

9-13 ABSOLUTE-VALUE OUTPUT CIRCUIT

Another important circuit used in waveshaping is the *absolute-value output circuit*. This circuit produces an output signal that swings positively only, regardless of the polarity of the input signal. Figure 9-30 shows an absolute-value output circuit with its input and output waveforms. Because of the nature of its output waveform, the circuit may be used as a full-wave rectifier if the input is a sine wave. During the positive half-cycle of v_{in}, diode D_1 is forward biased and D_2 is reverse biased; therefore, the equivalent circuit is as shown in Figure 9-30(b). In this circuit the voltage at the (+) input is

$$v_1 = \frac{V_P - V_{D1}}{2} \qquad (9\text{-}9\text{a})$$

where V_{D1}, the voltage drop across D_1, is 0.7 V. Similarly, voltage v_2 at the (−) input is

$$v_2 = \frac{v_o(+) - V_{D3}}{2} \qquad (9\text{-}9\text{b})$$

where $v_o(+)$ = output voltage during the positive half-cycle
V_{D3} = voltage drop across D_3 = 0.7 V

Since $v_{id} \cong 0$ V, $v_1 = v_2$. Therefore, from Equations (9-9a) and (9-9b),

$$\frac{V_P - V_{D1}}{2} = \frac{v_o(+) - V_{D3}}{2}$$

or

$$v_o(+) = V_P \qquad (9\text{-}10)$$

On the other hand, during the negative half-cycle of v_{in}, diode D_1 is reverse biased and D_2 is forward biased; hence the equivalent circuit is as shown in Figure 9-30(c). This circuit can further be simplified by Thévenizing to the left of (−) input. Thévenin's voltage and resistance are

$$V_{TH} = -\left(\frac{V_P - V_{D2}}{2}\right) \qquad (9\text{-}11\text{a})$$

$$R_{TH} \cong \frac{R}{2} \qquad (9\text{-}11\text{b})$$

[see Figure 9-30(c)]. Now output voltage $v_o(-)$ can be obtained by writing Kirchhoff's current equation at node v_2:

$$I_1 = I_2 + I_{B2}$$
$$I_1 \cong I_2 \quad \text{since } I_{B2} \cong 0 \text{ A}$$
$$\frac{[v_o(-) - V_{D3}] - v_2}{R} = \frac{v_2 - (V_{TH})}{R/2} \qquad (9\text{-}11\text{c})$$

However, $v_{id} = 0$ V and $v_1 = 0$. Therefore, $v_2 = 0$ and

$$\frac{v_o(-) - V_{D3}}{R} = \frac{-V_{TH}}{R/2}$$

(a)

(b)

(c)

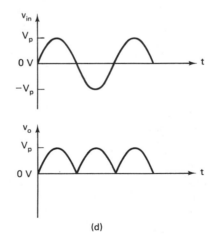

(d)

Figure 9–30 (a) Absolute-value output circuit, equivalent circuit during (b) positive half-cycle and (c) negative half-cycle. (d) Its input and output waveforms.

Substituting for V_{TH} from Equation (9-11a), we get

$$v_o(-) - V_{D3} = V_P - V_{D2}$$

$$v_o(-) = V_P \tag{9-12}$$

Thus, regardless of the polarity of the input signal, output is always positive going, hence the name *absolute-value output circuit*. Note that the gain of the circuit is 1; therefore, the positive peak amplitudes are the same as input peak amplitudes. In addition, diode D_3 compensates for the voltage drop across D_1 or D_2.

9-14 PEAK DETECTOR

Square, triangular, sawtooth, and pulse waves are typical examples of nonsinusoidal waveforms. A conventional *ac voltmeter* cannot be used to measure these nonsinusoidal waveforms because it is designed to measure the rms value of the pure sine wave. One possible solution to this problem is to measure the peak values of the nonsinusoidal waveforms.

Figure 9-31 shows a *peak detector* that measures the positive peak values of the square-wave input. During the positive half-cycle of v_{in}, the output of the op-amp drives D_1 on, charging capacitor C to the positive peak value V_P of the input voltage v_{in}. Thus, when D_1 is forward biased, the op-amp operates as a voltage follower. On the other hand, during the negative half-cycle of v_{in}, diode D_1 is reverse biased, and voltage across C is retained. The only discharge path for C is through R_L since the input bias current I_B is negligible. For proper operation of the circuit, charging time constant (CR_d) and discharging time constant (CR_L) must satisfy the following conditions:

$$CR_d \leq \frac{T}{10} \tag{9-13a}$$

(a)   (b)

Figure 9–31   (a) Peak detector circuit. (b) Its input and output waveforms.

where R_d = resistance of the forward-biased diode, 100 Ω, typically
   T = time period of the input waveform
and

$$CR_L \geq 10T \tag{9-13b}$$

where R_L is the load resistor.

   If R_L is very small so that Equation (9-13b) cannot be satisfied, use a buffer (voltage follower) between capacitor C and load resistor R_L. Although a 741-type op-amp is used in the circuit of Figure 9-31, a high-speed, precision-type op-amp such as the $\mu A771$ or $\mu A714$ may be desirable in critical applications. The resistor R is used to protect the op-amp against excessive discharge currents, especially when the power supply is switched off. The resistor $R_{OM} = R$ minimizes the offset problems caused by input currents. In addition, diode D_2 conducts during the negative half-cycle of v_{in} and hence prevents the op-amp from going into negative saturation. This in turn helps to reduce the recovery time of the op-amp. Negative peaks of input signal v_{in} can be detected simply by reversing diodes D_1 and D_2.

9-15 SAMPLE-AND-HOLD CIRCUIT

The sample-and-hold circuit, as its name implies, samples an input signal and holds on to its last sampled value until the input is sampled again. Figure 9-32 shows a sample-and-hold circuit using an op-amp with an E-MOSFET. In this circuit the E-MOSFET works as a switch that is controlled by the sample-and-hold control voltage V_S, and the capacitor C serves as a storage element. The circuit operates as follows. The analog signal v_{in} to be sampled is applied to the drain, and sample-and-hold control voltage V_S is applied to the gate of the E-MOSFET. During the positive portion of V_S, the E-MOSFET conducts and acts as a closed switch. This allows input voltage to charge capacitor C. In other words, input voltage appears across C and in turn at the output, as shown in Figure 9-32(b). On the other hand, when V_S is zero, the E-MOSFET is *off* (nonconductive) and acts as an open switch. The only discharge path for C is, therefore, through the op-amp. However, the input resistance of the op-amp voltage follower is also very high; hence the voltage across C is retained. The time periods T_S of the sample-and-hold control voltage V_S during which the voltage across the capacitor is equal to the input voltage are called *sample periods*. The time periods T_H of V_S during which the voltage across the capacitor is constant are called *hold periods* [see Figure 9-32(b)]. The output of the op-amp is usually processed/observed during hold periods. To obtain close approximation of the input waveform, the frequency of the sample-and-hold control voltage must be significantly higher than that of the input. In critical applications a precision and/or high-speed op-amp is helpful. If possible, choose a low-leakage capacitor such as Teflon or polyethylene.

   A significant reduction in size and improved performance can be achieved by using a specially designed sample-and-hold IC such as the LF398. The functional circuit of the LF398 shown in Figure 9-33 requires only an external storage capacitor.

   The sample-and-hold circuit is commonly used in digital interfacing and communications such as analog-to-digital and pulse modulation systems.

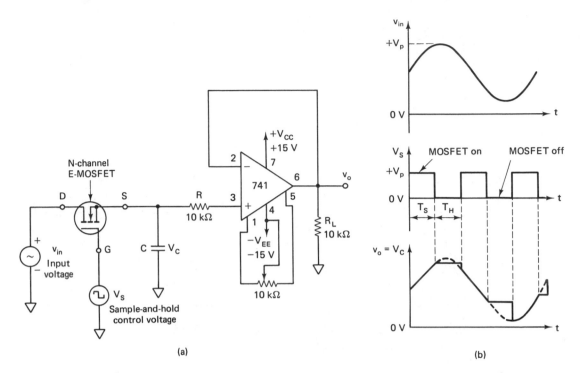

Figure 9–32 (a) Sample-and-hold circuit. (b) Its input and output waveforms.

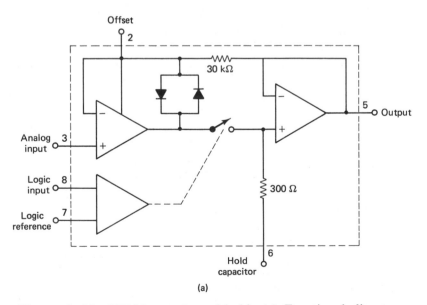

(a)

Figure 9–33 LF398 sample-and-hold. (a) Functional diagram.
(b) Typical connection diagram. (Courtesy of National Semiconductor.)

(b)

Figure 9–33   (Continued)

SUMMARY ━━━━━━━━━━━━━━━━━━━━━━━━━━━━━━━━━━━━━

1. Together with such functions as amplifier, filter, and oscillator, the op-amp can also be used as a comparator, limiter, detector, clipper, clamper, and converter.

2. In its simplest form, a comparator is nothing more than an open-loop op-amp with two analog inputs and a digital output, depending on which input is the larger. It is used in digital interfacing, Schmitt triggers, analog-to-digital converters, oscillators, and others.

3. An immediate application of the comparator is the zero-crossing detector, in which the reference voltage $V_{ref} = 0$ V.

4. The Schmitt trigger is a comparator with positive feedback that converts an irregular waveform to a square or pulse waveform. In the Schmitt trigger the input voltage triggers the output every time it exceeds certain voltage levels called upper threshold V_{ut} and lower threshold V_{lt}.

5. Switching speeds, accuracy, and compatibility of outputs are the limiting factors to the use of op-amp comparators in critical applications. The output of an op-amp can be limited to a predetermined value by using external components such as zeners and diodes. Such circuits are called voltage limiters.

6. The window detector uses two comparators and two threshold levels to determine when an unknown input is between these levels. When the input is between two predetermined limits, the output is high. However, when the input goes above or below the set limits, the output is low.

7. The Teledyne 9400 can be used as a V/F or F/V converter simply by using two external capacitors, three resistors, and a reference voltage. The V/F converter is used in such applications as temperature sensing and control, analog-to-digital converters, digital panel meters, and phase-locked loops. The F/V converter is used for applications such as frequency meters, speedometers, frequency multipliers/dividers, and motor controls.

8. The analog-to-digital (A/D) converter changes an analog input into a digital output, while a digital-to-analog (D/A) converter changes a digital input into an analog output. The D/A converter can be formed by using an op-amp and either binary-weighted resistors or an R and $2R$ ladder network. Integrated D/A converters are available as 8, 10, 12, 14, and 16 bits with voltage or current outputs. Microcomputer interfacing, CRT graphics generation, programmable power supplies, and digital filters are typical applications of D/A converters. On the other hand, the A/D converter can be a single-ramp or double-ramp integrating type, single-counter type, tracking type, or successive-approximation type. As with D/A converters, there are many monolithic/hybrid A/D converters. Typical applications of A/D converters include microprocessor interfacing, digital voltmeters, and LED/LCD displays.

9. A clipper is a circuit that removes certain parts of the input waveform. The clipper can be formed by using an op-amp with a rectifier diode. On the other hand, in an op-amp clamper a predetermined dc level is deliberately inserted in the output voltage. A half-wave rectifier is a form of clipper. The outputs of the amplifiers that operate on a single power supply are clamped to a desired value to avoid output distortion. Yet another waveshaping circuit is the absolute-value output circuit, in which the output swings positively only. Such a circuit is formed by using an op-amp with rectifier diodes.

10. The peak detector is generally used to detect the positive peak values of nonsinusoidal waveforms such as triangular, sawtooth, or pulse.

11. The sample-and-hold circuit samples an input signal and holds on to its last sampled value until the input is sampled again. It can be constructed by simply using an op-amp with an E-MOSFET as a switch. However, a monolithic sample-and-hold such as the LF398 requires only an external capacitor. The sample-and-hold circuit is used in digital interfacing and communications.

QUESTIONS

9-1. What is a comparator?

9-2. What is the difference between a basic comparator and the Schmitt trigger?

9-3. List the important characteristics of the comparator.

9-4. What is a voltage limiting, and why is it needed?

9-5. What is a window detector?

9-6. Name and then briefly describe one application of V/F and F/V converters.

9-7. Briefly describe the operation of the F/V ripple eliminator shown in Figure 9-16.

9-8. For the 9400 F/V converter, what is the limitation on input pulse widths?

9-9. What is the difference between A/D and D/A converters? Give one application of each.

9-10. What is the major difference between 1408 and 5018 D/A converters?

9-11. Define the following terms for D/A converters: resolution, settling time, conversion time.

9-12. What is the difference between clippers and clampers? Give one application of each.

9-13. What is an absolute-value output circuit? How can it be used as a full-wave rectifier?

9-14. What is the name of the circuit that is used to detect the peak value of the nonsinusoidal input waveforms? Briefly explain its operation.

9-15. What is a sample-and-hold circuit? Why is it needed?

PROBLEMS

9-1. For the basic comparator of Figure 9-1(a), $v_{in} = 2$ V pp sine wave at 1 kHz, $V_{ref} = 500$ mV, $R = 100$ Ω, and supply voltages $= \pm 15$ V. Draw the output waveform.

9-2. In the circuit of Figure 9-4(a), $R_1 = 150$ Ω, $R_2 = 68$ kΩ, $v_{in} = 500$ mV pp sine-wave, and the saturation voltages $= \pm 14$ V.
(a) Determine the threshold voltages V_{ut} and V_{lt}.
(b) What is the value of hysteresis voltage V_{hy}?

9-3. In the circuit of Figure 9-5, $v_{in} = 100$ mV peak sine wave at 100 Hz, $R = 1$ kΩ, and D_1 and D_2 are 6.2-V zeners. The op-amp is a 741 with supply voltages $= \pm 12$ V. Draw the output voltage waveform.

9-4. Repeat Problem 9-3 for the circuit shown in Figure 9-6(a). Assume that the voltage drop across the forward-biased diode is 0.7 V.

9-5. In the 9400 V/F converter shown in Figure 9-10(a), $V_{in} = 5$ V, $R_{in} = 1$ MΩ, $V_{ref} = -5$ V, and $C_{ref} = 180$ pF.
(a) Calculate the value of output frequency F_o.
(b) Draw the F_o and $F_o/2$ waveforms.
(c) To obtain bipolar outputs, what modification is required in the output circuitry?

9-6. Calculate the output voltage V_o for the 9400 F/V converter of Figure 9-14(a) if $F_{in} = 2.5$ kHz. What modifications are required in the external components if F_{in} is changed to 100 kHz?

9-7. Referring to the D/A converter with binary-weighted resistors shown in Figure 9-18(a), determine the size of each step if $R_F = 1.2$ kΩ. What is the output voltage when inputs b0 through b3 are at 5 V?

9-8. For the D/A converter using an R–$2R$ ladder network shown in Figure 9-19(a):
(a) Determine the size of each step if $R_F = 27$ kΩ.
(b) Calculate the output voltage when the inputs b0, b1, b2, and b3 are at 5 V.
(c) What is the advantage of this type of D/A converter over the one with binary-weighted resistors?

9-9. For the small-signal half-wave rectifier of Figure 9-25(a), draw the output wave-form if v_{in} is a 300-mV peak sine wave at 1 kHz.

9-10. Repeat Problem 9-9 with D_1 reversed.

9-11. For the peak clamper shown in Figure 9-29(a), draw the output voltage waveform if v_{in} = 500 mV pp sine wave at 100 Hz and V_{ref} = 50 mV.

9-12. In the peak detector of Figure 9-31, C = 0.01 μF, R_L = 1 MΩ, and v_{in} = 2 V pp square wave at 1 kHz. Draw the approximate output voltage waveform. (Assume that the resistance of the forward-biased diode R_d = 100 Ω.)

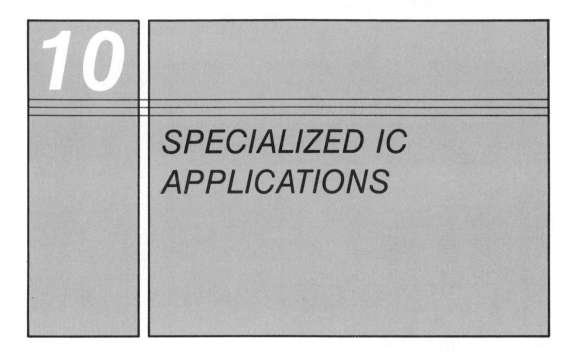

10

SPECIALIZED IC APPLICATIONS

10-1 INTRODUCTION

Integrated circuits with improved capabilities are appearing in ever-increasing numbers. Innovative design methods and fabrication procedures have not only helped produce a large variety of new integrated circuits but have also improved the old ones. Often the use of specialized ICs produces a simpler and more accurate circuit, such as Datel's universal filter FLT-U2, National's switched capacitor filter MF5, Signetics' phase-locked loop (PLL) SE/NE 565 and timer SE/NE 555, and others. This chapter presents a sampling of specialized integrated circuits and their applications to such devices as universal filters, timers, phase-locked loops, power amplifiers, voltage regulators, switching regulators, and voltage references.

10-2 UNIVERSAL ACTIVE FILTER

Various filter networks using general-purpose op-amps are discussed in Chapter 8. However, in critical applications, specially designed filter ICs are preferred. Besides being more accurate, the specially designed IC filters are simpler, easier to use, and more flexible. Datel's FLT-U2 is a typical example of such specialized IC filters. Datel's FLT-U2 is a universal filter that uses the state-variable active filter principle to implement second-order low-pass, high-pass, and band-pass output functions. These output functions are simultaneously available at the output of three committed op-amps (pins 3, 13, and 5), as shown in the FLT-U2 block diagram of Figure 10-1(a). A fourth uncommitted op-amp can be used as a gain stage or buffer amplifier or to raise the order of the low-pass, high-

Figure 10–1 (a) FLT-U2 universal filter block diagram. (b) Connection diagram. (c) Ceramic 16-pin DIP. (Courtesy of Datel-Intersil.)

pass, or band-pass functions. The uncommitted op-amp can also be used to realize the
notch and *all-pass* functions. The important characteristics of the FLT-U2 are the following:

Frequency range	0.001 Hz to 200 kHz
Figure of merit (Q) range	0.1 to 1000
Frequency accuracy	±5%
Voltage gain	0.1 to 1000
Input impedance	5 MΩ
Operating voltage range	±5 to ±18 V
Unity gain bandwidth	3 MHz
Slew rate	1 V/μs

Frequency tuning is accomplished by using two external resistors and Q tuning
by using a third resistor. In addition, by using the FLT-U2, any of the filter types, such
as Butterworth, Chebyshev, or Bessel, may be designed by proper selection of external components.

10-2.1 Design Procedure

In the design procedure that follows, it is assumed that the gain of the desired filter
function is unity. For simplicity, Q is assumed to be larger than 1. Figure 10-2 shows the
relative gains of the three simultaneous filter outputs assuming that the band-pass gain is
set to unity (0 dB). From this figure it is clear that the low-pass gain is always 10 dB higher
than band-pass gain, and high-pass gain is always 10 dB lower than band-pass gain.

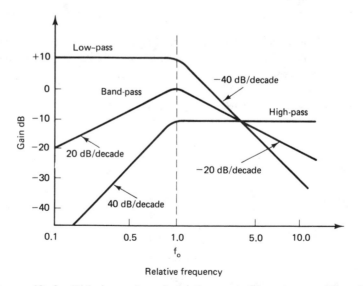

Figure 10–2   Relative gains of simultaneous filter outputs ($Q = 1$).

In following the design steps listed, refer to Figure 10-3.

1. For a desired function (low-pass, high-pass, or band-pass) use Table 10-1 to select an appropriate filter configuration for inverted or noninverted output.

TABLE 10-1 FILTER CONFIGURATIONS

Configuration	LP	HP	BP
Inverting input	Inverted	Inverted	Noninverted
Noninverting input	Noninverted	Noninverted	Inverted

Note that the low-pass, high-pass outputs are in phase.

2. For an inverting configuration, follow this step; otherwise, go to step 3. Using the desired value of Q, calculate R_1 and R_3 from Table 10-2. (R_2 is open.)

TABLE 10-2 INVERTING CONFIGURATIONS

	R_1	R_3
Low-pass	100 kΩ	$\dfrac{100\ \text{k}\Omega}{(3.80)(Q) - 1}$
High-pass	10 kΩ	$\dfrac{100\ \text{k}\Omega}{(6.64)(Q) - 1}$
Band-pass	$(Q)(31.6\ \text{k}\Omega)$	$\dfrac{100\ \text{k}\Omega}{(3.48)(Q)}$

3. Follow this step if a noninverting configuration is to be used; otherwise, go to step 4. Calculate R_2 and R_3 values from Table 10-3, using the desired value of Q. (R_1 is open.)

TABLE 10-3 NONINVERTING CONFIGURATIONS

	R_2	R_3
Low-pass	$\dfrac{316\ \text{k}\Omega}{Q}$	$\dfrac{100\ \text{k}\Omega}{(3.16)(Q) - 1}$
High-pass	$\dfrac{31.6\ \text{k}\Omega}{Q}$	$\dfrac{100\ \text{k}\Omega}{(0.316)(Q) - 1}$
Band-pass	100 kΩ	$\dfrac{100\ \text{k}\Omega}{(3.48)(Q) - 1}$

4. Using the desired value of resonant frequency f_1, which is the center frequency for band-pass and the cutoff frequency for low-pass or high-pass, calculate R_4 and R_5 values from the equation

Figure 10-3 FLT-U2 as second-order low-pass, high-pass, and band-pass filters. (Courtesy of Datel-Intersil.)

$$R_4 = R_5 = \frac{(5.03)\,(10^7)}{f_1} \quad \text{ohms} \tag{10-1}$$

where f_1 is either the center frequency or cutoff frequency expressed in hertz. For a band-pass filter the center frequency can be varied without affecting bandwidth by varying R_5, with R_4 fixed.

5. If $f_1 < 50$ Hz, the internal 1000-pF capacitors should be shunted with equal-value external capacitors across pins 5 and 7 and 13 and 14 (see Figure 10-3). R_4 and R_5 values are then computed from the equation

$$R_4 = R_5 = \frac{(5.03)\,(10^{10})}{f_1 C} \quad \text{ohms} \tag{10-2}$$

where C is the total capacitance in picofarads, that is, the sum of the external capacitance across pins 5 and 7 or across pins 13 and 14 in picofarads, and the internal 1000-pF capacitor. The value f_1 is in hertz.

6. The procedure described is based on unity gain output for the desired function. However, if additional gain is required, the fourth uncommitted op-amp should be used as an inverting or noninverting gain stage following the selected output (see Figure 10-4). Also, as shown in Figure 10-5, the order of the filter function can be raised by adding a capacitor to the gain stage.

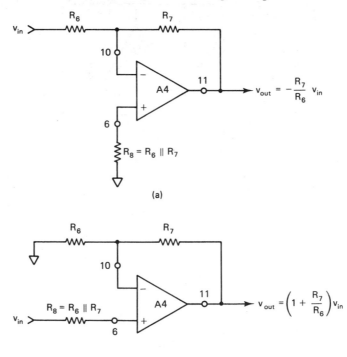

(a)

(b)

Figure 10-4   Uncommitted op-amp gain configurations. (a) Inverting. (b) Noninverting.

(a)

(b)

Figure 10–5 Using the uncommitted op-amp to raise the order of a low-pass function. (a) Inverting configuration. (b) Noninverting configuration.

The resonant frequency tuning is done by varying R_4 or R_5 resistors, while the figure of merit Q tuning is done by adjusting input resistors R_1, R_2, or R_3, depending on the filter configuration desired (see Figure 10-3). When tuning the filter and checking it over its frequency range, care should be exercised so that clipping does not occur in the filter. If clipping occurs, the input signal level should be reduced. For best performance, all external resistors should be 1% metal film, and capacitors should be a stable type such as tantalum or Mylar. Higher-order filters can be made by using cascaded FLT-U2 stages. To construct a notch filter, the simplest method is to use the FLT-U2 as an inverting band-pass filter and then to sum the output of the band-pass filter with the input signal by means of the uncommitted op-amp.

The FLT-U2 universal filter may be used in audio tone signaling, data acquisition, and feedback control systems.

EXAMPLE 10-1

The FLT-U2 is to be used as a second-order inverting Butterworth low-pass filter with a dc gain of 5, cutoff frequency of 2 kHz, and $Q = 10$. Determine the values of the external components.

SOLUTION   The values of external components can be determined by following these design steps:

1. According to Table 10-1, the inverting configuration would normally be used to give an inverting low-pass output. However, to obtain a gain of 5, an inverting uncommitted op-amp has to be used; hence the noninverting filter configuration must be used.
2. From Table 10-3, using $Q = 10$,

$$R_2 = \frac{316 \text{ k}\Omega}{10} = 31.6 \text{ k}\Omega$$

$$R_3 = \frac{100 \text{ k}\Omega}{(3.16)(10) - 1} = 3.27 \text{ k}\Omega \quad (\text{use } 3.3 \text{ k}\Omega)$$

$$R_1 = \text{open}$$

3. Substituting $f_1 = 2$ kHz in Equation (10-1), we get

$$R_4 = R_5 = \frac{(5.03)(10^7)}{(2)(10^3)} = 25.15 \text{ k}\Omega$$

(Use $R_4 = R_5 = 24.9$ kΩ.)

4. The final step is to use the uncommitted op-amp as an inverting amplifier with a gain of 5 [see Figure 10-4(a)]. Let $R_6 = 1.8$ kΩ; then $R_7 = (5)(1.8 \text{ k}\Omega) = 9.0$ kΩ (use a 10-kΩ potentiometer). $R_8 = R_6 \| R_7 = 1.5$ kΩ. The complete circuit is shown in Figure 10-6.

EXAMPLE 10-2

Using the FLT-U2, design a second-order inverting Butterworth band-pass filter with center frequency $f_1 = 5$ kHz and $Q = 10$.

SOLUTION   The values of external components needed to use the FLT-U2 as a band-pass filter can be determined by following these design steps:

1. According to Table 10-1, the noninverting configuration is chosen to obtain an inverted band-pass output.
2. Using $Q = 10$, the values of R_2 and R_3 from Table 10-3 are

$$R_2 = 100 \text{ k}\Omega$$

$$R_3 = \frac{100 \text{ k}\Omega}{(3.48)(10) - 1} = 2.96 \text{ k}\Omega$$

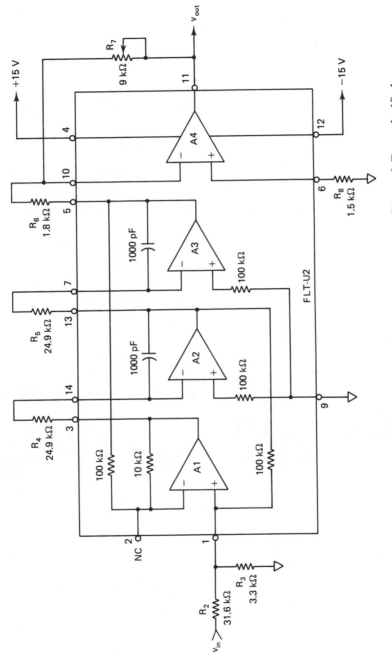

Figure 10-6 Second-order inverting Butterworth low-pass filter of Example 10-1.

(use $R_3 = 2.7$ kΩ). However,

$$R_1 = \text{open}$$

3. Since $f_1 = 5$ kHz, from Equation 10-1,

$$R_4 = R_5 = \frac{(5.03)(10^7)}{(5)(10^3)} \cong 10 \text{ kΩ}$$

For accurate circuit operation, R_4 may be a fixed 10-kΩ resistor, while R_5 may be a 10-kΩ potentiometer, so R_5 can be adjusted for the exact center frequency of 5 kHz.

The complete circuit is shown in Figure 10-7.

Figure 10–7 Band-pass filter of Example 10–2.

EXAMPLE 10-3

Using the FLT-U2, design a notch filter with 5-kHz notch-out frequency and $Q = 10$.

SOLUTION   As mentioned earlier, the FLT-U2 can be used as a notch filter by summing the inverted output of the band-pass filter designed in Example 10-2 with the input signal by means of the uncommitted op-amp (see Figure 10-8). The use of the band-pass filter of Example 10-2 is possible here because its 5-kHz center frequency and Q value are equal to the 5-kHz notch-out frequency and the Q value of the notch filter. Since no extra gain is required, the feedback resistance and the input resistances of the uncommitted op-amp used as the summing amplifier must be equal in value. Therefore, let

$$R_6 = R_7 = R_8 = 10 \text{ kΩ}$$

Then

$$R_9 \cong R_6 \| R_7 \| R_8 = 3.3 \text{ kΩ}$$

Figure 10-8 shows the complete diagram of the notch filter.

Figure 10-8   Notch filter of Example 10–3.

10-3 SWITCHED CAPACITOR FILTER

The switched capacitor filters have become popular because they require no external reactive components, capacitors or inductors. In fact, the operation of these filters is based on the ability of on-chip capacitors and MOS (metal oxide semiconductor) switches to simulate resistors. The values of these on-chip simulating capacitors can be closely matched to other capacitors on the IC, which results in integrated filters whose cutoff frequencies are proportional to and determined only by the external clock frequency. In addition, the cutoff or center frequencies of switched capacitor filters can be programmed to fall anywhere within an extremely wide range of frequencies, typically more than a 200,000:1 range. Also, the center frequencies can be very accurately determined. In short, switched capacitor filters offer the following advantages:

Low system cost
Low external component count
High accuracy
Excellent temperature stability

The only drawback of switched capacitor filters is that they generate more noise than standard active filter circuits.

10-3.1 Theory of Operation

In switched capacitor filters, a capacitor and a few switches can very closely approximate the behavior of a resistor. The value of this "simulated resistor" is inversely proportional

(a)   (b)

Figure 10–9   (a) Resistor to be simulated by switched capacitor. (b) Switched capacitor "resistor." Switches S_1 and S_2 are alternately opened and closed. (Courtesy of National Semiconductor Corporation.)

to the rate at which the switches are opened and closed. Let us consider the circuits shown in Figure 10-9. In Figure 10-9(a), a simple resistor is connected to ground. If a voltage V_{in} is applied to the resistor, a current

$$I = \frac{V_{in}}{R} \tag{10-3}$$

flows through the resistor. Now consider the circuit in Figure 10-9(b), which consists of a capacitor and two switches, S_1 and S_2. The switches are actually MOS transistors that are alternately opened and closed.

When S_1 is closed and S_2 is open, V_{in} is applied to the capacitor C. Therefore, the total charge on the capacitor is

$$Q = (V_{in})(C) \tag{10-4}$$

However, when S_1 is open and S_2 is closed, the charge Q flows to ground. If the switches are ideal, that is, they open and close instantaneously and have zero resistance when closed, C will charge and discharge instantly. Figure 10-10 shows the current into and out of the capacitor as a function of time for a constant input voltage. Remember that current is defined as the charge Q passing through a conductor per unit time. Note in Figure 10-10 that the capacitor current consists of short bursts at each switch closing.

If switches are opened and closed at a faster rate, the bursts of current will have the same amplitude but will occur more often. This means that the average current will be greater for a higher switching rate. Thus the average current flowing through the capacitor of Figure 10-9(b) is

$$I_{ave} = \frac{Q}{T}$$
$$= \frac{(V_{in})(C)}{T}$$
$$= (V_{in})(C)(f_{CLK}) \tag{10-5}$$

where T = time between closings of S_1 or S_2,

$$\frac{1}{T} = f_{CLK}$$

Since we know the average current (I_{ave}) and input voltage (V_{in}), the equivalent resistance is given by

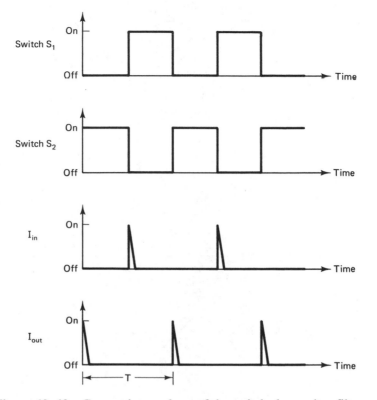

Figure 10-10 Current into and out of the switched capacitor filter as a function of time. (Courtesy of National Semiconductor Corporation.)

$$R = \frac{V_{in}}{I_{ave}}$$

$$R = \frac{V_{in}}{(V_{in})(C)(f_{CLK})}$$

$$R = \frac{1}{(C)(f_{CLK})} \qquad (10\text{-}6)$$

Equation (10-6) indicates that R is a function of C and f_{CLK}. However, C is a constant. This means the value of R can be adjusted by adjusting f_{CLK}. Note that to simulate a resistor it is necessary that V_{in} change at a rate much slower than f_{CLK}, especially when V_{in} is an ac signal.

10-3.2 Switched Capacitor Integrator

The discussion showed how a clock-tunable resistor can be built in IC form with just a capacitor and two switches. This concept can very easily be extended to an integrator,

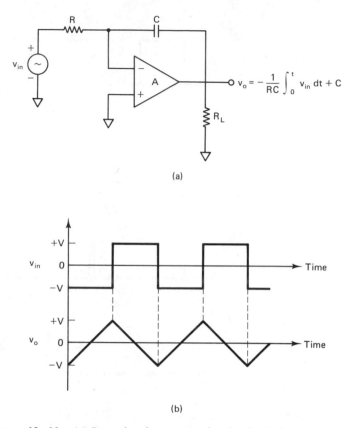

$$v_o = -\frac{1}{RC} \int_0^t v_{in} \, dt + C$$

(a)

(b)

Figure 10–11 (a) Inverting integrator circuit. (b) Its input and output waveforms.

which is the basis for a number of useful active filter configurations. Figure 10-11 shows an inverting integrator with its input and output waveforms. The same inverting integrator is built with a switched capacitor resistor as shown in Figure 10-12. In this figure, it is necessary that the switches S_1 and S_2 must never both be closed at the same time. In other words, the clock waveforms driving the MOS switches must not overlap if the filter is to operate properly.

  Next let us consider a specific switched capacitor filter IC.

10-3.3 Universal Monolithic Switched Capacitor Filters

There are several different brands of switched-capacitor filters. MF5 and MF10 are the most commonly used among them. These are manufactured by National Semiconductor Corporation and can be used to synthesize any of the filter types, band-pass, notch (or band-reject), low-pass, high-pass, and all-pass (or phase shift), simply by appropriate choice of a few external resistors. The values and placement of these resistors determine

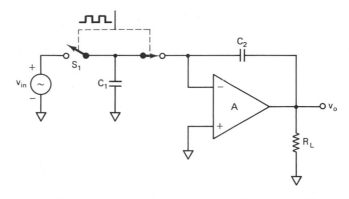

Figure 10–12 Inverting switched-capacitor integrator. (Courtesy of National Semiconductor Corporation.)

the basic shape of the amplitude and the phase response. However, the center or cutoff frequency is set by the external clock.

10-3.3(a) The MF5 filter.   The MF5 is a *universal* active filter in that it can be used to synthesize any of the normal filter types: band-pass, low-pass, high-pass, notch, and all-pass. It is a second-order circuit and can be cascaded to provide very steep attenuation slopes. The MF5 can realize virtually any response shape desired, including Butterworth, Chebyshev, Bessel, and elliptic.

Figure 10-13 is a block diagram of the MF5's internal circuitry. The basic filter consists of an operational amplifier, two positive integrators, and a summing node. A MOS switch, controlled by a logic voltage on pin 5 (S_A), connects one of the inputs of the first integrator either to ground or to the output of the second integrator, thus allowing more application flexibility. The MF5 includes a pin (9) that selects the clock (f_{CLK}) to center frequency (f_o) ratio at either 50:1 or 100:1. The maximum recommended clock frequency is 1 MHz, which results in a maximum center frequency f_o of 20 KHz at 50:1 and 10 KHz at 100:1, provided that (f_o) (Q) is less than 200 KHz. An extra uncommitted op-amp is available for additional signal processing.

In Figure 10-13, inputs and outputs of the summing amplifier and the two integrators are brought out to pins so that they can be connected to external resistors in various combinations. The way in which the external resistors are connected to the MF5 determines the filter's characteristics. Eight different connection schemes, called *modes,* are shown in the MF5 data sheet. For more information on modes, refer to Appendix D. Each of these modes has its own merits and shortcomings, which are summarized in the data sheet. For most applications, mode 3 is a good starting point because it gives low-pass, high-pass, and band-pass outputs and also allows independent adjustment of gain, Q, and the clock-to-center frequency ratio. The last feature can be especially handy if the only available clock has a frequency other than 50 or 100 times the desired center frequency, or if an application requires two or more filters, each with different center or cutoff frequencies.

(a)

(b)

Figure 10–13 MF5 universal monolithic switched-capacitor filter. (a) Block diagram. (b) Connection diagram. (Courtesy of National Semiconductor Corporation.)

Most of the modes have three outputs, with various combinations of low-pass, high-pass, band-pass, notch, and all-pass responses at the outputs. Two of the MF5's outputs are integrator outputs that are capable of driving a 5-kΩ load each to within 1 V of the power supply rails. The third summing amplifier output (pin 2) can produce somewhat more output current, driving a 3.5-kΩ load to within about 1 V of the supply rails. Lower load impedances than 5-kΩ for the integrators and 3.5-kΩ for the summing amplifier should be avoided. The MF5 can operate with a single or split power supply, but the total power supply must be between 8 and 14 V.

Pin Description.   For pin positions, refer to Figure 10-13.

LP (14), BP (1), N/AP/HP (2): The second order low-pass, band-pass, and notch/all-pass/high-pass outputs, respectively.

INV1 (3): The inverting input of the summing op-amp.

S1 (4): A signal input pin used in the all-pass filter configurations as shown in modes 4 and 5. (Refer to data sheet in Appendix D.) If not driven with a signal, S1 should be tied to AGND, pin 11.

S_A (5): When this pin is tied to V−, it activates a switch that connects one of the inputs of the filter's second summer to AGND. On the other hand, if S_A is tied to V+, it activates a switch that connects one of the inputs of the filter's second summer to the low-pass output.

50/100 (9): When this pin is tied to V+, an f_{CLK}/f_o ratio of about 50:1 is obtained. On the other hand, if tied to either AGND or V−, an f_{CLK}/f_o ratio of about 100:1 is obtained.

AGND (11): The analog ground pin. The supply voltages V+ and V− are measured with respect to this pin.

V+ (6), V− (10): The positive and negative supply voltage pins. Decoupling the supply pins with $0.1\text{-}\mu F$ capacitors is highly recommended.

CLK (8): The clock input pin for the filter. It can accept CMOS or TTL logic level clocks based on the status of *logic level shift,* pin 7.

L Sh (7): For a TTL logic level clock, this pin is tied to V−. If the L Sh pin is tied to AGND, the filter will accept a CMOS or TTL logic level clock at pin 8. However, for single supply operation, the L Sh pin should be tied to AGND for a CMOS logic level clock.

INV2 (12): This is the inverting input of the uncommitted op-amp, which behaves like a summing junction.

V_{02} (13): The output of the uncommitted op-amp.

Definitions of Terms.

f_{CLK}: The frequency of the external clock signal applied to pin 8.

f_o: Center frequency of the second-order band-pass filter.

f_{notch}: The frequency of minimum gain at the notch output.

Q: Quality factor of the second-order filter is equal to f_o divided by the −3-dB bandwidth of the band-pass filter. It determines the shape of the second-order filter responses.

$H_{OBP}, H_{OLP}, H_{OHP}, H_{ON}$: The voltage gains of the band-pass, low-pass, high-pass, and notch filters, respectively.

H_{ON1}: The gain of the notch output as $f \rightarrow 0$ Hz.

H_{ON2}: The gain of the notch output as $f \rightarrow f_{CLK}/2$.

The MF5 is a sampled-data filter. An important characteristic of sampled-data systems is their effect on signals at frequencies greater than one-half the sampling frequency (f_{CLK}). If a signal with a frequency greater than one-half the sampling frequency is applied to the input of a sampled-data system, it will be reflected to a frequency less than one-half the sampling frequency. For instance, an input signal whose frequency is ($f_{CLK}/2$) + 100 Hz will cause the system to respond as though the input frequency were ($f_{CLK}/2$) − 100 Hz. This phenomenon is known as *aliasing*. The aliasing can be reduced or eliminated by limiting the input signal spectrum to less than $f_{CLK}/2$.

Another characteristic of sampled-data circuits is that the output signal changes amplitude once every sampling period, resulting in *steps* in the output voltage that occur at the clock rate. If desired, these can be smoothed with a simple-*RC* low-pass filter at the MF5 output.

Finally, the ratio of f_{CLK} to f_o also affects the performance of the filter. A ratio of 100:1 reduces any aliasing problems and is usually recommended for wideband input signals. In noise-sensitive applications, however, a ratio of 50:1 may be better as it results in 3-dB lower output noise. The 50:1 ratio also results in lower dc offset voltages.

A very convenient feature of the MF5 is that f_o can be controlled independently of Q and passband gain. This allows f_o to be tuned simply by varying f_{CLK}, without altering the other characteristics. Another advantage is that choosing the external resistor values for the MF5 is a straightforward procedure. In short, the design procedure is much easier with the MF5 compared to typical *RC* active filters.

The major disadvantage of the MF5 is that its integrators have a higher equivalent input offset voltage than would be found in a typical active filter integrator. Larger offset voltages may cause clipping to occur at lower ac signal levels, and clipping at any of the outputs may cause gain nonlinearities and may change f_o and Q.

EXAMPLE 10-4

Using the MF5, design a second-order Butterworth low-pass filter with a cutoff frequency of 500 Hz and a passband gain of −2. Assume that a ±5-V power supply and a CMOS clock are used.

SOLUTION   For simplicity we will use mode 1, which has low-pass, band-pass, and notch outputs and inverts the signal polarity. It requires three external resistors that determine Q and the gain of the filter. The external resistors are connected as shown in Figure 10-14. For mode 1 the relationship between Q, H_{OLP} and the external resistors is given as follows (refer to the data sheet of MF5 in Appendix D):

$$Q = \frac{f_o}{\text{BW}} = \frac{R_3}{R_2}$$

and

$$H_{OLP} = -\frac{R_2}{R_1}$$

Figure 10–14   MF5 configured as the second-order Butterworth low-pass filter.

In this mode the input impedance of the filter is equal to R_1 because the input signal is applied to INV1 (pin 3) through R_1. See Figure 10-14. This means that we choose R_1 so that it will provide fairly high input impedance. Let $R_1 = 10\ k\Omega$. Therefore,

$$R_2 = -(R_1)(H_{OLP}) = (10\ k)(2) = 20\ k\Omega$$

Recall that, for the second-order Butterworth low-pass filter, $Q = 0.707$. Therefore,

$$R_3 = (Q)(R_2) = (0.707)(20\ k) = 14.14\ k\Omega, \quad \text{use } R_3 = 15\ k\Omega$$

Since a ± 5-V power supply is recommended, V+ (pin 6) is connected to +5 V, V– (pin 10) to -5 V, and AGND (pin 11) to ground. For "clean" power supplies the supply pins are decoupled with 0.1-μF capacitors as shown in Figure 10-14.

We will choose the f_{CLK} to f_o ratio equal to 50:1. Therefore, 50/100 (pin 9) must be connected to V+ (pin 6). Since the cutoff frequency is 500 Hz, the external clock frequency is $(500)(50) = 25\ kHz$. In addition, L.Sh (pin 7) should be connected to ground (pin 11) since the clock is a CMOS. Finally, the low-pass filter S_A (pin 5) is connected to V+ (pin 6), and S1 (pin 4) is connected to ground (pin 11). The complete circuit for the second-order Butterworth low-pass filter is shown in Figure 10-14.

10-3.3(b) The MF10 dual second-order universal filter.   The MF10 contains two of the second-order universal filter sections found in the MF5. Therefore, with the MF10, two second-order filters or one fourth-order filter can be built. As the MF5 and MF10 have similar filter sections, the design procedures for both are the same.

10-4 THE 555 TIMER

One of the most versatile linear integrated circuits is the 555 timer. Signetics Corporation first introduced this device as the SE/NE 555 in early 1970. Since its debut, the device has been used in a number of novel and useful applications. A sample of these applications includes monostable and astable multivibrators, dc–dc converters, digital logic probes, waveform generators, analog frequency meters and tachometers, temperature measurement and control, infrared transmitters, burglar and toxic gas alarms, voltage regulators, electric eyes, and many others. The 555 is a monolithic timing circuit that can produce accurate and highly stable time delays or oscillation. The timer basically operates in one of the two modes: either as a monostable (one-shot) multivibrator or as an astable (free-running) multivibrator. The device is available as an 8-pin metal can, an 8-pin mini DIP, or a 14-pin DIP. Figure 10-15 shows the connection diagram and the block diagram of the SE/NE 555 timer. The SE555 is designed for the operating temperature range from $-55°$ to $+125°C$, while the NE555 operates over a temperature range of $0°$ to $+70°C$. The important features of the 555 timer are these: it operates on $+5$ to $+18$ V supply voltage in both free-running (astable) and one-shot (monostable) modes; it has an adjustable duty cycle; timing is from microseconds through hours; it has a high current output; it can source or sink 200 mA; the output can drive TTL and has a temperature stability of 50 parts per million (ppm) per degree Celsius change in temperature, or equivalently 0.005%/°C. Like general-purpose op-amps, the 555 timer is reliable, easy to use, and low cost.

The next several sections explain the operation of the 555 timer as a monostable and astable multivibrator; a few simple applications using these two modes are then presented.

10-4.1 The 555 as a Monostable Multivibrator

A monostable multivibrator, often called a *one-shot* multivibrator, is a pulse-generating circuit in which the duration of the pulse is determined by the *RC* network connected externally to the 555 timer. In a stable or standby state the output of the circuit is approximately zero or at logic-low level. When an external trigger pulse is applied, the output is forced to go *high* ($\cong V_{CC}$). The time the output remains high is determined by the external *RC* network connected to the timer. At the end of the timing interval, the output automatically reverts back to its logic-low stable state. The output stays low until the trigger pulse is again applied. Then the cycle repeats. The monostable circuit has only one stable state (output low), hence the name *monostable*. Normally, the output of the monostable multivibrator is low.

Figure 10-16(a) shows the 555 configured for monostable operation. To better explain the circuit's operation, the internal block diagram is included in Figure 10-16(b).

Before proceeding with the operation of the 555 timer as a monostable multivibrator, it is important to examine its pin functions. The pin numbers used in the following discussion refer to the 8-pin mini DIP and 8-pin metal can packages [see Figure 10-15(a)].

Pin 1: Ground. All voltages are measured with respect to this terminal.

Pin 2: Trigger. The output of the timer depends on the amplitude of the external trigger pulse applied to this pin. The output is low if the voltage at this pin is greater than $2/3\ V_{CC}$. However, when a negative-going pulse of amplitude larger than $1/3\ V_{CC}$ is applied to this pin, the comparator 2 output goes low, which in turn switches the output of the timer high [see Figure 10-15(b)]. The output remains high as long as the trigger terminal is held at a low voltage.

(a)

(b)

Figure 10–15   (a) 555 timer connection diagram. (b) Block diagram. (Courtesy of Signetics Corporation.)

Pin 3: Output. There are two ways a load can be connected to the output terminal: either between pin 3 and ground (pin 1) or between pin 3 and supply voltage $+V_{CC}$ (pin 8) [see Figure 10-16(b)]. When the output is low, the load current flows through the load connected between pin 3 and $+V_{CC}$ into the output terminal and is called the *sink* current. However, the current through the grounded load is zero when the output is low. For this reason, the load connected between pin 3 and $+V_{CC}$ is called

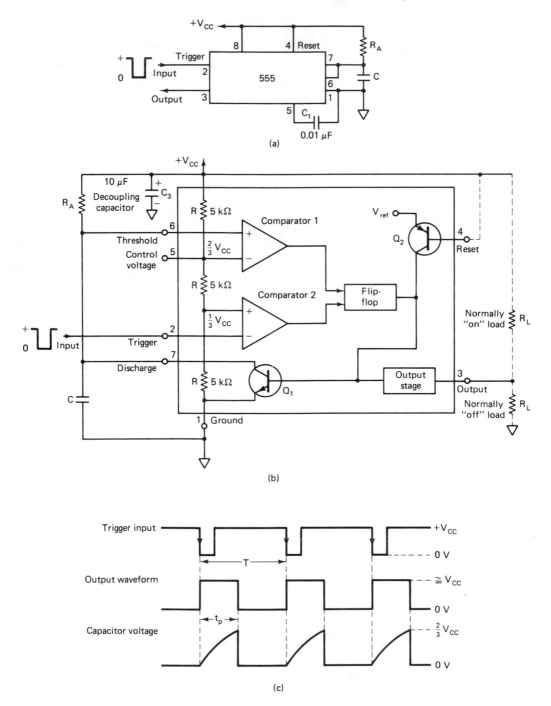

Figure 10–16   (a) and (b) 555 connected as a monostable multivibrator. (c) Input and output waveforms.

the *normally on load* and that connected between pin 3 and ground is called the *normally off load*. On the other hand, when the output is high, the current through the load connected between pin 3 and $+V_{CC}$ (normally on load) is zero. However, the output terminal supplies current to the normally off load. This current is called the *source* current. The maximum value of sink or source current is 200 mA.

Pin 4: Reset. The 555 timer can be reset (disabled) by applying a negative pulse to this pin. When the reset function is not in use, the reset terminal should be connected to $+V_{CC}$ to avoid any possibility of false triggering.

Pin 5: Control voltage. An external voltage applied to this terminal changes the threshold as well as the trigger voltage [see Figure 10-16(b)]. In other words, by imposing a voltage on this pin or by connecting a pot between this pin and ground, the pulse width of the output waveform can be varied. When not used, the control pin should be bypassed to ground with a 0.01-μF capacitor to prevent any noise problems.

Pin 6: Threshold. This is the noninverting input terminal of comparator 1, which monitors the voltage across the external capacitor [see Figure 10-16(b)]. When the voltage at this pin is $\geq$ threshold voltage $2/3 \ V_{CC}$, the output of comparator 1 goes high, which in turn switches the output of the timer low.

Pin 7: Discharge. This pin is connected internally to the collector of transistor Q_1, as shown in Figure 10-16(b). When the output is high, Q_1 is off and acts as an open circuit to the external capacitor C connected across it. On the other hand, when the output is low, Q_1 is saturated and acts as a short circuit, shorting out the external capacitor C to ground.

Pin 8: $+V_{CC}$. The supply voltage of $+5$ V to $+18$ is applied to this pin with respect to ground (pin 1).

10-4.1(a) Monostable operation.   According to Figure 10-16(b), initially when the output is low, that is, the circuit is in a stable state, transistor Q_1 is on and capacitor C is shorted out to ground. However, upon application of a negative trigger pulse to pin 2, transistor Q_1 is turned off, which releases the short circuit across the external capacitor C and drives the output high. The capacitor C now starts charging up toward V_{CC} through R_A. However, when the voltage across the capacitor equals $2/3 \ V_{CC}$, comparator 1's output switches from low to high, which in turn drives the output to its low state via the output of the flip-flop. At the same time, the output of the flip-flop turns transistor Q_1 on, and hence capacitor C rapidly discharges through the transistor. The output of the monostable remains low until a trigger pulse is again applied. Then the cycle repeats. Figure 10-16(c) shows the trigger input, output voltage, and capacitor voltage waveforms. As shown here, the pulse width of the trigger input must be smaller than the expected pulse width of the output waveform. Also, the trigger pulse must be a negative-going input signal with an amplitude larger than $1/3 \ V_{CC}$.

The time during which the output remains high is given by

$$t_p = 1.1R_A C \quad \text{seconds} \qquad (10\text{-}7)$$

where R_A is in ohms and C is in farads. Figure 10-17 shows a graph of the various combinations of R_A and C necessary to produce desired time delays. Note that this graph

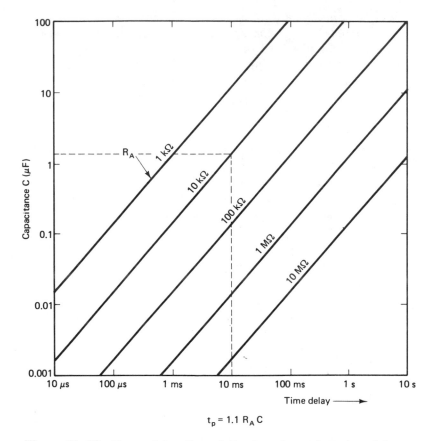

$$t_p = 1.1\ R_A\ C$$

Figure 10–17 Determining R_A and C values for various time delays.

can only be used as a guideline and gives only the approximate value of R_A and C for a given time delay.

    Once triggered, the circuit's output will remain in the high state until the set time t_p elapses. The output will not change its state even if an input trigger is applied again during this time interval t_p. However, the circuit can be reset during the timing cycle by applying a negative pulse to the reset terminal. The output will then remain in the low state until a trigger is again applied.

    Often in practice a decoupling capacitor (10 μF) is used between $+V_{CC}$ (pin 8) and ground (pin 1) to eliminate unwanted voltage spikes in the output waveform. Sometimes, to prevent any possibility of mistriggering the monostable multivibrator on positive pulse edges, a waveshaping circuit consisting of R, C_2, and diode D is connected between the trigger input pin 2 and V_{CC} pin 8, as shown in Figure 10-18. The values of R and C_2 should be selected so that the time constant RC_2 is smaller than the output pulse width t_p.

EXAMPLE 10-5

In the circuit of Figure 10-16(a), R_A = 10 kΩ, and the output pulse width t_p = 10 ms. Determine the value of C.

Figure 10–18   Monostable multivibrator with waveshaping network to prevent positive pulse edge triggering.

SOLUTION   Rearranging Equation (10-7), we get

$$C = \frac{(10)(10^{-3})}{(1.1)(10^4)} = 0.909 \ \mu F \cong 1 \ \mu F$$

(Notice that approximately the same value can be obtained for C from the time delay graph of Figure 10-17.)

10-4.2 Monostable Multivibrator Applications

10-4.2(a) Frequency divider.   The monostable multivibrator of Figure 10-16(a) can be used as a frequency divider by adjusting the length of the timing cycle t_p with respect to the time period T of the trigger input signal applied to pin 2. To use the monostable multivibrator as a divide-by-2 circuit, the timing interval t_p must be slightly larger than the time period T of the trigger input signal, as shown in Figure 10-19. By the

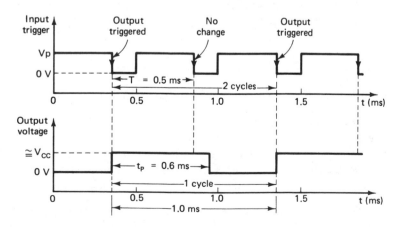

Figure 10–19   Input and output waveforms of the monostable multivibrator as a divide-by-2 network.

same concept, to use the monostable multivibrator as a divide-by-3 circuit, t_p must be slightly larger than twice the period of the input trigger signal, and so on.

The frequency-divider application is possible because the monostable multivibrator cannot be retriggered during the timing cycle.

EXAMPLE 10-6

The circuit of Figure 10-16(a) is to be used as a divide-by-2 network. The frequency of the input trigger signal is 2 kHz. If the value of $C = 0.01$ μF, what should be the value of R_A?

SOLUTION   For a divide-by-2 network, t_p should be slightly larger than T. Let

$$t_p = 1.2T$$

Therefore,

$$t_p = (1.2)\frac{1}{2 \text{ kHz}} = 0.6 \text{ ms}$$

Rearranging Equation (10-7) yields

$$R_A = \frac{(0.6)(10^{-3})}{(1.1)(10^{-8})} = 54.5 \text{ k}\Omega$$

A combination of a 10-kΩ fixed resistor and a 50-kΩ potentiometer may be used for R_A. The input and output waveforms are shown in Figure 10-19.

10-4.2(b) Pulse stretcher.   This application makes use of the fact that the output pulse width (timing interval) of the monostable multivibrator is of longer duration than the negative pulse width of the input trigger. As such, the output pulse width of the monostable multivibrator can be viewed as a stretched version of the narrow input pulse, hence the name *pulse stretcher*. Often, narrow-pulse-width signals are not suitable for driving an LED display, mainly because of their very narrow pulse widths. In other words, the LED may be flashing but is not visible to the eye because its *on* time is infinitesimally small compared to its *off* time. The 555 pulse stretcher can be used to remedy this problem.

Figure 10-20 shows a basic monostable used as a pulse stretcher with an LED indicator at the output. The LED will be on during the timing interval $t_p = 1.1R_AC$, which can be varied by changing the value of R_A and/or C.

10-4.3 The 555 as an Astable Multivibrator

An astable multivibrator, often called a *free-running* multivibrator, is a rectangular-wave-generating circuit. Unlike the monostable multivibrator, this circuit does not require an external trigger to change the state of the output, hence the name *free-running*. However, the time during which the output is either high or low is determined by the two resistors and a capacitor, which are externally connected to the 555 timer.

10-4.3(a) Astable operation.   Figure 10-21(a) shows the 555 timer connected as an astable multivibrator. Initially, when the output is high, capacitor C starts charging toward V_{CC} through R_A and R_B. However as soon as voltage across the capacitor equals

Figure 10–20   Monostable multi-vibrator as a pulse stretcher.

(a)

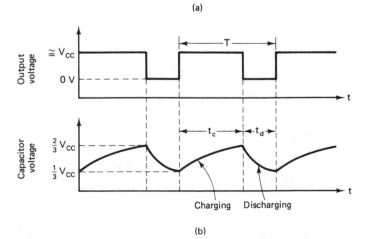

(b)

Figure 10–21   The 555 as an astable multivibrator. (a) Circuit. (b) Voltage across capacitor and output voltage waveforms.

$2/3\ V_{CC}$, comparator 1 triggers the flip-flop, and the output switches low [see Figure 10-21(b)]. Now capacitor C starts discharging through R_B and transistor Q_1. When the voltage across C equals $1/3\ V_{CC}$, comparator 2's output triggers the flip-flop, and the output goes high. Then the cycle repeats. The output voltage and capacitor voltage waveforms are shown in Figure 10-21(b).

As shown in this figure, the capacitor is periodically charged and discharged between $2/3\ V_{CC}$ and $1/3\ V_{CC}$, respectively. The time during which the capacitor charges from $1/3\ V_{CC}$ to $2/3\ V_{CC}$ is equal to the time the output is high and is given by

$$t_c = 0.69(R_A + R_B)C \qquad (10\text{-}8a)$$

where R_A and R_B are in ohms and C is in farads. Similarly, the time during which the capacitor discharges from $2/3\ V_{CC}$ to $1/3\ V_{CC}$ is equal to the time the output is low and is given by

$$t_d = 0.69(R_B)C \qquad (10\text{-}8b)$$

where R_B is in ohms and C is in farads. Thus the total period of the output waveform is

$$T = t_c + t_d = 0.69(R_A + 2R_B)C \qquad (10\text{-}9)$$

This, in turn, gives the frequency of oscillation as

$$f_o = \frac{1}{T} = \frac{1.45}{(R_A + 2R_B)C} \qquad (10\text{-}10)$$

Equation (10-10) indicates that the frequency f_o is independent of the supply voltage V_{CC}. The frequency of oscillation (free-running frequency) f_o can also be found by using Figure 10-22.

Often the term *duty cycle* is used in conjunction with the astable multivibrator. The duty cycle is the ratio of the time t_c during which the output is high to the total time period T. It is generally expressed as a percentage. In equation form,

$$\% \text{ duty cycle} = \frac{t_c}{T} \times 100$$

$$= \frac{R_A + R_B}{R_A + 2R_B}(100) \qquad (10\text{-}11)$$

The duty cycle of the square wave is 50%. This means that, according to Equation (10-11), the astable multivibrator shown in Figure 10-21(a) will not produce square-wave output unless the resistance $R_A = 0\ \Omega$. However, there is a danger in shorting resistance R_A to zero. With $R_A = 0\ \Omega$, terminal 7 is connected directly to $+V_{CC}$. When the capacitor discharges through R_B and Q_1 (pin 7), an extra current is supplied to Q_1 by V_{CC} through a short between terminal 7 and $+V_{CC}$, which may damage Q_1 and hence the timer. Fortunately, an alternative is available, which is explained in the section on astable multivibrator applications.

EXAMPLE 10-7

In the astable multivibrator of Figure 10-21(a), $R_A = 2.2\ \text{k}\Omega$, $R_B = 3.9\ \text{k}\Omega$, and $C = 0.1\ \mu\text{F}$. Determine the positive pulse width t_c, negative pulse width t_d, and free-running frequency f_o.

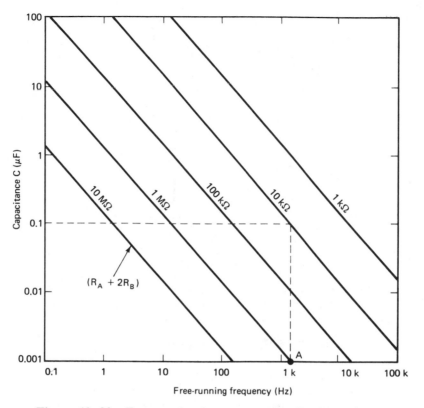

Figure 10–22   Free-running frequency versus R_A, R_B, and C.

SOLUTION   From Equations (10-8a) and (10-8b),

$$t_c = (0.69)\,(2.2\ \text{k}\Omega + 3.9\ \text{k}\Omega)\,(0.1)\,(10^{-6})$$
$$= 0.421\ \text{ms}$$
$$t_d = (0.69)\,(3.9\ \text{k}\Omega)\,(0.1)\,(10^{-6})$$
$$= 0.269\ \text{ms}$$

Therefore,

$$f_o = \frac{1}{(0.421 + 0.269)\,(10^{-3})} = 1.45\ \text{kHz}$$

Using $C = 0.1\ \mu\text{F}$ and $R_A + 2R_B = 10\ \text{k}\Omega$, the free-running frequency can also be found from the graph in Figure 10-22, as indicated by point A. Note that the graph uses the log–log scale; that is, vertical and horizontal axes are marked in logarithmic scales.

10-4.4 Astable Multivibrator Applications

10-4.4(a) Square-wave oscillator.   Without reducing $R_A = 0\ \Omega$, the astable multivibrator can be used to produce a square wave output simply by connecting diode D across resistor R_B, as shown in Figure 10-23. The capacitor C charges through R_A and diode D to approximately $2/3\ V_{CC}$ and discharges through R_B and terminal 7 (transistor

Figure 10–23　Astable multivibrator as a square wave oscillator.

Q_1) until the capacitor voltage equals approximately $1/3$ V_{CC}; then the cycle repeats. To obtain a square wave output (50% duty cycle), R_A must be a combination of a fixed resistor and potentiometer so that the potentiometer can be adjusted for the exact square wave.

10-4.4(b) Free-running ramp generator.
The astable multivibrator can be used as a free-running ramp generator when resistors R_A and R_B are replaced by a current mirror. Figure 10-24(a) shows an astable multivibrator configured to perform this function. The current mirror starts charging capacitor C toward V_{CC} at a constant rate. When voltage across C equals $2/3$ V_{CC}, comparator 1 turns transistor Q_1 on, and C rapidly discharges through transistor Q_1. Refer to Figure 10-15(b). However, when the discharge voltage across C is approximately equal to $1/3$ V_{CC}, comparator 2 switches transistor Q_1 off, and then capacitor C starts charging up again. Thus the charge–discharge cycle keeps repeating. The discharging time of the capacitor is relatively negligible compared to its charging time; hence, for all practical purposes, the time period of the ramp waveform is equal to the charging time and is approximately given by

$$T = \frac{V_{CC}C}{3I_C} \tag{10-12a}$$

where $I_C = (V_{CC} - V_{BE})/R$ = constant current in amperes and C is in farads. Therefore, the free-running frequency of the ramp generator is

$$f_o = \frac{3I_C}{V_{CC}C} \tag{10-12b}$$

Figure 10-24(b) shows the generator's output waveform. For the analysis and design of the current mirror, the reader may refer to Section 1-11.

EXAMPLE 10-8

Referring to the circuit of Figure 10-24(a), determine the frequency of the free-running ramp generator if R is set at 10 kΩ. Assume that the $V_{BE} = V_{D1} = 0.7$ V.

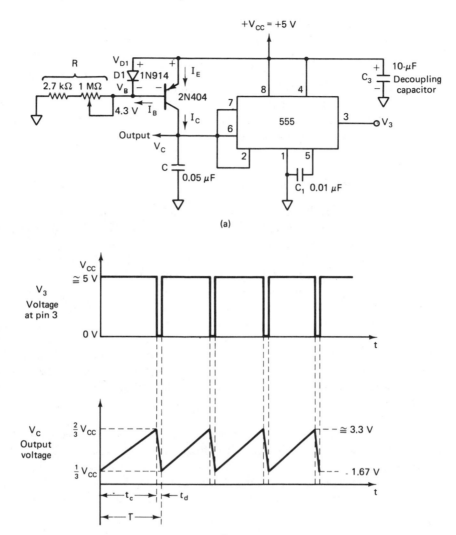

Figure 10–24   (a) Free-running ramp generator circuit. (b) Output waveform.

SOLUTION   By Equation (10-12b),

$$f_o = \frac{(3)\left[\dfrac{5 - 0.7}{10\text{ k}}\right]}{(5)(5)(10^{-8})} = 5.16 \text{ kHz}$$

10-5 PHASE-LOCKED LOOPS

Although the evolution of the phase-locked loop began in the early 1930s, its cost outweighed its advantages at first. With the rapid development of integrated-circuit technology, however, the phase-locked loop has emerged as one of the fundamental building blocks in electronics technology. The phase-locked-loop principle has been used in applications such as FM (frequency modulation) stereo decoders, motor speed controls, tracking filters, frequency synthesized transmitters and receivers, FM demodulators, frequency shift keying (FSK) decoders, and a generation of local oscillator frequencies in TV and in FM tuners. Today the phase-locked loop is even available as a single package, typical examples of which include the Signetics' SE/NE 560 series (the 560, 561, 562, 564, 565, and 567). However, for more economical operation, discrete ICs can be used to construct a phase-locked loop.

10-5.1 Operating Principles

Figure 10-25 shows the phase-locked loop (PLL) in its basic form. As illustrated in this figure, the phase-locked loop consists of (1) a phase detector, (2) a low-pass filter, and, (3) a voltage-controlled oscillator.

The phase detector, or comparator compares the input frequency f_{IN} with the feedback frequency f_{OUT}. The output of the phase detector is proportional to the phase difference between f_{IN} and f_{OUT}. The output voltage of a phase detector is a dc voltage and therefore is often referred to as the *error* voltage. The output of the phase detector is then applied to the low-pass filter, which removes the high-frequency noise and produces a dc level. This dc level, in turn, is the input to the voltage-controlled oscillator (VCO). The filter also helps in establishing the dynamic characteristics of the PLL circuit. The output frequency of the VCO is directly proportional to the input dc level. The VCO frequency is compared with the input frequencies and adjusted until it is equal to the input frequencies. In short, the phase-locked loop goes through three states: *free-running, capture,* and *phase lock*.

Before the input is applied, the phase-locked loop is in the free-running state. Once the input frequency is applied, the VCO frequency starts to change and the phase-locked loop is said to be in the capture mode. The VCO frequency continues to change until it equals the input frequency, and the phase-locked loop is then in the phase-locked state. When phase locked, the loop tracks any change in the input frequency through its repetitive action.

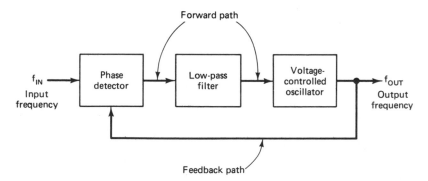

Figure 10–25   Block diagram of a phase-locked loop.

Before studying the specialized phase-locked-loop IC, we shall consider the discrete phase-locked loop, which may be assembled by combining a phase detector, a low-pass filter, and a voltage-controlled oscillator (see Figure 10-25).

10-5.1(a) Phase detector. The phase detector compares the input frequency and the VCO frequency and generates a dc voltage that is proportional to the phase difference between the two frequencies. Depending on the analog or digital phase detector used, the PLL is either called an analog or digital type, respectively. Even though most of the monolithic PLL integrated circuits use analog phase detectors, the majority of discrete phase detectors in use are of the digital type mainly because of its simplicity. For this reason, we shall consider only digital-type phase detectors here.

A double-balanced mixer is a classic example of an analog phase detector. On the other hand, examples of digital phase detectors are these:

1. Exclusive-OR phase detector
2. Edge-triggered phase detector
3. Monolithic phase detector (such as type 4044)

Figure 10-26(a) shows the exclusive-OR phase detector that uses an exclusive-OR gate such as CMOS type 4070. The output of the exclusive-OR gate is *high* only when f_{IN} *or* f_{OUT} is high, as shown in Figure 10-26(b).

In this figure, f_{IN} is leading f_{OUT} by ϕ (phi) degrees; that is, the phase difference between f_{IN} and f_{OUT} is ϕ degrees. The dc output voltage of the exclusive-OR phase detector is a function of the phase difference between its two inputs. Figure 10-26(c) shows dc output voltage as a function of the phase difference between f_{IN} and f_{OUT}. This graph indicates that the maximum dc output voltage occurs when the phase difference is π radians or 180°. The slope of the curve between 0 and π radians is the conversion gain k_p of the phase detector. For example, if the exclusive-OR gate uses a supply voltage $V_{CC} = 5$ V, the conversion gain k_p is

$$k_p = \frac{5 \text{ V}}{\pi} = 1.59 \text{ V/rad}$$

The exclusive-OR type of phase detector is generally used if the f_{IN} and f_{OUT} are square waves. The edge-triggered phase detector, on the other hand, is preferred if the f_{IN} and f_{OUT} are pulse waveforms with less than 50% duty cycles.

Figure 10-27(a) shows the edge-triggered type of phase detector using an R-S (reset–set) flip-flop. The R-S flip-flop, in turn, is formed from a pair of cross-coupled NOR gates, such as the CD4001. The R-S flip-flop is triggered; that is, the output of the detector changes its logic state on the positive (leading) edge of the inputs f_{IN} and f_{OUT} [see Figure 10-27(b)]. The graph of dc output voltage versus phase difference between f_{IN} and f_{OUT} is shown in Figure 10-27(c).

The advantages of the edge-triggered-type phase detector over the exclusive-OR type of detector are (1) the dc output voltage is linear over 2π radians or 360°, as opposed to π radians or 180° in the case of the exclusive-OR detector, and (2) the edge-triggered detector also exhibits better capture, tracking, and locking characteristics than the

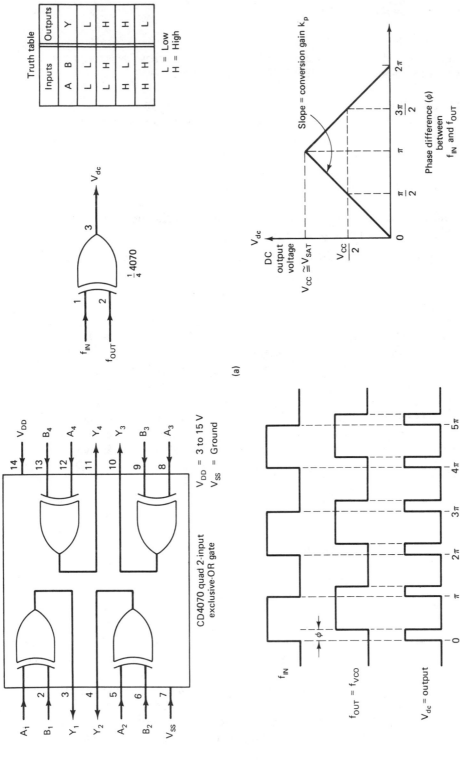

Truth table		
Inputs		Outputs
A	B	Y
L	L	L
L	H	H
H	L	H
H	H	L

L = Low
H = High

$\frac{1}{4}$4070

(a)

CD4070 quad 2-input
exclusive-OR gate

V_{DD} = 3 to 15 V
V_{SS} = Ground

Slope = conversion gain k_p

(c)

(b)

Figure 10–26 (a) Exclusive-OR phase detector connection and logic diagram. (Courtesy of National Semiconductor.) (b) Input and output waveforms. (c) Average output voltage versus phase difference between f_{IN} and f_{OUT} curve.

402

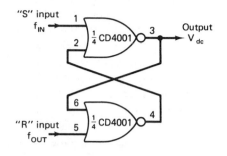

CD4001 Quad 2-input NOR gate

V_{DD} = 3 V to 15 V
V_{SS} = Ground

(a)

(b)

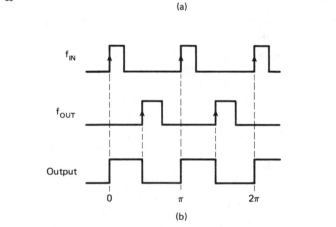

(c)

Figure 10–27   Edge-triggered type of phase detector. (a) NOR gate *R-S* flip-flop connection diagram. (b) Input and output waveforms. (c) DC output voltage versus phase difference between f_{IN} and f_{OUT}.

exclusive-OR detector. However, both types of detectors are sensitive to harmonics of the input signal and changes in the duty cycles of f_{IN} and f_{OUT}.

In a monolithic phase detector IC such as CMOS type 4044, the harmonic sensitivity and duty cycle problems are absent, since the circuit responds only to transitions in the input signals. In other words, the phase error and hence the output error voltage of the monolithic phase detector are independent of the variations in the amplitude and duty cycle of the input waveforms. Therefore, in critical applications the monolithic phase detector is preferred over both the exclusive-OR and edge-triggered types of phase detectors.

Figure 10-28(a) shows the block diagram of the MC4344/4044 phase detector. As shown in this figure, the MC4344/4044 consists of two digital phase detectors, a charge pump, and an amplifier. Phase detector 1 is used in applications that require zero frequency and phase difference at lock. On the other hand, if quadrature lock is desired, phase detector 2 can be used. When detector 1 is used in the main loop, detector 2 can also be used to indicate whether the main loop is in lock or out of lock.

The input/output transfer characteristic curve of phase detector 1 is shown in Figure 10-28(b). As shown here, the curve is linear over 4π radians or 720° and has a conversion gain k_p of 1.5 V/4π = 0.12 V/rad.

10-5.1(b) Low-Pass filter.   The second block shown in the PLL block diagram of Figure 10-25 is a low-pass filter. The function of the low-pass filter is to remove the high-frequency components in the output of the phase detector and to remove high-frequency noise. More important, the low-pass filter controls the dynamic characteristics of the phase-locked loop. These characteristics include capture and lock ranges, bandwidth, and transient response. The lock range is defined as the range of frequencies over which the PLL system follows the changes in the input frequency f_{IN}. An equivalent term for lock range is tracking range. On the other hand, the capture range is the frequency range in which the PLL acquires phase lock. Obviously, the capture range is always smaller than the lock range.

As the filter bandwidth is reduced, its response time increases. However, reduced bandwidth reduces the capture range of the PLL. Nevertheless, reduced bandwidth helps to keep the loop in lock through momentary losses of signal and also minimizes noise.

The loop filter used in the PLL may be one of the three types shown in Figure 10-29, page 406. With the passive filters of Figure 10-29(a) and (b), an amplifier is generally used for gain. On the other hand, the active filter of Figure 10-29(c) includes the gain.

10-5.1(c) Voltage-controlled oscillator.   A third section of the PLL is the voltage-controlled oscillator. The VCO generates an output frequency that is directly proportional to its input voltage. The block diagram of the VCO is shown in Figure 10-30, page 406.

While the SE/NE 566 VCO is discussed in Section 8-18, a typical connection diagram is repeated in Figure 10-31, page 407, for convenience. The maximum output frequency of the NE/SE 566 is 500 kHz. For higher output frequency, therefore, integrated circuits such as Motorola's MC4324/4024 and MC1648 may be used.

10-5.2 Monolithic Phase-Locked Loops

The Signetics SE/NE 560 series is monolithic phase-locked loops. The SE/NE 560, 561, 562, 564, 565, and 567 differ mainly in operating frequency range, power supply require-

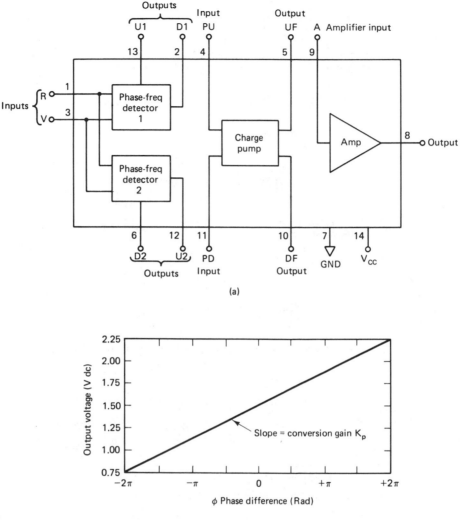

(a)

(b)

Figure 10–28 (a) MC4344/4044 phase detector block diagram. R, reference input; V, variable or feedback input; PU, pump-up signal; PD, pump-down signal,; UF, up-frequency output signal; DF, down-frequency output signal. (b) Its input/output transfer characteristic curve. (Courtesy of Motorola Semiconductor.)

ments, and frequency and bandwidth adjustment ranges. Only the SE/NE 565 phase-locked loop is discussed here because it is one of the most commonly used devices of the 560 series.

Figure 10-32 shows the block diagram and connection diagram of the 565 PLL. The device is available as a 14-pin DIP package and as 10-pin metal can package. The important electrical characteristics of the 565 PLL are:

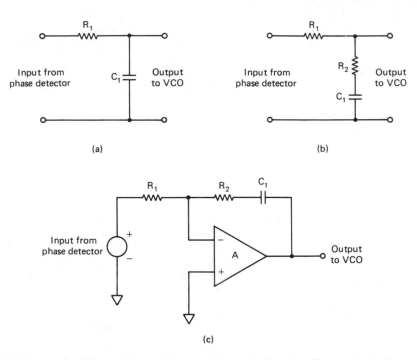

Figure 10–29 Low-pass filters. (a) and (b) Passive filters. (c) Active filter.

Figure 10–30 VCO block diagram.

Operating frequency range: 0.001 Hz to 500 kHz

Operating voltage range: ±6 to ±12 V

Input level required for tracking: 10 mV rms minimum to 3 V peak-to-peak maximum

Input impedance: 10 kΩ typically

Output sink current: 1 mA, typically

Output source current: 10 mA, typically

Drift in VCO center frequency (f_{OUT}) with temperature: 300 ppm/°C, typically

Drift in VCO center frequency with supply voltage: 1.5%/V maximum

Triangle wave amplitude: typically 2.4 V pp at ±6 V (see block diagram, pin 9)

Square wave amplitude: typically 5.4 V pp at ±6 V (see block diagram, pin 4)

Bandwidth adjustment range: ⟨±1 to⟩ ±60%

The center frequency of the PLL is determined by the free-running frequency of the VCO, which is given by the equation

Figure 10–31   SE566 VCO connection diagram.

$$f_{OUT} \cong \frac{1.2}{4R_1 C_1} \quad \text{Hz} \tag{10-13}$$

where R_1 and C_1 are an external resistor and a capacitor connected to pins 8 and 9, respectively [see block diagram of Figure 10-32(a).] The VCO free-running frequency f_{OUT} is adjusted externally with R_1 and C_1 to be at the center of the input frequency range. Although C_1 can be any value, R_1 must have a value between 2 kΩ and 20 kΩ. A capacitor C_2 connected between pin 7 and the positive supply (pin 10) forms a first-order low-pass filter with an internal resistance of 3.6 kΩ. The filter capacitor C_2 should be large enough to eliminate variations in the demodulated output voltage at pin 7 in order to stabilize the VCO frequency.

The 565 PLL can lock to and track an input signal over typically ±60% bandwidth with respect to f_{OUT} as the center frequency. The lock range f_L and capture range f_C of the PLL are given by the following equations:

$$f_L = \pm \frac{8f_{OUT}}{V} \quad \text{Hz} \tag{10-14}$$

where f_{OUT} = free-running frequency of VCO (Hz)
  $V = (+V) - (-V)$ (volts)

and

$$f_C = \pm \left[\frac{f_L}{(2\pi)(3.6)(10^3)(C_2)} \right]^{1/2} \tag{10-15}$$

where C_2 is in farads.

The lock range usually increases with an increase in input voltage but decreases with an increase in supply voltages. Pins 2 and 3 are the input terminals of the 565 PLL, and

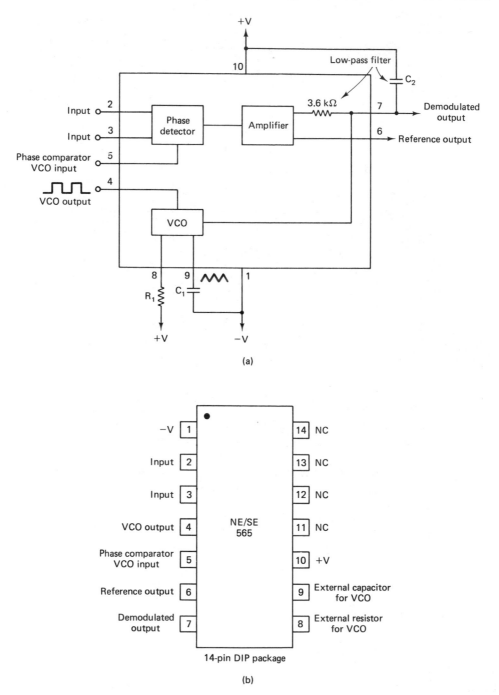

(a)

(b)

Figure 10–32   (a) NE/SE565 PLL block diagram. (b) Connection diagram. (Courtesy of Signetics Corporation.)

an input signal can be direct-coupled, provided that there is no dc voltage difference between the pins and the dc resistances seen from pins 2 and 3 are equal. A short between pins 4 and 5 connects the VCO output (f_{OUT}) to the phase comparator and enables the comparator to compare f_{OUT} with the input signal f_{IN}.

In frequency multiplication applications a digital frequency divider is inserted between pins 4 and 5, as will be shown later. A dc reference voltage at pin 6 is approximately equal to the dc potential of the demodulated output at pin 7. In applications such as frequency shift keying (FSK), the dc reference voltage at pin 6 is used as an input to the comparator. [Refer to Section 10-5.3(b).] The lock range of the PLL can be decreased with little change in the free-running frequency of the VCO by connecting a resistance between pins 6 and 7.

EXAMPLE 10-9

Referring to the circuit of Figure 10-33(a), determine the free-running frequency f_{OUT}, the lock range f_L, and the capture range f_C.

SOLUTION   From Equations (10-13) through (10-15),

$$f_{OUT} = \frac{1.2}{(4)(12)(10^3)(10^{-8})} = 2.5 \text{ kHz}$$

$$f_L = \pm \frac{8(2.5)(10^3)}{(20)} = \pm 1 \text{ kHz}$$

$$f_C = \pm \left[\frac{(10^3)}{(2\pi)(3.6)(10^3)(10)(10^{-6})} \right]^{1/2} = \pm 66.49 \text{ Hz}$$

The relationship between f_{OUT}, f_L, and f_C is depicted in Figure 10-33(b).

10-5.3 565 PLL Applications

10-5.3(a) Frequency multiplier.   Figure 10-34(a), page 411, is a block diagram of a frequency multiplier using the 565 PLL. As shown in this diagram, the frequency divider is inserted between the VCO and the phase comparator. Since the output of the divider is locked to the input frequency f_{IN}, the VCO is actually running at a multiple of the input frequency. The desired amount of multiplication can be obtained by selecting a proper divide-by-N network, where N is an integer. For example, to obtain the output frequency $f_{OUT} = 5f_{IN}$, a divide-by-$N = 5$ network is needed. Figure 10-34(b) shows this function performed by a 7490 (4-bit binary counter) configured as a divide-by-5 circuit. In this figure, transistor Q_1 is used as a driver stage to increase the driving capability of the NE565.

To verify the operation of the circuit, one must determine the input frequency range and then adjust the free-running frequency f_{OUT} of the VCO by means of R_1 and C_1 so that the output frequency of the 7490 divider is midway within the predetermined input frequency range. The output of the VCO now should be $5f_{IN}$. In Figure 10-34(b), the output frequency f_{OUT} can be adjusted from 1.5 kHz to 15 kHz by varying potentiometer R_1 [see Equation (10-13)]. This means that the input frequency f_{IN} range has to be within

Figure 10–33 (a) Circuit of Example 10–8. (b) Relationship between f_{OUT}, f_L, and f_C.

300 Hz to 3 kHz. In addition, the input waveform can either be sine or square and may be applied to input pins 2 or 3. Figure 10-34(c) shows the input–output waveforms for $f_{OUT} = 5f_{IN}$.

Even though supply voltages of ±10 V are used in Figure 10-34(b), the NE565 can be operated on ±5-V supply voltages instead. A small capacitor, typically 1000 pF, is connected between pins 7 and 8 to eliminate possible oscillations. Also, capacitor C_2 should be large enough to stabilize the VCO frequency.

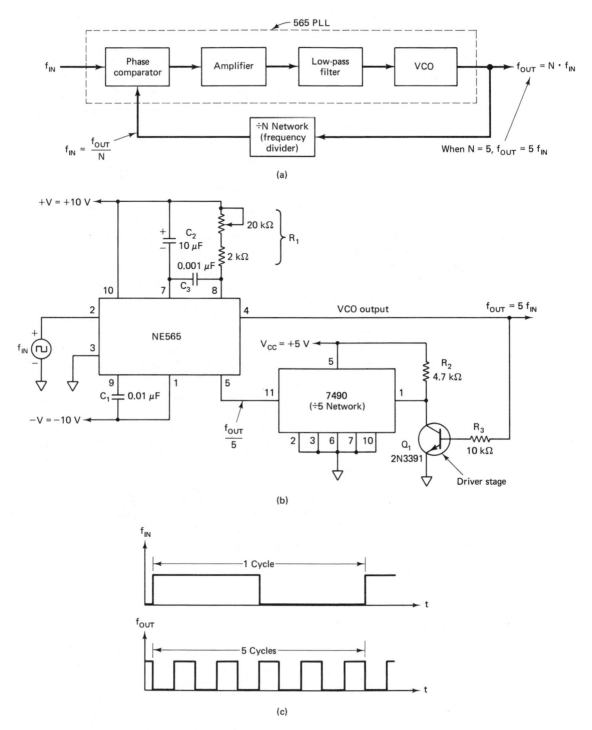

Figure 10–34 Frequency multiplier using the 565. (a) Block diagram. (b) Connection diagram for multiply-by-5. (c) Input–output waveforms.

10-5.3(b) Frequency shift keying (FSK) demodulator.   In computer periph-
eral and radio (wireless) communications, the binary data or code is transmitted by means
of a carrier frequency that is shifted between two preset frequencies. Since a carrier
frequency is shifted between two preset frequencies, the data transmission is said to use
a frequency shift keying (FSK) technique.

A very useful application of the 565 PLL is as a FSK demodulator. In the 565 PLL
the frequency shift is usually accomplished by driving a VCO with the binary data signal
so that the two resulting frequencies correspond to the logic 0 and logic 1 states of the
binary data signal. The frequencies corresponding to logic 1 and logic 0 states are commonly called the *mark* and *space* frequencies. Several standards are used to set the mark
and space frequencies. For example, when transmitting teletypewriter information using
a modulator–demodulator system (modem for short), a 1070 Hz–1270 Hz (mark–space)
pair represents the originate signal, while a 2025 Hz–2225 Hz (mark–space) pair represents the answer signal.

Figure 10-35 shows a typical block diagram of the AM (amplitude modulation) and
FM (frequency modulation) transmitters and receivers. A simplified diagram using only
an FSK generator and FSK demodulator is shown in Figure 10-36. The FSK generator is
formed by using a 555 as an astable multivibrator whose frequency is controlled by the
state of transistor Q_1. In other words, the output frequency of the FSK generator depends
on the logic state of the digital data input. One hundred and fifty hertz is one of the standard
frequencies at which the data are commonly transmitted. When the input is logic 1,
transistor Q_1 is off. Under these conditions, the 555 works in its normal mode as an astable
multivibrator; that is, capacitor C charges through R_A and R_B to $2/3$ V_{CC} and discharges
through R_B to $1/3$ V_{CC}. Thus capacitor C charges and discharges alternately between
$2/3$ V_{CC} and $1/3$ V_{CC} as long as the input is at a logic 1 state. The frequency of the output
waveform is given by Equation (10-10):

Figure 10–35   Block diagrams of AM/FM (a) transmitter and (b) receiver.

$$f_o = \frac{1.45}{(R_A + 2R_B)C} = 1070 \text{ Hz}$$

The values of R_A, R_B, and C are selected so that f_o represents a mark frequency (1070 Hz). On the other hand, when the input is logic 0, Q_1 is on (saturated), which in turn connects the resistance R_C across R_A. This action reduces the charging time of the capacitor and increases the output frequency, which is given by Equation (10-10):

$$f_o = \frac{1.45}{(R_A \| R_C + 2R_B)C} = 1270 \text{ Hz}$$

By proper selection of resistance R_C, this frequency is adjusted to equal the space frequency of 1270 Hz. The difference between FSK signals of 1070 Hz and 1270 Hz is 200 Hz; this difference is called the *frequency shift*.

As shown in Figure 10-36, the output of the 555 FSK generator is then applied to the 565 FSK demodulator. Capacitive coupling is used at the input to remove a dc level. As the signal appears at the input of the 565, the loop locks to the input frequency and tracks it between the two frequencies with a corresponding dc shift at the output. Resistor R_1 and capacitor C_1 determine the free-running frequency of the VCO, while C_2 is a loop filter capacitor that establishes the dynamic characteristics of the demodulator. Here C_2 must be chosen smaller than usual to eliminate overshoot on the output pulse. A three-stage RC ladder (low-pass) filter is used to remove the carrier component from the output. The high cutoff frequency ($f_H = 1/2\pi RC$) of the ladder filter is chosen to be approximately halfway between the maximum keying rate of 150 Hz and twice the input frequency, that is, approximately 2200 Hz. The output signal of 150 Hz can be made logic compatible by connecting a voltage comparator between the output of the ladder filter and pin 6 of the PLL. The VCO frequency is adjusted with R_1 so that at $f_{IN} = 1070$ Hz a slightly positive voltage is obtained at the output.

10-6 POWER AMPLIFIERS

Small-signal amplifiers are essentially voltage amplifiers that supply their loads with larger amplified signal voltages. On the other hand, large signal or power amplifiers supply a large-signal current to current-operated loads such as speakers and motors.

In previous discussion the general-purpose op-amp has been emphasized as a voltage amplifier because it is essentially a small-signal amplifier with a limited output current capability. For instance, the maximum output current supplied by the 741-type op-amp is equal to its *short-circuit* ($R_L = 0 \ \Omega$) current, that is, 25 mA. Obviously, the current supplied by the 741 will decrease with an increase in load. In audio applications, however, the amplifier is called upon to deliver much higher current than that supplied by general-purpose op-amps. This means that loads such as speakers and motors requiring substantial currents cannot be driven directly by the output of general-purpose op-amps. However, there are two possible solutions available. One method is to use discrete or monolithic power transistors called *power boosters* at the output of the op-amp; a second method is to use specialized ICs designed as power amplifiers.

Figure 10-36 The 565 as an FSK demodulator.

10-6.1 Power Amplifiers Using Power Boosters

A simple method of increasing the output current of a general-purpose op-amp is to connect a *power booster* in series with the op-amp. Obviously, the output characteristics of this arrangement, such as output current and output impedance, are then controlled by the choice of the power booster and the configuration in which it is used. Usually, a discrete power booster is an emitter-follower stage formed by using a power transistor, as in Figure 10-37(a). (Transistors with power dissipation larger than 1/2 W are called power transistors.) Because of its unity gain, the emitter follower helps to retain the voltage-gain characteristics of the general-purpose op-amp. In addition, the emitter follower offers a relatively high input impedance; therefore, it also works as an isolation stage (buffer) between the op-amp and the load. The input impedance and output current of the emitter follower may be further increased by the use of Darlington power transistors. For increased efficiency, the power booster is often configured as a push-pull Class B amplifier.

Figure 10-37(b) shows a typical connection diagram for the power amplifier using a general-purpose op-amp and the Burr-Brown 3329/03 monolithic power booster. The 3329/03, a unity-gain power booster amplifier that does not require any external components, is designed to be used inside the feedback loop of an op-amp. It operates over a power supply range of ± 12 to ± 18 V. The device has a Class B output stage, which provides output current of ± 100 mA at 20 V peak-to-peak output voltage swing when operated on supplies of ± 15 V. In addition, the device will operate over the temperature range from $-40°$ to $+85°C$ without a heat sink. Above all, when the 3329/03 power booster is used with an op-amp, it does not degrade the frequency response and output characteristics of the op-amp, mainly because of the following electrical characteristics:

Full power frequency = 1 MHz

Small-signal bandwidth = 5 MHz

Input impedance = 10 kΩ

Open-loop output impedance = 10 Ω

Another power booster with increased output current, slew rate, and bandwidth is the Burr-Brown 3553. It has an output current of ± 200 mA with a 300-MHz bandwidth and a slew rate of 2000 V/μs. The 3553 is therefore ideally suited for line driving applications (50-Ω load) in which fast pulses or wideband signals are involved. For applications such as stepper motor drivers, servomotor drivers, power supplies, and power DACs requiring output voltage up to ± 30 V and full output current ± 2 A, the Intersil ICL8063 (16-pin DIP) monolithic power transistor driver and amplifier is most suitable. The ICL8063 will operate from the outputs of most of op-amps and devices such as timers and comparators. When used in conjunction with general-purpose op-amps, external complementary power transistors, and 8 to 10 passive components (resistors and capacitors), the ICL8063 can deliver more than 50 W to external loads. It has built-in safe area protection, short-circuit proof protection, and built-in ± 13-V regulators.

10-6.2 Monolithic Power Amplifiers

A variety of monolithic as well as hybrid power amplifiers is commercially available. Although power amplifiers differ from general-purpose op-amps in delivering various

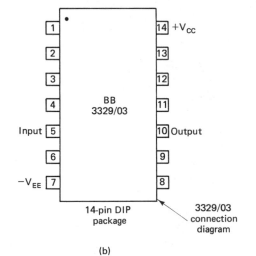

(b)

Figure 10–37 Power amplifiers using general-purpose op-amp input stage and (a) a discrete power booster and (b) a 3329/03 IC power booster. (Courtesy of Burr-Brown Research Corporation.)

amounts of power, they are nearly as compact. Because of its wide acceptance, National Semiconductor's LM380 audio power amplifier is introduced here and followed by a brief discussion of hybrid power amplifiers with output power up to 100 W.

10-6.2(a) LM380 power audio amplifier.   National Semiconductor's LM380 is a power audio amplifier designed to deliver a minimum of 2.5 W (rms) to an 8-Ω load and hence is ideal for consumer applications. Because of the following features, the LM380 requires a minimum number of external components:

> Internally fixed gain of 50 (34 dB)
>
> Output automatically self-centering to one-half of the supply voltage
>
> Output short-circuit proof with internal thermal limiting
>
> Unique input stage allowing inputs to be ground referenced or ac coupled

In addition, the LM380 has a wide supply voltage range (5 to 22 V), high peak current capability (1.3 A maximum), high impedance (150 kΩ), and low (0.2%) total harmonic distortion (THD) and is available in a standard DIP package. Also, the LM380 has a bandwidth of 100 kHz typically at $P_{OUT} = 2$ W and $R_L = 8$ Ω.

Figure 10-38 shows the connection diagram, block diagram, and schematic diagram of the LM380. A copper lead frame used with the center three pins (3, 4, 5, 10, 11, and 12) on either side of the DIP package comprises a heat sink [see Figure 10-38(a)]. Therefore, there is no need to use a separate heat sink for the audio amplifier.

LM380 Circuit Description.   The schematic diagram of the LM380 shown in Figure 10-38(c) is composed of four stages: PNP emitter follower, differential amplifier, common emitter, and quasi-complementary emitter follower. The input stage is an emitter follower composed of PNP transistors Q_1 and Q_2, which drives the PNP Q_3–Q_4 differential pair. The choice of PNP input transistors Q_1 and Q_2 allows the input to be referenced to ground; that is, input can be direct coupled to either the inverting (pin 6) or the non-inverting (pin 2) terminals of the amplifier.

The current in the PNP differential pair Q_3–Q_4 is established by Q_7, R_3, and $+$V. The current mirror formed by transistors Q_7, Q_8, and associated resistors then establishes the collector current of Q_9. Transistors Q_5 and Q_6 constitute collector loads for the PNP differential pair. The output of the differential amplifier is taken at the junction of Q_4 and Q_6 transistors and is applied as an input to the common-emitter voltage-gain stage.

The common-emitter amplifier stage is formed by transistor Q_9 with D_1, D_2, and Q_8 as a current source load. The capacitor C between the base and collector of Q_9 provides internal compensation and helps to establish the upper cutoff frequency of 100 kHz at 2 W for 8-Ω loads. Since Q_7 and Q_8 form a current mirror, the current through D_1 and D_2 is approximately the same as the current through R_3. In addition, D_1 and D_2 are temperature-compensating diodes for transistors Q_{10} and Q_{11} in that D_1 and D_2 have the same characteristics as the base–emitter junctions of Q_{11}. Therefore, the current through Q_{10} and (Q_{11}–Q_{12}) is approximately equal to the current through diodes D_1 and D_2.

The output stage is a quasi (false)-complementary pair emitter follower formed by NPN transistors Q_{10} and Q_{12}. In fact, the combination of PNP transistor Q_{11} and NPN

Figure 10–38 (a) LM380 power audio amplifier connection diagram. An asterisk indicates ground-heat sink pins. (b) Block diagram. (c) Schematic diagram. (Numbers in parentheses are pin numbers.) (Courtesy of National Semiconductor.)

transistor Q_{12} has the power capability of an NPN transistor but the characteristics of a PNP transistor.

Because of the arrangement of the output stage, the quiescent output voltage is half the supply voltage $(+V)$. Furthermore, the negative dc feedback applied through R_5 balances the differential amplifier so that the dc output voltage is stabilized at $+V/2$. To decouple the input stage from the supply voltage $+V$, a bypass capacitor on the order of microfarads should be connected between the bypass terminal (pin 1) and ground (pin 7). The overall internal voltage gain of the amplifier is fixed at 50. However, gain can be increased by using positive feedback, as illustrated in the applications discussed in the next section.

Another commonly used audio amplifier is the LM384. The schematic diagram of the LM384 is the same as that of the LM380, which is shown in Figure 10-38, except that the former is designed to deliver 5 W.

Applications. *Audio Power Amplifier.* Figure 10-39 shows the simplest and most basic application of the LM380—as an audio power amplifier. As shown in this figure, the amplifier requires very few external components because of the internal biasing, compensation, and the fixed gain. When the power amplifier is used in the noninverting configuration, the inverting terminal may be either shorted to ground, connected to ground through a resistor or capacitor, or left open as shown in Figure 10-39. Similarly, when the power amplifier is used in the inverting mode, the noninverting terminal may be either shorted to ground or returned to ground through a resistor or capacitor. Usually, a capacitor is connected between the inverting terminal and ground if the input has a high internal impedance. Nevertheless, in either configuration the supply voltage $+V$ should be decoupled by connecting a capacitor between the $+V$ terminal (pin 14) and ground. As a precautionary measure, an *RC* combination should be used at the output terminal (pin 8) to eliminate 5- to 10-MHz oscillation, especially in an RF-sensitive environment.

Although the gain of the LM380 is internally fixed at 50, it can be changed with the use of external components. Specifically, gains up to 300 are possible when positive feedback is used. Figure 10-40(a) shows the LM380 configured for a gain of 200 using

Figure 10–39 LM380 audio power amplifier. (Courtesy of National Semiconductor.)

(a)

(b)

Figure 10–40 (a) LM380 configured for a gain of 200 using positive feedback. (b) LM380 with a variable gain (volume control) up to 50 using a potentiometer across the two input terminals. (Courtesy of National Semiconductor.)

positive feedback. On the other hand, variable gains up to 50 are obtained with the use of a potentiometer across the two input terminals, as shown in Figure 10-40(b).

Bridge Power Audio Amplifier.   For applications requiring more power than is provided by the single LM380 amplifier, two LM380s can be used in the bridge configuration shown in Figure 10-41. In this arrangement the maximum output voltage swing will

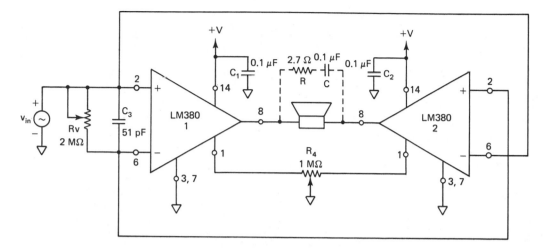

Figure 10–41   LM380s used in bridge configuration to provide more power. (*R* and *C* for stability with high-current loads.) (Courtesy of National Semiconductor.)

be twice that of a single LM380 amplifier; therefore, the power delivered to the load will be four times as much. For improved performances, potentiometer R_4 should be used to balance the output offset voltages of the LM380s.

Intercom System.   A simple and inexpensive intercom system can be formed by using the LM380 as shown in Figure 10-42. The speakers used in this figure are permanent-magnet types and hence act as microphones as well. The talk and listen modes are defined with reference to a master station. When the switch is in the talk position, the master speaker acts as the microphone [see Figure 10-42(a)]. On the other hand, when the switch is in the listen mode, the remote speaker acts as the microphone [see Figure 10-42(b)]. In either position the overall gain of the circuit is the same and depends on the turns ratio of the transformer *T* as well as the internal gain of the LM380. For example, if the turns ratio $N1/N2 = 20$, the overall gain of the circuit will be $50 \times 20 = 1000$. However, the internal gain of the LM380 can be controlled with the use of potentiometer R_V.

A sample of monolithic power audio amplifiers with their output power ratings is listed in Table 10-4. In these power ICs, the heat-sink requirements are met without the use of separate heat sinks. Because of their compact size, power audio amplifiers are more convenient than any of the discrete forms of audio amplifiers. Dual power amplifiers such as the LM377, which delivers 2 W/channel, are used for stereo phonographs, tape players and recorders, and AM–FM stereo receivers.

For applications requiring more output power than that supplied by the monolithic power ICs, hybrid power amplifiers are commonly used. For example, the Intersil's ICH8510/8520/8530 is a family of hybrid power amplifiers designed to deliver 1, 2, and 2.8 A, respectively, at 24-V output levels. All amplifiers are protected against inductive kickback with internal power limiting and against shorts to ground. Each device has a dc gain of 10^5 (100 dB), internal frequency compensation, and an electrically isolated pack-

Figure 10–42 Intercom systems using the LM380. (a) Talk mode. (b) Listen mode. (Courtesy of National Semiconductor.)

TABLE 10-4 TYPICAL MONOLITHIC POWER AUDIO AMPLIFIERS

Output power (W)	IC type
1	MC1454
2	LM377 (dual)
4	LM378 (dual)
5	μA706
6	LM379 (dual)
8	TDA2002

age style that allows easy heat sinking. These devices are used to drive electronic values, push-pull solenoids, and dc and ac motors.

Another classic example of a hybrid power amplifier is the Burr-Brown's 3573, which is designed to deliver 100 W peak or 40 W continuous output power. When operated from ±28-V power supplies, the device delivers ±5 A peak minimum at

±20 V (40 V peak to peak) to the load. The 3573 is internally frequency compensated with good input and distortion characteristics and is low in cost. It can be used to drive dc and ac motors, electronic valves, and push-pull solenoids.

10-7 VOLTAGE REGULATORS

A *voltage regulator* is a circuit that supplies a constant voltage regardless of changes in load currents. Although voltage regulators can be designed using op-amps, it is quicker and easier to use IC voltage regulators. Furthermore, IC voltage regulators are versatile and relatively inexpensive and are available with features such as a programmable output, current/voltage boosting, internal short-circuit current limiting, thermal shutdown, and floating operation for high-voltage applications. IC voltage regulators are of the following types:

Fixed output voltage regulators: positive and/or negative output voltage

Adjustable output voltage regulators: positive or negative output voltage

Switching regulators

Special regulators

Except for the switching regulators, all other types of regulators are called *linear regulators*. The impedance of a linear regulator's active element may be continuously varied to supply a desired current to the load. On the other hand, in the *switching regulator* a switch is turned on and off at a rate such that the regulator delivers the desired average current in periodic pulses to the load. Because the switching element dissipates negligible power in either the *on* or *off* state, the switching regulator is more efficient than the linear regulator. Nevertheless, in switching regulators the power dissipation is substantial during the switching intervals (on to off or off to on). In addition, most loads cannot accept the average current in periodic pulses. Therefore, most practical regulators are of the linear type.

Voltage regulators are commonly used for on-card regulation and laboratory-type power supplies. Voltage regulators, especially the switching type, are used as control circuits in pulse width modulation (PWD), push-pull bridges, and series-type switch mode supplies. Almost all power supplies use some type of voltage regulator IC because voltage regulators are simple to use, reliable, low in cost, and, above all, available in a variety of voltage and current ratings. A vast number of voltage regulators is available; data sheets and application notes provided by the manufacturers contain information on the design and use of these devices. The next section presents a selection of voltage regulators to illustrate the design and construction features of these devices.

10-7.1 Fixed Voltage Regulators

10-7.1(a) Positive voltage regulator series with seven voltage options.   The 7800 series consists of three-terminal positive voltage regulators with seven voltage options [see Figure 10-43(a)]. These ICs are designed as fixed voltage regulators and with adequate heat sinking can deliver output currents in excess of 1 A. Although these devices do not require external components, such components can be used to obtain adjustable voltages and currents. These ICs also have internal thermal overload protection

Device type	Output voltage (V)	Maximum input voltage (V)
7805	5.0	
7806	6.0	
7808	8.0	
7812	12.0	35
7815	15.0	
7818	18.0	
7824	24.0	40

(a)

(b)

(c)

*The two numbers XX
of the type number
indicate output voltage.

Figure 10–43   The 7800 series regulators. (a) Voltage options. (b) Package types. (c) Standard application. (Courtesy of Motorola Semiconductor.)

and internal short-circuit current limiting. As shown in Figure 10-43(c), proper operation requires a common ground between input and output voltages. In addition, the difference between input and output voltages ($V_{in} - V_o$), called *dropout voltage,* must be typically 2.0 V even during the low point on the input ripple voltage. Furthermore, the capacitor C_i is required if the regulator is located an appreciable distance from a power supply filter. Even though C_o is not needed, it may be used to improve the transient response of the regulator.

Typical performance parameters for voltage regulators are line regulation, load regulation, temperature stability, and ripple rejection. *Line or input regulation* is defined as the change in output voltage for a change in the input voltage and is usually expressed in millivolts or as a percentage of output voltage V_o. *Load regulation* is the change in output voltage for a change in load current and is also expressed in millivolts or as a percentage of V_o. *Temperature stability* or *average temperature coefficient of output voltage* (TCV$_o$) is the change in output voltage per unit change in temperature and is expressed in either millivolts/°C or parts per million (ppm)/°C. *Ripple rejection* is the measure of a regulator's ability to reject ripple voltages. It is usually expressed in decibels. The smaller the values of line regulation, load regulation, and temperature stability, the better the regulator.

The 7800 regulators can also be used as current sources. Figure 10-44 shows a typical connection diagram of the 7805C as a 0.5-A current source. The current supplied to the load is given by the equation

$$I_L = \frac{V_R}{R} + I_Q \qquad (10\text{-}16a)$$

where I_Q = quiscent current (amperes)
= 4.3 mA typically for the 7805C

Referring to Figure 10-44, $V_R = V_{23} = 5$ V and $R = 10\ \Omega$; therefore,

$$I_L \cong 0.5\ \text{A}$$

The output voltage V_o with respect to ground is

$$V_o = V_R + V_L \qquad (10\text{-}16b)$$

where $V_L = I_L R_L$. The load resistance $R_L = 10\ \Omega$; hence $V_L = 5$ V. Therefore, $V_o = 10$ V. Since the dropout voltage for the 7805C is 2 V, the minimum input voltage required is given by the equation

$$V_{\text{in}} = V_o + (\text{dropout voltage}) \qquad (10\text{-}16c)$$

$$= 12\ \text{V}$$

In short, a current source circuit using a voltage regulator can be designed for a desired

Figure 10–44   The 7805C as a 0.5-A current source.

value of load current (I_L) simply by selecting an appropriate value for R. Note, however, that V_{in} depends on the size of R_L and also the dropout voltage of the regulator [see Equation (10-16c)].

EXAMPLE 10-10

Using the 7805C voltage regulator, design a current source that will deliver a 0.25-A current to the 48-Ω, 10-W load.

SOLUTION   Determine the value of R and find the minimum value of V_{in} needed by using Equations (10-16a) through (10-16c):

$$R \cong \frac{5 \text{ V}}{0.25 \text{ A}} = 20 \ \Omega$$

$$V_o = 5.0 + (48)(0.25) = 17 \text{ V}$$

$$V_{in} = 17 + 2 = 19 \text{ V}$$

10-7.1(b) Negative voltage regulator series with nine voltage options.   The 7900 series of fixed output negative voltage regulators are complements to the 7800 series devices. These negative regulators are available in the same seven-voltage options as the 7800 devices. In addition, two extra voltage options, -2 V and -5.2 V, are also available in the negative 7900 series, as shown in Figure 10-45(a). Figure 10-45(b) shows the package types in which the 7900 series voltage regulators are available.

Device type	Output voltage (V)	Maximum input voltage (V)
7902	-2.0	
7905	-5.0	
7905.2	-5.2	
7906	-6.0	
7908	-8.0	-35
7912	-12.0	
7915	-15.0	
7918	-18.0	
7924	-24.0	-40

(a)

Figure 10–45   The 7900 series voltage regulators. (a) Voltage options. (b) Package types. (Courtesy of Motorola Semiconductor.)

10-7.2 Adjustable Voltage Regulators

The many manufacturers that used fixed voltage regulators, like the 7800 and 7900 series, in their line of products had to stock and hold in inventory quantities of each voltage in order to always have on hand a specific device for a particular system. Obviously, this approach increased inventory and proved to be very costly, especially when production was stopped due to the shortage of a particular voltage. Adjustable voltage regulators provided the answers to the excessive inventory and production costs because a single device satisfies many voltage requirements from 1.2 up to 47 V. In addition, they have the following performance and reliability advantages over the fixed types.

> Improved system performance by having line and load regulation of a factor of 10 or better.
>
> Improved overload protection allows greater output current over operating temperature range.
>
> Improved system reliability with each device being subjected to 100% thermal limit burn-in.

Thus, adjustable voltage regulators have become more popular because of versatility, performance, and reliability. The LM317 series is the most commonly used general-purpose adjustable voltage regulators.

10-7.2(a) Adjustable positive voltage regulators.

In this section we examine the LM317 series adjustable three-terminal positive voltage regulators. The different grades of regulators in the series are available with output voltage of 1.2 to 57 V and output current from 0.10 to 1.5 A as shown in Figure 10-46(a). The LM317 series regulators are available in standard transistor packages that are easily mounted and handled [see Figure 10-46(b)]. The three terminals are V_{IN}, V_{OUT} and ADJUSTMENT (ADJ). Figure 10-46(c) shows a typical connection diagram for the LM317 regulator. From this diagram it is obvious that the LM317 requires only two external resistors to set the output voltage. When configured as shown in Figure 10-46(c), the LM317 develops a nominal 1.25 V, referred to as the reference voltage V_{REF}, between the output and adjustment terminal. This reference voltage is impressed across resistor R_1 and, since the voltage is constant, the current I_1 is also constant for a given value of R_1. Because resistor R_1 sets current I_1, it is called the *current set* or *program resistor*. In addition to the current I_1, the current I_{ADJ} from the adjustment terminal also flows through the output set resistor R_2.

The LM317 is designed such that I_{ADJ} is very small and constant with line and load changes. The maximum value of adjustment pin current I_{ADJ} is 100 μA. Thus, referring to Figure 10-46(c), the output voltage V_o is

$$V_o = R_1 I_1 + R_2(I_1 + I_{ADJ}) \qquad (10\text{-}17a)$$

where $I_1 = \dfrac{V_{REF}}{R_1}$

R_1 = current (I_1) set resistor
R_2 = output (V_o) set resistor
I_{ADJ} = adjustment pin current

Device	Available V_o (V)	Output current (A)	V_{in} max (V)	Ripple rejection (dB)	Package
LM317	1.2 to 37	1.5	40	80	TO-39
LM317H	1.2 to 37	0.5	40	80	TO-39
LM317HV	1.2 to 57	1.5	60	80	TO-3
LM317HVH	1.2 to 37	0.50	40	80	TO-39
LM317L	1.2 to 37	0.10	40	65	TO-92
LM317M	1.2 to 37	0.50	40	80	TO-202

(a)

(b)

$$V_o = 1.25 \text{ V} \left(1 + \frac{R_2}{R_1}\right)$$

(c)

Figure 10–46 LM317. (a) Different grade regulators. (b) Standard package types. (c) A typical connection diagram. (Courtesy of National Semiconductor Corporation.)

Substituting the value of I_1 in Equation (10-17a) and rearranging, we get

$$V_o = V_{REF}\left(1 + \frac{R_2}{R_1}\right) + I_{ADJ}R_2 \qquad (10\text{-}17b)$$

where $V_{REF} = 1.25$ V = reference voltage between the output and adjustment terminals.

However, the current I_{ADJ} is very small (100 μA) and constant. Therefore, the voltage drop across R_2 due to I_{ADJ} is also very small and can be neglected. In short,

$$V_o = 1.25\left(1 + \frac{R_2}{R_1}\right) \qquad (10\text{-}17c)$$

Equation (10-17c) indicates that the output voltage V_o is a function of R_2 for a given value of R_1 and can be varied by adjusting the value of R_2. The current set resistor R_1 is usually 240 Ω, and to achieve good load regulation it should be tied directly to the output of the regulator rather than near the load.

Normally, no capacitors are needed unless the LM317 is situated far from the power supply filter capacitors, in which case an input bypass capacitor C_1 is needed. (see Figure 10-47). A 0.1-μF disc or 1-μF tantalum capacitor is suitable for input bypassing for almost all applications. Also, an optional output capacitor C_2 can be added to improve transient response. Output capacitors in the range of 1 to 1000 μF of aluminum or tantalum electrolytic are commonly used to provide improved impedance and rejection of transients. In addition, the adjustment terminal can be bypassed with C_3 to obtain very high ripple rejection ratios, which are difficult to achieve with standard three-terminal regulators. With a 10-μF bypass capacitor C_3, 80-dB ripple rejection is obtainable at any output level.

When external capacitors are used with the LM317, it is sometimes necessary to add protection diodes to prevent the capacitors from discharging through low current points into the regulator. However, there is no need to use diodes for output capacitors of 25 μF or less. Thus, protection diodes are included for use with outputs greater than 25 V and high values of output capacitance (see Figure 10-47).

Figure 10–47   LM317 with capacitors and protective diodes. (Courtesy of National Semiconductor Corporation.)

EXAMPLE 10-11

Design an adjustable voltage regulator to satisfy the following specifications:

Output voltage V_o = 5 to 12 V

Output current I_o = 1.0 A

Voltage regulator is LM317

SOLUTION   For the LM317, I_{ADJ} = 100 μA maximum. If we use R_1 = 240 Ω, then for V_o of 5 V the value of R_2 from Equation (10-17b) is

$$5 = 1.25\left(1 + \frac{R_2}{240}\right) + (10^{-4})R_2$$

or

$$R_2 = \frac{3.75}{(5.3)(10^{-3})}$$

$$= 0.71 \text{ k}\Omega$$

Similarly, for V_o = 12 V, the value of R_2 is

$$12 = 1.25\left(1 + \frac{R_2}{240}\right) + (10^{-4})R_2$$

or

$$R_2 = \frac{10.75}{(5.3)(10^{-3})}$$

$$= 2.01 \text{ k}\Omega$$

Thus, to obtain the output voltage of 5 to 12 V we need to vary R_2 from 0.71 to 2.01 kΩ, respectively. To accomplish this, we will use a 3-kΩ potentiometer for R_2. Also, to obtain I_o = 1.0 A, we should use a TO-3 package with 20-W power dissipation rating. We assume that the regulator is situated close to the power supply filter capacitors; therefore, we will not use an input bypass capacitor C_1 (refer to Figure 10-47). However, to provide improved impedance and rejection of transients and to obtain high ripple rejection, we will use an output capacitor C_2 and an adjustment terminal capacitor C_3 as shown in Figure 10-47 . As mentioned earlier, the value of C_2 = 1 μF and that of C_3 = 10 μF.

Finally, there is no need to use diodes because the output voltage is less than 25 V and capacitors are less than 25 μF. The complete circuit with component values is shown in Figure 10-48. Note for the LM317 that 3 V $\leq$ (V_{in} − V_o) $\leq$ 40 V; therefore, the input voltage V_{in} must be $\geq$ 15 V for an output voltage of 12 V.

10-7.2(b) Adjustable negative voltage regulators.   The LM337 series of adjustable voltage regulators is a complement to the LM317 series devices. These negative regulators are available in the same voltage and current options as the LM317 devices. Figure 10-49 shows the different grades of regulators in the series.

Figure 10–48   Adjustable voltage regulator for Example 10-11.

Device	Available V_o (V)	Output current (A)	V_{in} max (V)	Ripple rejection (dB)	Package
LM337	−1.2 to −37	1.5	40	77	TO-39
LM337H	−1.2 to −37	0.5	40	77	TO-39
LM337HV	−1.2 to −47	1.5	50	77	TO-3
LM337HVH	−1.2 to −47	0.5	50	77	TO-39
LM337LZ	−1.2 to −37	0.10	40	65	TO-92
LM337M	−1.2 to −37	0.50	40	77	TO-202

(a)

(b)

Figure 10–49   LM337. (a) Different grade regulators. (b) Standard package types. (Courtesy of National Semiconductor Corporation.)

10-7.3 Switching Regulators

An example of a general-purpose regulator is Motorola's MC1723. It can be used in many different ways, for example, as a fixed positive or negative output voltage regulator, variable output voltage regulator, or switching regulator. Because of its flexibility, it has become a standard type in the electronics industry. Although it is designed to deliver load current to 150 mA, the current capability can be increased to several amperes through the use of one or more external pass transistors. Figure 10-50 shows the connections for switching-regulator action of the 723-type regulator for positive output. The regulator requires an external transistor and a 1-mH choke. To minimize its power dissipation during switching, the external transistor used must be a switching power transistor. The 1-mH choke smooths out the current pulses delivered to the load, while capacitor C holds output voltage at a constant dc level.

Fixed voltage regulators such as the 78XX and 79XX and the adjustable regulators such as the LM317 are all called *series dissipative regulators*. This is because these regulators simulate a variable resistance between the input voltage and the load, and hence function in a linear mode. In fact, for a specified range of variation in the input voltage and load current, the linear regulator maintains a constant output voltage by dissipating the excess power as heat. In a series dissipative regulator, conversion efficiency decreases as the input/output voltage differential increases, or vice versa. For this reason, the linear series regulator is well suited for medium current applications with a small voltage differential, where the power dissipation can be handled with heat sinks.

Figure 10–50 Motorola MC1723 as a 5-V 1-A switching regulator. (Pin numbers adjacent to terminals are for the metal packages and those in parentheses are for the DIP packages.) (Courtesy of Motorola Semiconductor.)

To improve the efficiency of a regulator, the series-pass transistor is used as a *switch* (alternately turned on and off) rather than as a variable resistor as in the linear mode. A regulator constructed to operate in this manner is called a *series switching regulator*. In such regulators the series-pass transistor is switched between cutoff and saturation at a high frequency, which produces a pulse-width-modulated (PWM) square wave output. This output is then filtered throught a low-pass *LC* filter to produce an average dc output voltage. Thus the output voltage is proportional to the pulse width and frequency. The efficiency of a *series switching regulator* is independent of the input/output differential and can approach 95%.

Switching regulators come in various circuit configurations including the fly-back, feed-forward, push-pull, and nonisolated single-ended or single-polarity types. Also, the switching regulators can operate in any of three modes: step-down, step-up, or polarity inverting.

Theory of Switching Regulators.   A basic switching regulator consists of four major components:

1. Voltage source V_{in}
2. Switch $S1$
3. Pulse generator V_{pulse}
4. Filter $F1$

The interconnection between these four components is shown in Figure 10-51.

1. A *Voltage source* V_{in} may be any dc supply — a battery or an unregulated or a regulated voltage. The voltage source must satisfy the following requirements:

It must supply the required output power and the losses associated with the switching regulator.

It must be large enough to supply sufficient dynamic range for line and load variations.

It must be sufficiently high to meet the minimum requirement of the regulator system to be designed.

It may be required to store energy for a specified amount of time during power failures, especially if the system is designed for a computer power supply.

Figure 10–51 Basic switching regulator.

2. *Switch S*1 is typically a transistor or thyristor connected as a power switch and is operated in the saturated mode. The pulse generator output alternately turns the switch on and off.

3. *Pulse generator V*$_{\text{pulse}}$ produces an asymmetrical square wave varying in either frequency or pulse width called frequency modulation or pulse width modulation, respectively. The most effective frequency range for the pulse generator for optimum efficiency and component size is around 20 kHz. This frequency is inaudible to the human ear and also well within the switching speeds of most inexpensive transistors and diodes.

The duty cycle of the pulse waveform determines the relationship between the input and output voltages. The duty cycle is the ratio of the on-time, t_{on}, to the period T of the pulse waveform. In equation form

$$\text{duty cycle} = \frac{t_{\text{on}}}{t_{\text{on}} + t_{\text{off}}}$$

$$= \frac{t_{\text{on}}}{T} = t_{\text{on}}f \tag{10-18}$$

where t_{on} = on-time of the pulse waveform

t_{off} = off-time of the pulse waveform

$$T = \text{time period} = t_{\text{on}} + t_{\text{off}} = \frac{1}{\text{frequency}} \quad \text{or} \quad T = \frac{1}{f}$$

Typical operating frequencies of switching regulators range from 10 to 50 kHZ. However, there are certainly some tradeoffs. High operating frequencies reduce the ripple voltage at the expense of decreased efficiency and increased radiated electrical noise. On the other hand, lower operating frequencies improve efficiency and reduce electrical noise, but require larger filter components (inductors and capacitors).

4. *Filter F*1 converts the pulse from the output of the switch into a dc voltage. Since this switching mechanism allows a conversion similar to transformers, the switching regulator is often referred to as a *dc transformer*. The output voltage V_{o} of the switching regulator is a function of duty cycle and the input voltage V_{in}. In equation form, V_{o} is expressed as follows:

$$V_{\text{o}} = \frac{t_{\text{on}}}{t_{\text{on}} + t_{\text{off}}} \times V_{\text{in}} \tag{10-19}$$

Equation (10-19) indicates that, if time period T is constant, V_{o} is directly proportional to the on-time, t_{on}, for a given value of V_{in}. This method of changing the output voltage by varying t_{on} is referred to as *pulse width modulation*. Similarly, if t_{on} is held constant, the output voltage, V_{o}, is inversely proportional to the period T or directly proportional to the frequency f of the pulse waveform. This method of varying the output voltage is referred to as *frequency modulation*.

The frequency-modulated switching regulator is generally easier to design and build. However, the pulse-width-modulated switching regulator is most often used in commercial switching supplies with multiple outputs. The latter method is more complex and uses more components, but it is most suitable in high-current applications.

The filter is of major importance in the proper design of the switching regulator. The basic filter types are *RC*, *RL*, or *RLC*. While all three filters are used in switching regulators, the *RLC* filter is the most commonly used.

In Figure 10-51 the way the components (switch and filter) are connected determines the output mode: step-down, step-up, or polarity inverting.

The inductor in the *RL* or *RLC* filter is one of the most important components of the switching regulator. This is because there are several areas that are affected by the choice of inductor, including energy storage for the regulator, peak current limiting in a switch *S*1, output ripple, transient response, overshoot, size and cost limits, and radiated electric and magnetic fields. The energy stored by an inductor is directly proportional to the inductance. Therefore, increase in inductance increases the energy stored. Similarly, the transient response, overshoot, and size and cost increase as inductance increases because the inductance is directly proportional to the transient response, overshoot, and size of the inductor. However, the increase in inductance decreases peak current in the switch (transistor) and the output ripple because it is inversely proportional to the peak current and the output ripple. Finally, the effect of the inductor on radiated electrical and magnetic noise is a function of geometry, frequency, and size. In short, selecting the inductor requires careful consideration of the aforementioned tradeoffs.

The three most common techniques employed in the inductor design are:

1. Powdered permalloy toroids
2. Ferrite EI, U, and toroid cores
3. Silicon steel EI butt stacks

Powdered permalloy toroids have low leakage inductance, low core losses, and high permeability. Therefore, the powdered permalloy toroid yields the most stable and predictable inductor. The only disadvantage is the cost of manufacturing and mounting toroid inductors.

The ferrite EI, U, and toroid cores have low leakage inductance, low losses, low permeability, and poor high-temperature performance, but are as expensive to manufacture and mount as powdered permalloy toroids.

The silicon steel EI butt stack exhibits high permeability, high flux densities, and ease of construction and mounting. Therefore, it is most commonly used in low-voltage switching regulators.

Initially, due to the lack of high-speed switching transistors and low-loss inductors, the design of switching regulators was much more difficult. However, in recent years, due to advances in integrated circuit technology, that is, the monolithic switching regulators, which contain practically everything but the inductor, have made switching supplies more attractive and easy to design. The Fairchild μA78S40 and Texas Instrument TL497 are two such integrated regulators commonly used for switching supplies.

μA78S40 Switching Regulator: The μA78S40 consists of a temperature-compensated voltage reference, a duty-cycle controllable oscillator with an active current limit circuit, high-gain comparator, high-current, high-voltage output switch, a power-switching diode, and an uncommitted operational amplifier. The block diagram of the μA78S40 is shown in Figure 10-52.

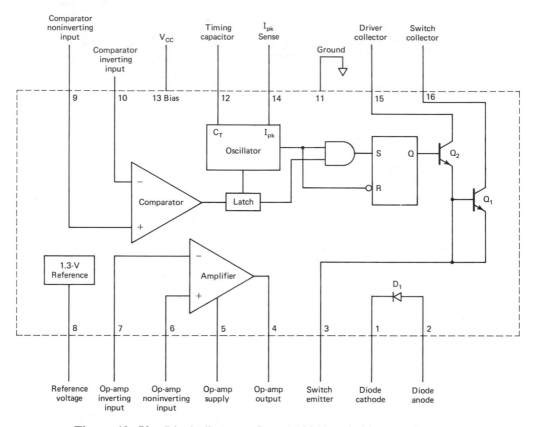

Figure 10–52 Block diagram of a μA78S40 switching regulator. (Copyright Fairchild Semiconductor Corporation — used by permission.)

The most important features of μA78S40 switching regulators are:

Step-up, step-down or inverting operation
Operation from 2.5- to 40-V input
80 dB line and load regulation
Output adjustable from 1.3 to 40 V
Peak currents to 1.5 A without external transistors
Variable frequency, variable duty-cycle device

The μA78S40 is a 16-pin DIP and is available as a ceramic or plastic package. The connection diagram of the μA78S40 is shown in Figure 10-53.

The initial switching frequency is set by the timing capacitor, C_T, connected between pin 12 and ground pin 11. The initial duty cycle is 6:1. The switching frequency and duty cycle can be modified by the current limit circuitry, I_{pk} sense, pin 14, and the comparator, pins 9 and 10. See Figure 10-52.

The comparator modifies the OFF time of the output switch transistors Q_1 and Q_2. In the step-up and step-down modes, the noninverting input (pin 9) of the comparator is

Figure 10–53 Connection diagram of the μA78S40 switching regulator. (Copyright Fairchild Semiconductor Corporation — used by permision.)

connected to the voltage reference of 1.3 V (pin 8), and the inverting input (pin 10) is connected to the output terminal via the sampling (voltage-divider) network. However, in the inverting mode, the noninverting input is connected to both the voltage reference and the output terminal through two separate resistors, and the inverting terminal is connected to ground. When the output voltage is correct (attains the desired value), the comparator output is in the *high* state and has no effect on the circuit operation. However, if the output voltage is too high and the voltage at the inverting terminal is higher than that at the noninverting terminal, then the comparator output goes *low*. In the *low* state the comparator inhibits the turn on of the output switching transistors. This means that, as long as the comparator output is *low*, the system is in *off* time. As the output current rises or the output voltage falls (inductor property), the *off* time of the system decreases. Consequently, as the output current nears its maximum, $I_{o\,max}$, the *off* time approaches its minimum value. In fact, the comparator can inhibit several on cycles, one on cycle, or any portion of an on cycle. However, once the on cycle has begun, the comparator cannot inhibit until the beginning of the next on cycle.

In all three operating modes, the current limit circuit is completed by connecting a sense resistor, R_{SC}, between I_{pk} sense, pin 14, and the V_{CC}, pin 13. The current limit (R_{SC}) modifies the *on* time of the switching transistors. The current limit circuit is activated when a 330 mV potential (350 mV max) appears across R_{SC}. R_{SC} is selected such that 330-mV appears across it when the desired peak current, I_{pk}, flows through it. When the peak current is reached, the current limit circuit is turned on. This action, however, terminates *on* time and immediately starts *off* time. Although the oscillator is free running, the current limit action resets the timing cycle. Increase in load results in more *on* time and less *off* time. Also, the increase in load current increases the switching frequency.

The forward voltage drop, V_D, across the internal power diode is used to determine the values of inductor L *off* time and the efficiency of the switching regulator. It is 1.25 V typical and 1.5 V maximum. However, if an external diode is used, the forward voltage drop across it must be used for V_D.

Another important quantity used in the design of a switching regulator is the saturation voltage V_S. The saturation voltage for the switch element, transistors Q_1 and Q_2, is listed on the data sheet. In the step-down mode, an "output saturation voltage 1" is 1.1 V typical, 1.3 V maximum as listed on the data sheet. The output saturation voltage 1 is defined as the voltage for transistors Q_1 and Q_2 in the Darlington configuration.

Similarly, in the step-up mode, the "output saturation voltage 2" is listed on the data sheet as 0.45 V typical and 0.7 maximum. It is defined as the switching element voltage for just Q_1 used as a transistor switch. Finally, for the inverting mode, the saturation voltage of the external transistor must be used for V_S. For convenience, the data sheet of the μA78S40 is included in Appendix D. The reader can obtain additional information on the μA78S40 by referring to the data sheet.

Design formulas for all three modes, step-down, step-up, and inverting, are given in Table 10-5.

EXAMPLE 10-12

Design a step-down switching regulator according to the following specifications:

Input voltage V_{in} = 12 Vdc
Output voltage V_o = 5 V at 500 mA maximum
Output ripple voltage V_{ripple} = 50 mV or 1% of V_o
Switching regulator: μA78S40

SOLUTION   Referring to the data sheet of the μA78S40:

Forward voltage drop across the power diode V_D = 1.25 V typical
Output saturation voltage V_S = 1.1 V typical
Reference voltage V_{REF} = 1.245 V typical
Comparator input bias current I_B = 35 nA typical and 200 nA maximum

Using the design formulas given in Table 10-5 for the step-down mode, we get:

1. $I_{pk} = 2I_{o\,max}$

$= 2(500 \text{ mA}) = 1 \text{ A}$

2. $R_{SC} = \dfrac{0.33}{I_{pk}}$

$R_{SC} = \dfrac{0.33}{1} = 0.33 \text{ ohm}, \dfrac{1}{2} \text{ W}$

TABLE 10-5 DESIGN FORMULAS FOR STEP-DOWN, STEP-UP, AND INVERTING MODES OF A 78S40 SWITCHING REGULATOR

Characteristic	Step down	Step up	Inverting	Unit
I_{pk}	$2I_{o(max)}$	$2I_{o(max)} \times \dfrac{V_o + V_D - V_S}{V_{in} - V_S}$	$2I_{o(max)} \times \dfrac{V_{in} + |V_o| + V_D - V_S}{V_{in} - V_S}$	A
R_{SC}	$0.33\,I_{pk}$	$0.33\,I_{pk}$	$0.33\,I_{pk}$	Ω
$\dfrac{t_{on}}{t_{off}}$	$\dfrac{V_o + V_D}{V_{in} - V_S - V_o}$	$\dfrac{V_o + V_D - V_{in}}{V_{in} - V_S}$	$\dfrac{|V_o| + V_D}{V_{in} - V_S}$	
L	$\dfrac{V_o + V_D}{I_{pk}} \times t_{off}$	$\dfrac{V_o + V_D - V_{in}}{I_{pk}} \times t_{off}$	$\dfrac{|V_o| + V_D}{I_{pk}} \times t_{off}$	μH
t_{off}	$\dfrac{I_{pk} \times L}{V_o + V_D}$	$\dfrac{I_{pk} \times L}{V_o + V_D - V_{in}}$	$\dfrac{I_{pk} \times L}{|V_o| + V_D}$	μs
C_T (μF)	$45 \times 10^{-5}\,t_{off}$ (μs)	$45 \times 10^{-5}\,t_{off}$ (μs)	$45 \times 10^{-5}\,t_{off}$ (μs)	
C_o	$\dfrac{I_{pk} \times (t_{on} + t_{off})}{8\,V_{ripple}}$	$\dfrac{(I_{pk} - I_o)^2 \times t_{off}}{2I_{pk} \times V_{ripple}}$	$\dfrac{(I_{pk} - I_o)^2 \times t_{off}}{2I_{pk} \times V_{ripple}}$	μF
Efficiency	$\dfrac{V_{in} - V_S + V_D}{V_{in}} \times \dfrac{V_o}{V_o + V_D}$	$\dfrac{V_{in} - V_S}{V_{in}} \times \dfrac{V_o}{V_o + V_D - V_S}$	$\dfrac{V_{in} - V_S}{V_{in}} \times \dfrac{|V_o|}{V_o + V_D}$	
$I_{in\,avg}$ (Max load condition)	$\dfrac{I_{pk}}{2} \times \dfrac{V_o + V_D}{V_{in} - V_S + V_D}$	$\dfrac{I_{pk}}{2}$	$\dfrac{I_{pk}}{2} \times \dfrac{|V_o| + V_D}{V_{in} + |V_o| + V_D - V_S}$	A

Courtesy of Fairchild Semiconductor Corporation — used by permission.

3. $\dfrac{t_{on}}{t_{off}} = \dfrac{V_o + V_D}{V_{in} - V_S - V_o}$

$\dfrac{t_{on}}{t_{off}} = \dfrac{5 + 1.25}{(12 - 1.1 - 5)} = 1.06$ or $t_{on} = 1.06\, t_{off}$

As mentioned earlier, for optimum efficiency and component size, the operating frequency of a switching regulator should be approximately 20 kHz. Also, it is recommended that neither t_{on} nor t_{off} should be less than 10 μs for the 78S40 regulator. Let us assume that the operating frequency is 20 kHz. Therefore,

$$T = 50\ \mu s$$

or

$$t_{on} + t_{off} = 50\ \mu s$$

Substituting $t_{on} = 1.06\, t_{off}$ in the preceding equation, we get

$$1.06\, t_{off} + t_{off} = 50\ \mu s$$

or

$$t_{off} = \frac{50\ \mu s}{2.06} = 24.27\ \mu s$$

and

$$t_{on} = 25.73\ \mu s$$

4. Using $t_{off} = 24.27\ \mu s$, the value of oscillator-timing capacitor C_T is

$$C_T = (45 \times 10^{-5})\, t_{off}$$
$$= (45 \times 10^{-5})\, (24.27)\, (10^{-6})$$
$$= 0.0109\ \mu F$$

Use a 0.01-μF standard capacitor.

5. $L = \dfrac{V_o + V_D}{I_{pk}} \cdot t_{off}$

$$= \frac{5 + 1.25}{1} \cdot (24.27)\, (10^{-6})$$
$$= 151.69\ \mu H$$

Use $L = 150\ \mu H$.

6. The output capacitor C_O is

$$C_O = I_{pk} \frac{t_{on} + t_{off}}{8 V_{ripple}}$$
$$= \frac{(1)\,(50)\,(10^{-6})}{(8)\,(50)\,(10^{-3})}$$
$$= 125\ \mu F$$

Use $C_O = 150\ \mu F$

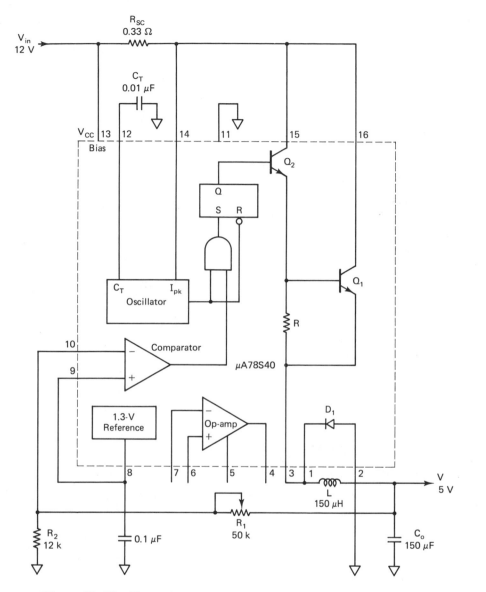

Figure 10–54 Five-volt step-down switching regulator using μA78S40.
(Copyright Fairchild Semiconductor Corporation — used by permission.)

7.   Finally, calculate the values of the resistors for the sampling network that is connected between the output terminal and the inverting input of the comparator (pin 10). Refer to Figure 10-54. The noninverting input of the comparator is connected to the voltage reference of 1.245 V typical; therefore, the threshold voltage for the inverting input is also 1.245 V. In addition, the input bias current, I_B, of the comparator is 35 nA typical but 200 nA maximum. Let us assume that the current through R_2 is 0.1 mA. Therefore,

$$R_2 = \frac{V_{REF}}{I_2}$$

$$= \frac{1.245}{(10)(10^{-5})}$$

$$= 12.45 \text{ k}\Omega$$

Use $R_2 = 12$ kΩ.

However, using the voltage-divider rule, we know

$$V_{R2} = \frac{R_2}{R_1 + R_2} \times V_o$$

Substituting the known values,

$$1.2 = \frac{(12)(10^3)}{R_1 + (12)(10^3)}(5)$$

or

$$R_1 = 36 \text{ k}\Omega$$

Use $R_1 = 50$-kΩ potentiometer.

The complete schematic diagram of the 5-V switching regulator with the component values is shown in Figure 10-54. Note that the 1.3-V voltage reference is bypassed by a 0.1-μF capacitor to protect it from noise and inductive spikes.

   8.   The efficiency of the switching regulator in Figure 10-54 is

$$\text{efficiency } (\eta) = \frac{V_{in} - V_S + V_D}{V_{in}} \times \frac{V_o}{V_o + V_D}$$

$$= \frac{12 - 1.1 + 1.25}{12} \times \frac{5}{5 + 1.25} \times 100$$

$$\eta = 81\%$$

The switching regulator in Figure 10-54 facilitates powering logic circuits when the only source available is the 12-V supply.

   In applications where the peak current is greater than 1.5 A or output voltage is greater than 40 V, an external diode and transistor must be used with the μA78S40, as in Example 10-13.

EXAMPLE 10-13
   Upgrade the switching regulator in Example 10-12 to provide +5 V at 3 A. Use the same specifications given in Example 10-12, except the output ratings.

SOLUTION   The output current requirement is 3 A at 5 V, which is much higher than the maximum current handling capacity (1.5 A peak) of the μA78S40. Therefore, we need to use an external transistor and diode to satisfy the requirement.

Since the output current is changed (from 0.5 to 3 A), the current-dependent components must also be changed, that is, R_{SC}, L, and C_O. In addition, we need to make necessary modifications in the circuit to incorporate an external transistor and diode. First, let us recalculate R_{SC}, L, and C_O:

1. $R_{SC} = \dfrac{0.33}{I_{pk}} = \dfrac{0.33}{6} = 0.055\ \Omega$

Use $R_{SC} = 0.05\ \Omega$, 1 W

2. $L = \dfrac{V_o + V_D}{I_{pk}} \times t_{off} = \dfrac{5 + 1.25}{6}(24.27)(10^{-6})$

$\qquad = 25.28\ \mu H$

Use $L = 25\ \mu H$.

3. $C_O = \dfrac{I_{pk}(t_{on} + t_{off})}{8 V_{ripple}} = \dfrac{(6)(50)(10^{-6})}{(8)(50)(10^{-3})}$

$\qquad = 750\ \mu F$

Use $C_O = 800\ \mu F$.

Next let us consider the modifications in the regulator of Figure 10-54 to include an external transistor and diode.

1. No connections to the internal diode pins 1 and 2.

2. Since an external transistor is used, the inductor L should not be connected to the switch emitter, pin 3. The output transistors, Q_1 and Q_2, should be connected in the Darlington configuration, and the switch emitter should be connected to the ground.

3. The external transistor is a *pass* transistor; therefore, it should be connected in series with R_{SC} as shown in Figure 10-55. Due to the nature of the configuration of Q_1 and Q_2, the external transistor must be a PNP. To properly bias the external transistor, Q_3, resistors R_3 and R_4 are used.

4. The inductor L (25 μH) is connected in the collector circuit of Q_3, and an external diode D_2 is connected between the collector of Q_3 and ground (pin 11). The function of the diode, D_2, is to bypass negative inductive spikes to ground and hence protect the regulator circuitry.

The values of R_3 and R_4 are not critical as long as the transistor Q_3 is properly biased and acts as a switch. However, note that the current through R_4 is equal to the current through R_3 plus the base current of Q_3. Therefore, R_4 must be a power resistor (1 W).

The transistor Q_3 must be a PNP high-speed switching power transistor with a continuous collector current, $I_C > 3$ A and $V_{CEO} > 12$ V. The 2N3791 satisfies these requirements and has the following specifications.

$$V_{CEO} = -60\ \text{V maximum}$$

$$I_C = -10\ \text{A continuous, maximum}$$

Figure 10–55 Five-volt, three-ampere step-down switching regulator of Example 10–12. (Copyright Fairchild Semiconductor Corporation — used by permission.)

$$h_{\text{FE}} = 30 \text{ minimum}$$

$$V_{\text{BE}} = -1.8 \text{ V maximum}$$

$$V_{\text{CE}}(\text{sat}) = -1 \text{ V maximum}$$

$$h_{\text{fe}} = 250 \text{ maximum}$$

Similarly, the MR821 is a power diode D_2 with the following specifications:

$$I_o = \text{average rectified forward current} = 5 \text{ A}$$

$$V_{\text{PRV}} = \text{peak reverse voltage} = 100 \text{ V}$$

Referring to Figure 10-55, we see that all other components and connections remain the same as in Figure 10-54.

Figure 10-56 shows a step-up regulator that converts $+10$ V to $+25$ V. The specifications for this regulator are as follows:

$$V_{\text{in}} = 10 \text{ V}$$

$$V_{\text{o}} = 25 \text{ V}$$

$$I_{\text{o max}} = 160 \text{ mA}$$

$$V_{\text{ripple}} = 30 \text{ mV}$$

The reader should be able to verify the values of the components using the formulas given in Table 10-5 for the step-up regulator.

Finally, an inverting regulator that converts $+12$ V to -15 V is shown in Figure 10-57, page 447. Again, the reader should be able to verify the values of the components if the following specifications are given:

$$V_{\text{in}} = 12 \text{ V}$$

$$V_{\text{o}} = -15 \text{ V}$$

$$I_{\text{o max}} = 160 \text{ mA}$$

$$V_{\text{ripple}} = 20 \text{ mV}$$

10-7.4 Special Regulators

In this section we will study special regulators: voltage references and voltage inverters.

10-7.4(a) Voltage references.   Voltage references are a special type of voltage regulator that is used for reference purposes as reference voltages. Typical applications of voltage references include D/A and A/D converters that do not have internal references, amplifier biasing, low-temperature-coefficient zener replacements, high-stability current references, comparator circuits, and voltmeter system references. Typical examples of voltage references include the following: The ICL8069 is a 1.2-V temperature-compensated voltage reference; the 9495 is a Teledyne 5-V reference; and the MC1403 is a Motorola 2.5-V reference.

MC1403 Used as a Voltage Reference for DAC MC1408.   Figure 10-58, page 448, shows a D/A converter using the MC1403 voltage reference. A stable current

Figure 10–56 A step-up regulator that converts $+10$ V to $+25$ V. (Copyright Fairchild Semiconductor Corporation — used by permission.)

reference of nominally 2.0 mA required for the MC1408 is obtained from the MC1403 with the addition of series resistors R_1 and R_2. Also, resistor R_3 improves temperature performance and capacitor C_o decouples any noise present on the reference line. A single MC1403 can provide the required current input for up to five of the D/A converters. For the complete connection diagram of the MC1408, refer to Figure 9-20(a).

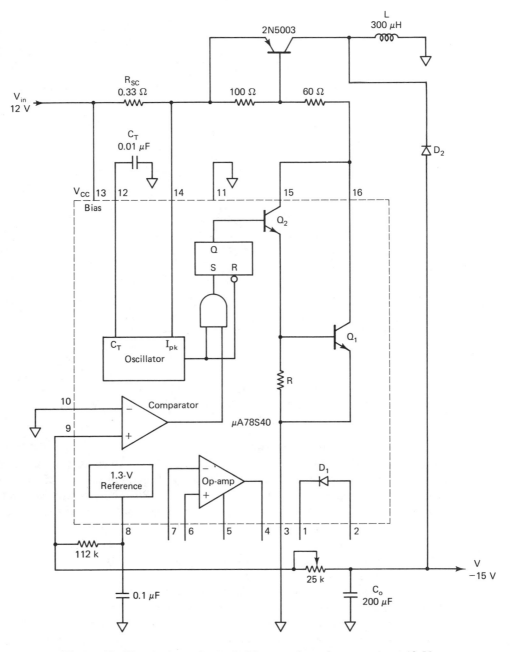

Figure 10–57   An inverting switching regulator that converts $+12$ V to -15 V. (Copyright Fairchild Semiconductor Corporation—used by permission.)

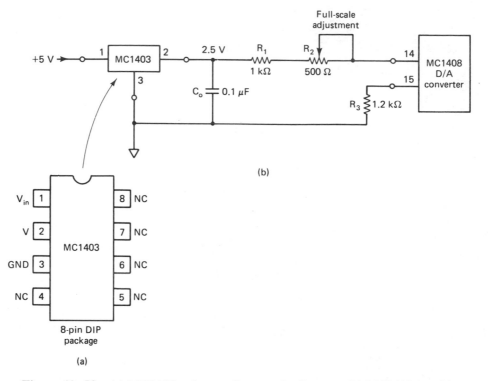

Figure 10–58 (a) MC1403 voltage reference pin diagram. (b) MC1403 providing the reference current for the MC1408. (Courtesy of Motorola Semiconductor.)

10-7.4(b) Voltage inverter. Datel's VI-7660 is a monolithic CMOS voltage inverter that provides −1.5 to −10 V from +1.5 to +10-V supplies with the addition of only two noncritical external capacitors. The block diagram of the VI-7660 is shown in Figure 10-59(a), which contains a dc voltage regulator, *RC* oscillator, voltage-level translator, four output-power MOS switches, and a logic network. The logic network senses the most negative voltage in the device and ensures that the output N-channel switches are not forward biased. This assures latch-up-free operation. When unloaded, the oscillator oscillates at a nominal frequency of 10 kHz for an input supply voltage of 5 V. Because of noise or other considerations, it may be desirable to increase the oscillator frequency in some applications. This may be done by overdriving the oscillator from an external clock. On the other hand, to maximize the conversion efficiency it may be necessary to lower the oscillator frequency. This is achieved by connecting an additional capacitor (typically 100 pF) between pins 7 and 8.

Typical applications for the VI-7660 include data acquisition and microprocessor-based systems in which a positive supply is available and an additional negative supply is required. The VI-7660 is also ideally suited as on-board negative supply for up to 64 dynamic RAMs (random-access memory ICs).

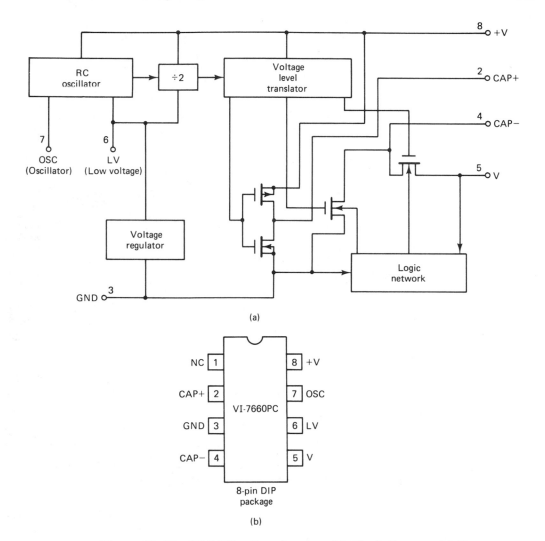

(a)

(b)

Figure 10–59 VI-7660 voltage inverter. (a) Block diagram. (b) Connection diagram. (Courtesy of Datel-Intersil, Inc.)

VI-7660 Applications *Simple negative converter.* Figure 10-60 shows typical connections when the VI-7660 is used as a simple negative converter. To improve the low-voltage (LV) operation, that is, when supply voltage $+V < 3.5$ V, the LV pin may be grounded. However, for $+V \geq 3.5$ V, the LV pin is left open to prevent device latch-up. Also, an additional diode D_1 must be included for proper operation at higher supply voltages ($+V > 6.5$ V) and/or elevated temperatures. When capacitor C_{OSC} is used to maximize the conversion efficiency of the device, the oscillator frequency decreases and hence the reactances of C_1 and C_2 will increase. Therefore, to overcome the increase in

Figure 10–60 VI-7660 as a simple negative converter. (Courtesy of Datel-Intersil, Inc.)

their reactances, the values of C_1 and C_2 must be increased by the same factor that the frequency has been reduced. For example, the addition of $C_{OSC} = 100$ pF between pins 7 and 8 will lower the oscillator frequency from 10 to 1 kHz and will thereby call for an increase in the values of C_1 and C_2 from 10 to 100 μF. The output voltage equation is

$$V_o = -(+V) \qquad \text{for } 1.5 \leq +V \leq 6.5 \text{ V} \qquad (10\text{-}20\text{a})$$

$$V_o = -(+V - V_{D1}) \qquad \text{for } 6.5 \leq +V \leq 10 \text{ V} \qquad (10\text{-}20\text{b})$$

where V_{D1} = forward voltage drop of diode D_1
    = 0.7 V typically

Negative-voltage multiplier. To produce larger negative multiplication of the initial supply voltage, the VI-7660s may be cascaded as shown in Figure 10-61. The practical limit to the number of devices that can be cascaded is 10 for light loads. The output voltage is given by the equation

$$V_o = -(n)(+V) \qquad (10\text{-}21)$$

where n = number of devices cascaded and must be ≤ 10
    $+V$ = input supply voltage

SUMMARY

1. Although a general-purpose op-amp may work satisfactorily in certain applications, the use of specialized ICs results in simpler, easier, more economical, more accurate circuits. In some other applications, general-purpose op-amps will not work at all. Under such circumstances, the designer is forced to use specialized ICs. The best source of such specialized ICs are the manufacturers' data books.

2. Datel's FLT-U2 is a universal filter that provides second-order low-pass, high-pass, and band-pass output functions simultaneously. It can also be used as a notch or all-pass

Figure 10–61 Cascading VI-7660s for increased negative output voltage. (Courtesy of Datel-Intersil, Inc.)

451

filter. An uncommitted op-amp available in the FLT-U2 can be used to either increase the overall gain of the filter or to raise the order of the low-pass or high-pass filter to the third order.

3. The MF5 is a switched-capacitor filter that can be used to synthesize any of the normal filter types: low-pass, high-pass, band-pass, notch, and all-pass. It is a second-order circuit and can be cascaded to provide higher-order filters. The MF5 is a digital filter as opposed to the FLT-U2, which is an analog filter.

4. Signetics' NE/SE 555 is a monolithic timing circuit that can produce accurate and highly stable time delays or oscillation. Thus the 555 may be used as either a monostable or an astable multivibrator. This device can be used in such applications as waveform generators, digital logic probes, infrared transmitters, burglar alarms, toxic gas alarms, and electric eyes.

5. The phase-locked loop (PLL) basically consists of a phase detector, an amplifier/ low-pass filter, and a VCO. The output f_{OUT} of the VCO is fed back to the phase detector, where it is compared with the input frequency f_{IN}. When $f_{IN} = f_{OUT}$, the loop is said to be locked. Signetics' SE/NE 560 series (560, 561, 562, 564, and 567) are monolithic PLLs. The PLLs are used in such applications as FM stereo decoders, motor speed controls, FSK decoders, and FM demodulators.

6. Power amplifiers are intended to supply a large-signal current to current-operated loads such as speakers and motors. The output current capability of general-purpose op-amps can be increased by using a power transistor called a power booster in series with it. In some applications, however, monolithic/hybrid power amplifiers are preferred, mainly because of their compact size and simpler and more reliable operation. National Semiconductor's LM380 is a monolithic power audio amplifier designed to deliver a minimum of 2.5 W (rms) to an 8-Ω load, while Burr-Brown's 3573 hybrid power amplifier is capable of delivering 100-W peak or 40-W continuous output power. Power audio amplifiers can be used for such applications as simple phonograph amplifiers, alarms, TV sound systems, and intercoms.

7. A voltage regulator is a circuit that supplies a constant voltage regardless of changes in load currents. Monolithic voltage regulators are available in a variety of different output voltage ratings and are also quicker and easier to use. The 7800 is a fixed-positive-voltage regulator series with seven voltage options, while the 7900 is a fixed-negative-voltage regulator series with nine voltage options. Adjustable voltage regulators such as the LM317 and LM337 are more popular because of their versatility, performance, and reliability. Switching regulators such as the μA78S40 and TL497 are most efficient compared to fixed or adjustable regulators. The efficiency of such regulators can approach 95%. In a switching regulator, to improve its efficiency the series-pass transistor is used as a *switch* rather than as a variable resistor as in the linear mode. Voltage references are special regulators designed for reference voltage purposes. A voltage inverter such as Datel's VI-7660 converts positive voltages from +1.5 to +10 V to −1.5 to −10 V. Voltage regulators are mainly used in power supplies, while voltage references are used with comparators, amplifiers, and D/A converters.

QUESTIONS ━━

10-1. What is the function of the FLT-U2? List important features of this device.

10-2. What maximum-order low-pass (or high-pass) filter can be realized using just one FLT-U2?

10-3. Can a notch filter be realized using the FLT-U2? Explain how.

10-4. To form a fourth-order low-pass, high-pass, or band-pass filter, how many FLT-U2s are needed? Explain the connection procedures required to form a fourth-order low-pass filter.

10-5. What is a switched-capacitor filter? How does it differ from an analog filter such as the FLT-U2?

10-6. List the advantages of the switched-capacitor filter.

10-7. List important features of the MF5 filter.

10-8. What is the major difference between the MF5 and the MF10 filters?

10-9. List important features of the 555 timer.

10-10. What are the two basic modes in which the 555 timer operates?

10-11. Briefly explain the differences between the two operating modes of the 555 timer.

10-12. What must the relationship be between the pulse width t_p and the period T of the input trigger signal if the 555 is to be used as a divide-by-4 network?

10-13. List one application each in which the 555 can be used as a monostable and astable multivibrator.

10-14. What is a phase-locked loop?

10-15. List the basic building blocks of the discrete PLL.

10-16. What is the major difference between digital and analog PLLs?

10-17. What is the major advantage of a monolithic phase detector such as the MC4344 over an exclusive-OR and an edge-triggered phase detector?

10-18. Briefly explain the roles of a low-pass filter and VCO in PLLs.

10-19. What are the advantages and disadvantages of monolithic PLLs over discrete PLLs?

10-20. List one application of the PLL and then briefly describe the role of the PLL in that application.

10-21. What is the major difference between small-signal and power amplifiers?

10-22. What is a power booster? Why is it needed?

10-23. List the differences between the monolithic power amplifiers and the power amplifiers using power boosters.

10-24. List important features of the LM380 power audio amplifier.

10-25. What is the major difference between monolithic and hybrid power amplifiers?

10-26. What is a voltage regulator? List four different types of voltage regulators.

10-27. What are the advantages of the adjustable voltage regulators over the fixed voltage regulators?

10-28. What is a switching regulator? List four major components of the switching regulator.

10-29. List the advantages of a switching regulator.

10-30. What is a voltage reference? Why is it needed?

10-31. List two important features of the VI-7660 voltage inverter.

PROBLEMS

10-1. Using the FLT-U2, design a second-order noninverting Butterworth low-pass filter with 10-kHz cutoff frequency, a gain of 10, $Q = 5$, and $V_S = \pm 15$ V.

10-2. Repeat Problem 10-1 for a high-pass filter.

10-3. Design a band-pass filter with 2-kHz center frequency, $Q = 20$, and noninverted output. Use the FLT-U2 with supply voltages $= \pm 15$ V.

10-4. Using the FLT-U2, design a notch filter with 2-kHz notch-out frequency and $Q = 20$.

10-5. Using the MF5, design a second-order Butterworth low-pass filter with 1-kHz cutoff frequency and a gain of -2. Assume that a ± 5-V supply and a TTL clock are used.

10-6. Repeat Problem 10-6 for a high-pass filter.

10-7. Using the MF5, design a notch filter with 2 kHz notch-out frequency and $Q = 20$.

10-8. Using the 555 timer, design a monostable multivibrator having an output pulse width of 100 ms. Verify the designed values of R_A and C with the graph of Figure 10-17.

10-9. In the circuit of Figure 10-16(a), $C = 0.1$ μF and the output pulse width is 1 ms. Determine the value of R_A.

10-10. The monostable multivibrator of Figure 10-16(a), is to be used as a divide-by-3 network. The frequency of the input trigger is 12 kHz. If the value of $C = 0.05$ μF, what should be the value of R_A?

10-11. For the astable multivibrator of Figure 10-21(a), $R_A = 4.7$ kΩ, $R_B = 1$ kΩ, and $C = 1$ μF. Determine the positive pulse width, the negative pulse width, and the free-running frequency. What is the duty cycle of the output waveform?

10-12. Design an astable multivibrator having an output frequency of 10 kHz with a duty cycle of 25%.

10-13. In the ramp generator of Figure 10-24(a), R is set at 100 kΩ, keeping all the other values the same. What is the approximate frequency of the output ramp?

10-14. Design a ramp generator having an output frequency of approximately 5 kHz.

10-15. Referring to Figure 10-33(a), determine f_{OUT}, f_L, and f_C if $C_1 = 0.001$ μF.

10-16. Repeat Problem 10-15 with $C_1 = 470$ pF.

10-17. Referring to the 565 PLL frequency multiplier of Figure 10-34(b), determine the range for the output frequency f_{OUT} if R_1 is varied from 2 to 20 kΩ.

10-18. In the circuit of Figure 10-34(b), if $f_{IN} = 500$ Hz, what is the value of R_1 needed?

10-19. In the intercom system of Figure 10-42, an overall gain of 200 is needed. What must be the turns ratio N_1/N_2 of the transformer if potentiometer R_V is open?

10-20. Using the 7805C voltage regulator, design a current source that will deliver 150-mA current to the 8-Ω, 10-W load.

10-21. **(a)** Using the LM317, design an adjustable voltage regulator to satisfy the following specifications: output voltage $V_o = 12$ to 15 V and output current $I_o = 0.50$ A.

(b) Draw the complete schematic diagram of the regulator designed in part (a).

10-22. Repeat Problem 10-21 for $V_o = 10$ to 12 V and $I_o = 200$ mA.

10-23. Using the μA78S40, design a step-down switching regulator to satisfy the following specifications: $V_{in} = 18$ V dc, $V_o = 12$ V at 1 A maximum, and $V_{ripple} = 1\%$ of V_o. Draw the complete schematic diagram for the regulator.

10-24. Using the μA78S40, design a step-up switching regulator to satisfy the following requirements: $V_{in} = 10$ V, $V_o = 15$ V at 500 mA, and $V_{ripple} = 1\%$ of V_o. Draw the complete schematic diagram for the regulator.

10-25. Using the μA78S40, design an inverting switching regulator that converts $+10$ V to -12 V. $I_o = 200$ mA maximum and $V_{ripple} = 1\%$ of V_o. Draw the complete schematic diagram for the regulator.

10-26. The VI-7660 is used as a simple negative converter, as shown in Figure 10-60. Determine the output voltage V_o if (a) $+V = 5$ V and (b) $+V = 9$ V.

10-27. VI-7660 voltage inverters are to be used to obtain an output voltage of -15 V. How many VI-7660s are needed if the supply voltage $+V = 5$ V? Draw the complete connection diagram.

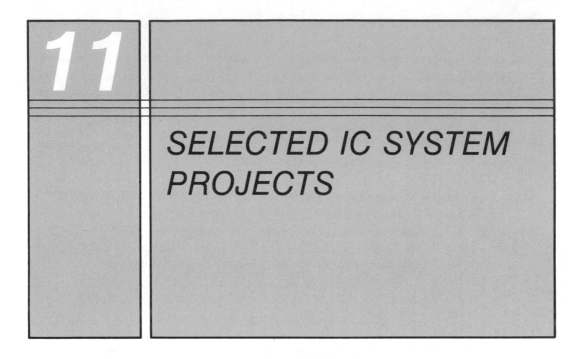

11

SELECTED IC SYSTEM PROJECTS

11-1 INTRODUCTION

The preceding chapters have emphasized the analysis and use of individual ICs. The next logical step would be to consider how some of these discrete ICs might be combined to form a system that serves some useful purpose. The aim of this chapter is to demonstrate some of the interesting and challenging projects devoted to constructing IC systems that serve a number of useful purposes.

To most people the word *system* means a very complex, cumbersome, and expensive combination of networks. On the contrary, a system may be simply a combination of a few ICs together with some discrete components, the effect of which is to perform a specific useful service. A system may even be built from a single IC and a few discrete components. The simpler system is increasingly more common today because of the increasing pace of refinement and advancement in IC technologies. Even though innumerable systems could be formed from combinations of the many ICs presented thus far, this chapter focuses on a limited number of systems only. These are simple enough to be discussed meaningfully at this level, and they are illustrative of IC systems designed for specific uses. This chapter introduces a power supply, an audio function generator, an LED temperature indicator, a digital dc motor speed control, an appliance timer, and a siren/alarm system. Most of these systems use a specially designed IC that requires very few external components. In addition, at the end of the chapter a collection of suggested projects is illustrated using block diagrams.

11-2 POWER SUPPLY: APPLICATION OF VOLTAGE REGULATORS

Since a power supply is a vital part of all electronic systems, it will be discussed first. Most digital ICs, including microprocessors and memory ICs, operate on a ± 5-V supply, while almost all linear ICs (op-amps and special-purpose ICs) require ± 15-V supplies. Therefore, the power supply presented in this section will have ± 5 and ± 15 V.

Figure 11-1 shows the block diagram of a typical power supply. The schematic diagram of the power supply that provides output voltages of ± 5 V at 1.0 A and ± 15 V at 0.500 A is shown in Figure 11-2. In this figure two separate transformers are used because they are readily available; however, it is possible to custom design a single transformer with the same specifications to replace the two. The ± 15-V supply voltages are obtained from a 26.8-V center-tapped (CT) transformer, and the ± 5-V supply voltages are obtained from the 12.6-V CT transformer. The output of these secondaries is then applied to the bridge rectifiers, which convert the sinusoidal inputs into full-wave rectified outputs. The filter capacitors at the output of the bridge rectifiers are charged to the peak value of the rectified output voltage whenever the diodes are forward biased. Since the diodes are not forward biased during the *entire* positive and negative half-cycle of the input waveform, the voltage across the filter capacitors is a pulsating dc that is a combination of dc and a ripple voltage. From the pulsating dc voltage, a regulated dc voltage is extracted by a regulator IC.

Consider first how the ± 15-V supply voltages are obtained in the circuit of Figure 11-2. As shown in Section 10-7, the 7815 is a $+15$-V regulator, the 7915 is a -15-V regulator, and both can deliver output current in excess of 1.0 A. They will hence perform satisfactorily in the circuit of Figure 11-2 by providing ± 15 V at 0.500 A. However, since the drop-out voltage ($V_{\text{in}} - V_o$) is 2 V, the input voltage for the 7815 must be at least $+17$ V and that for the 7915 must be at least -17 V. This means that the rectified peak voltage must be greater than $+17$ V and -17 V, which in turn implies that the secondary voltage must be larger than 34 V peak or 24 V rms. The voltage across the center-tapped secondary in Figure 11-2 is 26.8 V rms, thus satisfying the minimum voltage requirement of 24 V rms. Also, the peak voltage between either of the secondary terminals and the center-tap (ground terminal) is $13.4(\sqrt{2}) = 18.95$ V peak, which is less than the maximum peak voltages of $+35$ V and -35 V for the 7815 and 7915, respectively.

Note that the voltages across the two halves of the center-tapped secondary are equal in amplitude but opposite in phase. During the positive half-cycle of the input voltage, diode D_1 conducts and capacitor C_1 charges toward a positive peak value $\cong 18.95$ V. At the same time, diode D_3 is also conducting; hence capacitor C_3 charges toward a negative peak value $\cong -18.95$ V. This means that the voltage across nonconducting diodes D_2 and D_4 is 37.90 V peak, which implies that the peak-reverse-voltage (PRV) rating of the bridge rectifiers must be larger than 37.90 V peak or 26.8 V rms. The PRV rating of the bridge rectifier diodes, also known as a *working inverse voltage* (WIV), is specified on the data sheets. The bridge rectifier, MDA200 (Motorola's rectifier) in Figure 11-2, has

Figure 11-1   Block diagram of a power supply.

Figure 11-2 Power supply: ±5 V at 1.0 A and ±15 V at 0.5 A.

a PRV rating of 50 V, which is higher than needed. This bridge rectifier is, in fact, used here because it is readily available and more commonly used.

During the negative half-cycle of the input waveform, diodes D_2 and D_4 conduct and charge capacitors C_1 and C_3 toward the peak voltage of 18.95 V with indicated polarities. Note, however, that the diode pair that conducts during either the positive or negative half-cycle does not do so for the entire half-cycle. The diodes conduct only during the time when the anodes are positive with respect to the cathodes. In other words, when the diodes are forward biased, the capacitors are charged by current pulses. Data sheets give the maximum average rectified current $I_{o\,max}$ that the diode can safely handle. For the MDA200, $I_{o\,max}$ is 2.0 A. In addition, when the power supply is first turned on, the initial charging of the capacitor causes a large transient current called the *surge current* to pass through the diodes. The surge current I_{FS} flows only briefly and is therefore much larger than the maximum average current $I_{o\,max}$. The maximum surge-current I_{FSM} is normally included on the data sheets; it is 60 A for the MDA200.

Finally, the size of the filter capacitor depends on the secondary current rating of the transformer. As a rule of thumb, a 1500-μF capacitor should be used for each ampere of current. The working voltage rating (WVDC) of the capacitor, on the other hand, depends on the peak rectified output voltage and must be at least 20% higher than the peak value of the voltage it is expected to charge to. Capacitors C_1 and C_3 satisfy these requirements (see Figure 11-2). Capacitors C_2 and C_4 at the output of 7815 and 7915 regulators, respectively, help to improve the transient response and should be in the range of 1 μF.

Next consider the ±5-V supply. The circuit arrangement of the ±5-V supply is identical to that of the ±15-V supply except that here the specifications for the transformer T_2 secondary are different. Therefore, the operation and design considerations for the ±5-V supply are the same as those just presented for the ±15-V supply.

The voltage regulators in Figure 11-2 will require heat sinks. Let us examine why. The power dissipated by the 15-V regulators is as follows:

$$\text{power dissipated} = (\text{dropout voltage})(\text{current})$$
$$= (18.95 - 15)(0.5) = 1.98 \text{ W}$$

Similarly, the power dissipated by the 5-V regulators is

$$(8.91 - 5)(1.0) \cong 3.91 \text{ W}$$

Therefore, for the proper operation the regulators must be heat-sinked in order to keep their temperature down. If a regulator is a metal package (TO-3 type), the appropriate heat sink is mounted on the case of the package. However, if the regulator is an epoxy package, a silicon grease may be used on the back of the package, and then the package can be bolted to the chassis of the power supply cabinet with insulating hardware.

Besides the ±15 and ±5-V regulated supply voltages, there is often a need for a 60-Hz square-wave signal, which is used as a time base in scanning the digital displays and as a trigger for sequential and timing circuits. If needed, a 1-Hz (1-s) signal for the real-time clock can be readily obtained from the 60-Hz signal by using a divide-by-60 network. Although not commonly done, a higher-frequency signal can also be obtained from the 60-Hz signal by using a multiplier. For these reasons, in Figure 11-2 a 60-Hz square-wave signal is produced by using two small-signal diodes and a 555 timer as the Schmitt trigger.

Parts List

Transformer T_1	Primary:	117 V, 60 Hz: Hobart P-300
	Secondary:	26.8 V CT, 1.0 A
Transformer T_2	Primary:	117 V, 60 Hz: Hobart P-305
	Secondary:	12.6 V CT, 2.0 A
Bridge rectifiers	MDA200:	PRV = 50 V, $I_{o\,max}$ = 2.0 A, I_{FSM} = 60 A
	MDA970A1:	PRV = 50 V, $I_{o\,max}$ = 4.0 A, I_{FSM} = 100 A

NE555 timer
Two 1N914 signal diodes
10-kΩ resistor

Capacitors	3000 μF at 16 V(2):	Sprague TVA1175
	1500 μF at 50 V (2):	Sprague TVA1318
	1.0 μF at 50 V (4):	Sprague TVA1300

Regulators	+15 V:	MC7815
	−15 V:	MC7915
	+ 5 V:	MC7805
	− 5 V:	MC7905

Fuse   0.750 A slow blow

Switch   On–off toggle type

Silicon grease with insulating hardware or four heat sinks.

11-3 AUDIO FUNCTION GENERATOR: APPLICATION OF THE FUNCTION GENERATOR IC

A sine wave signal is generally used in testing linear circuits such as amplifiers and filters; on the other hand, a square wave input is often essential for testing digital circuits such as flip-flops, counters, and registers. In this section an audio function generator using a specially designed integrated circuit, ICL8038, will be presented. This IC produces not only sine and square waves but also the triangular wave.

As a first step, let us briefly consider the internal structure, operation, and electrical characteristics of the ICL8038. Figure 11-3 shows the functional diagram and connection diagram of the ICL8038. The 8038 function generator is a monolithic integrated circuit that utilizes advanced monolithic technology, such as thin-film resistors and Schottky-barrier diodes. As shown in Figure 11-3(a), it consists of two current sources, two comparators, two buffers, a flip-flop, and a sine converter.

The triangular wave is generated by alternately charging the external capacitor from one current source and then linearly discharging it with another. The triangular wave is then applied to comparators 1 and 2 and also simultaneously buffered to the sine converter. The comparators and the flip-flop together convert the triangular wave into a square wave, whereas the sine converter converts the triangular wave into a sine wave. The sine

(a)

(b)

Figure 11-3 ICL8038 waveform generator. (a) Block diagram. (b) Connection diagram. (Courtesy of Intersil, Inc.)

converter is composed of 16 transistors. These transistors function in a nonlinear manner by providing a decreasing shunt impedance as the triangular wave potential moves toward the positive and negative peaks. Figure 11-4 indicates the relationship between the sine, square, and triangular waveforms generated by the ICL8038.

As shown in the functional block diagram, both triangular and square-wave outputs are buffered so that the output impedance of each is 200 Ω. However, the sine wave output has a relatively high output impedance of 1 kΩ. For this reason the sine wave output is often fed into a separate noninverting amplifier that provides buffering, gain, and amplitude adjustment, as will be seen shortly.

Figure 11-3(b) shows that the 8038 function generator is a 14-pin DIP; it is available as a plastic or ceramic package. The pin functions are as follows:

Pins 1 and 12: Sine wave adjust. By adjusting a 100-kΩ potentiometer connected between pins 12 and 11 (negative supply or ground), sine wave distortion of less than 1% is achievable. However, to obtain a distortion close to 0.5%, two 100-kΩ potentiometers may be connected between pins 6 ($+V_{CC}$) and 11 ($-V_{EE}$ or ground) with the wiper of one potentiometer connected to pin 1 and that of the other to pin 12.

Pin 2: Sine wave out. The sine wave output is available at this pin, the amplitude of which is 0.22 V_S, where ±5 V $\leq V_S \leq \pm15$ V.

Pin 3: Triangle out. The triangular output is taken out at this pin. Like the sine wave, the output amplitude of the triangular wave is also a function of supply voltage V_S and is equal to 0.33 V_S, where ±5 V $\leq V_S \leq \pm15$ V.

Pins 4 and 5: Duty cycle/frequency adjust. The *symmetry* (50% duty cycle for square wave) of all waveforms can be adjusted by connecting two external resistors (generally equal in value) between $+V_{CC}$ and pins 4 and 5 (see the connection diagram shown in Figure 11-5). These external resistors, together with the external capacitor connected to pin 10, determine the frequency of the output waveforms. The recommended value of external timing resistors is between 1 kΩ and 1 MΩ so

Figure 11–4   Output waveforms of the ICL8038. (Courtesy of Intersil, Inc.)

that the resultant charging current is between 10 μA and 10 mA. The optimum performance is obtained for these charging currents.

Pin 6: $+V_{CC}$. Positive supply voltage is applied to this pin so that 10 V $\leq$ $V_{CC} \leq$ 30 V with respect to voltage at pin 11 ($-V_{EE}$ or ground).

Pin 7: FM bias. This pin is a junction of two resistors (10 and 40 kΩ) that form a voltage divider with the supply voltage(s). When this pin is shorted to pin 8, a high output frequency is obtained, as opposed to connecting pin 8 to the $+V_{CC}$ for the given values of external timing resistors and capacitor. This is because the output frequency of the waveform generator is a direct function of the dc voltage at pin 8 with respect to $+V_{CC}$. This procedure is generally used to test the ICL8038.

Pin 8: FM sweep input. For frequency sweeping or for larger FM deviations, the modulating (sweep) signal is applied between the $+V_{CC}$ and pin 8. In this way a very large sweep range is created. The sweep FM range is the ratio of maximum frequency to minimum frequency, which can be extended to 1000:1 for the 8038. However, for proper operation the sweep voltage should be within the range

$$\left(\frac{2}{3}V_{CC} + 2\text{ V}\right) < V_{\text{sweep}} < V_{CC}$$

where V_{CC} is the total supply voltage. The sweep frequency of FM is typically 10 kHz for the 8038. On the other hand, for small deviations ($\pm 10\%$ typically) the FM signal can be applied directly to pin 8.

Pin 9: Square wave output. The square wave output is available at this pin. It is an open collector output, in that this pin can be connected through a load resistor to a different power supply than that used for the function generator itself. However, the maximum value of the power supply used must be less than 30 V. This means that, to obtain a TTL-compatible square wave output, a load resistor (10 kΩ typically) simply needs to be connected to the $+5$-V supply. In addition, the duty cycle of the square wave can be varied from 2% to 98% by external components connected to pins 4, 5, and $+V_{CC}$.

Pin 10: Timing capacitor. An external capacitor connected to this pin, together with the timing resistors connected to pins 4 and 5, determines the frequency of output waveforms. Oscillations can be halted by connecting pin 10 to ground.

Pin 11: $-V_{EE}$/*ground.* If ($\pm$) supply voltages are used, a negative supply $-V_{EE}$ is connected to this pin. On the other hand, if a single supply is used, this pin is connected to ground. With ($\pm$) supply voltages, all output waveforms are symmetrical (bipolar) about the 0 V. Obviously, if a single supply is used, the output waveforms are unipolar; that is, their average voltage is equal to $+V_{CC}/2$.

Pins 13 and 14: NC. No connections.

The important features of the 8038 are these:

Simultaneous outputs: sine wave, square wave, and triangle wave
Frequency range of operation: 0.001 Hz to 500 kHz

Figure 11–5 ICL8038 as an audio function generator.

Low distortion: 1%

Low frequency drift with temperature: 50 ppm/°C maximum

Easy to use: requires very few external components

Now let us see how the 8038 can be used as an audio function generator. Figure 11-5 shows a function generator that produces the sine wave, square wave, and triangular wave and has a frequency range of less than 20 Hz to above 100 kHz. However, around 100 kHz the quality of output waveforms starts deteriorating, in that the sine and triangular waves distort and the square wave becomes a pulse waveform with <50% duty cycle.

Although the function generation of Figure 11-5 is not of a high laboratory standard and may not be used for calibration purposes or in critical design applications, it is nevertheless useful for beginning designers and hobbyists. It is also simple and low in cost.

In the circuit of Figure 11-5, switch $SW2$ helps to select the desired output waveform. For the output waveform chosen, the frequency select switch $SW1$ is then used to select an appropriate frequency range; a specific value of frequency is then obtained by adjusting potentiometer R_4. R_4 allows the potential at pin 8 to be varied from $+V_{CC}$ to $+V_{CC}/3$ with respect to ground, which in turn varies the frequency of the output waveform.

Potentiometer R_3 is used to vary the duty cycle of the square wave output and hence may be used to adjust the duty cycle of the square wave to 50%, especially at higher frequencies. Also, potentiometers R_7 and R_9, whose wipers are connected to pins 1 and 12, respectively, are adjusted to minimize sine wave distortion.

As mentioned earlier, the square wave output is uncommitted and is available at pin 9. Therefore, to obtain a TTL-compatible square wave output, pin 9 may be connected through a pull-up resistor to a +5-V supply if desired. However, in Figure 11-5, pin 9 is connected through resistor R_{10} and potentiometer R_{11} to +12 V. The output amplitude of the square wave can be adjusted to a desired value by varying R_{11}. Resistor R_{10} is used in series with potentiometer R_{11} so that the $+V_{CC}$ of 12 V is not directly applied to the (+) input of the op-amp (see Figure 11-5).

The triangular wave output is available at pin 3, the amplitude of which is $(0.33)(\pm12 \text{ V}) = \pm3.96$ V. However, potentiometer R_{12} is used to vary the amplitude of the triangular wave. Finally, the sine wave output is available at pin 2. Although the amplitude of the sine wave is $(0.22)(\pm12 \text{ V}) = \pm2.64$ V, it can be adjusted to a desired value with the help of potentiometer R_{13}.

Switch $SW2$ helps to select one of the three output waveforms, which is then applied to the output stage composed of the op-amp and a power booster. The output stage serves two functions: (1) it provides a low output impedance and (2) it increases the output power drive capability of the 8038. Recall that without the output stage the output resistance of the triangular and square waves is 200 Ω, while that of the sine wave is 1 kΩ. Because the output stage is configured as a noninverting amplifier, the output resistance is reduced to a negligibly small value. In addition, the use of the power booster MC1438R (or BB3553) inside the op-amp's feedback loop improves the output drive capability of the 8038 without degrading op-amp characteristics. The op-amp used here is a LF351 with a slew rate of 13 V/μs and a unity gain bandwidth of 4 MHz. It is used as a noninverting amplifier with a gain of 11 (see Figure 11-5).

For smaller-amplitude output ac signals, it may be necessary to reduce the output offset voltage to zero initially. This is accomplished by adjusting the offset null circuitry, that is, potentiometer R_p of the LF351 op-amp.

Parts List

ICL8038 function generator

LF351 op-amp

MC1438R power booster

Two 1-kΩ resistors

1-kΩ potentiometer

Two 3.9-kΩ resistors

Four 10-kΩ resistors

Three 10-kΩ potentiometers

20-kΩ resistor

Two 20-kΩ potentiometers

Two 100-kΩ potentiometers

470-pF capacitor

0.001-μF capacitor

0.01-μF capacitor

0.1-μF capacitor

Three 1.0-μF capacitors

10.0-μF capacitor

Five-position switch

Three-position switch

11-4 LED TEMPERATURE INDICATOR: APPLICATION OF THE V/F CONVERTER AND THE 555 TIMER

This project illustrates the use of a V/F converter in monitoring temperature in degrees Fahrenheit (°F). The block diagram of the temperature indicator is shown in Figure 11-6(a). The indicator is composed of a temperature sensor, amplifier, V/F converter, three-digit binary-coded-decimal (BCD) counter, time base, and LED display. In addition to the 9400 V/F converter, other ICs needed for this project include the LM334 temperature sensor, LF353 dual op-amp, NE555 timers, 74LS00 NAND gate, MC14553 three-digit BCD counter, MC14543 BCD-to-seven segment decoder/driver/latch, and three seven-segment (common anode or common cathode) LED displays with three PNP switching transistors.

Figure 11-6(b) shows the schematic diagram, which is designed to display temperatures from 0° to 100 °F. Operation of the circuit is as follows. The output of the temperature sensor changes linearly as a function of temperature (10 mV/K). This output is an input to the summing amplifier, which is used to calibrate the output of the tem-

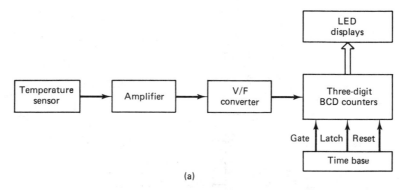

Figure 11–6 LED temperature indicator. (a) Block diagram.

Figure 11–6 (c) Timing diagram.

perature sensor for a desired temperature type (K,°C, or °F) and an intended range. That is, to display the temperature in either K,°C, or °F, potentiometer R_4 is adjusted accordingly so that a suitable voltage appears at the output of the summing amplifier. Since the output of the temperature sensor is directly proportional to temperature changes, R_4 needs to be adjusted at only one temperature. The output of the summing amplifier then drives the inverting amplifier. The purpose of the inverting amplifier is twofold: (1) to invert the input so that its output voltage is positive, which is necessary for the V/F converter, and (2) to provide a suitable gain, which depends on the voltage-to-frequency scaling used for the V/F converter.

Figure 11-6 (b) Schematic diagram.

The output of the inverting amplifier is the input to the V/F converter; therefore, the output frequency of the converter is directly proportional to the output voltage of the inverting amplifier. For example, as the temperature goes up the output voltage of the summing amplifier increases in the negative direction, whereas that of the inverting amplifier increases in the positive direction, which in turn causes the frequency of the V/F to increase in the positive direction.

The output frequency of the converter is then ANDed with the gating signal to produce the clock signal for the three-digit BCD counter. The BCD output of the counter drives the three LED displays sequentially via the BCD-to-seven segment decoder/latch/driver stage, and the temperature is displayed on the LEDs, depending on the relationship between the frequency of the V/F converter and the gate signal. The gate, latch, and reset signals are generated by the time-base circuit, which consists of a free-running multivibrator and two one-shot multivibrators.

Next let us examine the design considerations and procedures for each of the sections in the temperature indicator of Figure 11-6(b). The temperature sensor LM334 is a three-terminal adjustable current source whose current can be programmed from 1 μA to 10 mA with one external resistor R_{set}. The three terminals are labeled $+V$, R, and $-V$. The pin diagram and the connection diagram of the LM334 are shown separately in Figure 11-7.

The LM334 has a wide operating voltage range of 1 to 40 V. It can also withstand reverse voltage of up to 20 V (terminal $+V$ is negative with respect to $-V$). It is designed to operate over a temperature range of 0° to 70°C. For a wider temperature range, such as $-55°$ to 150°C, Intersil's AD590 temperature sensor is recommended.

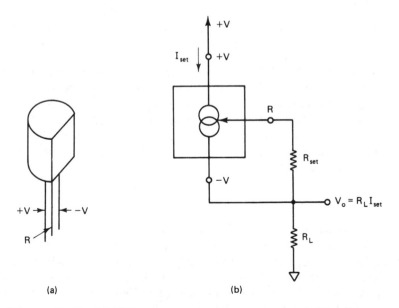

(a)   (b)

Figure 11-7 LM334 as a temperature sensor. (a) Pin diagram. (b) Connection diagram. (Courtesy of National Semiconductor.)

For the values indicated in Figure 11-6(b), the output of the LM334 changes 10 mV/K. This means that at 0 °F = 255.22 K the output of the sensor will be 2552.2 mV, which must be scaled down to 0 V so that the temperature displayed will be in degrees Fahrenheit. This is accomplished by the use of the summing amplifier. Specifically, potentiometer R_4 of the summing amplifier is adjusted so that the output is 0 V. The same procedure is used to calibrate the output of the summing amplifier at any other value of °F. Table 11-1 shows the relationship between K, °C, °F, and the output of the temperature sensor and the summing amplifier at corresponding values of temperature. Because the output of the sensor is directly proportional to the temperature, the output of the summing amplifier needs to be calibrated at the temperature at which the circuit is initially started up (refer to Table 11-1).

Note that the output of the summing amplifier is a negative dc voltage since the net input voltage is always positive for temperatures >0 °F (see Table 11-1).

TABLE 11-1 RELATIONSHIP BETWEEN DIFFERENT TEMPERATURE UNITS AND OUTPUTS OF THE SENSOR AND SUMMING AMPLIFIER

Kelvins (K)	Degrees Celsius (°C)	Degrees Fahrenheit (°F)	Output of the temperature sensor (mV)	Output of the summing amplifier (mV) to be adjusted to
255.22	−17.78	0	2552.2	0
273	0	32	2730	−177.8
298	25	77	2980	−427.8
310.78	37.78	100	310.7.8	−555.6

However, the 9400 V/F requires a positive input voltage. The summing amplifier must therefore be followed by an inverting amplifier. The gain of the inverting amplifier, however, depends on the voltage-to-frequency scaling of the converter. The V/F converter of Figure 11-6(b) is calibrated for a maximum frequency of 50 kHz, which represents a temperature of 100°F when the input voltage is 10 V maximum. For 9400 V/F calibration procedure, refer to Section 9-10.1(b). Since the output of the summing amplifier is −555.6 mV at 100°F, the gain of the inverting amplifier must be equal to

$$\frac{10\ V}{555.6\ mV} \cong 18$$

The output frequency of the V/F converter is then ANDed with the output frequency (called the gate signal) of the 555 free-running multivibrator to produce the clock signal for the three-digit BCD counter. Since the maximum 50-kHz output frequency of the converter represents 100 °F, the three-digit BCD counter must be clocked 100 times to display 100 °F. To accomplish this, the pulse width of the free-running multivibrator must be 100/50 K = 2 ms so that 100 pulses will be produced in 2 ms. At the end of 2 ms, the count of the counter is latched and displayed on the LED display. After the count is displayed as a temperature on the LEDs, the BCD counter is reset and the cycle repeats.

In other words, the counter continuously cycles through three states: count, latch, and reset. Therefore, the free-running multivibrator (gate signal) must provide for the time period required to count, latch, and reset the BCD counter. The latch enable and master reset pulses for the BCD counter MC14553 are produced by using two 555 one-shot multivibrators. Figure 11-6(c) indicates the timing diagram for the BCD counter MC14553, where the time period of the free-running multivibrator is approximately 12.5 ms with a pulse width of 2 ms. The pulse width of the latch enable pulse is approximately 10 ms, and the master reset pulse width is approximately 0.5 ms. To accomplish 2-, 10-, and 0.5-ms pulse widths, adjust potentiometers R_A, R_{A1}, and R_{A2}, respectively (Figure 11-6b).

The three-digit BCD counter used in Figure 11-6(b) is the MC14553. The block diagram and pin diagram of the MC14553 are shown in Figure 11-8. As shown in the block diagram, the MC14553 consists of three negative-edge-triggered BCD counters with a quad latch at the output of each counter, which enables the storage of any given count. The outputs of the latches are time multiplexed so that one BCD digit at a time is produced. The operation of the multiplexer output selector is controlled by the on-chip oscillator, whose frequency depends on the external capacitor.

The pin functions are as follows:

Pins 2, 1, and 15: $\overline{DS}1$, $\overline{DS}2$, and $\overline{DS}3$. These are the active-low digit-select outputs that sequentially control the three LED/LCD displays.

Pins 4 and 3: C1A and C1B. An external capacitor C_1 between these pins controls the frequency of the on-chip oscillator, which in turn drives the muliplexer output selector. If an external clock is to be used instead of the capacitor, the clock must be applied to pin 4.

Pins 9, 7, 6, and 5: Q_0, Q_1, Q_2, and Q_3. The BCD outputs are available at these pins and are active-high.

Pin 8: V_{SS}. This pin is connected to ground potential. The (+) supply voltage is also measured with respect to this pin.

Pin 10: LE. This is a latch-enable input pin. When the latch-enable input goes high, the information in the BCD counters is stored in the quad latches and is retained as long as the latch is high.

Pin 11: Dis. When an external input pulse applied to this pin goes high, the input clock is disabled, which in turn freezes the counters. However, the last count in the counters is retained. When not used, this pin should be connected to ground.

Pin 12: Clock. The input trigger source for the counters is connected to this pin. The counters are advanced on each high-to-low (logic 1 to 0) transition of the input trigger.

Pin 13: MR. When an input pulse applied to this pin goes high, the three BCD counters and the multiplexer scanning circuit are initialized (reset). In addition, the scan oscillator is inhibited, and digit-select outputs $\overline{DS}1$, $\overline{DS}2$, and $\overline{DS}3$ are disabled. For that reason, this pin is called *master reset* (MR).

Figure 11–8 MC14553 three-digit BCD counter. (a) Pin diagram.
(b) Block diagram. (Courtesy of Motorola Semiconductor.)

Pin 14: OF. This pin indicates the overflow condition of the counters. The voltage at pin 14 goes high for every 1000 counts and hence can be used to cascade multiple MC14553s.

Pin 16: V_{DD}. A (+) supply voltage up to +18 V dc can be applied to this pin with respect to pin 8 (V_{SS}).

As shown in Figure 11-6(b), the master reset (MR) and latch enable (LE) pulses for the MC14553 are produced by using two 555 one-shot multivibrators. In addition, the clock signal for the MC14553 is produced by ANDing the output of the V/F converter with the gating signal, which is obtained by using the 555 free-running multivibrator. (For more information on the 555 timer, refer to Section 10-4.) The digit select outputs $\overline{DS}1$, $\overline{DS}2$, and $\overline{DS}3$ sequentially drive the 2N1305 PNP transistors T_1, T_2, and T_3, which in turn control the three LED displays. The BCD outputs of the MC14553 are connected to the BCD inputs of the MC14543, which is a BCD-to-seven segment latch/decoder/driver. The seven-segment outputs of the MC14543 then drive the seven segments of the LED selected by the digit-select of the MC14553.

The MC14543 is designed to provide three functions: a 4-bit storage latch, an 8421 BCD-to-seven segment decoder, and a driver. The device is capable of driving LCD and LED displays. Figure 11-9(a) shows the pin diagram, with pin functions as follows:

Pin 1: LD. Since this pin is used to enable and disable the latches of the device, it is called *latch disable* (LD). When this pin is high (connected to V_{DD}), the BCD at the input is latched into the device. On the other hand, when this pin is low (connected to V_{SS}), the latches are disabled.

Pins 5, 3, 2, and 4: A, B, C, and D. These pins are the input pins to which the BCD inputs are applied.

Pin 6: Ph. This pin is called a *phase invert* (Ph) pin because it is used to invert the logic levels of the seven-segment output combinations. That is, when this pin is logic 0 (connected to V_{SS}), the MC14543 can drive common cathode LED readouts. On the other hand, for common anode LED readouts, this pin must be logic 1 (connected to V_{DD}).

Pin 7: BI. When this pin is logic 1, the display is blanked; hence it is called *blanking input* (BI). However, if this pin is logic 0, the display reads the BCD number corresponding to the code at pins 2, 3, 4, and 5.

Pin 8: V_{SS}. This is a ground pin, and (+) supply V_{DD} is measured with respect to it.

Pins 9, 10, 11, 12, 13, 15, and 14: a, b, c, d, e, f, and g. These pins represent the seven-segment outputs corresponding to the BCD code at the inputs A, B, C, and D [see the LED segment assignment shown in Figure 11-9(b)].

Pin 16: V_{DD}. A (+) voltage of 3.0 to 18 V dc can be applied to this pin with respect to pin 8 (V_{SS}).

As shown in Figure 11-6(b), the Ph pin 6 of the MC14543 is connected to V_{DD} (logic 1) because the LED displays are the common anode type. To limit the current through each of the LED segments, a separate resistor is used in series with each segment. Note, however, that only seven resistors are required for all the segments of the LEDs. This

(a)

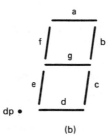

(b)

Figure 11–9   (a) MC14543 BCD-to-seven segment latch/decoder/
driver pin diagram. (Courtesy of Motorola Semiconductor.) (b) LED
segments.

is possible because the digit-select outputs of the MC14553 function sequentially. In
addition, the BCD outputs are also multiplexed, one BCD digit at a time.

Finally, since the accuracy of the temperature displayed depends mainly on the
frequency stability of the free-running multivibrator, all resistors must be of 1% or better
tolerance, and capacitors must be either Mylar or tantalum types.

Also, remember that the temperature indicator must be calibrated at the temperature
at which it is initially turned on.

Parts List

Teledyne 9400 V/F converter
LM334 temperature sensor
LF353 dual op-amp
Three NE/SE 555 timers

74LS00 NAND gate

MC14553 three-digit BCD counter

MC14543 BCD-to-seven segment decoder/driver/latch

Three seven-segment common anode LEDs: MAN72A or equivalent

Three 2N1305 switching transistors

Two 1N914 signal diodes

Seven 220-Ω resistors

Three 1-kΩ resistors

Two 1-kΩ potentiometers

3-kΩ potentiometer

Five 10-kΩ resistors

Two 10-kΩ potentiometers

15-kΩ resistor

20-kΩ resistor

Two 50-kΩ potentiometers

Four 100-kΩ resistors

180-kΩ resistor

510-kΩ resistor

1-MΩ potentiometer

100-pF capacitor

Two 0.001-μF capacitors

Five 0.01-μF capacitors

Three 1-μF capacitors

10-μF capacitor

All resistors have $\pm 1\%$ tolerance and all capacitors are either Mylar or tantalum.

11-5 DIGITAL DC MOTOR SPEED CONTROL: APPLICATION OF THE D/A AND F/V CONVERTERS

Various techniques can be used to control the speed of a dc motor, such as using the phase-locked-loop principles, digital inputs, or analog inputs. If desired, the speed of the motor may also be monitored with LED or LCD displays. This project illustrates the use of digital inputs to control the speed of a dc motor. To process the digital inputs, a D/A converter will be used, while a combination of a speed sensor and F/V converter will be used to sense and convert the speed into an appropriate voltage.

Figure 11-10 shows the block diagram of a digitally controlled dc motor. The output of the D/A converter is proportional to the binary equivalent of its digital inputs. The differential amplifier compares the D/A converter output with the output voltage of the F/V converter. The resulting difference voltage is an input to the power amplifier/driver stage. The output of the power amplifier/driver then drives the dc motor. The speed sensor

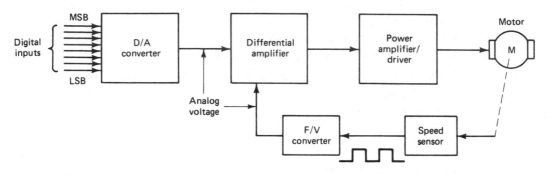

Figure 11–10   Block diagram of a digitally controlled dc motor.

converts the motor's speed into a pulse waveform, which is in turn converted into a proportional voltage by the F/V converter. Since the output of the F/V converter is processed using a negative feedback formed with the differential amplifier, the motor is kept at a constant speed corresponding to the setting of the digital inputs. In fact, the key to the operation of the circuit is that the differential amplifier maintains a specific difference between two input voltages so that motor speed is constant at the selected digital input setting.

Since the output of the D/A converter is directly proportional to the binary equivalent of its digital inputs, the output voltage of the D/A converter will be maximum positive when all the inputs are logic 1. This means that when all inputs are logic 1 the motor will run at a maximum speed.

Now suppose that the motor is initially running at a certain speed and digital inputs have just been set to lower the speed. This action will reduce the output voltage of the D/A converter, which in turn reduces the difference between the two input voltages of the differential amplifier, resulting in a reduced drive for the motor. Therefore, the speed of the motor will be lowered until the output of the F/V converter is such that a specific input difference voltage for the differential amplifier, which is required to keep the motor running at a constant speed, is reached. The difference voltage necessary to maintain the constant motor speed is a function of the physical dimensions and electrical characteristics of the motor. These include torque, speed, inertia, and current and voltage ratings of the motor. Thus the constant difference voltage and, in turn, a constant motor speed are maintained through the use of negative feedback.

The digital inputs may be calibrated in terms of revolutions per minute (rpm). In addition, the output of the speed sensor may be applied to the frequency meter/indicator to monitor the motor's speed.

Figure 11-11 shows the schematic diagram of a digitally controlled dc motor. As shown in the figure, the following ICs are used: SN74LS241 octal tri-state buffer, MC1408 8-bit D/A converter, MC1403 2.5-V voltage reference, LF353 dual op-amp, 9400 F/V converter, and the Hall-effect transducer. The desired digital inputs are selected by the use of a switch assembly. The eight LEDs indicate the state of the digital inputs applied to the D/A converter. When a switch is open, the corresponding LED lights up. The 74LS241 octal tristate buffer is used here because it isolates the LEDs from switches and also provides a current drive for the LEDs.

Figure 11-11 Digital dc motor speed control.

In Figure 11-11 the only device that has not been discussed so far is the *Hall-effect transducer*. A number of materials exhibit the Hall effect in that when a current-carrying semiconductor strip (usually silicon) is placed in the transverse (perpendicular) magnetic field, the combination produces an electromotive force (emf) between the opposite edges of the strip. This emf is proportional to the product of the current and field strength. On the other hand, when the magnetic field is zero, or of specific polarity, an emf between the opposite edges of the strip is also zero. Thus the Hall-effect transducer is a magnetically activated electronic switch that can be used for sensing a magnetic field.

Figure 11-12 shows the equivalent circuit, operating arrangement, and an electrical switching characteristic with hysteresis of Texas Instrument's TL170 bipolar Hall-effect switch. The TL170 is a three-terminal plastic package that consists of a silicon sensor, signal conditioning and hysteresis function, and an open-collector output stage integrated onto a monolithic chip [see Figure 11-12(a)]. The output of the device is compatible with bipolar or MOS logic circuits. Figure 11-12(b) shows the practical setup for the *on state*. The sensor is on (output voltage ≤ 0.04 V) when the magnetic field (B_{ON}) associated with the permanent-magnet north pole is perpendicular to the surface of the sensor and below a certain level, called the *operate point* or the *threshold*. On the other hand, the sensor is off when the magnetic field (B_{OFF}) emitted from the south pole of a permanent magnet is perpendicular to the surface of the sensor and above a certain level, called the *release point*. The TL170 has a typical operate point of ≤ -350 gauss and a release point of ≥ 350 gauss with a magnetic switching hysteresis ($B_{ON} - B_{OFF}$) of 200 gauss typically. The negative and positive magnetic fields are defined as those fields that are emitted from the north and south poles, respectively, of a permanent magnet. The magnetic switching hysteresis curve of the TL170 is shown in Figure 11-12(c). The sensor is designed so that its output stage can withstand up to 20 V in the off state and can sink up to 16 mA in the on state.

To operate, the TL170 sensor is positioned so that the plain surface of the sensor faces the permanent magnet. In addition, to obtain two samples per revolution and hence help to control the motor speed more accurately, four permanent magnets are used in Figure 11-11. These magnets are glued to the 4-in. diameter disk with alternately south and north poles up, as shown in Figure 11-12(d). The disk is then mounted on the motor's shaft. When the motor is running, the TL170 is turned on due to the magnetic field strength of the north pole and turned off due to the magnetic field strength of the south pole. Therefore, because of four permanent magnets the sensor will generate two cycles per revolution. The distance between the magnets and the sensor, however, depends on the field strength of the magnets. For the TL170 used in Figure 11-11, a magnetic field strength magnitude of ≥ 350 gauss is necessary. When the motor is running, the distance between the disk and sensor can be adjusted so that the output of the sensor is a pulse waveform. Remember that the output amplitude of the sensor depends on the supply voltage and is independent of the rpm of the motor.

Now let us reconsider the circuit of Figure 11-11. The D/A and F/V converters in this figure should be adjusted initially as follows:

1. With all the inputs high (logic 1), adjust R_{17} so that the output voltage of the *I*-to-*V* converter is 3.0 V.

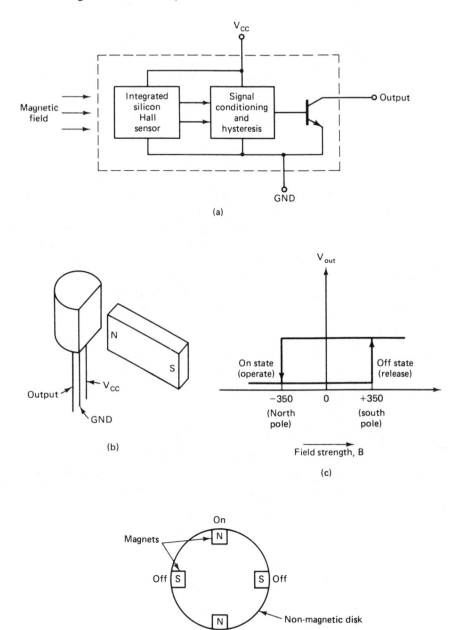

Figure 11–12 TL170 bipolar Hall-effect switch. (a) Equivalent circuit. (b) On-state arrangement. (c) Electrical switching characteristic. (d) Practical setup. (Courtesy of Texas Instruments.)

2. Disconnect pin 11 of the F/V converter from the junction of R_{27} and R_{26}. First, use the zero-adjust circuit connected to pin 2 to reduce the output voltage to zero. Second, apply 160-Hz, 5-V PP symmetrical square wave to pin 11 and adjust R_{int} until the output voltage is equal to 3.0 V.

Once the adjustments are performed in this order, they should not have to be repeated. Note that the F/V is calibrated for the maximum expected speed of the motor. In Figure 11-11, the motor speed is

$$4800 \text{ rpm} = \frac{(4800)\,(2)}{60} = 160 \text{ Hz}$$

The motor in Figure 11-11 initially starts running when the input binary code is $(00000110)_2$. Thereafter, the motor speed increases with the digital input until the motor attains a maximum speed at $(00111111)_2$. After $(00111111)_2$, however, the motor speed does not increase further even though the digital input is increased. In other words, we get 6-bit resolution instead of 8-bit. To obtain 8-bit resolution, an appropriate DAC with better resolution, a motor having favorable electromechanical specifications, and a differential amplifier with proper gain must be selected. The principles illustrated in the digital dc motor speed control of Figure 11-11 are used in the cruise control of automobiles.

Parts List

SN74LS241: tristate buffer

MC1408: 8-bit D/A converter

MC1403: 2.5-V voltage reference

LF353: dual op-amp

Teledyne 9400: F/V converter

TRW 405A100-1: dc motor

2N6569 and 2N6594: complementary power transistors ($I_C = 12$ A, $V_{CEO} = 40$ V)

TL170: bipolar Hall-effect switch

FLV 117: light-emitting diodes (8)

Miniature 8-slide or rocker DIP switch

Heat sinks for power transistors

Four small magnets with field strength magnitude ≥ 350 gauss

Nonmagnetic disk 4 in. in diameter

Eight 220-Ω resistors

Three 1-kΩ resistors

1.5-kΩ resistor

2.2-kΩ resistor

Eight 4.7-kΩ resistors

Six 100-kΩ resistors

1-kΩ potentiometer

20-kΩ potentiometer

1-MΩ potentiometer

Two 2.5-MΩ potentiometers

15-pF capacitor

75-pF capacitor

0.001-μF capacitor

0.1-μF capacitor

1.0-μF capacitor

11-6 APPLIANCE TIMER: APPLICATION OF THE 555 TIMERS

The previous project showed how the Hall-effect transducer works as a magnetic sensor. This project introduces another important type of transducer, an optical coupler. An optical coupler together with a square wave generator and counters can be used to construct a timer for appliances. That is, an appliance can be turned on at a desired time for a specific time interval. Figure 11-13 shows a block diagram of such a timer.

As shown in Figure 11-13, the output of the square wave generator is used as a clock (trigger signal) for the divide-by-N counters. A divide-by-N counter is a digital IC that produces a single output pulse for every N input pulses, where N is an integer. The integer N is commonly called the *modulus* of the counter. There are basically two types of divide-by-N counters: fixed and programmable. For example, the 7490 is the fixed modulus type and can be used as a divide-by-5 or divide-by-10 counter. On the other hand, the 74192 is a programmable counter that can be programmed to divide from 1 to 15. Often, to increase the resultant output count, the fixed-modulus-type counters are connected in series (cascaded). In this arrangement the output of the first counter is the input to the second, the output of the second counter is the input to the third, and so on. Using the cascading technique, time delays of hours, days, or even weeks can be achieved.

In Figure 11-13, the output of the divide-by-N counters is then applied to a one-shot (monostable) multivibrator, the output of which in turn drives the optical coupler. The output pulse width of the one-shot controls the time an optical coupler is on. The output

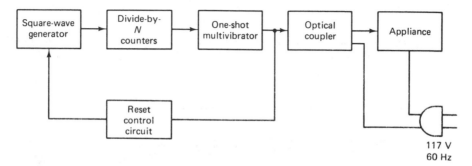

Figure 11–13   Block diagram of an appliance timer.

of the monostable multivibrator is also applied to the reset control circuit, which halts the square-wave generator and also resets the circuit for restart operation.

The resistor–capacitor combination used for the monostable determines its output pulse width. When the output pulse goes high (logic 0 to 1), the optical coupler is triggered and the appliance turns on. On the other hand, when the output pulse width of the monostable goes low (logic 1 to 0), the optical coupler is disabled and the appliance turns off. At the same time, the negative transition of the pulse also triggers the reset control circuit, which in turn shuts off the square wave generator. Thus the circuit is ready for the next cycle.

Before considering the schematic diagram of the appliance timer, examine the operation of Motorola's MOC3011, an optically isolated triac driver. Figure 11-14 shows the internal construction and the pin diagram of the MOC3011. The 3011 consists of an infrared emitting diode, optically coupled to a silicon bilateral switch, and is available as a 6-pin DIP. The bilateral switch, commonly known as a *diac,* is a member of the thyristor family. Other members of the thyristor family are the silicon-controlled rectifier (SCR) and the triac. A peculiarity of the thyristor is that it consists of four layers, P, N, P, and N, and is primarily designed for high-voltage and high-current switching applications. The diac consists of high-voltage/current diodes connected in opposite directions. On the other hand, the major difference between the SCR and triac is that the SCR conducts in only one direction, while the triac conducts in both directions. In fact, the triac is two SCRs connected in opposite directions. Both the SCR and triac are three-terminal devices (anode, cathode, and gate), whereas a diac is a two-terminal device (anode and cathode).

In the internal structure of the 3011, even though a symbolic notation for the diac is shown between output pins 4 and 6, this component actually works as a triac since its operation is controlled by the infrared-emitting diode. The maximum electrical ratings of an infrared-emitting diode are as follows:

1. Reverse voltage (V_R) = 3.0 V
2. Continuous forward current (I_F) = 50 mA at forward voltage (V_F) of 1.5 V

On the other hand, the output driver maximum electrical ratings are:

1. On-state output terminal voltage (V_{DRM}) = 250 V
2. On-state rms current (I_T) = 100 mA
3. Peak nonrepetitive surge current (I_{TSM}) = 1.2 A
4. Isolation surge voltage (an internal device dielectric breakdown rating) (V_{ISO}) = 7500 V peak

Figure 11–14 MOC3011 internal structure and pin diagram. (Courtesy of Motorola, Inc.)

In addition, the total device power dissipation is 330 mW, and the LED trigger current (the current required to latch output) is 5 mA typically.

Now let us see how the 3011 is used in the appliance timer. Figure 11-15 shows the schematic diagram of the timer, which is designed to switch an appliance on 23 minutes after the initial startup time. The appliance remains on for approximately 12 minutes. The 556, a dual 555 timer, is used to form the astable and monostable multivibrators. The time between the pulses as well as the output pulse width of the timers is limited by the size and cost of the timing components. In the circuit of Figure 11-15, the time period of the pulse output waveform of the astable multivibrator is $T = 0.69(R_A + 2R_B)C \cong$ 11.5 minutes, even though a maximum time period of approximately 20 minutes is possible. The output pulse waveform of the astable multivibrator works as a clock for the SN74177 4-bit binary counters/latches.

The SN74177, a 14-pin DIP, consists of four dc-coupled master–slave flip-flops, which accept frequencies of 0 to 35 MHz at the clock 1 input and 0 to 17.5 MHz at the clock 2 input (see Figure 11-15). It is fully programmable and triggers on the negative-going edge of the clock pulse. A, B, C, and D are the data inputs and Q_A, Q_B, Q_C, and Q_D are the outputs. When the count/load (pin 1) is low, the outputs directly follow the inputs. However, when the count/load is high (logic 1) and the clock inputs are inactive, the outputs remain unchanged. The *clear* terminal (pin 13), when taken low (logic 0), sets all outputs low regardless of the states of the clocks. In Figure 11-15, the SN74177 is used as a 4-bit ripple-through counter in which output Q_A must be externally connected to the clock 2 input. The output of the astable multivibrator is applied to the clock 1 input. In addition, the *count/load* is connected to +5 V and *clear* is taken low each time the timer is used. In fact, this arrangement enables the 74177 to perform simultaneous divisions by 2, 4, 8, and 16 at the Q_A, Q_B, Q_C, and Q_D outputs, respectively. That is, the outputs at Q_A, Q_B, Q_C, and Q_D will have timer periods of 23, 46, 92, and 184 minutes. However, only the Q_A output is used in Figure 11-15. To achieve longer time periods, output Q_D of the 74177 can be used to clock the next SN74177, and so on. In addition, by using all four outputs of the 74177, three more appliances can be controlled, although at different times.

The Q_A output of the 74177 with a time period of 23.0 minutes is applied to the monostable multivibrator, which has a pulse width of approximately 12 minutes. Note that the waveshaping network composed of R_1, C_1, and D_1 is used between trigger pin 8 and the V_{CC} pin 14. This arrangement assures proper operation of the monostable multivibrator, since the negative pulse width of the input trigger will be shorter than that of the output pulse width of the monostable multivibrator. Thus, in Figure 11-15, 23 minutes after the timer is initially set the appliance will come on and will remain on for approximately 12 minutes. At the end of 12 minutes, the output pulse width of the monostable goes low (logic 1 to 0), which disables the 3011 and the appliance turns off. At the same time, the negative transition in the output pulse of the monostable multivibrator also triggers the reset control circuit and shuts the astable multivibrator off.

The reset control circuit is composed of a single J-K negative-edge-triggered flip-flop 74H102 with *preset* and *clear*. The Q output of the flip-flop is connected to reset pin 4 of the astable multivibrator and is normally high (logic 1). However, when the output pulse of the monostable goes low, it switches the Q output low (logic 0), which in turn shorts out the reset terminal. This halts the astable operation and prevents further clocking

Figure 11-15 Appliance timer.

of the 74177 counters. The end result is that the appliance goes through one cycle (off–on–off) and then stops. Initially, to start the timer, apply the power to the timer by placing switch *SW*1 in the *on* position; then depress the START switch *SW*2. With *SW*2 depressed, the J-K flip-flop is preset ($Q = 1$), and the reset terminal of the astable multivibrator is pulled high (logic 1). The switch *SW*2 also clears the Q_A, Q_B, Q_C, and Q_D outputs of the 74177 counters. Now the timer is on and the appliance will come on in approximately 23 minutes and will remain on for approximately 12 minutes. At the end of 12 minutes the appliance turns off and the Q output of the J-K flip-flop switches low. This action shuts off the astable multivibrator and hence the timer. To restart the timer, simply depress the START switch *SW*1. When the timer is not in use, switch *SW*1 should be in the off position. The timing diagram for the appliance timer is shown in Figure 11-16. In the circuit of Figure 11-15, since the maximum *on* state rms current supplied by the MOC3011 is 100 mA, the power rating of an appliance whose operation is to be controlled must be ≤ 11 W $= (110$ V$)(100$ mA$)$. However, for appliances with higher power ratings (>11 W), a triac with appropriate electrical specifications (V and I) may be used between the MOC3011 and an appliance. For more information on using triacs with the MOC3011, refer to the MOC3011 data sheets. Finally, to verify the operation of the appliance timer, a 5-W light bulb may be used as the appliance.

The timer concepts discussed can be used to control the operation of a furnace or central air-conditioning unit so that it comes on at a desired time. For example, a timer may be used to shut off the furnace in the house after 10 P.M. and turn it back on at 5 A.M. This arrangement will help to save energy and money.

The concepts of appliance timers can also be used in switching house lights on and off at regular intervals, especially when the owner is on vacation. This provision should discourage burglars. In fact, when cyclic operation of the timer is desired, the reset control circuit shown in Figure 11-15 is not needed.

Parts List

NE556 timer

SN74177 4-bit binary counters/latches

MOC3011 optically isolated triac driver

SN74H102 single J-K flip-flop or 7476 dual J-K

1N914 signal diode

390-Ω resistor

4.7-kΩ resistor

5.1-kΩ resistor

Two 10-kΩ resistors

Two 5.1-MΩ resistors

Three 0.01-μF capacitors

1.0-μF capacitors

10-μF capacitor

100-μF capacitor

Figure 11–16   Timing diagram for the appliance timer.

130-μF capacitor
SPDT switch
Push-button switch (normally open)

11-7 SIREN/ALARM: APPLICATION OF THE LM380 POWER AMPLIFIER ▬

Devices such as burglar alarms and sirens, whose basic purpose is to monitor certain conditions, make enjoyable projects because of the variety of sounds they can generate. Figure 11-17 shows a simple siren/alarm circuit using a dual op-amp MC1458, audio

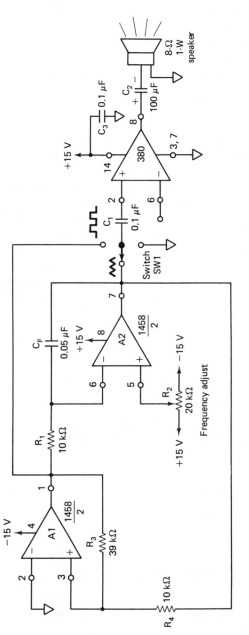

Figure 11-17 Siren/alarm schematic diagram.

amplifier LM380, and a 1-W speaker. The dual op-amp is used as a signal generator that produces square, pulse, and triangular or sawtooth waveforms as discussed in Section 8-17, while the operation of the audio power amplifier LM380 is explained in Section 10-6.

The operation of the circuit is as follows. The A_1 and A_2 op-amps make up a waveform generator in which the output of A_1 is a square wave or pulse waveform and that of A_2 is either a triangular or sawtooth waveform. The potentiometer R_2 controls the frequency as well as the type of output waveform of op-amps A_1 and A_2. The switch $SW1$ connects the output of A_1 or A_2 to the audio power amplifier LM380, which in turn drives the speaker.

Although not used in the circuit of Figure 11-17, a potentiometer may be connected between $(+)$ and $(-)$ inputs of the power amplifier to control its voltage gain, which in turn controls the sound volume. The sound level produced depends on the position of switch $SW1$, the wiper setting of potentiometer R_2, and the value of capacitor C_F. Therefore, sounds of varying intensities can be obtained by adjusting $SW1$, R_2, and C_F.

Recall that pulse and sawtooth waveforms can be generated by using the 555 timer, as shown in Section 10-4. Similarly, the VCO 566 presented in Section 8-18 also generates a square wave and a triangular wave. These circuits may therefore be used as the signal generator in the circuit of Figure 11-17, instead of an op-amp signal generator.

For higher output power (sound intensities), audio power amplifiers may be used in the bridge form as shown in Figure 10-41. This configuration will also require a higher wattage speaker.

Parts List

MC1458 dual op-amp

LM380 audio power amplifier

Three-position switch

1-W speaker

Two 10-kΩ resistors

39-kΩ resistor

20-kΩ potentiometer

0.05-μF capacitor

Two 0.1-μF capacitors

1000-μF capacitor

11-8 SUGGESTED IC PROJECTS ━━━━━━━━━━━━━━━━━━━━━━

The IC systems presented are a fair sample of practical applications of ICs discussed in the previous chapters. These IC projects demonstrate that a practical and useful system does not have to be complex or expensive. Also, the projects illustrated here should give you enough experience and confidence so that you might try your hand at a project of your own.

Although numerous practical IC systems can be formed with appropriate combinations of linear and/or digital ICs, block diagrams for suggested IC projects that should prove to be simple, interesting, practical, and yet challenging are shown in Figure 11-18. These are:

1. RPM speed indicator
2. Frequency meter
3. Motor speed control using phase-locked-loop principles
4. Digital light meter
5. Digital voltmeter
6. Sixteen-LED sequencer or roulette wheel
7. Sound generator
8. Amplifier with digitally controlled gain
9. Logic probe using window comparator

(a) RPM speed indicator.

(b) Frequency meter.

Figure 11–18   Block diagram for the IC projects.

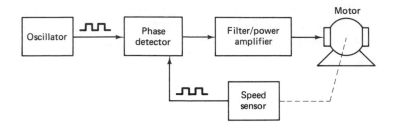

(c) Motor speed control using phase-locked loop principles.

(d) Digital light meter.

(e) Digital voltmeter.

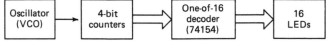

(f) Sixteen LED sequencer or roulette wheel.

(g) Sound generator.

Figure 11–18 (Continued)

(h) Amplifier with digitally controlled gain.

(i) Logic probe using window comparator.

Figure 11–18   (Continued)

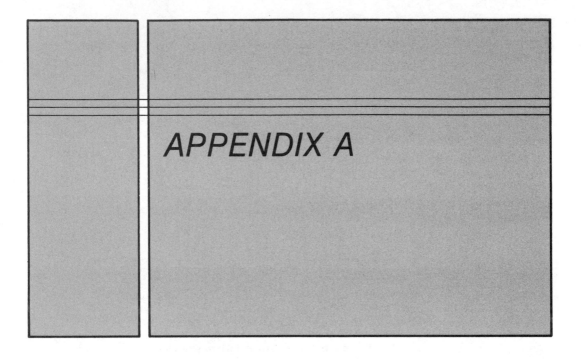

APPENDIX A

RESISTANCE CHART (10 Ω to 22 MΩ)[a]

Resistance (Ω)	Resistance (kΩ)	Resistance (kΩ)	Resistance (MΩ)
* 10	* 1.0	*100	* 1.0
11	1.1	110	1.1
* 12	* 1.2	*120	* 1.2
13	1.3	130	1.3
* 15	* 1.5	*150	* 1.5
16	1.6	160	1.6
* 18	* 1.8	*180	* 1.8
20	2.0	200	2.0
* 22	* 2.2	*220	* 2.2
24	2.4	240	2.4
* 27	* 2.7	*270	* 2.7
30	3.0	300	3.0
* 33	* 3.3	*330	* 3.3
36	3.6	360	3.6
* 39	* 3.9	*390	* 3.9
43	4.3	430	4.3
* 47	* 4.7	*470	* 4.7
51	5.1	510	5.1
* 56	* 5.6	*560	* 5.6
62	6.2	620	6.2
* 68	* 6.8	*680	* 6.8
75	7.5	750	7.5
* 82	* 8.2	*820	* 8.2
91	9.1	910	9.1
*100	*10.0		*10
110	11		11
*120	*12		*12
130	13		13
*150	*15		*15
160	16		16
*180	*18		*18
200	20		20
*220	*22		*22
240	24		
*270	*27		
300	30		
*330	*33		
360	36		
*390	*39		
430	43		
*470	*47		
510	51		
*560	*56		
620	62		
*680	*68		
750	75		
*820	*82		
910	91		

[a]Tolerance of $\pm 10\%$ available only in values with *; $\pm 5\%$ available for all.

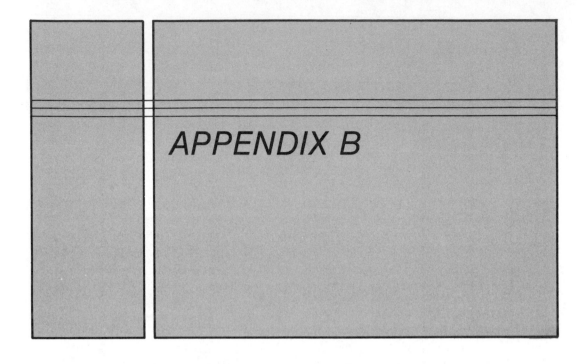

APPENDIX B

CAPACITANCE CHART (1 pF to 1000 μF)[a]

Capacitance (pF)	Capacitance (pF)	Capacitance (μF)	Capacitance (μF)	Capacitance (μF)	Capacitance (μF)
1	47	0.001	0.025	0.82	300
1.2	50	0.0012	0.027	1.0	400
1.5	56	0.0015	0.03	2	500
1.8	68	0.0018	0.033	3	600
2.2	75	0.002	0.035	4	750
2.7	82	0.0022	0.039	5	800
3.0	100	0.0025	0.04	6	900
3.3	120	0.0027	0.047	8	1000
3.9	150	0.003	0.05	10	
4.7	180	0.0033	0.056	12	
5.0	200	0.0039	0.06	15	
5.6	220	0.004	0.068	16	
6.8	270	0.0047	0.075	20	
8.2	300	0.005	0.082	25	
10	330	0.0056	0.1	30	
12	390	0.006	0.12	40	
15	430	0.0068	0.15	50	
18	470	0.0075	0.18	60	
20	500	0.008	0.22	75	
22	510	0.0082	0.25	80	
25	560	0.01	0.27	85	
27	620	0.012	0.33	100	
30	680	0.015	0.39	150	
33	750	0.018	0.47	160	
39	820	0.02	0.56	200	
	910	0.022	0.68	250	

[a]The standard WVDC ratings for the capacitors are: 1 V, 3 V, 6 V, 10 V, 12 V, 16 V, 20 V, 25 V, 50 V, 60 V, 75 V, 90 V, 100 V, 150 V, 175 V, 200 V, 250 V, 300 V, 350 V, 400 V, 450 V, 475 V, 500 V, 525 V, 550 V, 600 V, 1000 V.

APPENDIX C

IMPORTANT DERIVATIONS

7-2.2 AC Amplifier

To show:

$$f_L = \frac{1}{2\pi C_i (R_{iF} + R_o)} \tag{7-1}$$

Proof: The input circuit of the ac inverting amplifier of Figure 7-3(a) is drawn in Figure C-1. The voltage v_A is amplified by the ac inverting amplifier; hence it is necessary to determine v_A as a function of input voltage v_{in}. Using the voltage-divider rule, we get

$$v_A = \frac{(R_{iF})(v_{in})}{\left(R_o + \dfrac{1}{(j\omega C_i)} + R_{iF}\right)} \tag{C-1}$$

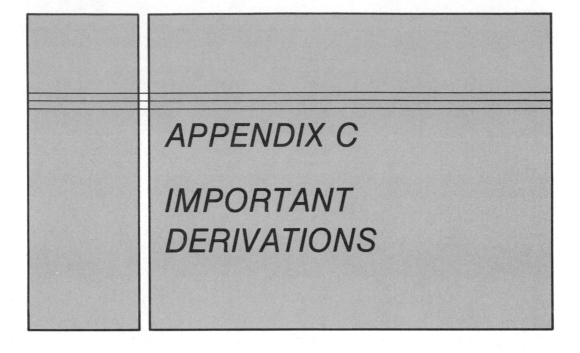

Figure C–1 Input circuit for the ac inverting amplifier of Figure 7–3(a).

where R_{iF} = input resistance of the inverting amplifier
R_o = source resistance, R_{in}

Rearranging Equation (C-1),

$$v_A = \frac{j(R_{iF}\omega C_i)(v_{in})}{j(R_{iF} + R_o)(\omega C_i) + 1}$$

$$= \frac{j(2\pi f C_i R_{iF})(v_{in})}{j(2\pi f C_i)(R_{iF} + R_o) + 1}$$

$$= \frac{j(2\pi f C_i R_{iF})(v_{in})}{j\left(\dfrac{f}{f_L}\right) + 1}$$

where f = input frequency (Hz)

$$f_L = \frac{1}{2\pi C_i(R_{iF} + R_o)} = \text{low frequency cutoff}$$

The output voltage of the ac inverting amplifier in Figure 7-3(a) is

$$v_o = -\left(\frac{R_F}{R_1}\right)(v_A)$$

$$v_o = -\left(\frac{R_F}{R_1}\right)\left[\frac{j(2\pi f C_i R_{iF})(v_{in})}{j\left(\dfrac{f}{f_L}\right) + 1}\right]$$

or

$$\frac{v_o}{v_{in}} = -\left(\frac{R_F}{R_1}\right)\left[\frac{j(2\pi f C_i R_{iF})}{j\left(\dfrac{f}{f_L}\right) + 1}\right] \qquad\text{(C-2)}$$

If we plot the frequency response (magnitude versus frequency plot) using Equation (C-2), the nature of the plot will be as shown in Figure C-2. This is because, as the input frequency f increases, the gain also increases. However, when $f = f_L$, the contribution due to the denominator of Equation (C-2) to the overall gain is -3.01 dB. Therefore, f_L is

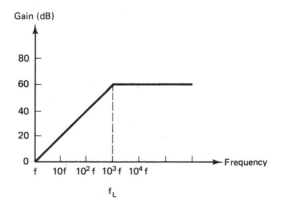

Figure C-2 Frequency response for the circuit of Figure C-1 (high-pass filter).

referred to as the break frequency or low-frequency cutoff. In fact, the frequency response of the amplifier is flat for $f > f_L$.

Equation (C-2) can also be derived by using the Laplace transform, and readers that are familiar with the Laplace transform are encouraged to do so.

7-12 The Integrator

To show:

$$f_b = \frac{1}{2\pi R_1 C_F} \tag{7-24}$$

Proof: Consider the basic integrator of Figure 7-23. To facilitate the derivation of Equation (7-24), the circuit is transformed into the S domain as shown in Figure C-3.

The gain of the integrator is

$$\frac{V_o(S)}{V_{in}(S)} = -\frac{Z_F(S)}{Z_1(S)}$$

where

$$Z_F(S) = \frac{1}{SC_F}$$

$$Z_1(S) = R_1$$

Therefore,

$$\frac{V_o(S)}{V_{in}(S)} = -\frac{1}{SR_1 C_F} \tag{C-3}$$

where $S = j\omega = j2\pi f$. Substituting $S = j\omega$ in Equation (C-3) and then obtaining the magnitude, we get

$$\left| \frac{V_o(j\omega)}{V_{in}(j\omega)} \right| = \frac{1}{\sqrt{(\omega R_1 C_F)^2}}$$

$$= \frac{1}{2\pi f R_1 C_F}$$

Figure C–3 Basic integrator circuit of Figure 7–23 transformed into the S domain.

The frequency ($f = f_b$) at which the magnitude of the gain of the basic integrator is 1 (or 0 dB) can be calculated by equating the preceding equation to 1. Hence

$$1 = \frac{1}{2\pi f R_1 C_F}$$

or

$$f_b = \frac{1}{2\pi R_1 C_F} \qquad (7\text{-}24)$$

To show:

$$f_a = \frac{1}{2\pi R_F C_F} \qquad (7\text{-}25)$$

Proof: Consider the practical integrator of Figure 7-25. For simplicity of derivation, the integrator is transformed into the S domain as shown in Figure C-4.

The voltage gain of the integrator is

$$\frac{V_o(S)}{V_{in}(S)} = -\frac{Z_F(S)}{Z_1(S)}$$

where $Z_F(S) = R_F \| \dfrac{1}{SC_F} = \dfrac{R_F}{R_F C_F S + 1}$

$$Z_1(S) = R_1$$

Therefore, substituting these values, we have

$$\frac{V_o(S)}{V_{in}(S)} = -\frac{R_F/R_1}{R_F C_F S + 1} \qquad (C\text{-}4)$$

where R_F/R_1 is the gain constant.

The break frequency ($f = f_a$) for the practical integrator at which the gain is $(0.707)(R_F/R_1)$ (or -3 dB down from its value of R_F/R_1) can be calculated by equating the denominator of Equation (C-4) to $\sqrt{2}$. That is,

Figure C–4   Practical integrator circuit of Figure 7–25 transformed into the S domain.

$$\sqrt{1 + (R_F C_F 2\pi f)^2} = \sqrt{2}$$

Solving for $f = f_a$, we get

$$f_a = \frac{1}{2\pi R_F C_F} \qquad (7\text{-}25)$$

7-13 The Differentiator
To show:

$$f_a = \frac{1}{2\pi R_F C_1} \qquad (7\text{-}28)$$

Proof: Consider the basic differentiator of Figure 7-27(a). To facilitate the derivation of Equation (7-28), the circuit is transformed into the S domain as shown in Figure C-5.
The ratio of $V_o(S)$ to $V_{in}(S)$ is

$$\frac{V_o(S)}{V_{in}(S)} = -\frac{Z_F(S)}{Z_1(S)}$$

where $Z_F(S) = R_F$

$$Z_1(S) = \frac{1}{SC_1}$$

$$S = j\omega$$

Therefore,

$$\frac{V_o(S)}{V_{in}(S)} = -(R_F C_1 S) \qquad (\text{C-5})$$

The frequency $(f = f_a)$ at which the magnitude of $[V_o(j\omega)/V_{in}(j\omega)] = 1$ is calculated by equating Equation (C-5) to 1. Thus,

$$1 = R_F C_1 (j2\pi f_a)$$

or

$$f_a = \frac{1}{2\pi R_F C_1} \qquad (7\text{-}28)$$

Figure C-5 Basic differentiator of Figure 7-27(a) transformed into the S domain.

Refer to Figure 7-27(b).

To show:

$$f_b = \frac{1}{2\pi R_1 C_1} \tag{7-29}$$

Proof: Consider the practical differentiator of Figure 7-28(a), which is transformed into the S domain and redrawn in Figure C-6.

The relationship between the output and the input voltages is

$$\frac{V_o(S)}{V_{in}(S)} = \frac{-Z_F(S)}{Z_1(S)}$$

where $Z_F(S) = R_F \| \dfrac{1}{SC_F} = \dfrac{R_F}{R_F C_F S + 1}$

$$Z_1(S) = R_1 + \frac{1}{SC_1} = \frac{R_1 C_1 S + 1}{C_1 S}$$

Therefore, substituting $Z_F(S)$ and $Z_1(S)$ values in the preceding equation, we get

$$\frac{V_o(S)}{V_{in}(S)} = -\frac{R_F C_1 S}{(R_F C_F S + 1)(R_1 C_1 S + 1)}$$

If $R_F C_F = R_1 C_1$, then

$$\frac{V_o(S)}{V_{in}(S)} = -\frac{R_F C_1 S}{(R_F C_F S + 1)^2} \tag{C-6}$$

The frequency ($f = f_b$) at which the frequency response of the practical differentiator starts decreasing at 20 dB/decade [see Figure 7-27(b)] can be determined by equating the denominator of Equation (C-6) to zero. Therefore,

$$(R_F C_F S + 1)^2 = 0$$

However, $S = j(2\pi f)$ and $f = f_a$ cannot be negative. Therefore,

Figure C–6 Practical differentiator of Figure 7–28(a) transformed into the S domain.

$$f_b = \frac{1}{2\pi R_1 C_1} \tag{7-29}$$

Since

$$R_1 C_1 = R_F C_F$$

CHAPTER 8

8-4 Second-Order Low-Pass Butterworth Filter

To show:

$$f_H = \frac{1}{2\pi \sqrt{R_2 R_3 C_2 C_3}} \tag{8-3}$$

Proof: To facilitate the analysis of Figure 8-4, we will use the Laplace transform. Consider the input circuitry of the second-order low-pass Butterworth filter as shown in Figure C-7. In this circuit all the components and the circuit parameters are expressed in the S domain where $S = j\omega$.

Writing Kirchhoff's current law at node $V_A(S)$,

$$I_1 = I_2 + I_3$$

or

$$\frac{V_{in} - V_A}{R_2} = \frac{V_A - V_o}{1/SC_2} + \frac{V_A - V_1}{R_3} \tag{C-7}$$

For simplicity in this equation, we have omitted S; for example $V_{in}(S)$ is written as V_{in}. Also, using the voltage-divider rule,

$$V_1 = \frac{1/SC_3}{R_3 + (1/SC_3)} V_A \qquad \text{since } R_{iF} \cong \infty, \quad I_B \cong 0 \text{ A}$$

$$= \frac{V_A}{R_3 C_3 S + 1}$$

or

$$V_A = (R_3 C_3 S + 1) V_1$$

Substituting the value of V_A in Equation (C-7) and solving for V_1, we get

Figure C–7   Input *RC* circuitry of the second-order low-pass filter of Figure 8–4 transformed into the *S* domain.

$$V_1 = \frac{(R_3)(V_{in}) + (R_3 R_2 C_2 S)(V_o)}{(R_3 C_3 S + 1)(R_2 + R_3 + R_3 R_2 C_2 S) - R_2}$$

However,

$$V_o = (A_F)V_1$$

where $A_F = 1 + (R_F/R_1)$.

Therefore,

$$V_o = \frac{(A_F)[(R_3)(V_{in}) + (R_3 R_2 C_2 S)(V_o)]}{(R_3 C_3 S + 1)(R_2 + R_3 + R_3 R_2 C_2 S) - R_2}$$

Solving this equation for V_o/V_{in}, we have

$$\frac{V_o}{V_{in}} = \frac{A_F}{S^2 + \dfrac{(R_3 C_3 + R_2 C_3 + R_2 C_2 - A_F R_2 C_2)S}{R_2 R_3 C_2 C_3} + \dfrac{1}{R_2 R_3 C_2 C_3}}$$

For frequencies above f_H, the gain of the second-order low-pass filter rolls off at the rate of -40 dB/decade. Therefore, the denominator quadratic in the gain (V_o/V_{in}) equation must have two real and equal roots. This means that

$$\omega_H^2 = \frac{1}{R_2 R_3 C_2 C_3}$$

or

$$\omega_H = \frac{1}{\sqrt{R_2 R_3 C_2 C_3}}$$

$$f_H = \frac{1}{2\pi\sqrt{R_2 R_3 C_2 C_3}} \tag{8-3}$$

8-12 Phase Shift Oscillator

To show:

$$f_o = \frac{1}{2\pi\sqrt{6}RC} \tag{8-22a}$$

$$\left|\frac{R_F}{R_1}\right| = 29 \tag{8-22b}$$

Proof: First consider the feedback circuit consisting of RC combinations of the phase shift oscillator. For simplicity we use the Laplace transform again. Thus, the circuit is represented in the S domain as shown in Figure C-8. Let us determine $V_f(S)/V_o(S)$ for the circuit.

Writing Kirchhoff's current law (KCL) at node $V_1(S)$, we get

$$I_1(S) = I_2(S) + I_3(S)$$

$$\frac{V_o(S) - V_1(S)}{1/SC} = \frac{V_1(S)}{R} + \frac{V_1(S) - V_2(S)}{1/SC}$$

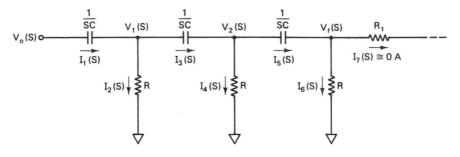

Figure C–8   *RC* network of the phase shift oscillator transformed into the *S* domain.

Solving for $V_1(S)$, we have

$$V_1(S) = \frac{[V_o(S) + V_2(S)](RCS)}{2RCS + 1} \tag{C-8}$$

Writing KCL at node $V_2(S)$,

$$I_3(S) = I_4(S) + I_5(S)$$

$$\frac{V_1(S) - V_2(S)}{1/SC} = \frac{V_2(S)}{R} + \frac{V_2(S) - V_f(S)}{1/SC}$$

Solving for $V_1(S)$,

$$V_1(S) = \frac{(2RCS + 1)V_2(S)}{(RCS)} - V_f(S) \tag{C-9}$$

If $R_1 \gg R$ in the circuit of Figure C-8, then $I_7(S) \cong 0$ A. This means that $I_5(S) = I_6(S)$. Therefore, using the voltage-divider rule,

$$V_f(S) = \frac{R}{R + (1/SC)} V_2(S)$$

or

$$V_2(S) = \frac{(RCS + 1)(V_f(S))}{RCS} \tag{C-10}$$

Substituting the value of $V_2(S)$ in Equation (C-8),

$$V_1(S) = \frac{(RCS)(V_o(S))}{2RCS + 1} + \frac{(RCS + 1)V_f(S)}{2RCS + 1} \tag{C-11}$$

Also, substituting the value of $V_2(S)$ in Equation (C-9), we get

$$V_1(S) = \frac{(2RCS + 1)(RCS + 1)V_f(S)}{(RCS)(RCS)} - V_f(S) \tag{C-12}$$

Equating Equations (C-11) and (C-12) and simplifying for $V_f(S)/V_o(S)$, we get

$$\frac{V_f(S)}{V_o(S)} = \frac{R^3C^3S^3}{(R^3C^3S^3 + 6R^2C^2S^2 + 5RCS + 1)} \tag{C-13}$$

$$= B$$

Figure C–9 Op-amp part of the phase shift oscillator of Figure 8–18.

Next, consider the op-amp part of the phase shift oscillator. For convenience the circuit is redrawn in Figure C-9. The voltage gain of the op-amp is

$$A_v = \frac{V_o(S)}{V_f(S)}$$

$$= -\left(\frac{R_F}{R_1}\right) \tag{C-14}$$

for an oscillator,

$$(A_v)(B) = 1$$

Therefore, using Equations (C-13) and (C-14),

$$-\frac{R_F}{R_1} \frac{R^3C^3S^3}{R^3C^3S^3 + 6R^2C^2S^2 + 5RCS + 1} = 1$$

Substituting $S = j\omega$ and equating real and imaginary parts, respectively, we get

$$\left(\frac{-R_F}{R_1}\right)(-jR^3C^3\omega^3) = (-jR^3C^3\omega^3) - (6R^2C^2\omega^2) + (j5RC\omega) + 1$$

$$0 = -6R^2C^2\omega^2 + 1 \qquad \text{(real part)}$$

or

$$f_o = \frac{1}{2\pi\sqrt{6}\,RC} \tag{8-22a}$$

$$\left(-\frac{R_F}{R_1}\right)(-jR^3C^3\omega^3) = (-jR^3C^3\omega^3) + (j5RC\omega) \qquad \text{(imaginary part)}$$

or

$$\left(-\frac{R_F}{R_1}\right) = 1 - \frac{5}{R^2C^2\omega^2}$$

Substituting the value of ω^2 in the preceding equation,

$$\frac{R_F}{R_1} = 29 \tag{8-22b}$$

8-13 Wien Bridge Oscillator

To show:

$$f_o = \frac{1}{2\pi RC} \tag{8-23a}$$

$$R_F = 2R_1 \tag{8-23b}$$

Proof: First consider the feedback circuit of the Wien bridge oscillator of Figure 8-19. The circuit is transformed in the S domain and redrawn in Figure C-10.
Using the voltage-divider rule,

$$V_f(S) = \frac{Z_P(S)V_o(S)}{Z_P(S) + Z_S(S)}$$

where $Z_P(S) = R \| \dfrac{1}{SC} = \dfrac{R}{RCS + 1}$

$$Z_S(S) = R + \frac{1}{SC} = \frac{RCS + 1}{SC}$$

Therefore, substituting $Z_P(S)$ and $Z_S(S)$ values, we get

$$V_f(S) = \frac{(RCS)V_o(S)}{(RCS + 1)^2 + RCS}$$

or

$$B = \frac{V_f(S)}{V_o(S)}$$

$$= \frac{RCS}{R^2C^2S^2 + 3RCS + 1} \tag{C-15}$$

Next, consider the op-amp part of the Wien bridge oscillator. The circuit is redrawn in Figure C-11.
The voltage gain A_v of the op-amp is

$$A_v = \frac{V_o(S)}{V_f(S)} = 1 + \frac{R_F}{R_1} \tag{C-16}$$

Figure C–10   Feedback circuit of the Wien bridge oscillator of Figure 8–19 represented in the S domain.

Figure C–11 Op-amp part of the Wien bridge oscillator.

Finally, the requirement for oscillation is

$$(A_v)(B) = 1$$

Therefore, using Equations (C-15) and (C-16), we have

$$\left(1 + \frac{R_F}{R_1}\right)\frac{RCS}{R^2C^2S^2 + 3RCS + 1} = 1$$

Substituting $S = j\omega$ in this equation and then equating the real and imaginary parts, we get the frequency of oscillation f_o and the gain required for oscillation, as follows:

$$\left(1 + \frac{R_F}{R_1}\right)jRC\omega = (-R^2C^2\omega^2) + j(3RC\omega) + 1$$

$$\omega^2 = \frac{1}{R^2C^2} \qquad \text{(real part)}$$

or

$$f_o = \frac{1}{2\pi RC} \tag{8-23a}$$

and

$$\left(1 + \frac{R_F}{R_1}\right)jRC\omega = j(3RC\omega) \qquad \text{(imaginary part)}$$

or

$$1 + \frac{R_F}{R_1} = 3$$

$$R_F = 2R_1 \tag{8-23b}$$

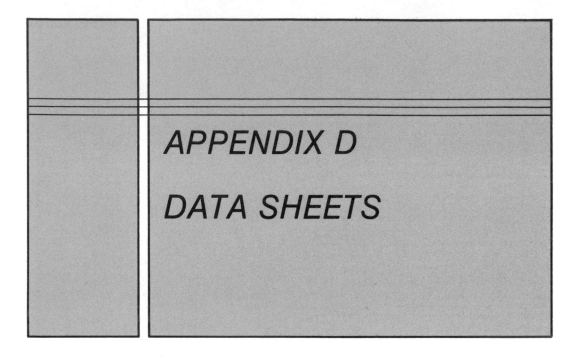

APPENDIX D

DATA SHEETS

The following data sheets are reprinted with permission, copyright Fairchild Semiconductor Corporation: μA714, 78S40.

The following data sheets are reprinted with the permission of National Semiconductor Corporation: MF5, LM380.

The following data sheets are reprinted with the permission of Telydyne Semiconductor: TSC 9400.

The following data sheets are reprinted with the permission of Signetics Corporation: NE/SE566.

The following data sheets are reprinted with the permission of Motorola Corporation: MC1408, MC7800, MC7900, MOC3009, MOC3010, MOC3011.

The following data sheets are reprinted with permission of Datel-Intersil Corporation: 1CL8038.

The following data sheets are reprinted with the permission of Texas Instruments Incorporated: TL170C.

μA714
Precision
Operational Amplifier

Linear Division Operational Amplifiers

Description

The μA714 is a monolithic instrumentation operational amplifier constructed using the Fairchild Planar Epitaxial process. It is intended for precise, low level signal amplification applications where low noise, low drift and accurate closed loop gain are required. The offset null capability, low power consumption, very high voltage gain as well as wide power supply voltage range provide superior performance for a wide range of instrumentation applications.

- **Low Offset Voltage — 75 μV**
- **Low Offset Voltage Drift — 1.0 μV/°C Typically**
- **Low Bias Current — ± 2.6 nA**
- **Low Input Noise Current — 0.12 pA/ $\sqrt{Hz}$ at 1.0 kHz Typically**
- **High Open Loop Gain — 500 K Typically**
- **Low Input Offset Current — 2.8 nA**
- **High Common Mode Rejection — 110 dB**
- **Wide Power Supply Range — ± 3.0 To ± 22 V**
- **Plug-In Replacement For Op-07**

Absolute Maximum Ratings

Storage Temperature Range
Metal Can	−65°C to +175°C
Molded DIP and SO-8	−65°C to +150°C

Operating Temperature Range
Extended (μA714M)	−55°C to +125°C
Commercial	
(μA714C, μA714EC, μA714LC)	0°C to +70°C

Lead Temperature
Metal Can (soldering, 60 s)	300°C
Molded DIP and SO-8	
(soldering, 10 s)	265°C

Internal Power Dissapation[1, 2]
8L-Metal Can	1.00 W
8L-Molded DIP	0.93 W
SO-8	0.81 W

Supply Voltage
μA714, μA714C, μA714E	± 22 V
μA714L	± 18 V
Differential Input Voltage	± 30 V

Input Voltage[3]
μA714, μA714C, μA714E	± 22 V
μA714L	± 18 V

Notes

1. $T_{J\ Max}$ = 150°C for the Molded DIP and SO-8, and 175°C for the Metal Can.
2. Ratings apply to ambient temperature at 25°C. Above this temperature, derate the 8L-Metal Can at 6.7 mW/°C, the 8L-Molded DIP at 7.5 mW/°C, and the SO-8 at 6.5 mW/°C.
3. For supply voltage less than ± 22 V, the absolute maximum input voltage is equal to the supply voltage.

Connection Diagram
8-Lead Metal Package
(Top View)

Lead 4 connected to case.

Order Information

Device Code	Package Code	Package Description
μA714HM	5W	Metal
μA714HC	5W	Metal
μA714EHC	5W	Metal
μA714LHC	5W	Metal

Connection Diagram
8-Lead DIP and SO-8 Package
(Top View)

Order Information

Device Code	Package Code	Package Description
μA714SC	KC	Molded Surface Mount
μA714TC	9T	Molded DIP
μA714LSC	KC	Molded Surface Mount
μA714LTC	9T	Molded DIP

μA714

Equivalent Circuit

EQ00161F

Note

1. R2A and R2B are electronically adjusted on chip at the factory for minimum offset voltage

μA714

μA714
Electrical Characteristics $T_A = 25°C$, $V_{CC} = \pm 15$ V

Symbol	Characteristic	Condition	Min	Typ	Max	Unit
V_{IO}	Input Offset Voltage	$R_S = 50\ \Omega$, $V_{CM} = 0$ V		30	75	μV
S	Long Term Input Offset Voltage Stability	$R_S = 50\ \Omega$, $V_{CM} = 0$ V		0.2		μV/mo
$V_{IO\ adj}$	Input Offset Voltage Adjustment Range	$R_O = 20$ kΩ		±4.0		mV
I_{IO}	Input Offset Current	$V_{CM} = 0$ V		0.4	2.8	nA
I_{IB}	Input Bias Current	$V_{CM} = 0$ V		1.0	3.0	nA
Z_I	Input Impedance		20	60		MΩ
P_c	Power Consumption	$V_O = 0$ V		75	120	mW
		$V_{CC} = \pm 3.0$ V, $V_O = 0$ V		4.0	6.0	
CMR	Common Mode Rejection	$V_{CM} = \pm 13$ V, $R_S = 50\ \Omega$	110	126		dB
V_{IR}	Input Voltage Range		±13.0	±14.0		V
PSRR	Power Supply Rejection Ratio	$V_{CC} = \pm 3.0$ V to ±18 V, $R_S = 50\ \Omega$	100	110		dB
A_{VS}	Large Signal Voltage Gain	$R_L \geqslant 2.0$ kΩ, $V_O = \pm 10$ V	200	500		V/mV
		$R_L \geqslant 500\ \Omega$, $V_O = \pm 0.5$ V $V_{CC} = \pm 3.0$ V	150	500		
V_{OP}	Output Voltage Swing	$R_L = 10$ kΩ	±12.5	±13.0		V
		$R_L = 2.0$ kΩ	±12.0	±12.8		
		$R_L = 1.0$ kΩ	±10.5	±12.0		
BW	Bandwidth	$A_V = 1.0$		0.6		MHz
SR	Slew Rate	$R_L = 2.0$ kΩ, $A_V = 1.0$		0.17		V/μs
e_n	Input Noise Voltage	0.1 Hz to 1.0 kHz		0.35	0.6	μVp—p
	Input Noise Voltage Density	$f_o = 10$ Hz		10.3	18.0	nV/√Hz
		$f_o = 100$ Hz		10.0	13.0	
		$f_o = 1000$ Hz		9.6	11.0	
i_n	Input Noise Current	0.1 Hz to 1.0 kHz		14		pA p—p
	Input Noise Current Density	$f_o = 10$ Hz		0.32	0.80	pA/√Hz
		$f_o = 100$ Hz		0.14	0.23	
		$f_o = 1000$ Hz		0.12	0.17	

µA714

µA714 (Cont.)
Electrical Characteristics $T_A = 25°C$, $V_{CC} = \pm 15$ V

Symbol	Characteristic		Condition	Min	Typ	Max	Unit
The following specifications apply for $V_{CC} = \pm 15$ V, $-55°C \leqslant T_A \leqslant +125°C$								

Symbol	Characteristic		Condition	Min	Typ	Max	Unit
V_{IO}	Input Offset Voltage		$R_S = 50$ Ω, $V_{CM} = 0$ V		60	200	µV
$\Delta V_{IO}/\Delta T$	Input Offset Voltage Temperature Sensitivity[1]	Without External Trim	$R_S = 50$ Ω, $V_{CM} = 0$ V		0.3	1.3	µV/°C
		With External Trim	$R_O = 20$ kΩ, $R_S = 50$ Ω		0.3	1.3	
I_{IO}	Input Offset Current		$V_{CM} = 0$ V		1.2	5.6	nA
$\Delta I_{IO}/\Delta T$	Input Offset Current Temperature Sensitivity[1]		$V_{CM} = 0$ V		8.0	50	pA/°C
I_{IB}	Input Bias Current		$V_{CM} = 0$ V		2.0	6.0	nA
$\Delta I_{IB}/\Delta T$	Input Bias Current Temperature Sensitivity[1]		$V_{CM} = 0$ V		13	50	pA/°C
CMR	Common Mode Rejection		$V_{CM} = \pm 13$ V, $R_S = 50$ Ω	106	123		dB
V_{IR}	Input Voltage Range			± 13.0	± 13.5		V
PSRR	Power Supply Rejection Ratio		$V_{CC} = \pm 3.0$ V to ± 18 V, $R_S = 50$ Ω	94	106		dB
A_{VS}	Large Signal Voltage Gain		$R_L \geqslant 2.0$ kΩ, $V_O = \pm 10$ V	150	400		V/mV
V_{OP}	Output Voltage Swing		$R_L = 2.0$ kΩ	± 12.0	± 12.6		V

µA714E
Electrical Characteristics $T_A = 25°C$, $V_{CC} = \pm 15$ V

Symbol	Characteristic	Condition	Min	Typ	Max	Unit
V_{IO}	Input Offset Voltage	$R_S = 50$ Ω, $V_{CM} = 0$ V		30	75	µV
S	Long Term Input Offset Voltage Stability	$R_S = 50$ Ω, $V_{CM} = 0$ V		0.3		µV/mo
$V_{IO\ adj}$	Input Offset Voltage Adjustment Range	$R_O = 20$ kΩ		± 4.0		mV
I_{IO}	Input Offset Current	$V_{CM} = 0$ V		0.5	3.8	nA
I_{IB}	Input Bias Current	$V_{CM} = 0$ V		1.2	4.0	nA
Z_I	Input Impedance		15	50		MΩ
P_c	Power Consumption	$V_O = 0$ V		75	120	mW
		$V_{CC} = \pm 3.0$ V, $V_O = 0$ V		4.0	6.0	
CMR	Common Mode Rejection	$V_{CM} = \pm 13$ V, $R_S = 50$ Ω	106	123		dB

μA714

μA714E (Cont.)

Electrical Characteristics $T_A = 25°C$, $V_{CC} = \pm 15$ V

Symbol	Characteristic	Condition	Min	Typ	Max	Unit
V_{IR}	Input Voltage Range		± 13.0	± 14.0		V
PSRR	Power Supply Rejection Ratio	$V_{CC} = \pm 3.0$ V to ± 18 V, $R_S = 50\ \Omega$	94	107		dB
A_{VS}	Large Signal Voltage Gain	$R_L \geqslant 2.0$ kΩ, $V_O = \pm 10$ V	200	500		V/mV
		$R_L \geqslant 500\ \Omega$, $V_O = \pm 0.5$ V $V_{CC} = \pm 3.0$ V	150	500		
V_{OP}	Output Voltage Swing	$R_L = 10$ kΩ	± 12.5	± 13.0		V
		$R_L = 2.0$ kΩ	± 12.0	± 12.8		
		$R_L = 1.0$ kΩ	± 10.5	± 12.0		
BW	Bandwidth	$A_V = 1.0$		0.6		MHz
SR	Slew Rate	$R_L = 2.0$ kΩ, $A_V = 1.0$		0.17		V/μs
e_n	Input Noise Voltage[1]	0.1 Hz to 1.0 kHz		0.35	0.6	μV p—p
	Input Noise Voltage Density[1]	$f_o = 10$ Hz		10.3	18.0	nV/$\sqrt{Hz}$
		$f_o = 100$ Hz		10.0	13.0	
		$f_o = 1000$ Hz		9.6	11.0	
i_n	Input Noise Current[1]	0.1 Hz to 1.0 kHz		14	30	pA p—p
	Input Noise Current Density[1]	$f_o = 10$ Hz		0.32	0.80	pA/$\sqrt{Hz}$
		$f_o = 100$ Hz		0.14	0.23	
		$f_o = 1000$ Hz		0.12	0.17	

The following specifications apply for $V_{CC} = \pm 15$ V, $0°C \leqslant T_A \leqslant 70°C$

Symbol	Characteristic		Condition	Min	Typ	Max	Unit
V_{IO}	Input Offset Voltage		$R_S = 50\ \Omega$, $V_{CM} = 0$ V		45	130	μV
$\Delta V_{IO}/\Delta T$	Input Offset Voltage Temperature Sensitivity[1]	Without External Trim	$R_S = 50\ \Omega$, $V_{CM} = 0$ V		0.3	1.3	μV/°C
		With External Trim	$R_O = 20$ kΩ, $R_S = 50\ \Omega$		0.3	1.3	
I_{IO}	Input Offset Current		$V_{CM} = 0$ V		0.9	5.3	nA
$\Delta I_{IO}/\Delta T$	Input Offset Current Temperature Sensitivity[1]		$V_{CM} = 0$ V		8.0	35	pA/°C
I_{IB}	Input Bias Current		$V_{CM} = 0$ V		1.5	5.5	nA
$\Delta I_{IB}/\Delta T$	Input Bias Current Temperature Sensitivity[1]		$V_{CM} = 0$ V		13	35	pA/°C
CMR	Common Mode Rejection		$V_{CM} = \pm 13$ V, $R_S = 50\ \Omega$	103	123		dB

513

μA714

Electrical Characteristics $T_A = 25°C$, $V_{CC} = \pm 15$ V

Symbol	Characteristic	Condition	Min	Typ	Max	Unit
V_{IR}	Input Voltage Range		± 13.0	± 13.5		V
PSRR	Power Supply Rejection Ratio	$V_{CC} = \pm 3.0$ V to ± 18 V, $R_S = 50$ Ω	90	104		dB
A_{VS}	Large Signal Voltage Gain	$R_L \geqslant 2.0$ kΩ, $V_O = \pm 10$ V	180	450		V/mV
V_{OP}	Output Voltage Swing	$R_L = 2.0$ kΩ	± 12.0	± 12.6		V

μA714C

Electrical Characteristics $T_A = 25°C$, $V_{CC} = 15$ V

Symbol	Characteristic	Condition	Min	Typ	Max	Unit
V_{IO}	Input Offset Voltage	$R_S = 50$ Ω, $V_{CM} = 0$ V		60	150	μV
S	Long Term Input Offset Voltage Stability	$R_S = 50$ Ω, $V_{CM} = 0$ V		0.4	2.0	μV/mo
$V_{IO\ adj}$	Input Offset Voltage Adjustment Range	$R_O = 20$ kΩ		± 4.0		mV
I_{IO}	Input Offset Current	$V_{CM} = 0$ V		0.8	6.0	nA
I_{IB}	Input Bias Current	$V_{CM} = 0$ V		1.8	7.0	nA
Z_I	Input Impedance		8.0	33		MΩ
P_c	Power Consumption	$V_O = 0$ V		80	150	mW
		$V_{CC} = \pm 3.0$ V, $V_O = 0$ V		4.0	8.0	
CMR	Common Mode Rejection	$V_{CM} = \pm 13$ V, $R_S = 50$ Ω	100	120		dB
V_{IR}	Input Voltage Range		± 13.0	± 14.0		V
PSRR	Power Supply Rejection Ratio	$V_{CC} = \pm 3.0$ V to ± 18 V, $R_S = 50$ Ω	90	104		dB
A_{VS}	Large Signal Voltage Gain	$R_L \geqslant 2.0$ kΩ, $V_O = \pm 10$ V	120	400		V/mV
		$R_L \geqslant 500$ Ω, $V_O = \pm 0.5$ V $V_{CC} = \pm 3.0$ V	100	400		
V_{OP}	Output Voltage Swing	$R_L = 10$ kΩ	± 12.0	± 13.0		V
		$R_L = 2.0$ kΩ	± 11.5	± 12.8		
		$R_L = 1.0$ kΩ		± 12.0		
BW	Bandwidth	$A_V = 1.0$		0.6		MHz
SR	Slew Rate	$R_L = 2.0$ kΩ, $A_V = 1.0$		0.17		V/μs

μA714C (Cont.)
Electrical Characteristics $T_A = 25°C$, $V_{CC} = 15$ V

Symbol	Characteristic		Condition	Min	Typ	Max	Unit
e_n	Input Noise Voltage[1]		0.1 Hz to 1.0 kHz		0.38	0.65	μVp — p
	Input Noise Voltage Density[1]		f_o = 10 Hz		10.5	20.0	nV/$\sqrt{Hz}$
			f_o = 100 Hz		10.2	13.5	
			f_o = 1000 Hz		9.8	11.5	
i_n	Input Noise Current[1]		0.1 Hz to 1.0 kHz		0.15	35	μVp — p
	Input Noise Current Density[1]		f_o = 10 Hz		0.35	0.90	pA/$\sqrt{Hz}$
			f_o = 100 Hz		0.15	0.27	
			f_o = 1000 Hz		0.13	0.18	

The following specifications apply for $V_{CC} = \pm 15$ V, $0°C \leqslant T_A \leqslant 70°C$

Symbol	Characteristic		Condition	Min	Typ	Max	Unit
V_{IO}	Input Offset Voltage		R_S = 50 Ω, V_{CM} = 0 V		85	250	μV
$\Delta V_{IO}/\Delta T$	Input Offset Voltage Temperature Sensitivity[1]	Without External Trim	R_S = 50 Ω, V_{CM} = 0 V		0.5	1.8	μV/°C
		With External Trim	R_O = 20 kΩ, R_S = 50 Ω		0.4	1.6	
I_{IO}	Input Offset Current		V_{CM} = 0 V		1.6	8.0	nA
$\Delta I_{IO}/\Delta T$	Input Offset Current Temperature Sensitivity[1]		V_{CM} = 0 V		12	50	pA/°C
I_{IB}	Input Bias Current		V_{CM} = 0 V		2.2	9.0	nA
$\Delta I_{IB}/\Delta T$	Input Bias Current Temperature Sensitivity[1]		V_{CM} = 0 V		18	50	pA/°C
CMR	Common Mode Rejection		$V_{CM} = \pm 13$ V, R_S = 50 Ω	97	120		dB
V_{IR}	Input Voltage Range			± 13.0	± 13.5		V
PSRR	Power Supply Rejection Ratio		$V_{CC} = \pm 3.0$ V to ± 18 V, R_S = 50 Ω	86	100		dB
A_{VS}	Large Signal Voltage Gain		$R_L \geqslant 2.0$ kΩ, $V_O = \pm 10$ V	100	400		V/mV
V_{OP}	Output Voltage Swing		R_L = 2.0 kΩ	± 11.0	± 12.6		V

µA714

µA714L
Electrical Characteristics $T_A = 25°C$, $V_{CC} = \pm 15$ V

Symbol	Characteristic	Condition	Min	Typ	Max	Unit
V_{IO}	Input Offset Voltage	$R_S = 50$ Ω, $V_{CM} = 0$ V		100	250	µV
S	Long Term Input Offset Voltage Stability	$R_S = 50$ Ω, $V_{CM} = 0$ V		0.5	3.0	µV/mo
$V_{IO\ adj}$	Input Offset Voltage Adjustment Range	$R_O = 20$ kΩ		±4.0		mV
I_{IO}	Input Offset Current	$V_{CM} = 0$ V		5.0	20	nA
I_{IB}	Input Bias Current	$V_{CM} = 0$ V		6.0	30	nA
Z_I	Input Impedance		8.0	33		MΩ
P_c	Power Consumption	$V_O = 0$ V		100	180	mW
		$V_{CC} = \pm 3.0$ V, $V_O = 0$ V		5.0	12	
CMR	Common Mode Rejection	$V_{CM} = \pm 13$ V, $R_S = 50$ Ω	100	120		dB
V_{IR}	Input Voltage Range		±13.0	±14.0		V
PSRR	Power Supply Rejection Ratio	$V_{CC} = \pm 3.0$ V to ±18 V, $R_S = 50$ Ω	90	104		dB
A_{VS}	Large Signal Voltage Gain	$R_L \geqslant 2.0$ kΩ, $V_O = \pm 10$ V	100	300		V/mV
		$R_L \geqslant 500$ Ω, $V_O = \pm 0.5$ V, $V_{CC} = \pm 3.0$ V	50	150		
V_{OP}	Output Voltage Swing	$R_L = 10$ kΩ	±12.0	±13.0		V
		$R_L = 2.0$ kΩ	±11.0	±12.8		
		$R_L = 1.0$ kΩ		±12.0		
BW	Bandwidth	$A_V = 1.0$		0.6		MHz
SR	Slew Rate	$R_L = 2.0$ kΩ, $A_V = 1.0$		0.17		V/µs
e_n	Input Noise Voltage[1]	0.1 Hz to 1.0 kHz		0.5		µV p—p
	Input Noise Voltage Density[1]	$f_o = 10$ Hz		10.5		nV/√Hz
		$f_o = 100$ Hz		10.2		
		$f_o = 1000$ Hz		9.8		
i_n	Input Noise Current[1]	0.1 Hz to 1.0 kHz		0.15		pA p—p
	Input Noise Current Density[1]	$f_o = 10$ Hz		0.35		pA/√Hz
		$f_o = 100$ Hz		0.15		
		$f_o = 1000$ Hz		0.13		

µA714L (Cont.)
Electrical Characteristics 0°C $\leqslant T_A \leqslant$ +70°C, V_{CC} = ± 15 V

Symbol	Characteristic		Condition	Min	Typ	Max	Unit
V_{IO}	Input Offset Voltage		R_S = Ω, V_{CM} = 0 V			400	µV
$\Delta V_{IO}/\Delta T$	Input Offset Voltage Temperature Sensitivity[1]	Without External Trim	R_S = 50 Ω, V_{CM} = 0 V		1.0	3.0	µV/°C
		With External Trim	R_O = 20 kΩ, R_S = 50 Ω		1.3		
I_{IO}	Input Offset Current		V_{CM} = 0 V		8.0	40	nA
$\Delta I_{IO}/\Delta T$	Input Offset Current Temperature Sensitivity[1]		V_{CM} = 0 V		20	100	pA/°C
I_{IB}	Input Bias Current		V_{CM} = 0 V		15	60	nA
$\Delta I_{IB}/\Delta T$	Input Bias Current Temperature Sensitivity[1]		V_{CM} = 0 V		35	150	pA/°C
CMR	Common Mode Rejection		V_{CM} = ± 13 V, R_S = 50 Ω	94	120		dB
V_{IR}	Input Voltage Range			± 13.0	± 13.5		V
PSRR	Power Supply Rejection Ratio		V_{CC} = ± 3.0 V to ± 18 V, R_S = 50 Ω	83	100		dB
A_{VS}	Large Signal Voltage Gain		$R_L \geqslant$ 2.0 kΩ, V_O = ± 10 V	80	400		V/mV
V_{OP}	Output Voltage Swing		R_L = 2.0 kΩ	± 10.0	± 12.6		V

Note
1. Parameter is not 100% tested; 90% of the units meet this specification.

Typical Performance Curves

Untrimmed Offset Voltage vs Temperature

Trimmed Offset Voltage vs Temperature

Offset Voltage Stability vs Time

Voltage Gain vs Temperature

Offset Voltage Change Due to Thermal Shock

Warm-Up Drift

Maximum Error vs Source Resistance

Maximum Error vs Source Resistance

Maximum Error vs Source Resistance

Typical Performance Curves (Cont.)

Input Bias Current vs Temperature

PC05870F

Input Bias Current vs Differential Input Voltage

PC05891F

Input Offset Current vs Temperature

PC05880F

Input Noise Voltage vs Frequency

PC05901F

Input Wideband Noise vs Bandwidth (0.1 Hz to Frequency Indicated)

PC05911F

CMR vs Frequency

PC05821F

PSRR vs Frequency

PC05931F

Voltage Gain vs Supply Voltage

PC05941F

Open Loop Frequency Response

PC05951F

μA714

Typical Performance Curves (Cont.)

Frequency Response For Various Closed Loop Gains

PC05961F

Maximum Undistorted Output vs Frequency

PC05971F

Output Voltage vs Load Resistance

PC05981F

Power Consumption vs Supply Voltage

PC05991F

Short Circuit Current vs Time

PC06001F

Test Circuits

Offset Voltage Test Circuit

200 kΩ

50 Ω

2

3

μA714

6

V$_O$

$V_{IO} = \dfrac{V_O}{4000}$

CR01440F

Optional Offset Nulling Circuit

R$_O$
20 kΩ

V+

1

2

8

7

IN

3

μA714

6

OUT

4

V−

CR01451F

Low Frequency Noise Test Circuit

+15V

100 Ω

2

7

100 Ω

3

μA714

6

3.3 kΩ

OUT

4

4.7 μF

−15 V

2.5 MΩ

(≈ 10 Hz Filter)

CR01461F

Input Referred Noise = $\dfrac{V_O}{25,000}$

TELEDYNE SEMICONDUCTOR

TSC9400 (0.05%)
TSC9401 (0.01%)
TSC9402 (0.25%)
Voltage to Frequency/ Frequency to Voltage Converters

General Description

The TSC9400/9401/9402 are low cost Voltage-to-Frequency converters combining Bipolar and CMOS technology on the same substrate. The converters accept a variable analog input signal and generate an output pulse train whose frequency is linearly proportional to the input voltage.

The devices can also be used as highly accurate Frequency-to-Voltage converters, accepting virtually any input frequency waveform and providing a linearly proportional voltage output.

A complete V/F or F/V system requires the addition of two capacitors , three resistors and reference voltage.

Applications

• Voltage-to-Frequency

— Temperature Sensing and Control
— μP Data Acquisition
— Instrumentation
— 13-Bit A/D Converters
— Digital Panel Meters
— Analog Data Transmission and Recording
— Phase Locked Loops
— Medical Isolation
— Transducer Encoding
— Alternate to 555 Astable Timer

• Frequency-to-Voltage

— Frequency Meters/Tachometer
— Speedometers
— Analog Data Transmission and Recording
— Medical Isolation
— Motor Control
— RPM Indicator
— FM Demodulation
— Frequency Multiplier/Divider
— Flow Measurement and Control

Features

• Voltage-to-Frequency

— 1Hz to 100kHz Operation
— Choice of Guaranteed Linearity:
  TSC9401   .01%
  TSC9400   .05%
  TSC9402   .25%
— ±25ppm/°C Typ. Gain Temperature Stability
— Open Collector Output
— Output can Interface with any Form of Logic
— Pulse and Square Wave Outputs
— Programmable Scale Factor
— Low Power Dissipation (27mW Typ.)
— Single Supply Operation (8V to 15V)
— Dual Supply Operation (±4V to ±7.5V)
— Current or Voltage Input

• Frequency-to-Voltage

— DC to 100kHz Operation
— Choice of Guaranteed Linearity:
  TSC9401   0.02%
  TSC9400   0.05%
  TSC9402   0.25%
— Op Amp Output
— Programmable Scale Factor
— High Input Impedance (>10MΩ)
— Accepts any Voltage Wave Shape

HANDLING PRECAUTIONS: The 9400 Series are CMOS Bipolar devices and must be handled correctly to prevent damage. Package and store only in conductive foam, anti-static tubes or other conductive material. Use proper anti-static handling procedures. Do not connect in circuits under "power on" conditions, as high transients may cause permanent damage.

Ordering Information

Part No.	Linearity (V/F)	Package	Temperature Range
TSC9400CJ	0.05%	14-Pin Plastic Dip	0°C to +70°C
TSC9400CL	0.05%	14-Pin CerDIP	-40°C to +85°C
TSC9401CJ	0.01%	14-Pin Plastic Dip	0°C to +70°C
TSC9401CL	0.01%	14-Pin CerDIP	-40°C to +85°C
TSC9402CJ	0.25%	14-Pin Plastic Dip	0°C to +70°C
TSC9402CL	0.25%	14-Pin CerDIP	-40°C to +85°C

Pin Configuration

NOTE: PIN 1 INDICATED BY ADJACENT DOT (N PACKAGE), INDENT OR END NOTCH (J PACKAGE).

TELEDYNE SEMICONDUCTOR

9400, 9401, 9402

Absolute Maximum Ratings

Storage Temperature .... -65°C to +150°C
Operating Temperature
  J Package .... 0°C to 70°C
  L Package .... -40°C to +85°C
$V_{DD} - V_{SS}$ .... 18 V

I_{IN} .... 10 mA
V_{OUT} Max — V_{OUT} Common .... 25 V
$V_{REF} - V_{SS}$ .... -1.5 V
Package Dissipation .... 500 mW
Lead Temperature (Soldering, 10 sec) .... 300°C

Electrical Characteristics, V/F Mode

Unless otherwise specified, $V_{DD} = +5V$, $V_{SS} = -5V$, $V_{GND} = 0$, $V_{REF} = -5V$, $R_{BIAS} = 100K\Omega$, Full Scale = 10KHz. $T_A = 25°C$ unless
Full Temp. Range is specified (—40°C to +85°C for L package, 0°C to 70°C for J package).

VOLTAGE-TO-FREQUENCY		TSC9401			TSC9400			TSC9402				
Parameter	Definition	Min.	Typ.	Max.	Min.	Typ.	Max.	Min.	Typ.	Max.	Units	Notes
Accuracy Linearity (10KHz)	Output Deviation from Straight Line between Normalized Zero and Full Scale Input		0.004	**0.01**		0.01	**0.05**		0.05	**0.25**	% Full Scale	
Linearity (100KHz)			0.04	0.08		0.1	0.25		0.25	0.50	% Full Scale	
Gain Temperature Drift	Variation in Gain (A) due to Temperature Change		±25	±40		±25	±40		±50	±100	ppm/°C Full Scale	1
Gain Variance	Variation from Exact A (Compensate by Trimming R_{IN}, V_{REF}, or C_{REF})		±10			±10			±10		% of Nominal	
Zero Offset	Correction at Zero Adjust for Zero Output When Input is Zero		±10	±50		±10	±50		±20	±100	mV	2
Zero Temperature Drift	Variation in Zero Offset Due to Temperature Change		±25	±50		±25	±50		±50	±100	μV/°C	1
Analog Inputs I_{IN} Full Scale	Full Scale Analog Input Current to Achieve Specified Accuracy		10			10			10		μA	
I_{IN} Overrange	Overrange Current			50			50			50	μA	
Response Time	Settling Time to 0.01% Full Scale		2			2			2		Cycles	
Digital Outputs V_{SAT} @ $I_{OL} = 10\mu A$	Logical "0" Output Voltage			0.4			0.4			0.4	V	3
V_{OUT} Max. — V_{OUT} Common	Voltage Range between Output and Common			18.0			18.0			18.0	V	4
Pulse Frequency Output Width			3.0			3.0			3.0		μsec	
Supply Current I_{DD} Quiescent (L Package) (J Package)	Current Required from Positive Supply During Operation		2.0 2.0	4.0 6.0		2.0 2.0	4.0 6.0		3.0	10.0	mA mA	9
I_{SS} Quiescent (L Package) (J Package)	Current Required from Negative Supply During Operation		-1.5 -1.5	-4.0 -6.0		-1.5 -1.5	-4.0 -6.0		-3.0	-10.0	mA mA	10
V_{DD} Supply	Operating Range of Positive Supply	4.0		7.5	4.0		7.5	4.0		7.5	V	
V_{SS} Supply	Operating Range of Negative Supply	-4.0		-7.5	-4.0		-7.5	-4.0		-7.5	V	
Reference Voltage $V_{REF} - V_{SS}$	Range of Voltage Reference Input	-1.0			-1.0			-1.0			V	

NOTES:
1. Full temperature range.
2. $I_{IN} = 0$.
3. Full temperature range. $I_{OUT} = 10mA$.
4. $I_{OUT} = 10\mu A$.
5. 10Hz to 100KHz.
6. 5μs min. positive pulse width and 0.5μs min. negative pulse width.

7. $T_r = t_f = 20ns$.
8. $R_L \geq 2K\Omega$.
9. Full temperature range, $V_{IN} = -0.1V$.
10. $V_{IN} = -0.1V$.
11. I_{IN} connects the summing junction of an operational amplifier. Voltage sources cannot be attached directly but must be buffered by external resistors.

☀ TELEDYNE SEMICONDUCTOR

V/F Circuit Description

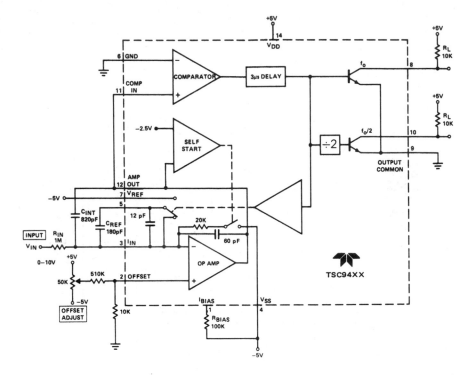

Figure 1. 10Hz to 10KHz V/F Converter

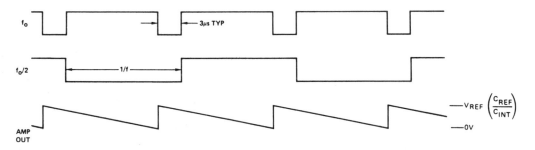

1. To adjust f_{min}, set V_{IN} =10mV and adjust the 50K offset for 10Hz out.

2. To adjust f_{max}, set V_{IN} =10V and adjust R_{IN} or V_{REF} for 10KHz out.

3. To increase f_{OUT}MAX to 100KHz change C_{REF} to 27pF and C_{INT} to 75pF.

4. For high performance applications use high stability components for R_{IN}, C_{REF}, V_{REF} (metal film resistors and glass film capacitors). Also separate the output ground (Pin 9) from the input ground (Pin 6).

Figure 2. Output Waveforms

9400, 9401, 9402

V/F Circuit Description (Contd.)

The Teledyne 9400 V/F Converter operates on the principal of charge balancing. The input voltage (V_{IN}) is converted to a current (I_{IN}) by the input resistor. This current is then converted to a charge by the integrating capacitor and shows up as a linearly decreasing voltage at the output of the op amp. The zero crossing of the output is sensed by the comparator causing the reference voltage to be applied to the reference capacitor for a time period long enough to virtually charge the capacitor to the reference voltage. This action reduces the charge on the integrating capacitor by a fixed amount ($q = C_{REF} \times V_{REF}$) causing the op amp output to step up a finite amount.

At the end of the charging period, C_{REF} is shorted out dissipating the stored reference charge so that when the output again crosses zero, the system is ready to recycle. In this manner, the continued discharging of the integrating capacitor by the input is balanced out by fixed charges from the reference voltage. As the input voltage is increased, the number of reference pulses required to maintain balance increases causing the output frequency to also increase. Since each charge increment is fixed the increase in frequency with voltage is linear. In addition, the accuracy of the output pulses does not directly effect the linearity of the V/F. It must simply be long enough for full charge transfer to take place.

The 9400 contains a "self-start" circuit to assure that the V/F will always operate properly when power is first applied. In the event that during "Power-on" the op amp output is below the comparator threshold and C_{REF} is already charged, a positive voltage step will not occur. The op amp output will continue to decrease until it crosses the −2.5 volt threshold of the "self-start" comparator. When this happens a resistor is connected to the op amp input causing the output to quickly go positive until the 9400 is once again in its normal operating mode.

The 9400 utilizes both bipolar and MOS transistors on the same substrate, taking advantage of the best features of each. MOS transistors are used at the inputs to reduce offset and bias currents. Bipolar transistors are used in the op amp, for high gain, and on all outputs for excellent current driving capabilities. CMOS logic is used throughout to minimize power consumption.

Pin Functions

Comparator Input − In the V/F mode, this input is connected to the amplifier output (pin 12) and triggers the 3μsec pulse delay when the input voltage passes its threshold. In the F/V mode, the input frequency is applied to the comparator input.

Pulse Freq Out − This output is an open-collector bipolar transistor providing a pulse waveform whose frequency is proportional to the input voltage. This output requires a pull up resistor and interfaces directly with MOS, CMOS and TTL logic.

Freq/2 Out − This output is an open-collector bipolar transistor providing a square wave that is one-half the frequency of the pulse frequency output. This output requires a pull up resistor and interfaces directly with MOS, CMOS, and TTL logic.

Output Common − The emitters of both the freq/2 out and the pulse freq out are connected to this pin. An output level swing from the collector voltage to ground or to the V_{SS} supply may be obtained by connecting to the appropriate point.

R_{BIAS} − Specifications for the 9400 are based on R_{BIAS} = 100K ±10% unless otherwise noted. R_{BIAS} may be varied between the range of $82K \le R_{BIAS} \le 120K$.

Amplifier Out − The output stage of the operational amplifier. A negative going ramp signal is available at this pin in the V/F mode. In the F/V mode a voltage proportional to the frequency input is generated.

Zero Adjust − The non-inverting input of the operational amplifier. The low frequency set point is detemined by adjusting the voltage at this pin.

I_{IN} − The inverting input of the operational amplifier and the summing junction when connected in the V/F mode. An input current of 10μA is specified for nominal full scale but an over range current up to 50μA can be used without detrimental effect to the circuit operation.

V_{REF} − A reference voltage from either a precision source or the V_{SS} supply may be applied to this pin. Accuracy will be dependent on the voltage regulation and temperature characteristics of the circuitry.

V_{REF} OUT − The charging current for C_{REF} is derived from the internal circuitry and switched by the break-before-make switch to this pin.

V/F Design Information

Input/Output Relationships − The output frequency is related to the analog input voltage (V_{IN}) by the transfer equation:

$$\text{Frequency Out} = \frac{V_{IN}}{R_{IN}} \times \frac{1}{(V_{REF})(C_{REF})} = f_o$$

External Component Selection

R_{IN} − The value of this component is chosen to give a full scale input current of approximately 10μA.

Example:

$$R_{IN} \cong \frac{V_{IN} \text{ FULL SCALE}}{10\mu A} \qquad R_{IN} \cong \frac{10V}{10\mu A} = 1M\Omega$$

☆ TELEDYNE SEMICONDUCTOR

V/F Design Information (Contd.)

Note that the value is an approximation, and the exact relationship is defined by the transfer equation. In practice, the value of R_{IN} typically would be trimmed to obtain full scale frequency at V_{IN} FULL SCALE (see adjustment procedure). Metal film resistors with 1% tolerance or better are recommended for high accuracy applications because of their thermal stability and low noise generation.

C_{INT} — Exact value not critical but is related to C_{REF} by the relationship:

$$3C_{REF} \leq C_{INT} \leq 10C_{REF}$$

Improved stability and linearity is obtained when $C_{INT} \geq 4C_{REF}$. Low leakage types are recommended although mica and ceramic devices can be used in applications where their temperature limits are not exceeded. Locate as close as possible to pins 12 and 3.

C_{REF} — Exact value not critical and may be used to trim the full scale frequency (see input/output relation). Glass film or air trimmer capacitors are recommended because of their stability and low leakage. Locate as close as possible to pins 5 and 3.

V_{DD}, V_{SS} — Power supplies of ±5V are recommended. For high accuracy requirements 0.05% line and load regulation and 0.1μF disc decoupling capacitors located near the pins are recommended.

Adjustment Procedure — Figure 1 shows a circuit for trimming the zero location. Full scale may be trimmed by adjusting R_{IN}, V_{REF}, or C_{REF}. Recommended procedure is as follows for a 10KHz full scale frequency.

1. Set V_{IN} to 10mV and trim the zero adjust circuit to obtain a 10Hz output frequency.

2. Set V_{IN} to 10.000V and trim either R_{IN}, V_{REF}, or C_{REF} to obtain a 10KHz output frequency.

If adjustments are performed in this order, there should be no interaction and they should not have to be repeated.

Figure 3. Recommended C_{REF} vs V_{REF}

V/F Single Supply Operation

NOTE:
See Also the TSC7660
Data Sheet;

Figure 4. Fixed Voltage—Single Supply Operation

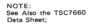

Figure 5. Variable Voltage—Single Supply Operation

9400, 9401, 9402

Figure 6. Single Variable Supply Voltage with Offset and Gain Adjust

Electrical Characteristics, F/V Mode

Unless otherwise specified, V_{DD} = +5V, V_{SS} = —5V, V_{GND} = 0, V_{REF} = —5V, R_{BIAS} = 100KΩ, Full Scale = 10KHz. T_A = 25°C unless Full Temp. Range is specified —40°C to +85°C for **L** package, 0°C to 70°C for J package

FREQUENCY-TO-VOLTAGE		TSC9401			TSC9400			TSC9402				
Parameter	Definition	Min.	Typ.	Max.	Min.	Typ.	Max.	Min.	Typ.	Max.	Units	Notes
Accuracy Non-Linearity	Deviation from Ideal Transfer Function as a Percentage of Full Scale Voltage		0.01	**0.02**		0.02	**0.05**		0.05	**0.25**	% Full Scale	5
Input Frequency Range	Frequency Range for Specified Non-Linearity	10		100K	10		100K	10		100K	Hz	6
Frequency Inputs Positive Excursion	Voltage Required to Turn Comparator On	0.4		V_{DD}	0.4		V_{DD}	0.4		V_{DD}	V	7
Negative Excursion	Voltage Required to Turn Comparator Off	-0.4		-2V	-0.4		-2V	-0.4		-2V	V	7
Min. Positive Pulse Width	Time between Threshold Crossings		5.0			5.0			5.0		μs	7
Min. Negative Pulse Width	Time between Threshold Crossings		0.5			0.5			0.5		μs	7
Input Impedance		10			10			10			MΩ	
Analog Outputs Output Voltage	Voltage Range of Op Amp Output for Specified Non-Linearity		V_{DD}-1			V_{DD}-1			V_{DD}-1		V	8
Output Loading	Resistive Loading at Output of Op Amp	2K			2K			2K			Ω	
Supply Current I_{DD} Quiescent (L Package) (J Package)	Current Required from Positive Supply During Operation		2.0 2.0	4.0 6.0		2.0 2.0	4.0 6.0		3.0	10.0	mA mA	9
I_{SS} Quiescent (L Package) (J Package)	Current Required from Negative Supply During Operation		-1.5 -1.5	-4.0 -6.0		-1.5 -1.5	-4.0 -6.0		-3.0	-10.0	mA mA	10
V_{DD} Supply	Operating Range of Positive Supply	4.0		7.5	4.0		7.5	4.0		7.5	V	
V_{SS} Supply	Operating Range of Negative Supply	-4.0		-7.5	-4.0		-7.5	-4.0		-7.5	V	
Reference Voltage V_{REF}—V_{SS}	Range of Voltage Reference Input	-1.0			-1.0			-1.0			V	

NOTES:
1. Full temperature range.
2. I_{IN} = 0.
3. Full temperature range, I_{OUT} = 10mA.
4. I_{OUT} = 10μA.
5. 10Hz to 100KHz.
6. 5μs min. positive pulse width and 0.5μs min. negative pulse width.

7. T_r = t_f = 20ns.
8. R_L ≥ 2KΩ.
9. Full temperature range, V_{IN} = -0.1V.
10. V_{IN} = -0.1V.
11. I_{IN} connects the summing junction of an operational amplifier. Voltage sources cannot be attached directly but must be buffered by external resistors.

𝍏 TELEDYNE SEMICONDUCTOR

527

F/V Circuit Description

The 9400, when used as a frequency to voltage converter, generates an output voltage which is linearly proportional to the input frequency waveform.

Each zero crossing at the comparator's input causes a precise amount of charge ($q = C_{REF} \times V_{REF}$) to be dispensed into the op amp's summing junction. This charge in turn flows through the feedback resistor generating voltage pulses at the output of the op amp. A capacitor (C_{INT}) across R_{INT} averages these pulses into a DC voltage which is linearly proportional to the input frequency.

Figure 7. DC — 10KHz F/V Converter

F/V Design Information

Input/Output Relationships — The output voltage is related to the input frequency (F_{IN}) by the transfer equation:

$$V_{OUT} = \left[V_{REF}\, C_{REF}\, R_{INT} \right] F_{IN}$$

The response time to a change in F_{IN} is equal to $(R_{INT}\, C_{INT})$. The amount of ripple on V_{OUT} is inversely proportional to C_{INT} and the Input Frequency.

C_{INT} can be increased to lower the ripple. $1\mu F$ to $100\mu F$ are perfectly acceptable values for low frequencies.

When 9400 is used in the single supply mode, V_{REF} is defined as the voltage difference between Pin 7 and Pin 2.

Input Voltage Levels — The input signal must cross through zero in order to trip the comparator. In order to overcome the hysteresis the amplitude must be greater than $\pm 200mV$.

If only a unipolar input signal (F_{IN}) is available, it is recommended that either an offset circuit using resistor be used or that the signal be coupled in via a capacitor.

Note: C_{REF} should be increased for lower F_{IN} max. Adjust C_{REF} so that V_0 is approximately 2.5 to 3.0 volts for the maximum input frequency. When F_{IN} max is less than 1 kHz, the duty cycle should be greater than 20% to insure that C_{REF} is fully charged and discharged.

✵ TELEDYNE SEMICONDUCTOR

9400, 9401, 9402
F/V Design Information (Contd.)

For 100KHz maximum input R_{INT} should be decreased to 100KΩ.

Input Buffer — f_O and $f_O/2$ are not used in the F/V mode. However, these outputs may be useful for some applications, such as a buffer to feed additional circuitry. f_O will then follow the input frequency waveform; except that f_O will go high 3μs after F_{IN} goes high. $f_O/2$ will be square wave with a frequency of one half f_O.

If these outputs are not used, then Pins 8, 9 and 10 may be left floating or connected to ground.

Figure 8. F/V Digital Outputs

The sawtooth ripple which is on the output of an F/V can be eliminated without affecting the F/V's response time by using the circuit in Figure 10. The circuit has a DC gain of +1. Any AC components such as a ripple are amplified both positively, via the lower path, and negatively, via the upper path. When both paths have the same gain, the AC ripple is cancelled. The amount of cancellation is directly proportional to gain matching. If the two paths are matched within 10%, then the ripple will be lowered by 1/10. For 1% matching, the ripple is lowered by 1/100. The 10K potentiometer is used to make the gain equal in both paths. This circuit is insensitive to both frequency changes and to signal wave shape.

Package Information

NOTE:
See Also the TSC7660 Data Sheet.

1. The input is now referenced to 6.2 V (Pin 6). The input signal must therefore be restricted to be greater than 4 volts (Pin 6 −2V) and less than 10 to 15V (V_{DD}).

If the signal is AC coupled then a resistor (100K to 10MΩ) must be placed between the input (Pin 11) and Pin 6.

2. The output will now be referenced to Pin 6 which is at 6.2 V (V_Z). For frequency meter applications a 1mA meter with a series scaling resistor can be placed across Pins 6 and 12.

Figure 9. F/V Single Supply

Figure 10. F/V Ripple Eliminator

(Package #6)
14-Pin Plastic Dip (J Package)

(Package #7)
14-Pin CerDIP (L Package)

⚡ TELEDYNE SEMICONDUCTOR

MOTOROLA

Specifications and Applications Information

EIGHT-BIT MULTIPLYING DIGITAL-TO-ANALOG CONVERTER

. . . designed for use where the output current is a linear product of an eight-bit digital word and an analog input voltage.

- Eight-Bit Accuracy Available in Both Temperature Ranges
  Relative Accuracy: ±0.19% Error maximum
  (MC1408L8, MC1408P8, MC1508L8)
- Seven and Six-Bit Accuracy Available with MC1408 Designated by 7 or 6 Suffix after Package Suffix
- Fast Settling Time — 300 ns typical
- Noninverting Digital Inputs are MTTL and CMOS Compatible
- Output Voltage Swing — +0.4 V to –5.0 V
- High-Speed Multiplying Input
  Slew Rate 4.0 mA/μs
- Standard Supply Voltages: +5.0 V and –5.0 V to –15 V

EIGHT-BIT MULTIPLYING DIGITAL-TO-ANALOG CONVERTER

SILICON MONOLITHIC INTEGRATED CIRCUIT

L SUFFIX
CERAMIC PACKAGE
CASE 620-02

P SUFFIX
PLASTIC PACKAGE
CASE 648-05

D SUFFIX
PLASTIC PACKAGE
CASE 751B-01
SO-16*

*Surface Mount pin-out is different than standard package pin-out for this device. Please refer to Chapter 7 for details.

FIGURE 1 – D-to-A TRANSFER CHARACTERISTICS

FIGURE 2 – BLOCK DIAGRAM

TYPICAL APPLICATIONS

- Tracking A-to-D Converters
- Successive Approximation A-to-D Converters
- 2 1/2 Digit Panel Meters and DVM's
- Waveform Synthesis
- Sample and Hold
- Peak Detector
- Programmable Gain and Attenuation
- CRT Character Generation
- Audio Digitizing and Decoding
- Programmable Power Supplies
- Analog-Digital Multiplication
- Digital-Digital Multiplication
- Analog-Digital Division
- Digital Addition and Subtraction
- Speech Compression and Expansion
- Stepping Motor Drive

MC1408, MC1508

MAXIMUM RATINGS (T$_A$ = +25°C unless otherwise noted.)

Rating	Symbol	Value	Unit
Power Supply Voltage	V$_{CC}$ V$_{EE}$	+5.5 −16.5	Vdc
Digital Input Voltage	V$_5$ thru V$_{12}$	0 to +5.5	Vdc
Applied Output Voltage	V$_O$	+0.5,−5.2	Vdc
Reference Current	I$_{14}$	5.0	mA
Reference Amplifier Inputs	V$_{14}$,V$_{15}$	V$_{CC}$,V$_{EE}$	Vdc

Operating Temperature Range
    MC1508
    MC1408 Series
	T$_A$	
		−55 to +125
		0 to +75
			^{O}C

Storage Temperature Range	T$_{stg}$	−65 to +150	^{O}C

ELECTRICAL CHARACTERISTICS (V$_{CC}$ = +5.0 Vdc, V$_{EE}$ = −15 Vdc, $\frac{Vref}{R14}$ = 2.0 mA, MC1508L8: T$_A$ = −55°C to +125°C.
MC1408L Series: T$_A$ = 0 to +75°C unless otherwise noted. All digital inputs at high logic level.)

Characteristic	Figure	Symbol	Min	Typ	Max	Unit

Relative Accuracy (Error relative to full scale I$_O$)
  MC1508L8, MC1408L8, MC1408P8
  MC1408P7, MC1408L7, See Note 1
  MC1408P6, MC1408L6, See Note 1
	4	E$_r$	
			−
			−
			−
				
				−
				−
				−
					
					±0.19
					±0.39
					±0.78
						%

Settling Time to within ±1/2 LSB [includes t$_{PLH}$](T$_A$=+25°C)See Note 2	5	t$_S$	−	300	−	ns
Propagation Delay Time T$_A$ = +25°C	5	t$_{PLH}$,t$_{PHL}$	−	30	100	ns
Output Full Scale Current Drift		TCI$_O$	−	−20	−	PPM/^{O}C

Digital Input Logic Levels (MSB)
  High Level, Logic "1"
  Low Level, Logic "0"
	3	
		V$_{IH}$
		V$_{IL}$
			
			2.0
			−
				
				−
				−
					
					−
					0.8
						Vdc

Digital Input Current (MSB)
  High Level, V$_{IH}$ = 5.0 V
  Low Level, V$_{IL}$ = 0.8 V
	3	
		I$_{IH}$
		I$_{IL}$
			
			−
			−
				
				0
				−0.4
					
					0.04
					−0.8
						mA

Reference Input Bias Current (Pin 15)	3	I$_{15}$	−	−1.0	−5.0	µA

Output Current Range
  V$_{EE}$ = −5.0 V
  V$_{EE}$ = −15 V, T$_A$ = 25°C
	3	I$_{OR}$	
			0
			0
				
				2.0
				2.0
					
					2.1
					4.2
						mA

Output Current V$_{ref}$ = 2.000 V, R14 = 1000 Ω	3	I$_O$	1.9	1.99	2.1	mA
Output Current (All bits low)	3	I$_{O(min)}$	−	0	4.0	µA

Output Voltage Compliance (E$_r$ ≤ 0.19% at T$_A$ = +25°C)
  Pin 1 grounded
  Pin 1 open, V$_{EE}$ below −10 V
	3	V$_O$	
			−
			−
				
				−
				−
					
					−0.55, +0.4
					−5.0, +0.4
						Vdc

Reference Current Slew Rate	6	SR I$_{ref}$	−	4.0	−	mA/µs
Output Current Power Supply Sensitivity		PSRR(−)	−	0.5	2.7	µA/V
Power Supply Current (All bits low)	3	I$_{CC}$ I$_{EE}$	− −	+13.5 −7.5	+22 −13	mA
Power Supply Voltage Range (T$_A$ = +25°C)	3	V$_{CCR}$ V$_{EER}$	+4.5 −4.5	+5.0 −15	+5.5 −16.5	Vdc

Power Dissipation
  All bits low
    V$_{EE}$ = −5.0 Vdc
    V$_{EE}$ = −15 Vdc
  All bits high
    V$_{EE}$ = −5.0 Vdc
    V$_{EE}$ = −15 Vdc
	3	P$_D$	
			
			−
			−
			
			−
			−
				
				105
				190
				
				90
				160
					
					170
					305
					
					−
					−
						mW

Note 1.   All current switches are tested to guarantee at least 50% of rated output current.
Note 2.   All bits switched.

TEST CIRCUITS

FIGURE 3 – NOTATION DEFINITIONS TEST CIRCUIT

Typical Values: R14 = R15 = 1 k
V_{ref} = +2.0 V
C = 15 pF

V_I and i_j apply to inputs A1 thru A8

The resistor tied to pin 15 is to temperature compensate the bias current and may not be necessary for all applications.

$$I_O = K \left\{ \frac{A1}{2} + \frac{A2}{4} + \frac{A3}{8} + \frac{A4}{16} + \frac{A5}{32} + \frac{A6}{64} + \frac{A7}{128} + \frac{A8}{256} \right\}$$

where $K \cong \dfrac{V_{ref}}{R14}$

and A_N = "1" if A_N is at high level
A_N = "0" if A_N is at low level

FIGURE 4 – RELATIVE ACCURACY TEST CIRCUIT

FIGURE 5 – TRANSIENT RESPONSE and SETTLING TIME

TEST CIRCUITS (continued)

FIGURE 6 — REFERENCE CURRENT SLEW RATE MEASUREMENT

FIGURE 7 — POSITIVE V$_{ref}$

See text for values of C.

FIGURE 8 — NEGATIVE V$_{ref}$

See text for values of C.

MOTOROLA LINEAR/INTERFACE DEVICES

**FIGURE 9 — MC1408, MC1508 SERIES EQUIVALENT
CIRCUIT SCHEMATIC**

CIRCUIT DESCRIPTION

The MC1408 consists of a reference current amplifier, an R-2R ladder, and eight high-speed current switches. For many applications, only a reference resistor and reference voltage need be added.

The switches are noninverting in operation, therefore a high state on the input turns on the specified output current component. The switch uses current steering for high speed, and a termination amplifier consisting of an active load gain stage with unity gain feedback. The termination amplifier holds the parasitic capacitance of the ladder at a constant voltage during switching, and provides a low impedance termination of equal voltage for all legs of the ladder.

The R-2R ladder divides the reference amplifier current into binarily-related components, which are fed to the switches. Note that there is always a remainder current which is equal to the least significant bit. This current is shunted to ground, and the maximum output current is 255/256 of the reference amplifier current, or 1.992 mA for a 2.0 mA reference amplifier current if the NPN current source pair is perfectly matched.

MOTOROLA LINEAR/INTERFACE DEVICES

MC1408, MC1508

GENERAL INFORMATION

Reference Amplifier Drive and Compensation

The reference amplifier provides a voltage at pin 14 for converting the reference voltage to a current, and a turn-around circuit or current mirror for feeding the ladder. The reference amplifier input current, I14, must always flow into pin 14 regardless of the setup method or reference voltage polarity.

Connections for a positive reference voltage are shown in Figure 7. The reference voltage source supplies the full current I14. For bipolar reference signals, as in the multiplying mode, R15 can be tied to a negative voltage corresponding to the minimum input level. It is possible to eliminate R15 with only a small sacrifice in accuracy and temperature drift. Another method for bipolar inputs is shown in Figure 25.

The compensation capacitor value must be increased with increases in R14 to maintain proper phase margin; for R14 values of 1.0, 2.5 and 5.0 kilohms, minimum capacitor values are 15, 37, and 75 pF. The capacitor should be tied to V_{EE} as this increases negative supply rejection.

A negative reference voltage may be used if R14 is grounded and the reference voltage is applied to R15 as shown in Figure 8. A high input impedance is the main advantage of this method. Compensation involves a capacitor to V_{EE} on pin 16, using the values of the previous paragraph. The negative reference voltage must be at least 3.0-volts above the V_{EE} supply. Bipolar input signals may be handled by connecting R14 to a positive reference voltage equal to the peak positive input level at pin 15.

When a dc reference voltage is used, capacitive bypass to ground is recommended. The 5.0-V logic supply is not recommended as a reference voltage. If a well regulated 5.0-V supply which drives logic is to be used as the reference, R14 should be decoupled by connecting it to +5.0 V through another resistor and bypassing the junction of the two resistors with 0.1 µF to ground. For reference voltages greater than 5.0 V, a clamp diode is recommended between pin 14 and ground.

If pin 14 is driven by a high impedance such as a transistor current source, none of the above compensation methods apply and the amplifier must be heavily compensated, decreasing the overall bandwidth.

Output Voltage Range

The voltage on pin 4 is restricted to a range of -0.55 to +0.4 volts at +25°C, due to the current switching methods employed in the MC1408. When a current switch is turned "off", the positive voltage on the output terminal can turn "on" the output diode and increase the output current level. When a current switch is turned "on", the negative output voltage range is restricted. The base of the termination circuit Darlington transistor is one diode voltage below ground when pin 1 is grounded, so a negative voltage below the specified safe level will drive the low current device of the Darlington into saturation, decreasing the output current level.

The negative output voltage compliance of the MC1408 may be extended to -5.0 V volts by opening the circuit at pin 1. The negative supply voltage must be more negative than -10 volts. Using a full scale current of 1.992 mA and load resistor of 2.5 kilohms between pin 4 and ground will yield a voltage output of 256 levels between 0 and -4.980 volts. Floating pin 1 does not affect the converter speed or power dissipation. However, the value of the load resistor determines the switching time due to increased voltage swing. Values of R_L up to 500 ohms do not significantly affect performance, but a 2.5-kilohm load increases "worst case" settling time to 1.2 µs (when all bits are switched on).

Refer to the subsequent text section on Settling Time for more details on output loading.

If a power supply value between -5.0 V and -10 V is desired, a voltage of between 0 and -5.0 V may be applied to pin 1. The value of this voltage will be the maximum allowable negative output swing.

Output Current Range

The output current maximum rating of 4.2 mA may be used only for negative supply voltages typically more negative than -8.0 volts, due to the increased voltage drop across the 350-ohm resistors in the reference current amplifier.

Accuracy

Absolute accuracy is the measure of each output current level with respect to its intended value, and is dependent upon relative accuracy and full scale current drift. Relative accuracy is the measure of each output current level as a fraction of the full scale current. The relative accuracy of the MC1408 is essentially constant with temperature due to the excellent temperature tracking of the monolithic resistor ladder. The reference current may drift with temperature, causing a change in the absolute accuracy of output current. However, the MC1408 has a very low full scale current drift with temperature.

The MC1408/MC1508 Series is guaranteed accurate to within ±1/2 LSB at +25°C at a full scale output current of 1.992 mA. This corresponds to a reference amplifier output current drive to the ladder network of 2.0 mA, with the loss of one LSB = 8.0 µA which is the ladder remainder shunted to ground. The input current to pin 14 has a guaranteed value of between 1.9 and 2.1 mA, allowing some mismatch in the NPN current source pair. The accuracy test circuit is shown in Figure 4. The 12-bit converter is calibrated for a full scale output current of 1.992 mA. This is an optional step since the MC1408 accuracy is essentially the same between 1.5 and 2.5 mA. Then the MC1408 circuits' full scale current is trimmed to the same value with R14 so that a zero value appears at the error amplifier output. The counter is activated and the error band may be displayed on an oscilloscope, detected by comparators, or stored in a peak detector.

Two 8-bit D-to-A converters may not be used to construct a 16-bit accurate D-to-A converter. 16-bit accuracy implies a total error of ±1/2 of one part in 65,536, or ±0.00076%, which is much more accurate than the ±0.19% specification provided by the MC1408x8.

Multiplying Accuracy

The MC1408 may be used in the multiplying mode with eight-bit accuracy when the reference current is varied over a range of 256:1. The major source of error is the bias current of the termination amplifier. Under "worst case" conditions, these eight amplifiers can contribute a total of 1.6 µA extra current at the output terminal. If the reference current in the multiplying mode ranges from 16 µA to 4.0 mA, the 1.6 µA contributes an error of 0.1 LSB. This is well within eight-bit accuracy referenced to 4.0 mA.

A monotonic converter is one which supplies an increase in current for each increment in the binary word. Typically, the MC1408 is monotonic for all values of reference current above 0.5 mA. The recommended range for operation with a dc reference current is 0.5 to 4.0 mA.

GENERAL INFORMATION (Continued)

Settling Time

The "worst case" switching condition occurs when all bits are switched "on", which corresponds to a low-to-high transition for all bits. This time is typically 300 ns for settling to within ±1/2 LSB, for 8-bit accuracy, and 200 ns to 1/2 LSB for 7 and 6-bit accuracy. The turn off is typically under 100 ns. These times apply when $R_L \leqslant 500$ ohms and $C_O \leqslant 25$ pF.

The slowest single switch is the least significant bit, which turns "on" and settles in 250 ns and turns "off" in 80 ns. In applications where the D-to-A converter functions in a positive-going ramp mode, the "worst case" switching condition does not occur, and a settling time of less than 300 ns may be realized. Bit A7 turns "on" in 200 ns and "off" in 80 ns, while bit A6 turns "on" in 150 ns and "off" in 80 ns.

The test circuit of Figure 5 requires a smaller voltage swing for the current switches due to internal voltage clamping in the MC-1408. A 1.0-kilohm load resistor from pin 4 to ground gives a typical settling time of 400 ns. Thus, it is voltage swing and not the output RC time constant that determines settling time for most applications.

Extra care must be taken in board layout since this is usually the dominant factor in satisfactory test results when measuring settling time. Short leads, 100 μF supply bypassing for low frequencies, and minimum scope lead length are all mandatory.

TYPICAL CHARACTERISTICS
(V_{CC} = +5.0 V, V_{EE} = –15 V, T_A = +25°C unless otherwise noted.)

FIGURE 10 – LOGIC INPUT CURRENT versus INPUT VOLTAGE

FIGURE 11 – TRANSFER CHARACTERISTIC versus TEMPERATURE
(A5 thru A8 thresholds lie within range for A1 thru A4)

FIGURE 12 – OUTPUT CURRENT versus OUTPUT VOLTAGE
(See text for pin 1 restrictions)

FIGURE 13 – OUTPUT VOLTAGE versus TEMPERATURE
(Negative range with pin 1 open is –5.0 Vdc over full temperature range)

MOTOROLA LINEAR/INTERFACE DEVICES

MC1408, MC1508

TYPICAL CHARACTERISTICS (continued)

(V_{CC} = +5.0 V, V_{EE} = –15 V, T_A = +25°C unless otherwise noted.)

FIGURE 14 – REFERENCE INPUT FREQUENCY RESPONSE

FIGURE 15 – TYPICAL POWER SUPPLY CURRENT
versus TEMPERATURE (all bits low)

FIGURE 16 – TYPICAL POWER SUPPLY CURRENT
versus V_{EE} (all bits low)

Unless otherwise specified:

R14 = R15 = 1.0 kΩ
C = 15 pF, pin 16 to V_{EE}
R_L = 50 Ω, pin 4 to GND

Curve A: Large Signal Bandwidth
Method of Figure 7
V_{ref} = 2.0 V(p-p) offset 1.0 V above GND

Curve B: Small Signal Bandwidth
Method of Figure 7 R_L = 250 Ω
V_{ref} = 50 mV(p-p) offset 200 mV above GND

Curve C: Large and Small Signal Bandwidth
Method of Figure 25 (no op-ampl, R_L = 50 Ω)
R_S = 50 Ω
V_{ref} = 2.0 V
V_S = 100 mV(p-p) centered at 0 V

APPLICATIONS INFORMATION

FIGURE 17 – OUTPUT CURRENT TO VOLTAGE CONVERSION

V_{ref} = 2.0 Vdc
R14 = R15 = 1.0 kΩ
R_O = 5.0 kΩ

Theoretical V_O

$$V_O = \frac{V_{ref}}{R14}(R_O)\left[\frac{A1}{2} + \frac{A2}{4} + \frac{A3}{8} + \frac{A4}{16} + \frac{A5}{32} + \frac{A6}{64} + \frac{A7}{128} + \frac{A8}{256}\right]$$

Adjust V_{ref}, R14 or R_O so that V_O with all digital inputs at high level is equal to 9.961 volts.

$$V_O = \frac{2 \text{ V}}{1 \text{ k}}(5 \text{ k})\left[\frac{1}{2} + \frac{1}{4} + \frac{1}{8} + \frac{1}{16} + \frac{1}{32} + \frac{1}{64} + \frac{1}{128} + \frac{1}{256}\right]$$

$$= 10 \text{ V}\left[\frac{255}{256}\right] = 9.961 \text{ V}$$

MOTOROLA LINEAR/INTERFACE DEVICES

MC1408, MC1508

APPLICATIONS INFORMATION (continued)

Voltage outputs of a larger magnitude are obtainable with this circuit which uses an external operational amplifier as a current to voltage converter. This configuration automatically keeps the output of the MC1408 at ground potential and the operational amplifier can generate a positive voltage limited only by its positive supply voltage. Frequency response and settling time are primarily determined by the characteristics of the operational amplifier. In addition, the operational amplifier must be compensated for unity gain, and in some cases overcompensation may be desirable.

Note that this configuration results in a positive output voltage only, the magnitude of which is dependent on the digital input.

The following circuit shows how the MLM301AG can be used in a feedforward mode resulting in a full scale settling time on the order of 2.0 μs.

FIGURE 18

An alternative method is to use the MC1539G and input compensation. Response of this circuit is also on the order of 2.0 μs. See Motorola Application Note AN-459 for more details on this concept.

FIGURE 19

The positive voltage range may be extended by cascading the output with a high beta common base transistor, Q1, as shown.

FIGURE 20 – EXTENDING POSITIVE VOLTAGE RANGE

The output voltage range for this circuit is 0 volts to BV_{CBO} of the transistor. If pin 1 is left open, the transistor base may be grounded, eliminating both the resistor and the diode. Variations in beta must be considered for wide temperature range applications. An inverted output waveform may be obtained by using a load resistor from a positive reference voltage to the collector of the transistor. Also, high-speed operation is possible with a large output voltage swing, because pin 4 is held at a constant voltage. The resistor (R) to V_{EE} maintains the transistor emitter voltage when all bits are "off" and insures fast turn-on of the least significant bit.

Combined Output Amplifier and Voltage Reference

For many of its applications the MC1408 requires a reference voltage and an operational amplifier. Normally the operational amplifier is used as a current to voltage converter and its output need only go positive. With the popular MC1723G voltage regulator both of these functions are provided in a single package with the added bonus of up to 150 mA of output current. See Figure 21. The MC1723G uses both a positive and negative power supply. The reference voltage of the MC1723G is then developed with respect to the negative voltage and appears as a common-mode signal to the reference amplifier in the D-to-A converter. This allows use of its output amplifier as a classic current-to-voltage converter with the non-inverting input grounded.

Since ±15 V and +5.0 V are normally available in a combination digital-to-analog system, only the –5.0 V need be developed. A resistor divider is sufficiently accurate since the allowable range on pin 5 is from –2.0 to –8.0 volts. The 5.0 kilohm pulldown resistor on the amplifier output is necessary for fast negative transitions.

Full scale output may be increased to as much as 32 volts by increasing R_O and raising the +15 V supply voltage to 35 V maximum. The resistor divider should be altered to comply with the maximum limit of 40 volts across the MC1723G. C_O may be decreased to maintain the same R_OC_O product if maximum speed is desired.

MOTOROLA LINEAR/INTERFACE DEVICES

MC1408, MC1508

Programmable Power Supply

The circuit of Figure 21 can be used as a digitally programmed power supply by the addition of thumbwheel switches and a BCD-to-binary converter. The output voltage can be scaled in several ways, including 0 to +25.5 volts in 0.1-volt increments, ±0.05 volt; or 0 to 5.1 volts in 20 mV increments, ±10 mV.

FIGURE 21 – COMBINED OUTPUT AMPLIFIER and VOLTAGE REFERENCE CIRCUIT

$$V_O = V_{ref} = \frac{R_O}{R14}\{A\}$$

Settling time for a 10-volt step $\cong$ 1.0 μs

Bipolar or Negative Output Voltage

The circuit of Figure 22 is a variation from the standard voltage output circuit and will produce bipolar output signals. A positive current may be sourced into the summing node to offset the output voltage in the negative direction. For example, if approximately 1.0 mA is used a bipolar output signal results which may be described as 8-bit "1's" complement offset binary. V_{ref} may be used as this auxiliary reference. Note that R_O has been doubled to 10 kilohms because of the anticipated 20 V(p-p) output range.

FIGURE 22 – BIPOLAR OR NEGATIVE OUTPUT VOLTAGE CIRCUIT

$$V_O = \frac{V_{ref}}{R14}(R_O)\left[\frac{A1}{2}+\frac{A2}{4}+\frac{A3}{8}+\frac{A4}{16}+\frac{A5}{32}+\frac{A6}{64}+\frac{A7}{128}+\frac{A8}{256}\right] - \frac{V_{ref}}{R_B}(R_O)$$

FIGURE 23 – BIPOLAR OR INVERTED NEGATIVE OUTPUT VOLTAGE CIRCUIT

Decrease R_O to 2.5 kΩ for a 0 to –5.0-volt output range. This application provides somewhat lower speed, as previously discussed in the Output Voltage Range section of the General Information.

APPLICATIONS INFORMATION (continued)

Polarity Switching Circuit, 8-Bit Magnitude Plus Sign D-to-A Converter

Bipolar outputs may also be obtained by using a polarity switching circuit. The circuit of Figure 24 gives 8-bit magnitude plus a sign bit. In this configuration the operational amplifier is switched between a gain of +1.0 and −1.0. Although another operational amplifier is required, no more space is taken when a dual operational amplifier such as the MC1558G is used. The transistor should be selected for a very low saturation voltage and resistance.

FIGURE 24 – POLARITY SWITCHING CIRCUIT
(8-Bit Magnitude Plus Sign D-to-A Converter)

Programmable Gain Amplifier or Digital Attenuator

When used in the multiplying mode the MC1408 can be applied as a digital attenuator. See Figure 25. One advantage of this technique is that if R_S = 50 ohms, no compensation capacitor is needed. The small and large signal bandwidths are now identical and are shown in Figure 14.

The best frequency response is obtained by not allowing I_{14} to reach zero. However, the high impedance node, pin 16, is clamped to prevent saturation and insure fast recovery when the current through R14 goes to zero. R_S can be set for a ±1.0 mA variation in relation to I_{14}. I_{14} can never be negative.

The output current is always unipolar. The quiescent dc output current level changes with the digital word which makes ac coupling necessary.

FIGURE 25 – PROGRAMMABLE GAIN AMPLIFIER OR DIGITAL ATTENUATOR CIRCUIT

Panel Meter Readout

The MC1408 can be used to read out the status of BCD or binary registers or counters in a digital control system. The current output can be used to drive directly an analog panel meter. External meter shunts may be necessary if a meter of less than 2.0 mA full scale is used. Full scale calibration can be done by adjusting R14 or V_{ref}.

FIGURE 26 – PANEL METER READOUT CIRCUIT

FIGURE 27 – DC COUPLED DIGITAL ATTENUATOR and DIGITAL SUBTRACTION

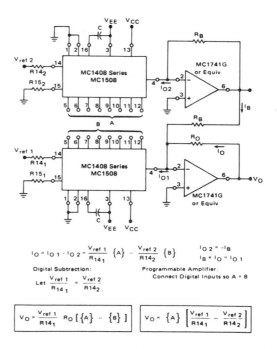

MOTOROLA LINEAR/INTERFACE DEVICES

MC1408, MC1508

APPLICATIONS INFORMATION (continued)

FIGURE 36 — TWO-DIGIT BCD CONVERSION

Two 8-bit, D-to-A converters can be used to build a two digit BCD D-to-A or A-to-D converter. If both outputs feed the virtual ground of an operational amplifier, 10:1 current scaling can be achieved with a resistive current divider. If current output is desired, the units may be operated at full scale current levels of 4.0 mA and 0.4 mA with the outputs connected to sum the currents. The error of the D-to-A converter handling the least significant bits will be scaled down by a factor of ten and thus an MC1408L6 may be used for the least significant word.

FIGURE 37 — DIGITAL QUOTIENT OF TWO ANALOG VARIABLES
or ANALOG-TO-DIGITAL CONVERSION

The circuit shown is a simple counter-ramp converter. An UP/DOWN counter and dual threshold comparator can be used to provide faster operation and continuous conversion.

$$C = \frac{V_{in}/R_O}{V_{ref}/R14}$$

DESCRIPTION

The NE/SE 566 Function Generator is a voltage controlled oscillator of exceptional linearity with buffered square wave and triangle wave outputs. The frequency of oscillation is determined by an external resistor and capacitor and the voltage applied to the control terminal. The oscillator can be programmed over a ten to one frequency range by proper selection of an external resistance and modulated over a ten to one range by the control voltage, with exceptional linearity.

FEATURES

- Wide range of operating voltage (up to 24 volts)
- High linearity of modulation
- Highly stable center frequency (200 ppm/°C typical)
- Highly linear triangle wave output
- Frequency programming by means of a resistor or capacitor, voltage or current
- Frequency adjustable over 10 to 1 range with same capacitor

APPLICATIONS

- Tone generators
- Frequency shift keying
- FM modulators
- Clock generators
- Signal generators
- Function generators

PIN CONFIGURATIONS

H PACKAGE*

ORDER PART NO.
NE/SE566H

*Metal cans (H) not recommended for new designs

N PACKAGE

ORDER PART NO.
NE/SE566N

F PACKAGE

ORDER PART NO.
NE/SE566F

BLOCK DIAGRAM

EQUIVALENT SCHEMATIC

ABSOLUTE MAXIMUM RATINGS

PARAMETER	RATING	UNIT
Maximum operating voltage	26	V
Input voltage	3	V_{P-P}
Storage temperature	−65 to +150	°C
Operating temperature range		
NE566	0 to +70	°C
SE566	−55 to +125	°C
Power dissipation	300	mW

ELECTRICAL CHARACTERISTICS T_A = 25°C; V_{CC} = 12V unless otherwise specified.

PARAMETER	SE566			NE566			UNIT
	Min	Typ	Max	Min	Typ	Max	
GENERAL							
Operating temperature range	–55		125	0		70	°C
Operating supply voltage			24			24	V
Operating supply current		7	12.5		7	12.5	mA
VCO[1]							
Maximum operating frequency		1			1		MHz
Frequency drift with temperature		200			300		ppm/°C
Frequency drift with supply voltage		1			2		%/V
Control terminal input impedance[2]		1			1		MΩ
FM distortion (±10% deviation)		0.2	0.75		0.4	1.5	%
Maximum sweep rate		1			1		MHz
Sweep range		10:1			10:1		
OUTPUT							
Triangle wave output							
Impedance		50			50		Ω
Voltage	1.9	2.4		1.9	2.4		Vpp
Linearity		0.2			0.5		%
Square wave input							
Impedance		50			50		Ω
Voltage	5	5.4		5	5.4		Vpp
Duty Cycle	45	50	55	40	50	60	%
Rise time		20			20		ns
Fall Time		50			50		ns

NOTES

1. The external resistance for frequency adjustment (R_1) must have a value between 2kΩ and 20KΩ.
2. The bias voltage (V_c) applied to the control terminal (pin 5) should be in the range
   $3/4V^+ \leq V_c \leq V^+$.

TYPICAL PERFORMANCE CHARACTERISTICS

NORMALIZED FREQUENCY AS A FUNCTION OF CONTROL VOLTAGE

NORMALIZED FREQUENCY AS A FUNCTION OF RESISTANCE (R1)

CHANGE IN FREQUENCY AS A FUNCTION OF TEMPERATURE

Signetics

TYPICAL PERFORMANCE CHARACTERISTICS (Cont'd)

| POWER SUPPLY CURRENT AS A FUNCTION OF SUPPLY VOLTAGE | FREQUENCY AS A FUNCTION OF CAPACITANCE (C1) | VCO OUTPUT WAVEFORMS |

OPERATING INSTRUCTIONS

The SE/NE 566 Function Generator is a general purpose voltage controlled oscillator designed for highly linear frequency modulation. The circuit provides simultaneous square wave and triangle wave outputs at frequencies up to 1MHz. A typical connection diagram is shown in Figure 1. The control terminal (pin 5) must be biased externally with a voltage (V_C) in the range

$$3/4 \, V^+ \leq V_C \leq V^+$$

where V_{CC} is the total supply voltage. In Figure 1, the control voltage is set by the voltage divider formed with R_2 and R_3. The modulating signal is then ac coupled with the capacitor C_2. The modulating signal can be direct coupled as well, if the appropriate dc bias voltage is applied to the control terminal. The frequency is given approximately by

$$f_o \simeq \frac{2[(V^+) - (V_C)]}{R_1 C_1 V^+}$$

and R_1 should be in the range $2k\Omega < R_1 < 20k\Omega$.

A small capacitor (typically $0.001\mu f$) should be connected between pins 5 and 6 to eliminate possible oscillation in the control current source.

If the VCO is to be used to drive standard logic circuitry, it may be desirable to use a dual supply of ±5 volts as shown in Figure 2. In this case the square wave output has the proper dc levels for logic circuitry. RTL can be driven directly from pin 3. For DTL or T^2L gates, which require a current sink of more than 1mA, it is usually necessary to connect a 5kΩ resistor between pin 3 and negative supply. This increases the current sinking capability to 2mA. The third type of interface shown uses a saturated transistor between the 566 and the logic circuitry. This scheme is used primarily for T^2L circuitry which requires a fast fall time (<50ns) and a large current sinking capability.

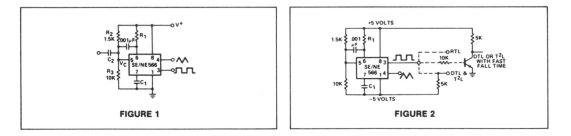

FIGURE 1

FIGURE 2

National Semiconductor

MF5 Universal Monolithic Switched Capacitor Filter

General Description

The MF5 consists of an extremely easy to use, general purpose CMOS active filter building block and an uncommitted op amp. The filter building block, together with an external clock and a few resistors, can produce various second order functions. The filter building block has 3 output pins. One of the output pins can be configured to perform highpass, allpass or notch functions and the remaining 2 output pins perform bandpass and lowpass functions. The center frequency of the filter can be directly dependent on the clock frequency or it can depend on both clock frequency and external resistor ratios. The uncommitted op amp can be used for cascading purposes, for obtaining additional allpass and notch functions, or for various other applications. Higher order filter functions can be obtained by cascading several MF5s or by using the MF5 in conjuction with the MF10 (dual switched capacitor filter building block). The MF5 is functionally compatible with the MF10. Any of the classical filter configurations (such as Butterworth, Bessel, Cauer and Chebyshev) can be formed.

Features

- Low cost
- 14-pin DIP
- Easy to use
- Clock to center frequency ratio accuracy ±0.6%
- Filter cutoff frequency stability directly dependent on external clock quality
- Low sensitivity to external component variations
- Separate highpass (or notch or allpass), bandpass, lowpass outputs
- $f_0 \times Q$ range up to 200 kHz
- Operation up to 30 kHz (typical)
- Additional uncommitted op-amp

Block and Connection Diagrams

TL/H/5066–1

Dual-In-Line Package

	MF5	
BP — 1		14 — LP
N/AP/HP — 2		13 — V02
INV1 — 3		12 — INV2
S1 — 4		11 — AGND
SA — 5		10 — V⁻
V⁺ — 6		9 — 50/100
L Sh — 7		8 — CLK

Top View

Order Number MF5J, N
See NS Packages J14A, N14A

TL/H/5066–2

Absolute Maximum Ratings

Supply Voltage (V⁺ − V⁻)	14V	Input Voltage (any pin)	$V^- \leq V_{in} \leq V^+$
Power Dissipation T_A = 25°C (note 1)	500 mW	Operating Temp. Range	$T_{MIN} \leq T_A \leq T_{MAX}$
Storage Temp.	150°C	MF5BN, MF5CN	$0°C \leq T_A \leq 70°C$
Lead Temp. (soldering, 10 seconds)	300°C		

Electrical Characteristics

V^+ = 5V ± 0.5%, V^- = −5V ± 0.5% unless otherwise noted. **Boldface limits apply over temperature, $T_{MIN} \leq T_A \leq T_{MAX}$. For all other limits T_A = 25°C.**

Parameter		Conditions	Typical (Note 6)	Tested Limit (Note 7)	Design Limit (Note 8)	Units
Supply Voltage (V⁺ − V⁻)	Min				**8**	V
	Max				**14**	V
Maximum Supply Current		Clock applied to Pin 8 No Input Signal	4.5	6.0		mA
Clock Feedthrough	Filter Output		10			mV
	Op-amp Output		10			mV

Filter Electrical Characteristics

V^+ = 5V ± 0.5%, V^- = −5V ± 0.5% unless otherwise noted. **Boldface limits apply over temperature, $T_{MIN} \leq T_A \leq T_{MAX}$. For all other limits T_A = 25°C.**

Parameter		Conditions		Typical (Note 6)	Tested Limit (Note 7)	Design Limit (Note 8)	Units
Center Frequency Range (f_o)	Max			30		**20**	kHz
	Min			0.1		**0.2**	Hz
Clock Frequency Range (f_{CLK})	Max			1.5		**1.0**	MHz
	Min			5.0		**10**	Hz
Clock to Center Frequency Ratio (f_{CLK}/f_o)	MF5BN	Ideal Q=10 Mode 1	V_{pin9} = +5V F_{CLK}=250 kHz	50.11 ± 0.2%	50.11 ± 0.6%		
	MF5CN			50.11 ± 0.2%	50.11 ± 1.5%		
	MF5BN		V_{pin9} = −5V F_{CLK}=500 kHz	100.04 ± 0.2%	100.04 ± 0.6%		
	MF5CN			100.04 ± 0.2%	100.04 ± 1.5%		
f_{CLK}/f_o Temp. Coefficient		V_{pin9} = +5V (50:1 CLK ratio)		± 10			ppm/°C
		V_{pin9} = −5V (100:1 CLK ratio)		± 20			ppm/°C
Q Accuracy (Max) (Note 2)	MF5BN	Ideal Q=10 Mode 1	V_{pin9} = +5V F_{CLK}=250 kHz		± 5		%
	MF5CN				± 6		%
	MF5BN		V_{pin9} = −5V F_{CLK}=500 kHz		± 4		%
	MF5CN				± 6		%
Q Temperature Coefficient		V_{pin9} = +5V (50:1 CLK ratio)		− 200			ppm/°C
		V_{pin9} = −5V (100:1 CLK ratio)		− 70			ppm/°C
DC Lowpass Gain Accuracy (Max)		Mode 1 R1 = R2 = 10 kΩ			± 0.2		dB
DC Offset Voltage (Max)	V_{os1}			± 5.0			mV
	V_{os2}	V_{pin9} = +5V		− 185			mV
	V_{os3}	(50:1 CLK ratio)		+ 115			mV
(Note 3)	V_{os2}	V_{pin9} = −5V		− 310			mV
	V_{os3}	(100:1 CLK ratio)		+ 240			mV
Output Swing (Min)	BP, LP pins	RL = 5 kΩ		± 4.0	± 3.8		V
	N/AP/HP pin	RL = 3.5 kΩ		± 4.2	± 3.8		V

Filter Electrical Characteristics

$V^+ = 5V \pm 0.5\%$, $V^- = -5V \pm 0.5\%$ unless otherwise noted. **Boldface limits apply over temperature, $T_{MIN} \leq T_A \leq T_{MAX}$. For all other limits $T_A = 25°C$. (Continued)**

Parameter		Conditions	Typical (Note 6)	Tested Limit (Note 7)	Design Limit (Note 8)	Units
Dynamic Range (Note 4)		$V_{pin9} = +5V$ (50:1 CLK ratio)	83			dB
		$V_{pin9} = -5V$ (100:1 CLK ratio)	80			dB
Maximum Output Short Circuit Current (Note 5)	Source		20			mA
	Sink		3.0			mA

OP-AMP Electrical Characteristics

$V^+ = +5V \pm 0.5\%$, $V^- = -5V \pm 0.5\%$ unless other noted. **Boldface limits apply over temperature, $T_{MIN} \leq T_A \leq T_{MAX}$. For all other limits $T_A = 25°C$.**

Parameter		Conditions	Typical (Note 6)	Tested Limit (Note 7)	Design Limit (Note 8)	Units
Gain Bandwidth Product			2.5			MHz
Output Voltage Swing (Min)		$RL = 3.5 k\Omega$	± 4.2	± 3.8		V
Slew Rate			7.0			V/μs
DC Open-Loop Gain			80			db
Input Offset Voltage (Max)			± 5.0	± 20		mV
Input Bias Current			10			pA
Maximum Output Short Circuit Current (Note 5)	Source		20			mA
	Sink		3.0			mA

Logic Input Characteristics

Boldface limits apply over temperature, $T_{MIN} \leq T_A \leq T_{MAX}$. All other limits $T_A = 25°C$.

Parameter		Conditions	Typical (Note 6)	Tested Limit (Note 7)	Design Limit (Note 8)	Units
CMOS Clock Input	Min Logical "1" Input Voltage	$V^+ = +5V, V^- = -5V$, $V_{L.Sh.} = 0V$		3.0		V
	Max Logical "0" Input Voltage			-3.0		V
	Min Logical "1" Input Voltage	$V^+ = +10V, V^- = 0V$, $V_{L.Sh.} = +5V$		8.0		V
	Max Logical "0" Input Voltage			2.0		V
TTL Clock Input	Min Logical "1" Input Voltage	$V^+ = +5V, V^- = -5V$, $V_{L.Sh.} = 0V$		2.0		V
	Max Logical "0" Input Voltage			0.8		V

Note 1: The typical junction-to-ambient thermal resistance (θ_{JA}) of the 14 pin N package is 160°C/W.

Note 2: The accuracy of the Q value is a function of the center frequency (f_0). This is illustrated in the curves under the heading "Typical Performance Characteristics".

Note 3: V_{OS1}, V_{OS2}, and V_{OS3} refer to the internal offsets as discussed in the Application Information section 3.4.

Note 4: For $\pm 5V$ supplies the dynamic range is referenced to 2.82V rms (4V peak) where the wideband noise over a 20 kHz bandwidth is typically 200 μV rms for the MF5 with a 50:1 CLK ratio and 280 μV rms for the MF5 with a 100:1 CLK ratio.

Note 5: The short circuit source current is measured by forcing the output that is being tested to its maximum positive voltage swing and then shorting that output to the negative supply. The short circuit sink current is measured by forcing the output that is being tested to its maximum negative voltage swing and then shorting that output to the positive supply. These are the worst case conditions.

Note 6: Typicals are at 25°C and represent most likely parametric norm.

Note 7: Guaranteed and 100% tested.

Note 8: Guaranteed, but not 100% tested. These limits are not used to calculate outgoing quality levels.

Pin Description

LP(14), BP(1), N/AP/HP(2):	The second order lowpass, bandpass, and notch/allpass/highpass outputs. The LP and BP outputs can typically sink 1 mA and source 3 mA. The N/AP/HP output can typically sink 1.5 mA and source 3 mA. Each output typically swings to within 1V of each supply.
INV1(3):	The inverting input of the summing op amp of the filter. This is a high impedance input, but the non-inverting input is internally tied to AGND, making INV1 behave like a summing junction (low impedance current input).
S1(4):	S1 is a signal input pin used in the allpass filter configurations (see modes 4 and 5). The pin should be driven with a source impedance of less than 1 kΩ. If S1 is not driven with a signal it should be tied to AGND (mid-supply).
SA(5):	This pin activates a switch that connects one of the inputs of the filter's second summer to either AGND (SA tied to $V-$) or to the lowpass (LP) output (SA tied to $V+$). This offers the flexibility needed for configuring the filter in its various modes of operation.
50/100(9):	This pin is used to set the internal clock to center frequency ratio (f_{CLK}/f_o) of the filter. By tying the pin to $V+$ an f_{CLK}/f_o ratio of about 50:1 (typically 50.11 ± 0.2%) is obtained. Tying the 50/100 pin to either AGND or $V-$ will set the f_{CLK}/f_o ratio to about 100:1 (typically 100.04 ± 0.2%).
AGND(11):	This is the analog ground pin. This pin should be connected to the system ground for dual supply operation or biased to mid-supply for single supply operation. For a further discussion of mid-supply biasing techniques see the Applications Information (Section 3.2). For optimum filter performance a "clean" ground must be provided.

$V+$(6), $V-$(10):	These are the positive and negative supply pins. The MF5 will operate over a total supply range of 8V to 14V. Decoupling the supply pins with 0.1 μF capacitors is highly recommended.
CLK(8):	This is the clock input for the filter. CMOS or TTL logic level clocks can be accomodated by setting the L. Sh pin to the levels described in the L. Sh pin description. For optimum filter performance a 50% duty cycle clock is recommended for clock frequencies greater than 200 kHz. This gives each op amp the maximum amount of time to settle to a new sampled input.
L. Sh(7):	This pin allows the MF5 to accommodate either CMOS or TTL logic level clocks. For dual supply operation (i.e., ±5V), a CMOS or TTL logic level clock can be accepted if the L. Sh pin is tied to mid-supply (AGND), which should be the system ground. For single supply operation the L. Sh pin should be tied to mid-supply (AGND) for a CMOS logic level clock. The mid-supply bias should be a very low impedance node. See Applications Information for biasing techniques. For a TTL logic level clock the L. Sh pin should be tied to $V-$ which should be the system ground.
INV2(12):	This is the inverting input of the uncommitted op amp. This is a very high impedance input, but the non-inverting input is internally tied to AGND, making INV2 behave like a summing junction (low-impedance current input).
Vo2(13):	This is the output of the uncommitted op amp. It will typically sink 1.5 mA and source 3.0 mA. It will typically swing to within 1V of each supply.

Typical Performance Characteristics

Deviation of $\frac{F_{CLK}}{F_o}$ vs Nominal Q

Deviation of $\frac{F_{CLK}}{F_o}$ vs Nominal Q

OPAMP Output Voltage Swing vs Temperature

TL/H/5066–3

548

Typical Performance Characteristics (Continued)

Supply Current vs Temperature

TL/H/5066–4

1.0 Definitions of Terms

f_{CLK}: the frequency of the external clock signal applied to pin 8.

f_0: center frequency of the second order function complex pole pair. f_0 is measured at the bandpass output of the MF5, and is the frequency of maximum bandpass gain. (*Figure 1*).

f_{notch}: the frequency of minimum (ideally zero) gain at the notch output.

f_z: the center frequency of the second order complex zero pair, if any. If f_z is different from f_0 and if Q_z is high, it can be observed as the frequency of a notch at the allpass output. (*Figure 10*).

Q: "quality factor" of the 2nd order filter. Q is measured at the bandpass output of the MF5 and is equal to f_0 divided by the −3dB bandwidth of the 2nd order bandpass filter (*Figure 1*). The value of Q determines the shape of the 2nd order filter responses as shown in *Figure 6*.

Q_z: the quality factor of the second order complex zero pair, if any. Q_z is related to the allpass characteristic, which is written:

$$H_{AP}(s) = \frac{H_{OAP}\left(s^2 - \dfrac{s\omega_0}{Q_z} + \omega_0{}^2\right)}{s^2 + \dfrac{s\omega_0}{Q} + \omega_0{}^2}$$

where $Q_z = Q$ for an all-pass response.

H_{OBP}: the gain (in V/V) of the bandpass output at $f = f_0$.

H_{OLP}: the gain (in V/V) of the lowpass output as $f \rightarrow 0$ Hz (*Figure 2*).

H_{OHP}: the gain (in V/V) of the highpass output as $f \rightarrow f_{clk}/2$ (*Figure 3*).

H_{ON}: the gain (in V/V) of the notch output as $f \rightarrow 0$ Hz and as $f \rightarrow f_{clk}/2$, when the notch filter has equal gain above and below the center frequency (*Figure 4*). When the low-frequency gain differs from the high-frequency gain, as in modes 2 and 3a (*Figures 11 and 8*), the two quantities below are used in place of H_{ON}.

H_{ON1}: the gain (in V/V) of the notch output as $f \rightarrow 0$ Hz.

H_{ON2}: the gain (in V/V) of the notch output as $f \rightarrow f_{clk}/2$.

FIGURE 1. 2nd-Order Bandpass Response

$$H_{BP}(s) = \frac{H_{OBP}s}{s^2 + \dfrac{s\omega_0}{Q} + \omega_0{}^2}$$

$$Q = \frac{f_0}{f_H - f_L}; \quad f_0 = \sqrt{f_L f_H}$$

$$f_L = f_0\left(\frac{-1}{2Q} + \sqrt{\left(\frac{1}{2Q}\right)^2 + 1}\right)$$

$$f_H = f_0\left(\frac{1}{2Q} + \sqrt{\left(\frac{1}{2Q}\right)^2 + 1}\right)$$

$$\omega_0 = 2\pi f_0$$

TL/H/5066–5   TL/H/5066–6

FIGURE 2. 2nd-Order Low-Pass Response

$$H_{LP}(s) = \frac{H_{OLP}\omega_0{}^2}{s^2 + \dfrac{s\omega_0}{Q} + \omega_0{}^2}$$

$$f_c = f_0 \times \sqrt{\left(1 - \frac{1}{2Q^2}\right) + \sqrt{\left(1 - \frac{1}{2Q^2}\right)^2 + 1}}$$

$$f_p = f_0\sqrt{1 - \frac{1}{2Q^2}}$$

$$H_{OP} = H_{OLP} \times \frac{1}{\dfrac{1}{Q}\sqrt{1 - \dfrac{1}{4Q^2}}}$$

TL/H/5066–7   TL/H/5066–8

FIGURE 3. 2nd-Order High-Pass Response

$$H_{HP}(s) = \frac{H_{OHP}s^2}{s^2 + \dfrac{s\omega_0}{Q} + \omega_0{}^2}$$

$$f_c = f_0 \times \left[\sqrt{\left(1 - \frac{1}{2Q^2}\right) + \sqrt{\left(1 - \frac{1}{2Q^2}\right)^2 + 1}}\right]^{-1}$$

$$f_p = f_0 \times \left[\sqrt{1 - \frac{1}{2Q^2}}\right]^{-1}$$

$$H_{OP} = H_{OHP} \times \frac{1}{\dfrac{1}{Q}\sqrt{1 - \dfrac{1}{4Q^2}}}$$

TL/H/5066–9   TL/H/5066–10

1.0 Definition of Terms (Continued)

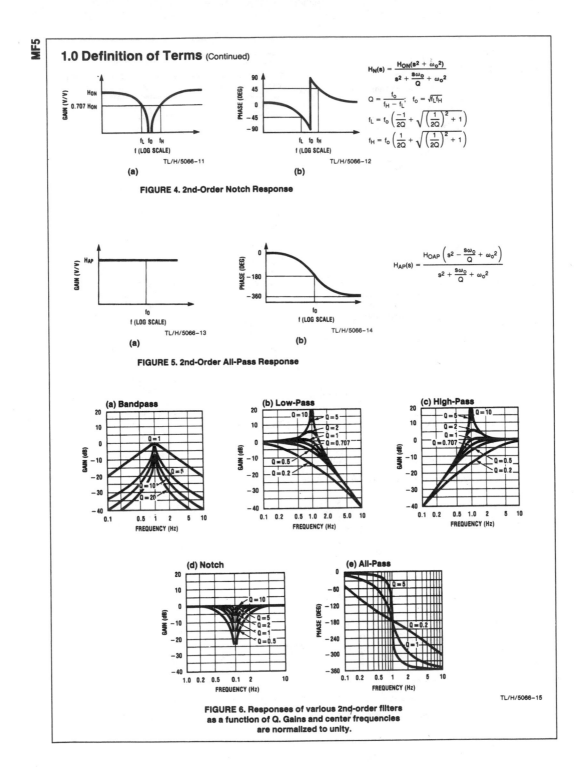

$$H_N(s) = \frac{H_{ON}(s^2 + \omega_o^2)}{s^2 + \frac{s\omega_o}{Q} + \omega_o^2}$$

$$Q = \frac{f_o}{f_H - f_L}; \quad f_o = \sqrt{f_L f_H}$$

$$f_L = f_o \left(\frac{-1}{2Q} + \sqrt{\left(\frac{1}{2Q}\right)^2 + 1} \right)$$

$$f_H = f_o \left(\frac{1}{2Q} + \sqrt{\left(\frac{1}{2Q}\right)^2 + 1} \right)$$

TL/H/5066–11

(a)

TL/H/5066–12

(b)

FIGURE 4. 2nd-Order Notch Response

$$H_{AP}(s) = \frac{H_{OAP}\left(s^2 - \frac{s\omega_o}{Q} + \omega_o^2\right)}{s^2 + \frac{s\omega_o}{Q} + \omega_o^2}$$

TL/H/5066–13

(a)

TL/H/5066–14

(b)

FIGURE 5. 2nd-Order All-Pass Response

TL/H/5066–15

FIGURE 6. Responses of various 2nd-order filters as a function of Q. Gains and center frequencies are normalized to unity.

2.0 Modes of Operation

The MF5 is a switched capacitor (sampled data) filter. To fully describe its transfer functions, a time domain approach is appropriate. Since this is cumbersome, and since the MF5 closely approximates continuous filters, the following discussion is based on the well known frequency domain. Each MF5 can produce a full 2nd order function. See Table 1 for a summary of the characteristics of the various modes.

MODE 1: Notch 1, Bandpass, Lowpass Outputs:

$f_{notch} = f_o$ (See *Figure 7*)

f_o = center frequency of the complex pole pair

$= \dfrac{f_{CLK}}{100}$ or $\dfrac{f_{CLK}}{50}$

f_{notch} = center frequency of the imaginary zero pair $= f_o$.

H_{OLP} = Lowpass gain (as $f \rightarrow 0$) $= -\dfrac{R2}{R1}$

H_{OBP} = Bandpass gain (at $f = f_o$) $= -\dfrac{R3}{R1}$

H_{ON} = Notch output gain as $\left.\begin{array}{l} f \rightarrow 0 \\ \\ f \rightarrow f_{CLK}/2 \end{array}\right\} = \dfrac{-R_2}{R_1}$

$Q = \dfrac{f_o}{BW} = \dfrac{R3}{R2}$

BW = the -3 dB bandwidth of the bandpass output.

Circuit dynamics:

$$H_{OLP} = \dfrac{H_{OBP}}{Q} \text{ or } H_{OBP} = H_{OLP} \times Q = H_{ON} \times Q.$$

$$H_{OLP(peak)} \cong Q \times H_{OLP} \text{ (for high Q's)}$$

MODE 1a: Non-Inverting BP, LP (See *Figure 8*)

$f_o = \dfrac{f_{CLK}}{100}$ or $\dfrac{f_{CLK}}{50}$

$Q = \dfrac{R3}{R2}$

$H_{OLP} = -1$; $H_{OLP(peak)} \cong Q \times H_{OLP}$ (for high Q's)

$H_{OBP_1} = -\dfrac{R3}{R2}$

$H_{OBP_2} = 1$ (non-inverting)

Circuit dynamics: $H_{OBP_1} = Q$

Note: V_{IN} should be driven from a low impedance (<1 kΩ)

FIGURE 7. MODE 1

TL/H/5066–16

FIGURE 8. MODE 1a

TL/H/5066–17

2.0 Modes of Operation (Continued)

MODE 2: Notch 2, Bandpass, Lowpass: $f_{notch} < f_o$
(See Figure 9)

f_o = center frequency

$$= \frac{f_{CLK}}{100}\sqrt{\frac{R2}{R4} + 1} \text{ or } \frac{f_{CLK}}{50}\sqrt{\frac{R2}{R4} + 1}$$

f_{notch} = $\dfrac{f_{CLK}}{100}$ or $\dfrac{f_{CLK}}{50}$

Q = quality factor of the complex pole pair

$$= \frac{\sqrt{R2/R4 + 1}}{R2/R3}$$

H_{OLP} = Lowpass output gain (as $f \rightarrow 0$)

$$= -\frac{R2/R1}{R2/R4 + 1}$$

H_{OBP} = Bandpass output gain (at $f = f_o$) = $-R3/R1$

H_{ON_1} = Notch output gain (as $f \rightarrow 0$)

$$= -\frac{R2/R1}{R2/R4 + 1}$$

H_{ON_2} = Notch output gain $\left(\text{as } f \rightarrow \dfrac{f_{CLK}}{2}\right)$ = $-R2/R1$

Filter dynamics: $H_{OBP} = Q\sqrt{H_{OLP} H_{ON_2}} = Q\sqrt{H_{ON_1} H_{ON_2}}$

MODE 3: Highpass, Bandpass, Lowpass Outputs
(See Figure 10)

f_o = $\dfrac{f_{CLK}}{100} \times \sqrt{\dfrac{R2}{R4}}$ or $\dfrac{f_{CLK}}{50} \times \sqrt{\dfrac{R2}{R4}}$

Q = quality factor of the complex pole pair

$$= \sqrt{\frac{R2}{R4}} \times \frac{R3}{R2}$$

H_{OHP} = Highpass gain $\left(\text{as } f \rightarrow \dfrac{f_{CLK}}{2}\right)$ = $-\dfrac{R2}{R1}$

H_{OBP} = Bandpass gain (at $f = f_o$) = $-\dfrac{R3}{R1}$

H_{OLP} = Lowpass gain (as $f \rightarrow 0$) = $-\dfrac{R4}{R1}$

Circuit dynamics: $\dfrac{R2}{R4} = \dfrac{H_{OHP}}{H_{OLP}}$; $H_{OBP} = \sqrt{H_{OHP} \times H_{OLP}} \times Q$

$H_{OLP(peak)} \cong Q \times H_{OLP}$ (for high Q's)

$H_{OHP(peak)} \cong Q \times H_{OHP}$ (for high Q's)

TL/H/5066–18

FIGURE 9. MODE 2

TL/H/5066–19

FIGURE 10. MODE 3

*In Mode 3, the feedback loop is closed around the input summing amplifier; the finite GBW product of this op amp causes a slight Q enhancement. If this is a problem, connect a small capacitor (10 pF–100 pF) across R4 to provide some phase lead.

2.0 Modes of Operation (Continued)

MODE 3a: HP, BP, LP and Notch with External Op amp (See Figure 11)

$$f_o = \frac{f_{CLK}}{100} \times \sqrt{\frac{R2}{R4}} \text{ or } \frac{f_{CLK}}{50} \times \sqrt{\frac{R2}{R4}}$$

$$Q = \sqrt{\frac{R2}{R4}} \times \frac{R3}{R2}$$

$$H_{OHP} = -\frac{R2}{R1}$$

$$H_{OBP} = -\frac{R3}{R1}$$

$$H_{OLP} = -\frac{R4}{R1}$$

$$f_n = \text{notch frequency} = \frac{f_{CLK}}{100}\sqrt{\frac{R_h}{R_l}} \text{ or } \frac{f_{CLK}}{50}\sqrt{\frac{R_h}{R_l}}$$

$$H_{on} = \text{gain of notch at } f = f_o = \left\| Q\left(\frac{R_g}{R_l}H_{OLP} - \frac{R_g}{R_h}H_{OHP}\right)\right\|$$

$$H_{n1} = \text{gain of notch (as } f \rightarrow 0) = \frac{R_g}{R_l} \times H_{OLP}$$

$$H_{n2} = \text{gain of notch }\left(\text{as } f \rightarrow \frac{f_{CLK}}{2}\right) = -\frac{R_g}{R_h} \times H_{OHP}$$

MODE 4: Allpass, Bandpass, Lowpass Outputs (See Figure 12)

$$f_o = \text{center frequency}$$

$$= \frac{f_{CLK}}{100} \text{ or } \frac{f_{CLK}}{50};$$

f_z = center frequency of the complex zero pair $\approx f_o$

$$Q = \frac{f_o}{BW} = \frac{R3}{R2};$$

$$Q_z = \text{quality factor of complex zero pair} = \frac{R3}{R1}$$

For AP output make R1 = R2

$$H^*_{OAP} = \text{Allpass gain}\left(\text{at } 0 < f < \frac{f_{CLK}}{2}\right) = -\frac{R2}{R1} = -1$$

$$H_{OLP} = \text{Lowpass gain (as } f \rightarrow 0)$$

$$= -\left(\frac{R2}{R1} + 1\right) = -2$$

$$H_{OBP} = \text{Bandpass gain (at } f = f_o)$$

$$= -\frac{R3}{R2}\left(1 + \frac{R2}{R1}\right) = -2\left(\frac{R3}{R2}\right)$$

Circuit dynamics: $H_{OBP} = (H_{OLP}) \times Q = (H_{OAP} + 1) Q$

*Due to the sampled data nature of the filter, a slight mismatch of f_z and f_o occurs causing a 0.4 dB peaking around f_o of the allpass filter amplitude response (which theoretically should be a straight line). If this is unacceptable, Mode 5 is recommended.

FIGURE 11. MODE 3a

TL/H/5066-20

FIGURE 12. MODE 4

TL/H/5066-21

553

2.0 Modes of Operation (Continued)

MODE 5: Numerator Complex Zeros, BP, LP
(See *Figure 13*)

$$f_o = \sqrt{1 + \frac{R2}{R4}} \times \frac{f_{CLK}}{100} \text{ or } \sqrt{1 + \frac{R2}{R4}} \times \frac{f_{CLK}}{50}$$

$$f_z = \sqrt{1 - \frac{R1}{R4}} \times \frac{f_{CLK}}{100} \text{ or } \sqrt{1 - \frac{R1}{R4}} \times \frac{f_{CLK}}{50}$$

$$Q = \sqrt{1 + R2/R4} \times \frac{R3}{R2}$$

$$Q_z = \sqrt{1 - R1/R4} \times \frac{R3}{R1}$$

$$H_{0z1} = \text{gain at C.Z. output (as } f \rightarrow 0 \text{ Hz)} = \frac{-R2\,(R4-R1)}{R1\,(R4+R2)}$$

$$H_{0z2} = \text{gain at C.Z. output}\left(\text{as } f \rightarrow \frac{f_{CLK}}{2}\right) = \frac{-R2}{R1}$$

$$H_{OBP} = -\left(=\frac{R2}{R1}+1\right) \times \frac{R3}{R2}$$

$$H_{OLP} = -\left(\frac{R2+R1}{R2+R4}\right) \times \frac{R4}{R1}$$

MODE 6a: Single Pole, HP, LP Filter (See *Figure 14*)

$$f_c = \text{cutoff frequency of LP or HP output}$$

$$= \frac{R2}{R3}\frac{f_{CLK}}{100} \text{ or } \frac{R2}{R3}\frac{f_{CLK}}{50}$$

$$H_{OLP} = -\frac{R3}{R1}$$

$$H_{OHP} = -\frac{R2}{R1}$$

MODE 6b: Single Pole LP Filter (Inverting and Non-Inverting) (See *Figure 15*)

$$f_c = \text{cutoff frequency of LP outputs}$$

$$\cong \frac{R2}{R3}\frac{f_{CLK}}{100} \text{ or } \frac{R2}{R3}\frac{f_{CLK}}{50}$$

$$H_{OLP_1} = 1 \text{ (non-inverting)}$$

$$H_{OLP_2} = -\frac{R3}{R2}$$

TL/H/5066-22

FIGURE 13. MODE 5

TL/H/5066-23

FIGURE 14. MODE 6a

TL/H/5066-24

FIGURE 15. MODE 6b

2.0 Modes of Operation (Continued)

TABLE I. Summary of Modes. Realizable filter types (e.g. low-pass) denoted by asterisks. Unless otherwise noted, gains of various filter outputs are inverting and adjustable by resistor ratios.

Mode	BP	LP	HP	N	AP	Number of resistors	Adjustable f_{CLK}/f_O	Notes
1	*	*		*		3	No	
1a	(2) $H_{OBP1} = -Q$ $H_{OBP2} = +1$	$H_{OLP} = +1$				2	No	May need input buffer. Poor dynamics for high Q.
2	*	*		*		3	Yes (above $f_{CLK}/50$ or $f_{CLK}/100$)	
3	*	*	*			4	Yes	Universal State-Variable Filter. Best general-purpose mode.
3a	*	*	*	*		7	Yes	As above, but also includes resistor-tuneable notch.
4	*	*			*	3	No	Gives Allpass response with $H_{OAP} = -1$ and $H_{OLP} = -2$.
5	*	*			*	4		Gives flatter allpass response than above if $R_1 = R_2 = 0.02R_4$.
6a		*	*			3		Single pole.
6b		(2) $H_{OLP} = +1$ $H_{OLP2} = \dfrac{-R3}{R2}$				2		Single pole

3.0 Applications Information

The MF5 is a general-purpose second-order state variable filter whose center frequency is proportional to the frequency of the square wave applied to the clock input (f_{CLK}). By connecting pin 9 to the appropriate DC voltage, the filter center frequency f_O can be made equal to either $f_{CLK}/100$ or $f_{CLK}/50$. f_O can be very accurately set (within ±0.6%) by using a crystal clock oscillator, or can be easily varied over a wide frequency range by adjusting the clock frequency. If desired, the f_{CLK}/f_O ratio can be altered by external resistors as in *Figures 9, 10, 11, 13, 14,* and *15*. The filter Q and gain are determined by external resistors.

All of the five second-order filter types can be built using the MF5. These are illustrated in *Figures 1* through *5* along with their transfer functions and some related equations. *Figure 6* shows the effect of Q on the shapes of these curves. When filter orders greater than two are desired, two or more MF5s can be cascaded. The MF5 also includes an uncommitted CMOS operational amplifier for additional signal processing applications.

3.1 DESIGN EXAMPLE

An example will help illustrate the MF5 design procedure. For the example, we will design a 2nd order Butterworth low-pass filter with a cutoff frequency of 200 Hz, and a passband gain of −2. The circuit will operate from a ±5V power supply, and the clock amplitude will be ±5v (CMOS levels).

From the specifications, the filter parameters are: $f_O = 200$ Hz, $H_{OLP} = -2$, and, for Butterworth response, $Q = 0.707$.

In section 2.0 are several modes of operation for the MF5, each having different characteristics. Some allow adjustment of f_{CLK}/f_O, others produce different combinations of filter types, some are inverting while others are non-inverting, etc. These characteristics are summarized in Table I. To keep the example simple, we will use mode 1, which has notch, bandpass, and lowpass outputs, and inverts the signal polarity. Three external resistors determine the filter's Q and gain. From the equations accompanying *Figure 7*, $Q = R_3/R_2$ and the passband gain $H_{OLP} = -R_2/R_1$. Since the input signal is driving a summing junction through R_1, the input impedance will be equal to R_1. Start by choosing a value for R_1. 10k is convenient and gives a reasonable input impedance. For $H_{OLP} = -2$, we have:

$$R_2 = -R_1 H_{OLP} = 10k \times 2 = 20k.$$

For $Q = 0.707$ we have:

$$R_3 = R_2 Q = 20k \times 0.707 = 14.14k. \text{ Use } 15k.$$

For operation on ±5V supplies, V^+ is connected to +5V, V^- to −5V, and AGND to ground. The power supplies should be "clean" (regulated supplies are preferred) and 0.1 µF bypass capacitors are recommended.

3.0 Applications Information (Continued)

FIGURE 16. 2nd-Order Butterworth Low-Pass Filter of Design

Example. For $\dfrac{f_{CLK}}{f_0} = 50$**, Connect Pin 9 to** $+5V$**, and**

Change Clock Frequency to 10 kHz.

TL/H/5066-25

TL/H/5066-26

FIGURE 17. Butterworth Low-Pass Circuit of Example, but Designed for Single-Supply Operation

3.0 Applications Information (Continued)

TL/H/5066–27

(a) Resistive Divider with Decoupling Capacitor

TL/H/5066–28

(b) Voltage Regulator

TL/H/5066–29

(c) Operational Amplifier with Divider

FIGURE 18. Three Ways of Generating $\frac{V^+}{2}$ for Single-supply Operation

For a cutoff frequency of 200 Hz, the external clock can be either 10 kHz with pin 9 connected to V^+ (50:1) or 20 kHz with pin 9 tied to A_{GND} or V^- (100:1). The voltage on the Logic Level Shift pin (7) determines the logic threshold for the clock input. The threshold is approximately 2V higher than the voltage applied to pin 7. Therefore, when pin 7 is grounded, the clock logic threshold will be 2V, making it compatible with 0–5 volt TTL logic levels and ±5 volt CMOS levels. Pin 7 should be connected to a clean, low-impedance (less than 1000Ω) voltage source.

The complete circuit of the design example is shown for a 100:1 clock ratio in *Figure 16*.

3.2 SINGLE SUPPLY OPERATION

The MF5 can also operate with a single-ended power supply. *Figure 17* shows the example filter with a single-ended power supply. V^+ is again connected to the positive power supply (8 to 14 volts), and V^- is connected to ground. The A_{GND} pin must be tied to $V^+/2$ for single supply operation. This half-supply point should be very "clean", as any noise appearing on it will be treated as an input to the filter. It can be derived from the supply voltage with a pair of resistors and a bypass capacitor (*Figure 18a*), or a low-impedance half-supply voltage can be made using a three-terminal voltage regulator or an operational amplifier (*Figures 18b and 18c*). The passive resistor divider with a bypass capacitor is sufficient for many applications, provided that the time constant is long enough to reject any power supply noise. It is also important that the half-supply reference present a low impedance to the clock frequency, so at very low clock frequencies the regulator or op-amp approaches may be preferable because they will require smaller capacitors to filter the clock frequency. The main power supply voltage should be clean (preferably regulated) and bypassed with 0.1 μF.

3.3 DYNAMIC CONSIDERATIONS

The maximum signal handling capability of the MF5, like that of any active filter, is limited by the power supply voltages and speed. The amplifiers in the MF5 are able to swing to within about 1 volt of the supplies, so the input signals must be kept small enough that none of the outputs will exceed

these limits. If the MF5 is operating on ±5 volts, for example, the outputs will clip at about $8V_{p-p}$. The maximum input voltage multiplied by the filter gain should therefore be less than $8V_{p-p}$.

Note that if the filter has high Q, the gain at the lowpass or highpass outputs will be much greater than the nominal filter gain (*Figure 6*). As an example, a lowpass filter with a Q of 10 will have a 20 dB peak in its amplitude response at f_0. If the nominal gain of the filter H_{OLP} is equal to 1, the gain at f_0 will be 10. The maximum input signal at f_0 must therefore be less than 800 mV_{p-p} when the circuit is operated on ±5 volt supplies.

Also note that one output can have a reasonable small voltage on it while another is saturated. This is most likely for a circuit such as the notch in Mode 1 (*Figure 7*). The notch output will be very small at f_0, so it might appear safe to apply a large signal to the input. However, the bandpass will have its maximum gain at f_0 and can clip if overdriven. If one output clips, the performance at the other outputs will be degraded, so avoid overdriving any filter section, even ones whose outputs are not being directly used. Accompanying *Figures 7* through *15* are equations labeled "circuit dynamics", which relate the Q and the gains at the various outputs. These should be consulted to determine peak circuit gains and maximum allowable signals for a given application.

3.4 OFFSET VOLTAGE

The MF5's switched capacitor integrators have a higher equivalent input offset voltage than would be found in a typical continuous-time active filter integrator. *Figure 19* shows an equivalent circuit of the MF5 from which the output dc offsets can be calculated. Typical values for these offsets are:

V_{os1} = opamp offset = ±5mV

V_{os2} = −185mV @ 50:1   −310mV @ 100:1

V_{os3} = +115mV @ 50:1   +240mV @ 100:1

The dc offset at the BP output is equal to the input offset of the lowpass integrator (V_{os3}). The offsets at the other outputs depend on the mode of operation and the resistor ratios, as described in the following expressions.

3.0 Applications Information (Continued)

Mode 1 and Mode 4

$$V_{OS(N)} = V_{OS1}\left(\frac{1}{Q} + 1 + \|H_{OLP}\|\right) - \frac{V_{OS3}}{Q}$$

$$V_{OS(BP)} = V_{OS3}$$
$$V_{OS(LP)} = V_{OS(N)} - V_{OS2}$$

Mode 1a

$$V_{OS(N.INV.BP)} = \left(1 + \frac{1}{Q}\right)V_{OS1} - \frac{V_{OS3}}{Q}$$

$$V_{OS(INV.BP)} = V_{OS3}$$
$$V_{OS(LP)} = V_{OS(N.INV.BP)} - V_{OS2}$$

Mode 2 and Mode 5

$$V_{OS(N)} = \left(\frac{R2}{Rp} + 1\right)V_{OS1} \times \frac{1}{1 + R2/R4}$$
$$+ V_{OS2}\frac{1}{1 + R4/R2} - \frac{V_{OS3}}{Q\sqrt{1 + R2/R4}}:$$
$$R_p = R1//R2//R4$$

$$V_{OS(BP)} = V_{OS3}$$
$$V_{OS(LP)} = V_{OS(N)} - V_{OS2}$$

Mode 3

$$V_{OS(HP)} = V_{OS2}$$
$$V_{OS(BP)} = V_{OS3}$$
$$V_{OS(LP)} = -\frac{R4}{R2}\left(\frac{R2}{R3}V_{OS3} + V_{OS2}\right) +$$
$$-\frac{R4}{R2}\left(1 + \frac{R2}{Rp}\right)V_{OS1}; R_p = R1//R3//R4$$

FIGURE 19. Block Diagram Showing MF5
Offset Voltage Sources

TL/H/5066-30

FIGURE 20. Method for Trimming V_OS,
See Text, Section 3.4

TL/H/5066-31

3.0 Applications Information (Continued)

For most applications, the outputs are AC coupled and DC offsets are not bothersome unless large signals are applied to the filter input. However, larger offset voltages will cause clipping to occur at lower ac signal levels, and clipping at any of the outputs will cause gain nonlinearities and will change f_o and Q. When operating in Mode 3, offsets can become excessively large if R_2 and R_4 are used to make f_{CLK}/f_o significantly higher than the nominal value, especially if Q is also high. An extreme example is a bandpass filter having unity gain, a Q of 20, and f_{CLK}/f_o = 250 with pin 9 tied to V$^-$ (100:1 nominal). R_4/R_2 will therefore be equal to 6.25 and the offset voltage at the lowpass output will be about +1.9V. Where necessary, the offset voltage can be adjusted by using the circuit of Figure 20. This allows adjustment of V_{os1}, which will have varying effects on the different outputs as described in the above equations. Some outputs cannot be adjusted this way in some modes, however ($V_{os(BP)}$ in modes 1a and 3, for example).

3.5 SAMPLED DATA SYSTEM CONSIDERATIONS

The MF5 is a sampled data filter, and as such, differs in many ways from conventional continuous-time filters. An important characteristic of sampled-data systems is their effect on signals at frequencies greater than one-half the sampling frequency. (The MF5's sampling frequency is the same as its clock frequency). If a signal with a frequency greater than one-half the sampling frequency is applied to the input of a sampled data system, it will be "reflected" to a frequency less than one-half the sampling frequency. Thus, an input signal whose frequency is $f_s/2$ + 100 Hz will cause the system to respond as though the input frequency was $f_s/2$ - 100 Hz. This phenomenon is known as "alias-

ing", and can be reduced or eliminated by limiting the input signal spectrum to less than $f_s/2$. This may in some cases require the use of a bandwidth-limiting filter ahead of the MF5 to limit the input spectrum. However, since the clock frequency is much higher than the center frequency, this will often not be necessary.

Another characteristic of sampled-data circuits is that the output signal changes amplitude once every sampling period, resulting in "steps" in the output voltage which occur at the clock rate. (Figure 21) If necessary, these can be "smoothed" with a simple R-C low-pass filter at the MF5 output.

The ratio of f_{CLK} to f_c (normally either 50:1 or 100:1) will also affect performance. A ratio of 100:1 will reduce any aliasing problems and is usually recommended for wideband input signals. In noise sensitive applications, however, a ratio of 50:1 may be better as it will result in 3 dB lower output noise. The 50:1 ratio also results in lower DC offset voltages, as discussed in 3.4.

The accuracy of the f_{CLK}/f_o ratio is dependent on the value of Q. This is illustrated in the curves under the heading "Typical Performance Characteristics". As Q is changed, the true value of the ratio changes as well. Unless the Q is low, the error in f_{CLK}/f_o will be small. If the error is too large for a specific application, use a mode that allows adjustment of the ratio with external resistors.

It should also be noted that the product of Q and f_o should be limited to 300 kHz when f_o < 5 kHz, and to 200 kHz for f_o > 5 kHz.

100:1

50:1

TL/H/5066-32

FIGURE 21. The Sampled-Data Output Waveform

Physical Dimensions inches (millimeters)

Cavity Dual-In-Line Package (J)
Order Number MF5J
NS Package Number J14A

Molded Dual-In-Line Package (N)
Order Number MF5N
NS Package Number N14A

National Semiconductor

LM380 Audio Power Amplifier

General Description

The LM380 is a power audio amplifier for consumer application. In order to hold system cost to a minimum, gain is internally fixed at 34 dB. A unique input stage allows inputs to be ground referenced. The output is automatically self centering to one half the supply voltage.

The output is short circuit proof with internal thermal limiting. The package outline is standard dual-in-line. A copper lead frame is used with the center three pins on either side comprising a heat sink. This makes the device easy to use in standard p-c layout.

Uses include simple phonograph amplifiers, intercoms, line drivers, teaching machine outputs, alarms, ultrasonic drivers, TV sound systems, AM–FM radio, small servo drivers, power converters, etc.

A selected part for more power on higher supply voltages is available as the LM384. For more information see AN-69.

Features

- Wide supply voltage range
- Low quiescent power drain
- Voltage gain fixed at 50
- High peak current capability
- Input referenced to GND
- High input impedance
- Low distortion
- Quiescent output voltage is at one-half of the supply voltage
- Standard dual-in-line package

Connection Diagrams (Dual-In-Line Packages, Top View)

BYPASS 1 — 14 V_S
NON-INVERTING INPUT 2 — 13 NC
3 — 12
*GND { 4 — 11 } GND
5 — 10
INVERTING INPUT 6 — 9 NC
GND 7 — 8 V_{OUT}

*Heatsink Pins

Order Number LM380N
See NS Package N14A

NC 1 — 8 BYPASS
NON-INVERTING INPUT 2 — 7 V_S
INVERTING INPUT 3 — 6 V_{OUT}
GND 4 — 5 GND

Order Number LM380N-8
See NS Package N08B

Block and Schematic Diagrams

LM380N

LM380N-8

Absolute Maximum Ratings

Supply Voltage	22V
Peak Current	1.3A
Package Dissipation 14-Pin DIP (Notes 6 and 7)	10W
Input Voltage	±0.5V
Storage Temperature	$-65°C$ to $+150°C$
Operating Temperature	$0°C$ to $+70°C$
Junction Temperature	$+150°C$
Lead Temperature (Soldering, 10 sec)	$+300°C$

Electrical Characteristics (Note 1)

PARAMETER	SYMBOL	CONDITIONS	MIN	TYP	MAX	UNITS
Output Power	$P_{OUT(RMS)}$	(Notes 3, 4) $R_L = 8\Omega$, THD = 3%	2.5			W
Gain	A_V		40	50	60	V/V
Output Voltage Swing	V_{OUT}	$R_L = 8\Omega$		14		V_{p-p}
Input Resistance	Z_{IN}			150k		Ω
Total Harmonic Distortion	THD	(Note 4, 5)		0.2		%
Power Supply Rejection Ratio	PSRR	(Note 2)		38		dB
Supply Voltage	V_S	(Note 8)	10		22	V
Bandwidth	BW	$P_{OUT} = 2W$, $R_L = 8\Omega$		100k		Hz
Quiescent Supply Current	I_Q			7	25	mA
Quiescent Output Voltage	V_{OUTQ}		8	9.0	10	V
Bias Current	I_{BIAS}	Inputs Floating		100		nA
Short Circuit Current	I_{SC}			1.3		A

Note 1: V_S = 18V and $T_A = 25°C$ unless otherwise specified.
Note 2: Rejection ratio referred to the output with C_{BYPASS} = 5 µF.
Note 3: With device Pins 3, 4, 5, 10, 11, 12 soldered into a 1/16'' epoxy glass board with 2 ounce copper foil with a minimum surface of 6 square inches.
Note 4: If oscillation exists under some load conditions, add 2.7Ω and 0.1 µfd series network from Pin 8 to Gnd.
Note 5: C_{BYPASS} = 0.47 µfd on Pin 1.
Note 6: The maximum junction temperature of the LM380 is 150°C.
Note 7: The package is to be derated at 12°C/W junction to heat sink pins.
Note 8: Can select for 8V operation.

Heat Sink Dimensions

COPPER WINGS 2 REQUIRED SOLDERED TO PINS 3, 4, 5, 10, 11, 12 THICKNESS 0.04 INCHES

Typical Performance Characteristics

Typical Applications

Phono Amplifier

Bridge Amplifier

Intercom

*FOR STABILITY WITH
HIGH CURRENT LOADS

Phase Shift Oscillator

MC7800 Series

Advance Information

3-TERMINAL POSITIVE VOLTAGE REGULATORS

These voltage regulators are monolithic integrated circuits designed as fixed-voltage regulators for a wide variety of applications including local, on-card regulation. These regulators employ internal current limiting, thermal shutdown, and safe-area compensation. With adequate heatsinking they can deliver output currents in excess of 1.0 ampere. Although designed primarily as a fixed voltage regulator, these devices can be used with external components to obtain adjustable voltages and currents.

- Output Current in Excess of 1.0 Ampere
- No External Components Required
- Internal Thermal Overload Protection
- Internal Short-Circuit Current Limiting
- Output Transistor Safe-Area Compensation
- Output Voltage Offered in 2% and 4% Tolerance

K SUFFIX
METAL PACKAGE
CASE 1
(TO-3 TYPE)

PIN	1.	INPUT
	2.	OUTPUT
CASE		GROUND

T SUFFIX
PLASTIC PACKAGE
CASE 221A
TO-220 TYPE

PIN	1.	INPUT
	2.	GROUND
	3.	OUTPUT

SCHEMATIC DIAGRAM

Pin 2 is ground
for Case 221A.
Case is ground
for Case 1.

STANDARD APPLICATION

A common ground is required between the input and the output voltages. The input voltage must remain typically 2.0 V above the output voltage even during the low point on the input ripple voltage.

XX = these two digits of the type number indicate voltage.

* = C_{in} is required if regulator is located an appreciable distance from power supply filter.

** = C_O is not needed for stability; however, it does improve transient response.

XX indicates nominal voltage

ORDERING INFORMATION

Device	Output Voltage Tolerance	Temperature Range	Package
MC78XXK	4%	−55 to +150°C	Metal Power
MC78XXAK	2%		
MC78XXCK	4%	0 to +125°C	
MC78XXACK	2%		
MC78XXCT	4%		Plastic Power
MC78XXACT	2%		

XX Indicates Nominal Voltage
This is advance information and specifications are subject to change without notice.

TYPE NO'/VOLTAGE			
MC7805	5.0 Volts	MC7815	15 Volts
MC7806	6.0 Volts	MC7818	18 Volts
MC7808	8.0 Volts	MC7824	24 Volts
MC7812	12 Volts		

MC7800 Series MAXIMUM RATINGS (T_A = +25°C unless otherwise noted.)

Rating	Symbol	Value	Unit
Input Voltage (5.0 V – 18 V) (24 V)	V_{in}	35 40	Vdc
Power Dissipation and Thermal Characteristics Plastic Package T_A = +25°C Derate above T_A = +25°C Thermal Resistance, Junction to Air T_C = +25°C Derate above T_C = +95°C (See Figure 1) Thermal Resistance, Junction to Case Metal Package T_A = +25°C Derate above T_A = +25°C Thermal Resistance, Junction to Air T_C = +25°C Derate above T_C = +65°C (See Figure 2) Thermal Resistance, Junction to Case	P_D 1/θ_{JA} θ_{JA} P_D 1/θ_{JC} θ_{JC} P_D 1/θ_{JA} θ_{JA} P_D 1/θ_{JC} θ_{JC}	Internally Limited 15.4 65 Internally Limited 200 5.0 Internally Limited 22.5 45 Internally Limited 182 5.5	Watts mW/$^{\circ}$C $^{\circ}$C/W Watts mW/$^{\circ}$C $^{\circ}$C/W Watts mW/$^{\circ}$C $^{\circ}$C/W Watts mW/$^{\circ}$C $^{\circ}$C/W
Storage Junction Temperature Range	T_{stg}	−65 to +150	$^{\circ}$C
Operating Junction Temperature Range MC7800, A MC7800C, AC	T_J	−55 to +150 0 to +150	$^{\circ}$C

DEFINITIONS

Line Regulation – The change in output voltage for a change in the input voltage. The measurement is made under conditions of low dissipation or by using pulse techniques such that the average chip temperature is not significantly affected.

Load Regulation – The change in output voltage for a change in load current at constant chip temperature.

Maximum Power Dissipation – The maximum total device dissipation for which the regulator will operate within specifications.

Quiescent Current – That part of the input current that is not delivered to the load.

Output Noise Voltage – The rms ac voltage at the output, with constant load and no input ripple, measured over a specified frequency range.

Long Term Stability – Output voltage stability under accelerated life test conditions with the maximum rated voltage listed in the devices' electrical characteristics and maximum power dissipation.

MC7800 Series

MC7805, C
ELECTRICAL CHARACTERISTICS (V_{in} = 10V, I_O = 500 mA, T_J = T_{low} to T_{high} (Note 1) unless otherwise noted).

Characteristic	Symbol	MC7805			MC7805C			Unit
		Min	Typ	Max	Min	Typ	Max	
Output Voltage (T_J = +25°C)	V_O	4.8	5.0	5.2	4.8	5.0	5.2	Vdc
Output Voltage (5.0 mA ≤ I_O ≤ 1.0 A, P_O ≤ 15W) 7.0 Vdc ≤ V_{in} ≤ 20 Vdc 8.0 Vdc ≤ V_{in} ≤ 20 Vdc	V_O	— 4.65	5.0 5.0	— 5.35	4.75 —	5.0 —	5.25 —	Vdc
Line Regulation (T_J = +25°C, Note 2) 7.0 Vdc ≤ V_{in} ≤ 25 Vdc 8.0 Vdc ≤ V_{in} ≤ 12 Vdc	Reg_{in}	— —	2.0 1.0	50 25	— —	7.0 2.0	100 50	mV
Load Regulation (T_J = +25°C, Note 2) 5.0 mA ≤ I_O ≤ 1.5 A 250 mA ≤ I_O ≤ 750 mA	Reg_{load}	— —	25 8.0	100 25	— —	40 15	100 50	mV
Quiescent Current (T_J = +25°C)	I_B	—	3.2	6.0	—	4.3	8.0	mA
Quiescent Current Change 7.0 Vdc ≤ V_{in} ≤ 25 Vdc 8.0 Vdc ≤ V_{in} ≤ 25 Vdc 5.0 mA ≤ I_O ≤ 1.0 A	ΔI_B	— — —	— 0.3 0.04	— 0.8 0.5	— — —	— — —	1.3 — 0.5	mA
Ripple Rejection 8.0 Vdc ≤ V_{in} ≤ 18 Vdc, f = 120 Hz	RR	68	75	—	—	68	—	dB
Dropout Voltage (I_O = 1.0 A, T_J = +25°C)	V_{in} · V_O	—	2.0	2.5	—	2.0	—	Vdc
Output Noise Voltage (T_A = +25°C) 10 Hz ≤ f ≤ 100 kHz	V_n	—	10	40	—	10	—	µV/V_O
Output Resistance f = 1.0 kHz	R_O	—	17	—	—	17	—	mΩ
Short-Circuit Current Limit (T_A = +25°C) V_{in} = 35 Vdc	I_{sc}	—	0.2	1.2	—	0.2	—	A
Peak Output Current (T_J = +25°C)	I_{max}	1.3	2.5	3.3	—	2.2	—	A
Average Temperature Coefficient of Output Voltage	TCV_O	—	±0.6	—	—	–1.1	—	mV/°C

MC7805A, AC
ELECTRICAL CHARACTERISTICS (V_{in} = 10 V, I_O = 1.0 A, T_J = T_{low} to T_{high} (Note 1) unless otherwise noted)

Characteristics	Symbol	MC7805A			MC7805AC			Unit
		Min	Typ	Max	Min	Typ	Max	
Output Voltage (T_J = +25°C)	V_O	4.9	5.0	5.1	4.9	5.0	5.1	Vdc
Output Voltage (5.0 mA ≤ I_O ≤ 1.0 A, P_O ≤ 15 W) 7.5 Vdc ≤ V_{in} ≤ 20 Vdc	V_O	4.8	5.0	5.2	4.8	5.0	5.2	Vdc
Line Regulation (Note 2) 7.5 Vdc ≤ V_{in} ≤ 25 Vdc, I_O = 500 mA 8.0 Vdc ≤ V_{in} ≤ 12 Vdc 8.0 Vdc ≤ V_{in} ≤ 12 Vdc, T_J = +25°C 7.3 Vdc ≤ V_{in} ≤ 20 Vdc, T_J = +25°C	Reg_{in}	— — — —	2.0 3.0 1.0 2.0	10 10 4.0 10	— — — —	7.0 10 2.0 7.0	50 50 25 50	mV
Load Regulation (Note 2) 5.0 mA ≤ I_O ≤ 1.5 A 5.0 mA ≤ I_O ≤ 1.0 A 5.0 mA ≤ I_O ≤ 1.5 A, T_J = +25°C 250 mA ≤ I_O ≤ 750 mA	Reg_{load}	— — — —	25 — — 8.0	50 — — 25	— — — —	— 25 25 8.0	— 100 100 50	mV
Quiescent Current T_J = +25°C	I_B	— —	— 3.2	5.0 4.0	— —	— 4.3	6.0 6.0	mA
Quiescent Current Change 8.0 Vdc ≤ V_{in} ≤ 25 Vdc, I_O = 500 mA 7.5 Vdc ≤ V_{in} ≤ 20 Vdc, T_J = +25°C 5.0 mA ≤ I_O ≤ 1.0 A	ΔI_B	— — —	0.3 0.2 0.04	0.5 0.5 0.2	— — —	— — —	0.8 0.8 0.5	mA
Ripple Rejection 8.0 Vdc ≤ V_{in} ≤ 18 Vdc, f = 120 Hz, T_J = +25°C 8.0 Vdc ≤ V_{in} ≤ 18 Vdc, f = 120 Hz, I_O = 500 mA	RR	68 68	75 75	— —	— —	— 68	— —	dB
Dropout Voltage (I_O = 1.0 A, T_J = +25°C)	V_{in}·V_O	—	2.0	2.5	—	2.0	—	Vdc
Output Noise Voltage (T_A = +25°C) 10 Hz ≤ f ≤ 100 kHz	V_n	—	10	40	—	10	—	µV/V_O
Output Resistance (f = 1.0 kHz)	R_O	—	17	—	—	17	—	mΩ
Short-Circuit Current Limit (T_A = +25°C) V_{in} = 35 Vdc	I_{sc}	—	0.2	1.2	—	0.2	—	A
Peak Output Current (T_J = +25°C)	I_{max}	1.3	2.5	3.3	—	2.2	—	A
Average Temperature Coefficient of Output Voltage	TCV_O	—	±0.6	—	—	–1.1	—	mV/°C

Notes: 1. T_{low} = –55°C for MC78XX, A
          = 0°C for MC78XXC, AC

   T_{high} = +150°C for MC78XX, A
           = +125°C for MC78XXC, AC

2. Load and line regulation are specified at constant junction temperature. Changes in V_O due to heating effects must be taken into account separately. Pulse testing with low duty cycle is used.

MC7806, C
ELECTRICAL CHARACTERISTICS (V_{in} = 11 V, I_O = 500 mA, T_J = T_{low} to T_{high} (Note 1) unless otherwise noted.)

Characteristic	Symbol	MC7806			MC7806C			Unit
		Min	Typ	Max	Min	Typ	Max	
Output Voltage (T_J = +25°C)	V_O	5.75	6.0	6.25	5.75	6.0	6.25	Vdc
Output Voltage (5.0 mA ≤ I_O ≤ 1.0 A, P_O ≤ 15W)	V_O							Vdc
8.0 Vdc ≤ V_{in} ≤ 21 Vdc		—	—	—	5.7	6.0	6.3	
9.0 Vdc ≤ V_{in} ≤ 21 Vdc		5.65	6.0	6.35	—	—	—	
Line Regulation (T_J = +25°C, Note 2)	Reg_{in}							mV
8.0 Vdc ≤ V_{in} ≤ 25 Vdc		—	3.0	60	—	9.0	120	
9.0 Vdc ≤ V_{in} ≤ 13 Vdc		—	2.0	30	—	3.0	60	
Load Regulation (T_J = +25°C, Note 2)	Reg_{load}							mV
5.0 mA ≤ I_O ≤ 1.5 A		—	27	100	—	43	120	
250 mA ≤ I_O ≤ 750 mA		—	9.0	30	—	16	60	
Quiescent Current (T_J = +25°C)	I_B	—	3.2	6.0	—	4.3	8.0	mA
Quiescent Current Change	ΔI_B							mA
8.0 Vdc ≤ V_{in} 25 Vdc		—	—	—	—	—	1.3	
9.0 Vdc ≤ V_{in} 25 Vdc		—	0.3	0.8	—	—	—	
5.0 mA ≤ I_O ≤ 1.0 A		—	0.04	0.5	—	—	0.5	
Ripple Rejection	RR							dB
9.0 Vdc ≤ V_{in} ≤ 19 Vdc, f = 120 Hz		65	73	—	—	65	—	
Dropout Voltage (I_O = 1.0 A, T_J = +25°C)	V_{in}-V_O	—	2.0	2.5	—	2.0	—	Vdc
Output Noise Voltage (T_A = +25°C) 10 Hz ≤ f ≤ 100 kHz	V_n	—	10	40	—	10	—	μV/V_O
Output Resistance (f = 1.0 kHz)	R_O	—	17	—	—	17	—	mΩ
Short-Circuit Current Limit (T_A = +25°C) V_{in} = 35 Vdc	I_{sc}	—	0.2	1.2	—	0.2	—	A
Peak Output Current (T_J = +25°C)	I_{max}	1.3	2.5	3.3	—	2.2	—	A
Average Temperature Coefficient of Output Voltage	TCV_O	—	±0.7	—	—	0.8	—	mV/°C

MC7806A, AC
ELECTRICAL CHARACTERISTICS (V_{in} = 11 V, I_O = 1.0 A, T_J = T_{low} to T_{high} (Note 1) unless otherwise noted.)

Characteristic	Symbol	MC7806A			MC7806AC			Unit
		Min	Typ	Max	Min	Typ	Max	
Output Voltage (T_J = +25°C)	V_O	5.88	6.0	6.12	5.88	6.0	6.12	Vdc
Output Voltage (5.0 mA ≤ I_O ≤ 1.0 A, P_O ≤ 15W) 8.6 Vdc ≤ V_{in} ≤ 21 Vdc	V_O	5.76	6.0	6.24	5.76	6.0	6.24	Vdc
Line Regulation (Note 2)	Reg_{in}							mV
8.6 Vdc ≤ V_{in} ≤ 25 Vdc, I_O = 500 mA		—	3.0	11	—	9.0	60	
9.0 Vdc ≤ V_{in} ≤ 13 Vdc		—	5.0	15	—	11	60	
9.0 Vdc ≤ V_{in} ≤ 13 Vdc, T_J = +25°C		—	2.0	5.0	—	3.0	30	
8.3 Vdc ≤ V_{in} ≤ 21 Vdc, T_J = +25°C		—	4.0	11	—	9.0	60	
Load Regulation (Note 2)	Reg_{load}							mV
5.0 mA ≤ I_O ≤ 1.5A		—	27	50	—	—	—	
5.0 mA ≤ I_O ≤ 1.0 A		—	—	—	—	43	100	
5.0 mA ≤ I_O ≤ 1.5 A, T_J = +25°C		—	—	—	—	43	100	
250 mA ≤ I_O ≤ 750 mA		—	9.0	25	—	16	50	
Quiescent Current	I_B							mA
T_J = +25°C		—	—	5.0	—	—	6.0	
		—	3.2	4.0	—	4.3	6.0	
Quiescent Current Change	ΔI_B							mA
9.0 Vdc ≤ V_{in} ≤ 25 Vdc, I_O = 500 mA		—	0.3	0.5	—	—	0.8	
8.6 Vdc ≤ V_{in} ≤ 21 Vdc, T_J = +25°C		—	0.2	0.5	—	—	0.8	
5.0 mA ≤ I_O ≤ 1.0 A		—	0.04	0.2	—	—	0.5	
Ripple Rejection	RR							dB
9.0 Vdc ≤ V_{in} ≤ 19 Vdc, f = 120 Hz, T_J = +25°C		65	73	—	—	—	—	
9.0 Vdc ≤ V_{in} ≤ 19 Vdc, f = 120 Hz, I_O = 500 mA		65	73	—	—	65	—	
Dropout Voltage (I_O = 1.0 A, T_J = +25°C)	V_{in}-V_O	—	2.0	2.5	—	2.0	—	Vdc
Output Noise Voltage (T_A = +25°C) 10 Hz ≤ f ≤ 100 kHz	V_n	—	10	40	—	10	—	μV/V_O
Output Resistance (f = 1.0 kHz)	R_O	—	17	—	—	17	—	mΩ
Short-Circuit Current Limit (T_A = +25°C) V_{in} = 35 Vdc	I_{sc}	—	0.2	1.2	—	0.2	—	A
Peak Output Current (T_J = +25°C)	I_{max}	1.3	2.5	3.3	—	2.2	—	A
Average Temperature Coefficient of Output Voltage	TCV_O	—	±0.7	—	—	-0.8	—	mV/°C

Notes: 1. T_{low} = −55°C for MC78XX, A
  = 0°C for MC78XXC, AC

 T_{high} = +150°C for MC78XX, A
  = +125°C for MC78XXC, AC

2. Load and line regulation are specified at constant junction temperature. Changes in V_O due to heating effects must be taken into account separately. Pulse testing with low duty cycle is used.

MC7800 Series

MC7808, C
ELECTRICAL CHARACTERISTICS (V_{in} = 14 V, I_O = 500 mA, T_J = T_{low} to T_{high} (Note 1) unless otherwise noted.)

Characteristic	Symbol	MC7808			MC7808C			Unit
		Min	Typ	Max	Min	Typ	Max	
Output Voltage (T_J = +25°C)	V_O	7.7	8.0	8.3	7.7	8.0	8.3	Vdc
Output Voltage	V_O							Vdc
(5.0 mA ≤ I_O ≤ 1.0 A, P_O ≤ 15W)								
10.5 Vdc ≤ V_{in} ≤ 23 Vdc		−	−	−	7.6	8.0	8.4	
11.5 Vdc ≤ V_{in} ≤ 23 Vdc		7.6	8.0	8.4	−	−	−	
Line Regulation (T_J = +25°C, Note 2)	Reg_{in}							mV
10.5 Vdc ≤ V_{in} ≤ 25 Vdc		−	3.0	80	−	12	160	
11 Vdc ≤ V_{in} ≤ 17 Vdc		−	2.0	40	−	5.0	80	
Load Regulation (T_J = +25°C, Note 2)	Reg_{load}							mV
5.0 mA ≤ I_O ≤ 1.5 A		−	28	100	−	45	160	
250 mA ≤ I_O ≤ 750 mA		−	9.0	40	−	16	80	
Quiescent Current (T_J = +25°C)	I_B	−	3.2	6.0	−	4.3	8.0	mA
Quiescent Current Change	ΔI_B							mA
10.5 Vdc ≤ V_{in} 25 Vdc		−	−	−	−	−	1.0	
11.5 Vdc ≤ V_{in} 25 Vdc		−	0.3	0.8	−	−	−	
5.0 mA ≤ I_O ≤ 1.0 A		−	0.04	0.5	−	−	0.5	
Ripple Rejection	RR							dB
11.5 Vdc ≤ V_{in} ≤ 21.5 Vdc, f = 120 Hz		62	70	−	−	62	−	
Dropout Voltage (I_O = 1.0 A, T_J = +25°C)	V_{in}-V_O	−	2.0	2.5	−	2.0	−	Vdc
Output Noise Voltage (T_A = +25°C)	V_n							μV/V_O
10 Hz ≤ f ≤ 100 kHz		−	10	40	−	10	−	
Output Resistance (f = 1.0 kHz)	R_O	−	18	−	−	18	−	mΩ
Short-Circuit Current Limit (T_A = +25°C)	I_{sc}							A
V_{in} = 35 Vdc		−	0.2	1.2	−	0.2	−	
Peak Output Current (T_J = +25°C)	I_{max}	1.3	2.5	3.3	−	2.2	−	A
Average Temperature Coefficient of Output Voltage	TCV_O	−	±1.0	−	−	−0.8	−	mV/°C

MC7808A, AC
ELECTRICAL CHARACTERISTICS (V_{in} = 14 V, I_O = 1.0 A, T_J = T_{low} to T_{high} (Note 1) unless otherwise noted.)

Characteristic	Symbol	MC7808A			MC7808AC			Unit
		Min	Typ	Max	Min	Typ	Max	
Output Voltage (T_J = +25°C)	V_O	7.84	8.0	8.16	7.84	8.0	8.16	Vdc
Output Voltage	V_O							Vdc
(5.0 mA ≤ I_O ≤ 1.0 A, P_O ≤ 15W)								
10.6 Vdc ≤ V_{in} ≤ 23 Vdc		7.7	8.0	8.3	7.7	8.0	8.3	
Line Regulation (Note 2)	Reg_{in}							mV
10.6 Vdc ≤ V_{in} ≤ 25 Vdc, I_O = 500 mA		−	4.0	13	−	12	80	
11 Vdc ≤ V_{in} ≤ 17 Vdc		−	6.0	20	−	15	80	
11 Vdc ≤ V_{in} ≤ 17 Vdc, T_J = +25°C		−	2.0	6.0	−	5.0	40	
10.4 Vdc ≤ V_{in} ≤ 23 Vdc, T_J = +25°C		−	4.0	13	−	12	80	
Load Regulation (Note 2)	Reg_{load}							mV
5.0 mA ≤ I_O ≤ 1.5A		−	28	50	−	−	−	
5.0 mA ≤ I_O ≤ 1.0 A		−	−	−	−	45	100	
5.0 mA ≤ I_O ≤ 1.5 A, T_J = +25°C		−	−	−	−	45	100	
250 mA ≤ I_O ≤ 750 mA		−	9.0	25	−	16	50	
Quiescent Current	I_B							mA
T_J = +25°C		−	−	5.0	−	−	6.0	
		−	3.2	4.0	−	4.3	6.0	
Quiescent Current Change	ΔI_B							mA
11 Vdc ≤ V_{in} ≤ 25 Vdc, I_O = 500 mA		−	0.3	0.5	−	−	0.8	
10.6 Vdc ≤ V_{in} ≤ 23 Vdc, T_J = +25°C		−	0.2	0.5	−	−	0.8	
5.0 mA ≤ I_O ≤ 1.0 A		−	0.04	0.2	−	−	0.5	
Ripple Rejection	RR							dB
11.5 Vdc ≤ V_{in} ≤ 21.5 Vdc, f = 120 Hz, T_J = +25°C		62	70	−	−	−	−	
11.5 Vdc ≤ V_{in} ≤ 21.5 Vdc, f = 120 Hz, I_O = 500 mA		62	70	−	−	62	−	
Dropout Voltage (I_O = 1.0 A, T_J = +25°C)	V_{in}-V_O	−	2.0	2.5	−	2.0	−	Vdc
Output Noise Voltage (T_A = +25°C)	V_n							μV/V_O
10 Hz ≤ f ≤ 100 kHz		−	10	40	−	10	−	
Output Resistance (f = 1.0 kHz)	R_O	−	18	−	−	18	−	mΩ
Short-Circuit Current Limit (T_A = +25°C)	I_{sc}							A
V_{in} = 35 Vdc		−	0.2	1.2	−	0.2	−	
Peak Output Current (T_J = +25°C)	I_{max}	1.3	2.5	3.3	−	2.2	−	A
Average Temperature Coefficient of Output Voltage	TCV_O	−	±1.0	−	−	−0.8	−	mV/°C

Notes: 1. T_{low} = −55°C for MC78XX, A
         = 0°C for MC78XXC, AC

    T_{high} = +150°C for MC78XX, A
          = +125°C for MC78XXC, AC

2. Load and line regulation are specified at constant junction temperature. Changes in V_O due to heating effects must be taken into account separately. Pulse testing with low duty cycle is used.

MC7800 Series

MC7812, C
ELECTRICAL CHARACTERISTICS (V_{in} = 19 V, I_O = 500 mA, $T_J = T_{low}$ to T_{high} (Note 1) unless otherwise noted.)

Characteristic	Symbol	MC7812			MC7812C			Unit
		Min	Typ	Max	Min	Typ	Max	
Output Voltage (T_J = +25°C)	V_O	11.5	12	12.5	11.5	12	12.5	Vdc
Output Voltage (5.0 mA ≤ I_O ≤ 1.0 A, P_O ≤ 15W)	V_O							Vdc
14.5 Vdc ≤ V_{in} ≤ 27 Vdc		—	—	—	11.4	12	12.6	
15.5 Vdc ≤ V_{in} ≤ 27 Vdc		11.4	12	12.6	—	—	—	
Line Regulation (T_J = +25°C, Note 2)	Reg_{in}							mV
14.5 Vdc ≤ V_{in} ≤ 30 Vdc		—	5.0	120	—	13	240	
16 Vdc ≤ V_{in} ≤ 22 Vdc		—	3.0	60	—	6.0	120	
Load Regulation (T_J = +25°C, Note 2)	Reg_{load}							mV
5.0 mA ≤ I_O ≤ 1.5 A		—	30	120	—	46	240	
250 mA ≤ I_O ≤ 750 mA		—	10	60	—	17	120	
Quiescent Current (T_J = +25°C)	I_B	—	3.4	6.0	—	4.4	8.0	mA
Quiescent Current Change	ΔI_B							mA
14.5 Vdc ≤ V_{in} ≤ 30 Vdc		—	—	—	—	—	1.0	
15 Vdc ≤ V_{in} ≤ 30 Vdc		—	0.3	0.8	—	—	—	
5.0 mA ≤ I_O ≤ 1.0 A		—	0.04	0.5	—	—	0.5	
Ripple Rejection	RR							dB
15 Vdc ≤ V_{in} ≤ 25 Vdc, f = 120 Hz		61	68	—	—	60	—	
Dropout Voltage (I_O = 1.0 A, T_J = +25°C)	V_{in}-V_O	—	2.0	2.5	—	2.0	—	Vdc
Output Noise Voltage (T_A = +25°C) 10 Hz ≤ f ≤ 100 kHz	V_n	—	10	40	—	10	—	µV/V_O
Output Resistance (f = 1.0 kHz)	R_O	—	18	—	—	18	—	mΩ
Short-Circuit Current Limit (T_A = +25°C) V_{in} = 35 Vdc	I_{sc}	—	0.2	1.2	—	0.2	—	A
Peak Output Current (T_J = +25°C)	I_{max}	1.3	2.5	3.3	—	2.2	—	A
Average Temperature Coefficient of Output Voltage	TCV_O	—	±1.5	—	—	−1.0	—	mV/°C

MC7812A, AC
ELECTRICAL CHARACTERISTICS (V_{in} = 19 V, I_O = 1.0 A, $T_J = T_{low}$ to T_{high} (Note 1) unless otherwise noted.)

Characteristic	Symbol	MC7812A			MC7812AC			Unit
		Min	Typ	Max	Min	Typ	Max	
Output Voltage (T_J = +25°C)	V_O	11.75	12	12.25	11.75	12	12.25	Vdc
Output Voltage (5.0 mA ≤ I_O ≤ 1.0 A, P_O ≤ 15W) 14.8 Vdc ≤ V_{in} ≤ 27 Vdc	V_O	11.5	12	12.5	11.5	12	12.5	Vdc
Line Regulation	Reg_{in}							mV
14.8 Vdc ≤ V_{in} ≤ 30 Vdc, I_O = 500 mA		—	5.0	18	—	13	120	
16 Vdc ≤ V_{in} ≤ 22 Vdc		—	8.0	30	—	16	120	
16 Vdc ≤ V_{in} ≤ 22 Vdc, T_J = +25°C		—	3.0	9.0	—	6.0	60	
14.5 Vdc ≤ V_{in} ≤ 27 Vdc, T_J = +25°C		—	5.0	18	—	13	120	
Load Regulation (Note 2)	Reg_{load}							mV
5.0 mA ≤ I_O ≤ 1.5A		—	30	50	—	—	—	
5.0 mA ≤ I_O ≤ 1.0 A		—	—	—	—	46	100	
5.0 mA ≤ I_O ≤ 1.5 A, T_J = +25°C		—	—	—	—	46	100	
250 mA ≤ I_O ≤ 750 mA		—	10	25	—	17	50	
Quiescent Current	I_B							mA
T_J = +25°C		—	—	5.0	—	—	6.0	
		—	3.4	4.0	—	4.4	6.0	
Quiescent Current Change	ΔI_B							mA
15 Vdc ≤ V_{in} ≤ 30 Vdc, I_O = 500 mA		—	0.3	0.5	—	—	0.8	
14.8 Vdc ≤ V_{in} ≤ 27 Vdc, T_J = +25°C		—	0.2	0.5	—	—	0.8	
5.0 mA ≤ I_O ≤ 1.0 A		—	0.04	0.2	—	—	0.5	
Ripple Rejection	RR							dB
15 Vdc ≤ V_{in} ≤ 25 Vdc, f = 120 Hz, T_J = +25°C		61	68	—	—	—	—	
15 Vdc ≤ V_{in} ≤ 25 Vdc, f = 120 Hz, I_O = 500 mA		61	68	—	—	60	—	
Dropout Voltage (I_O = 1.0 A, T_J = +25°C)	V_{in}·V_O	—	2.0	2.5	—	2.0	—	Vdc
Output Noise Voltage (T_A = +25°C) 10 Hz ≤ f ≤ 100 kHz	V_n	—	10	40	—	10	—	µV/V_O
Output Resistance (f = 1.0 kHz)	R_O	—	18	—	—	18	—	mΩ
Short-Circuit Current Limit (T_A = +25°C) V_{in} = 35 Vdc	I_{sc}	—	0.2	1.2	—	0.2	—	A
Peak Output Current (T_J = +25°C)	I_{max}	1.3	2.5	3.3	—	2.2	—	A
Average Temperature Coefficient of Output Voltage	TCV_O	—	±1.5	—	—	−1.0	—	mV/°C

Notes: 1. T_{low} = −55°C for MC78XX, A
      = 0°C for MC78XXC, AC

     T_{high} = +150°C for MC78XX, A
          = +125°C for MC78XXC, AC

    2. Load and line regulation are specified at constant junction temperature. Changes in V_O due to heating effects must be taken into account separately. Pulse testing with low duty cycle is used.

MC7800 Series

MC7815, C
ELECTRICAL CHARACTERISTICS
(V_{in} = 23 V, I_O = 500 mA, T_J = T_{low} to T_{high} (Note 1) unless otherwise noted.)

Characteristic	Symbol	MC7815			MC7815C			Unit
		Min	Typ	Max	Min	Typ	Max	
Output Voltage (T_J = +25°C)	V_O	14.4	15	15.6	14.4	15	15.6	Vdc
Output Voltage (5.0 mA ≤ I_O ≤ 1.0 A, P_O ≤ 15W)	V_O							Vdc
17.5 Vdc ≤ V_{in} ≤ 30 Vdc		—	—	—	14.25	15	15.75	
18.5 Vdc ≤ V_{in} ≤ 30 Vdc		14.25	15	15.75	—	—	—	
Line Regulation (T_J = +25°C, Note 2)	Reg_{in}							mV
17.5 Vdc ≤ V_{in} ≤ 30 Vdc		—	6.0	150	—	13	300	
20 Vdc ≤ V_{in} ≤ 26 Vdc		—	3.0	75	—	6.0	150	
Load Regulation (T_J = +25°C, Note 2)	Reg_{load}							mV
5.0 mA ≤ I_O ≤ 1.5 A		—	32	150	—	52	300	
250 mA ≤ I_O ≤ 750 mA		—	10	75	—	20	150	
Quiescent Current (T_J = +25°C)	I_B	—	3.4	6.0	—	4.4	8.0	mA
Quiescent Current Change	ΔI_B							mA
17.5 Vdc ≤ V_{in} ≤ 30 Vdc		—	—	—	—	—	1.0	
18.5 Vdc ≤ V_{in} ≤ 30 Vdc		—	0.3	0.8	—	—	—	
5.0 mA ≤ I_O ≤ 1.0 A		—	0.04	0.5	—	—	0.5	
Ripple Rejection	RR							dB
18.5 Vdc ≤ V_{in} ≤ 28.5 Vdc, f = 120 Hz		60	66	—	—	58	—	
Dropout Voltage (I_O = 1.0 A, T_J = +25°C)	V_{in}-V_O	—	2.0	2.5	—	2.0	—	Vdc
Output Noise Voltage (T_A = +25°C) 10 Hz ≤ f ≤ 100 kHz	V_n	—	10	40	—	10	—	μV/V_O
Output Resistance (f = 1.0 kHz)	R_O	—	19	—	—	19	—	mΩ
Short-Circuit Current Limit (T_A = +25°C) V_{in} = 35 Vdc	I_{sc}	—	0.2	1.2	—	0.2	—	A
Peak Output Current (T_J = +25°C)	I_{max}	1.3	2.5	3.3	—	2.2	—	A
Average Temperature Coefficient of Output Voltage	TCV_O	—	±1.8	—	—	-1.0	—	mV/°C

MC7815A, AC
ELECTRICAL CHARACTERISTICS
(V_{in} = 23 V, I_O = 1.0 A, T_J = T_{low} to T_{high} (Note 1) unless otherwise noted.)

Characteristic	Symbol	MC7815A			MC7815AC			Unit
		Min	Typ	Max	Min	Typ	Max	
Output Voltage (T_J = +25°C)	V_O	14.7	15	15.3	14.7	15	15.3	Vdc
Output Voltage (5.0 mA ≤ I_O ≤ 1.0 A, P_O ≤ 15W) 17.9 Vdc ≤ V_{in} ≤ 30 Vdc	V_O	14.4	15	15.6	14.4	15	15.6	Vdc
Line Regulation (Note 2)	Reg_{in}							mV
17.9 Vdc ≤ V_{in} ≤ 30 Vdc, I_O = 500 mA		—	6.0	22	—	13	150	
20 Vdc ≤ V_{in} ≤ 26 Vdc		—	9.0	30	—	16	150	
20 Vdc ≤ V_{in} ≤ 26 Vdc, T_J = +25°C		—	3.0	10	—	6.0	75	
17.5 Vdc ≤ V_{in} ≤ 30 Vdc, T_J = +25°C		—	6.0	22	—	13	150	
Load Regulation (Note 2)	Reg_{load}							mV
5.0 mA ≤ I_O ≤ 1.5A		—	32	50	—	—	—	
5.0 mA ≤ I_O ≤ 1.0 A		—	—	—	—	52	100	
5.0 mA ≤ I_O ≤ 1.5 A, T_J = +25°C		—	—	—	—	52	100	
250 mA ≤ I_O ≤ 750 mA		—	10	25	—	20	50	
Quiescent Current	I_B							mA
T_J = +25°C		—	—	5.5	—	—	6.0	
		—	3.4	4.5	—	4.4	6.0	
Quiescent Current Change	ΔI_B							mA
17.5 Vdc ≤ V_{in} ≤ 30 Vdc, I_O = 500 mA		—	0.3	0.5	—	—	0.8	
17.5 Vdc ≤ V_{in} ≤ 30 Vdc, T_J = +25°C		—	0.2	0.5	—	—	0.8	
5.0 mA ≤ I_O ≤ 1.0 A		—	0.04	0.2	—	—	0.5	
Ripple Rejection	RR							dB
18.5 Vdc ≤ V_{in} ≤ 28.5 Vdc, f = 120 Hz, T_J = +25°C		60	66	—	—	—	—	
18.5 Vdc ≤ V_{in} ≤ 28.5 Vdc, f = 120 Hz, I_O = 500 mA		60	66	—	—	58	—	
Dropout Voltage (I_O = 1.0 A, T_J = +25°C)	V_{in}-V_O	—	2.0	2.5	—	2.0	—	Vdc
Output Noise Voltage (T_A = +25°C) 10 Hz ≤ f ≤ 100 kHz	V_n	—	10	40	—	10	—	μV/V_O
Output Resistance (f = 1.0 kHz)	R_O	—	19	—	—	19	—	mΩ
Short-Circuit Current Limit (T_A = +25°C) V_{in} = 35 Vdc	I_{sc}	—	0.2	1.2	—	0.2	—	A
Peak Output Current (T_J = +25°C)	I_{max}	1.3	2.5	3.3	—	2.2	—	A
Average Temperature Coefficient of Output Voltage	TCV_O	—	±1.8	—	—	-1.0	—	mV/°C

Notes: 1. T_{low} = −55°C for MC78XX, A
= 0°C for MC78XXC, AC

T_{high} = +150°C for MC78XX, A
= +125°C for MC78XXC, AC

2. Load and line regulation are specified at constant junction temperature. Changes in V_O due to heating effects must be taken into account separately. Pulse testing with low duty cycle is used.

MC7800 Series

MC7818, C
ELECTRICAL CHARACTERISTICS (V_{in} = 27 V, I_O = 500 mA, T_J = T_{low} to T_{high} (Note 1) unless otherwise noted.)

Characteristic	Symbol	MC7818			MC7818C			Unit
		Min	Typ	Max	Min	Typ	Max	
Output Voltage (T_J = +25°C)	V_O	17.3	18	18.7	17.3	18	18.7	Vdc
Output Voltage (5.0 mA ≤ I_O ≤ 1.0 A, P_O ≤ 15W)	V_O							Vdc
21 Vdc ≤ V_{in} ≤ 33 Vdc		—	—	—	17.1	18	18.9	
22 Vdc ≤ V_{in} ≤ 33 Vdc		17.1	18	18.9	—	—	—	
Line Regulation (T_J = +25°C, Note 2)	Reg_{in}							mV
21 Vdc ≤ V_{in} ≤ 33 Vdc		—	7.0	180	—	25	360	
24 Vdc ≤ V_{in} ≤ 30 Vdc		—	4.0	90	—	10	180	
Load Regulation (T_J = +25°C, Note 2)	Reg_{load}							mV
5.0 mA ≤ I_O ≤ 1.5 A		—	35	180	—	55	360	
250 mA ≤ I_O ≤ 750 mA		—	12	90	—	22	180	
Quiescent Current (T_J = +25°C)	I_B	—	3.5	6.0	—	4.5	8.0	mA
Quiescent Current Change	ΔI_B							mA
21 Vdc ≤ V_{in} ≤ 33 Vdc		—	—	—	—	—	1.0	
22 Vdc ≤ V_{in} ≤ 33 Vdc		—	0.3	0.8	—	—	—	
5.0 mA ≤ I_O ≤ 1.0 A		—	0.04	0.5	—	—	0.5	
Ripple Rejection 22 Vdc ≤ V_{in} ≤ 32 Vdc, f = 120 Hz	RR	59	65	—	—	57	—	dB
Dropout Voltage (I_O = 1.0 A, T_J = +25°C)	$V_{in}\cdot V_O$	—	2.0	2.5	—	2.0	—	Vdc
Output Noise Voltage (T_A = +25°C) 10 Hz ≤ f ≤ 100 kHz	V_n	—	10	40	—	10	—	µV/V_O
Output Resistance (f = 1.0 kHz)	R_O	—	19	—	—	19	—	mΩ
Short-Circuit Current Limit (T_A = +25°C) V_{in} = 35 Vdc	I_{sc}	—	0.2	1.2	—	0.2	—	A
Peak Output Current (T_J = +25°C)	I_{max}	1.3	2.5	3.3	—	2.2	—	A
Average Temperature Coefficient of Output Voltage	TCV_O	—	±2.3	—	—	−1.0	—	mV/°C

MC7818A, AC
ELECTRICAL CHARACTERISTICS (V_{in} = 27 V, I_O = 1.0 A, T_J = T_{low} to T_{high} (Note 1) unless otherwise noted.)

Characteristic	Symbol	MC7818A			MC7818AC			Unit
		Min	Typ	Max	Min	Typ	Max	
Output Voltage (T_J = +25°C)	V_O	17.64	18	18.36	17.64	18	18.36	Vdc
Output Voltage (5.0 mA ≤ I_O ≤ 1.0 A, P_O ≤ 15W) 21 Vdc ≤ V_{in} ≤ 33 Vdc	V_O	17.3	18	18.7	17.3	18	18.7	Vdc
Line Regulation (Note 2)	Reg_{in}							mV
21 Vdc ≤ V_{in} ≤ 33 Vdc, I_O = 500 mA		—	7.0	31	—	25	180	
24 Vdc ≤ V_{in} ≤ 30 Vdc		—	12	45	—	28	180	
24 Vdc ≤ V_{in} ≤ 30 Vdc, T_J = +25°C		—	4.0	15	—	10	90	
20.6 Vdc ≤ V_{in} ≤ 33 Vdc, T_J = +25°C		—	7.0	31	—	25	180	
Load Regulation (Note 2)	Reg_{load}							mV
5.0 mA ≤ I_O ≤ 1.5A		—	35	50	—	—	—	
5.0 mA ≤ I_O ≤ 1.0 A		—	—	—	—	55	100	
5.0 mA ≤ I_O ≤ 1.5 A, T_J = +25°C		—	—	—	—	55	100	
250 mA ≤ I_O ≤ 750 mA		—	12	25	—	22	50	
Quiescent Current T_J = +25°C	I_B	— —	— 3.4	5.5 4.5	— —	— 4.5	6.0 6.0	mA
Quiescent Current Change	ΔI_B							mA
21 Vdc ≤ V_{in} ≤ 33 Vdc, I_O = 500 mA		—	0.3	0.5	—	—	0.8	
21 Vdc ≤ V_{in} ≤ 33 Vdc, T_J = +25°C		—	0.2	0.5	—	—	0.8	
5.0 mA ≤ I_O ≤ 1.0 A		—	0.04	0.2	—	—	0.5	
Ripple Rejection	RR							dB
22 Vdc ≤ V_{in} ≤ 32 Vdc, f = 120 Hz, T_J = +25°C		59	65	—	—	—	—	
22 Vdc ≤ V_{in} ≤ 32 Vdc, f = 120 Hz, I_O = 500 mA		59	65	—	—	57	—	
Dropout Voltage (I_O = 1.0 A, T_J = +25°C)	$V_{in}\cdot V_O$	—	2.0	2.5	—	2.0	—	Vdc
Output Noise Voltage (T_A = +25°C) 10 Hz ≤ f ≤ 100 kHz	V_n	—	10	40	—	10	—	µV/V_O
Output Resistance (f = 1.0 kHz)	R_O	—	19	—	—	19	—	mΩ
Short-Circuit Current Limit (T_A = +25°C) V_{in} = 35 Vdc	I_{sc}	—	0.2	1.2	—	0.2	—	A
Peak Output Current (T_J = +25°C)	I_{max}	1.3	2.5	3.3	—	2.2	—	A
Average Temperature Coefficient of Output Voltage	TCV_O	—	±2.3	—	—	−1.0	—	mV/°C

Notes: 1. T_{low} = −55°C for MC78XX, A
= 0°C for MC78XXC, AC

T_{high} = +150°C for MC78XX, A
= +125°C for MC78XXC, AC

2. Load and line regulation are specified at constant junction temperature. Changes in V_O due to heating effects must be taken into account separately. Pulse testing with low duty cycle is used.

MC7800 Series

MC7824, C
ELECTRICAL CHARACTERISTICS (V_{in} = 33 V, I_O = 500 mA, T_J = T_{low} to T_{high} (Note 1) unless otherwise noted.)

Characteristic	Symbol	MC7824			MC7824C			Unit
		Min	Typ	Max	Min	Typ	Max	
Output Voltage (T_J = +25°C)	V_O	23	24	25	23	24	25	Vdc
Output Voltage (5.0 mA ≤ I_O ≤ 1.0 A, P_O ≤ 15W)	V_O							Vdc
27 Vdc ≤ V_{in} ≤ 38 Vdc		–	–	–	22.8	24	25.2	
28 Vdc ≤ V_{in} ≤ 38 Vdc		22.8	24	25.2	–	–	–	
Line Regulation (T_J = +25°C, Note 2)	Reg_{in}							mV
27 Vdc ≤ V_{in} ≤ 38 Vdc		–	10	240	–	31	480	
30 Vdc ≤ V_{in} ≤ 36 Vdc		–	5.0	120	–	14	240	
Load Regulation (T_J = +25°C, Note 2)	Reg_{load}							mV
5.0 mA ≤ I_O ≤ 1.5 A		–	40	240	–	60	480	
250 mA ≤ I_O ≤ 750 mA		–	15	120	–	25	240	
Quiescent Current (T_J = +25°C)	I_B	–	3.6	6.0	–	4.6	8.0	mA
Quiescent Current Change	ΔI_B							mA
27 Vdc ≤ V_{in} ≤ 38 Vdc		–	–	–	–	–	1.0	
28 Vdc ≤ V_{in} ≤ 38 Vdc		–	0.3	0.8	–	–	–	
5.0 mA ≤ I_O ≤ 1.0 A		–	0.04	0.5	–	–	0.5	
Ripple Rejection 28 Vdc ≤ V_{in} ≤ 38 Vdc, f = 120 Hz	RR	56	62	–	–	54	–	dB
Dropout Voltage (I_O = 1.0 A, T_J = +25°C)	$V_{in}\cdot V_O$	–	2.0	2.5	–	2.0	–	Vdc
Output Noise Voltage (T_A = +25°C) 10 Hz < f < 100 kHz	V_n	–	10	40	–	10	–	$\mu V/V_O$
Output Resistance (f = 1.0 kHz)	R_O	–	20	–	–	20	–	$m\Omega$
Short-Circuit Current Limit (T_A = +25°C) V_{in} = 35 Vdc	I_{sc}	–	0.2	1.2	–	0.2	–	A
Peak Output Current (T_J = +25°C)	I_{max}	1.3	2.5	3.3	–	2.2	–	A
Average Temperature Coefficient of Output Voltage	TCV_O	–	±3.0	–	–	–1.5	–	mV/°C

MC7824A, AC
ELECTRICAL CHARACTERISTICS (V_{in} = 33 V, I_O = 1.0 A, T_J = T_{low} to T_{high} (Note 1) unless otherwise noted.)

Characteristic	Symbol	MC7824A			MC7824AC			Unit
		Min	Typ	Max	Min	Typ	Max	
Output Voltage (T_J = +25°C)	V_O	23.5	24	24.5	23.5	24	24.5	Vdc
Output Voltage (5.0 mA ≤ I_O ≤ 1.0 A, P_O ≤ 15W) 27.3 Vdc ≤ V_{in} ≤ 38 Vdc	V_O	23	24	25	23	24	25	Vdc
Line Regulation (Note 2)	Reg_{in}							mV
27 Vdc ≤ V_{in} ≤ 38 Vdc, I_O = 500 mA		–	10	36	–	31	240	
30 Vdc ≤ V_{in} ≤ 36 Vdc		–	15	60	–	35	240	
30 Vdc ≤ V_{in} ≤ 36 Vdc, T_J = +25°C		–	5.0	19	–	14	120	
26.7 Vdc ≤ V_{in} ≤ 38 Vdc, T_J = +25°C		–	10	36	–	31	240	
Load Regulation (Note 2)	Reg_{load}							mV
5.0 mA ≤ I_O ≤ 1.5 A		–	40	50	–	–	–	
5.0 mA ≤ I_O ≤ 1.0 A		–	–	–	–	60	100	
5.0 mA ≤ I_O ≤ 1.5 A, T_J = +25°C		–	–	–	–	60	100	
250 mA ≤ I_O ≤ 750 mA		–	15	25	–	25	50	
Quiescent Current	I_B							mA
		–	–	6.0	–	–	6.0	
T_J = +25°C		–	3.6	5.0	–	4.6	6.0	
Quiescent Current Change	ΔI_B							mA
27.3 Vdc ≤ V_{in} ≤ 38 Vdc, I_O = 500 mA		–	0.3	0.5	–	–	0.8	
27.3 Vdc ≤ V_{in} ≤ 38 Vdc, T_J = +25°C		–	0.2	0.5	–	–	0.8	
5.0 mA ≤ I_O ≤ 1.0 A		–	0.04	0.2	–	–	0.5	
Ripple Rejection	RR							dB
28 Vdc ≤ V_{in} ≤ 38 Vdc, f = 120 Hz, T_J = +25°C		56	62	–	–	–	–	
28 Vdc ≤ V_{in} ≤ 38 Vdc, f = 120 Hz, I_O = 500 mA		56	62	–	–	54	–	
Dropout Voltage (I_O = 1.0 A, T_J = +25°C)	$V_{in}\cdot V_O$	–	2.0	2.5	–	2.0	–	Vdc
Output Noise Voltage (T_A = +25°C) 10 Hz < f < 100 kHz	V_n	–	10	40	–	10	–	$\mu V/V_O$
Output Resistance (f = 1.0 kHz)	R_O	–	20	–	–	20	–	$m\Omega$
Short-Circuit Current Limit (T_A = +25°C) V_{in} = 35 Vdc	I_{sc}	–	0.2	1.2	–	0.2	–	A
Peak Output Current (T_J = +25°C)	I_{max}	1.3	2.5	3.3	–	2.2	–	A
Average Temperature Coefficient of Output Voltage	TCV_O	–	±3.0	–	–	–1.5	–	mV/°C

Notes: 1. T_{low} = –55°C for MC78XX, A
= 0°C for MC78XXC, AC

T_{high} = +150°C for MC78XX, A
= +125°C for MC78XXC, AC

2. Load and line regulation are specified at constant junction temperature. Changes in V_O due to heating effects must be taken into account separately. Pulse testing with low duty cycle is used.

MC7800 Series

TYPICAL CHARACTERISTICS
(T_A = +25°C unless otherwise noted.)

**FIGURE 1 — WORST CASE POWER DISSIPATION
versus AMBIENT TEMPERATURE (Case 221A)**

**FIGURE 2 — WORST CASE POWER DISSIPATION
versus AMBIENT TEMPERATURE (Case 1)**

**FIGURE 3 — INPUT OUTPUT DIFFERENTIAL AS A
FUNCTION OF JUNCTION TEMPERATURE
(MC78XXC, AC)**

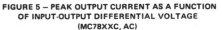

**FIGURE 4 — INPUT OUTPUT DIFFERENTIAL AS A
FUNCTION OF JUNCTION TEMPERATURE
(MC78XX, A)**

**FIGURE 5 — PEAK OUTPUT CURRENT AS A FUNCTION
OF INPUT-OUTPUT DIFFERENTIAL VOLTAGE
(MC78XXC, AC)**

**FIGURE 6 — PEAK OUTPUT CURRENT AS A
FUNCTION OF INPUT-OUTPUT DIFFERENTIAL
VOLTAGE (MC78XX, A)**

TYPICAL CHARACTERISTICS (continued)
(T_A = 25°C unless otherwise noted.)

FIGURE 7 – RIPPLE REJECTION AS A FUNCTION
OF OUTPUT VOLTAGES
(MC78XXC, AC)

f = 120 Hz
I_O = 20 mA
ΔV_{in} = 1.0 V(RMS)

PART #	V_{in}
MC7805C	10 V
MC7806C	11 V
MC7808C	14 V
MC7812C	19 V
MC7815C	23 V
MC7818C	27 V
MC7824C	33 V

FIGURE 8 – RIPPLE REJECTION AS A FUNCTION
OF FREQUENCY
(MC78XXC, AC)

MC78XX, A
V_{in} = 8.0 to 18 Vdc
V_O = 5.0 V
I_O = 1.0 A

MC78XXC, AC
V_{in} = 10 V
V_O = 5 V
I_O = 20 mA

FIGURE 9 – OUTPUT VOLTAGE AS A FUNCTION
OF JUNCTION TEMPERATURE (MC78XXC, AC)

V_{in} = 11 V
V_O = 6.0 V
I_O = 20 mA

FIGURE 10 – OUTPUT IMPEDANCE AS A
FUNCTION OF OUTPUT VOLTAGE (MC78XXC, AC)

f = 120 Hz
I_O = 500 mA
C_L = 0 μF

FIGURE 11 – QUIESCENT CURRENT AS A
FUNCTION OF TEMPERATURE (MC78XXC, AC)

MC78XXC, AC
V_{in} = 10V
V_O = 5V
I_O = 20 mA

MC78XX, A
V_{in} = 10V
V_O = 5V
I_O = 5 mA

FIGURE 12 – DROPOUT CHARACTERISTICS
(MC78XX, A)

LM140K-5.0
T_J = 25°C

I_{out} = 10 mA
I_{out} = 500 mA
I_{out} = 1.0 A

MC7800 Series

APPLICATIONS INFORMATION

Design Considerations

The MC7800 Series of fixed voltage regulators are designed with Thermal Overload Protection that shuts down the circuit when subjected to an excessive power overload condition, Internal Short-Circuit Protection that limits the maximum current the circuit will pass, and Output Transistor Safe-Area Compensation that reduces the output short-circuit current as the voltage across the pass transistor is increased.

In many low current applications, compensation capacitors are not required. However, it is recommended that the regulator input be bypassed with a capacitor if the regulator is connected to the power supply filter with long wire lengths, or if the output load capacitance is large. An input bypass capacitor should be selected to provide good high-frequency characteristics to insure stable operation under all load conditions. A 0.33 µF or larger tantalum, mylar, or other capacitor having low internal impedance at high frequencies should be chosen. The bypass capacitor should be mounted with the shortest possible leads directly across the regulators input terminals. Normally good construction techniques should be used to minimize ground loops and lead resistance drops since the regulator has no external sense lead.

FIGURE 13 – CURRENT REGULATOR

The MC7800 regulators can also be used as a current source when connected as above. In order to minimize dissipation the MC7805C is chosen in this application. Resistor R determines the current as follows:

$$I_O = \frac{5 \text{ V}}{R} + I_Q$$

$I_Q \cong 1.5$ mA over line and load changes

For example, a 1-ampere current source would require R to be a 5-ohm, 10-W resistor and the output voltage compliance would be the input voltage less 7 volts.

FIGURE 14 – ADJUSTABLE OUTPUT REGULATOR

V_O, 7.0 V to 20 V
$V_{IN} \cdot V_O \geqslant 2.0$ V

The addition of an operational amplifier allows adjustment to higher or intermediate values while retaining regulation characteristics. The minimum voltage obtainable with this arrangement is 2.0 volts greater than the regulator voltage.

FIGURE 15 – CURRENT BOOST REGULATOR

XX = 2 digits of type number indicating voltage.

The MC7800 series can be current boosted with a PNP transistor. The MJ2955 provides current to 5.0 amperes. Resistor R in conjunction with the V_{BE} of the PNP determines when the pass transistor begins conducting; this circuit is not short-circuit proof. Input-output differential voltage minimum is increased by V_{BE} of the pass transistor.

FIGURE 16 – SHORT-CIRCUIT PROTECTION

XX – 2 digits of type number indicating voltage.

The circuit of Figure 15 can be modified to provide supply protection against short circuits by adding a short-circuit sense resistor, R_{SC}, and an additional PNP transistor. The current sensing PNP must be able to handle the short-circuit current of the three-terminal regulator. Therefore, a four-ampere plastic power transistor is specified.

MC7900C SERIES THREE-TERMINAL NEGATIVE VOLTAGE REGULATORS

The MC7900C Series of fixed output negative voltage regulators are intended as complements to the popular MC7800C Series devices. These negative regulators are available in the same seven-voltage options as the MC7800C devices. In addition, two extra voltage options commonly employed in MECL systems are also available in the negative MC7900C Series.

Available in fixed output voltage options from –2.0 to –24 volts, these regulators employ current limiting, thermal shutdown, and safe-area compensation — making them remarkably rugged under most operating conditions. With adequate heat-sinking they can deliver output currents in excess of 1.0 ampere.

- No External Components Required
- Internal Thermal Overload Protection
- Internal Short-Circuit Current Limiting
- Output Transistor Safe-Area Compensation
- Packaged in the Plastic Case 221A and Case 1 (TO-220 and Hermetic TO-3)

THREE-TERMINAL NEGATIVE FIXED VOLTAGE REGULATORS

K SUFFIX
METAL PACKAGE
CASE 1
(TO-3 TYPE)

① ②
Gnd Output

Case
Input

(bottom view)

T SUFFIX
PLASTIC PACKAGE
CASE 221A

Pin 1. Ground
2. Input
3. Output

(Heatsink surface connected to Pin2)

STANDARD APPLICATION

Input ——→ MC79XXC ——→ Output

C_{in}^* 0.33 μF

C_O^{**} 1.0 μF

A common ground is required between the input and the output voltages. The input voltage must remain typically 2.0 V more negative even during the high point on the input ripple voltage.

XX = these two digits of the type number indicate voltage.

* = C_{in} is required if regulator is located an appreciable distance from power supply filter.

** = C_O improves stability and transient response.

SCHEMATIC DIAGRAM

DEVICE TYPE/NOMINAL OUTPUT VOLTAGE

MC7902C – 2.0 Volts	MC7906C – 6.0 Volts	MC7915C – 15 Volts
MC7905C – 5.0 Volts	MC7908C – 8.0 Volts	MC7918C – 18 Volts
MC7905.2C – 5.2 Volts	MC7912C – 12 Volts	MC7924C – 24 Volts

ORDERING INFORMATION		
DEVICE	TEMPERATURE RANGE	PACKAGE
MC79XXCK	$T_J = 0^0$ C to +150^0 C	Metal Power
MC79XXCT	$T_J = 0^0$ C to +150^0 C	Plastic Power
XX indicates nominal voltage		

MC7900C Series

MC7900C Series MAXIMUM RATINGS (T_A = +25°C unless otherwise noted.)

Rating	Symbol	Value	Unit
Input Voltage (2.0 V – 18 V)	V_I	–35	Vdc
(24 V)		–40	

Power Dissipation
Plastic Package
  T_A = +25°C
  Derate above T_A = +25°C
	P_D
	1/$R_{\theta JA}$
		Internally Limited
		15.4
			Watts
			mW/$^{\circ}$C

T_C = +25°C Derate above T_C = +95°C (See Figure 1)	P_D 1/$R_{\theta JC}$	Internally Limited 200	Watts mW/$^{\circ}$C

Metal Package
  T_A = +25°C
  Derate above T_A = +25°C
	P_D
	1/$R_{\theta JA}$
		Internally Limited
		22.2
			Watts
			mW/$^{\circ}$C

T_C = +25°C Derate above T_C = +65°C	P_D 1/$R_{\theta JC}$	Internally Limited 182	Watts mW/$^{\circ}$C
Storage Temperature Range	T_{stg}	–65 to +150	$^{\circ}$C
Junction Temperature Range	T_J	0 to +150	$^{\circ}$C

THERMAL CHARACTERISTICS

Characteristic		Symbol	Max	Unit
Thermal Resistance, Junction to Ambient	– Plastic Package	$R_{\theta JA}$	65	$^{\circ}$C/W
	– Metal Package		45	
Thermal Resistance, Junction to Case	– Plastic Package	$R_{\theta JC}$	5.0	$^{\circ}$C/W
	– Metal Package		5.5	

MC7902C ELECTRICAL CHARACTERISTICS (V_I = –10 V, I_O = 500 mA, 0°C < T_J < +125°C unless otherwise noted.)

Characteristic	Symbol	Min	Typ	Max	Unit
Output Voltage (T_J = +25°C)	V_O	–1.92	–2.00	–2.08	Vdc

Line Regulation
  (T_J = +25°C, I_O = 100 mA)
  –7.0 Vdc $\geqslant V_I \geqslant$ –25 Vdc
  –8.0 Vdc $\geqslant V_I \geqslant$ –12 Vdc
  (T_J = +25°C, I_O = 500 mA)
  –7.0 Vdc $\geqslant V_I \geqslant$ –25 Vdc
  –8.0 Vdc $\geqslant V_I \geqslant$ –12 Vdc
	Reg$_{line}$	
		
		–
		–
		
		–
		–
			
			8.0
			4.0
			
			18
			8.0
				
				20
				10
				
				40
				20
					mV

Load Regulation
  T_J = +25°C, 5.0 mA $\leqslant I_O \leqslant$ 1.5 A
  250 mA $\leqslant I_O \leqslant$ 750 mA
	Reg$_{load}$	
		–
		–
			
			70
			20
				
				120
				60
					mV

Output Voltage –7.0 Vdc $\geqslant V_I \geqslant$ –20 Vdc, 5.0 mA $\leqslant I_O \leqslant$ 1.0 A, P $\leqslant$ 15 W	V_O	–1.90	–	–2.10	Vdc
Input Bias Current (T_J = +25°C)	I_{IB}	–	4.3	8.0	mA

Input Bias Current Change
  –7.0 Vdc $\geqslant V_I \geqslant$ –25 Vdc
  5.0 mA $\leqslant I_O \leqslant$ 1.5 A
	ΔI_{IB}	
		–
		–
			
			–
			–
				
				1.3
				0.5
					mA

Output Noise Voltage (T_A = +25°C, 10 Hz $\leqslant$ f $\leqslant$ 100 kHz)	e_{on}	–	40	–	μV
Long-Term Stability	$\Delta V_O / \Delta t$	–	–	20	mV/1.0 k Hrs
Ripple Rejection (I_O = 20 mA, f = 120 Hz)	RR	–	65	–	dB
Input-Output Voltage Differential I_O = 1.0 A, T_J = +25°C	$|V_I - V_O|$	–	3.5	–	Vdc
Average Temperature Coefficient of Output Voltage I_O = 5.0 mA, 0°C $\leqslant T_A \leqslant$ +125°C	$\Delta V_O / \Delta T$	–	–1.0	–	mV/$^{\circ}$C

MC7900C Series

MC7905C ELECTRICAL CHARACTERISTICS (V_I = -10 V, I_O = 500 mA, $0^oC < T_J < +125^oC$, unless otherwise noted.)

Characteristic	Symbol	Min	Typ	Max	Unit
Output Voltage (T_J = +25°C)	V_O	-4.8	-5.0	-5.2	Vdc
Line Regulation (T_J = +25°C, I_O = 100 mA) -7.0 Vdc $\geqslant V_I \geqslant$ -25 Vdc -8.0 Vdc $\geqslant V_I \geqslant$ -12 Vdc (T_J = +25°C, I_O = 500 mA) -7.0 Vdc $\geqslant V_I \geqslant$ -25 Vdc -8.0 Vdc $\geqslant V_I \geqslant$ -12 Vdc	Reg$_{line}$	 - - - -	 7.0 2.0 35 8.0	 50 25 100 50	mV
Load Regulation T_J = +25°C, 5.0 mA $\leqslant I_O \leqslant$ 1.5 A 250 mA $\leqslant I_O \leqslant$ 750 mA	Reg$_{load}$	 - -	 11 4.0	 100 50	mV
Output Voltage -7.0 Vdc $\geqslant V_I \geqslant$ -20 Vdc, 5.0 mA $\leqslant I_O \leqslant$ 1.0 A, P $\leqslant$ 15 W	V_O	-4.75	-	-5.25	Vdc
Input Bias Current (T_J = +25°C)	I_{IB}	-	4.3	8.0	mA
Input Bias Current Change -7.0 Vdc $\geqslant V_{in} \geqslant$ -25 Vdc 5.0 mA $\leqslant I_O \leqslant$ 1.5 A	$\triangle I_{IB}$	 - -	 - -	 1.3 0.5	mA
Output Noise Voltage (T_A = +25°C, 10 Hz $\leqslant$ f $\leqslant$ 100 kHz)	e_{on}	-	40	-	μV
Long-Term Stability	$\Delta V_O/\Delta t$	-	-	20	mV/1.0k Hrs
Ripple Rejection (I_O = 20 mA, f = 120 Hz)	RR	-	70	-	dB
Input-Output Voltage Differential I_O = 1.0 A, T_J = +25°C	$|V_I - V_O|$	-	2.0	-	Vdc
Average Temperature Coefficient of Output Voltage I_O = 5.0 mA, $0^oC \leqslant T_A \leqslant +125^oC$	$\Delta V_O/\Delta T$	-	-1.0	-	mV/^{o}C

MC7905.2C ELECTRICAL CHARACTERISTICS (V_I = -10 V, I_O = 500 mA, $0^oC < T_J < +125^oC$, unless otherwise noted.)

Characteristic	Symbol	Min	Typ	Max	Unit
Output Voltage (T_J = +25°C)	V_O	-5.0	-5.2	-5.4	Vdc
Line Regulation (T_J = +25°C, I_O = 100 mA) -7.2 Vdc $\geqslant V_I \geqslant$ -25 Vdc -8.0 Vdc $\geqslant V_I \geqslant$ -12 Vdc (T_J = +25°C, I_O = 500 mA) -7.2 Vdc $\geqslant V_I \geqslant$ -25 Vdc -8.0 Vdc $\geqslant V_I \geqslant$ -12 Vdc	Reg$_{line}$	 - - - -	 8.0 2.2 37 8.5	 52 27 105 52	mV
Load Regulation T_J = +25°C, 5.0 mA $\leqslant I_O \leqslant$ 1.5 A 250 mA $\leqslant I_O \leqslant$ 750 mA	Reg$_{load}$	 - -	 12 4.5	 105 52	mV
Output Voltage -7.2 Vdc $\geqslant V_I \geqslant$ -20 Vdc, 5.0 mA $\leqslant I_O \leqslant$ 1.0 A, P $\leqslant$ 15 W	V_O	-4.94	-	-5.46	Vdc
Input Bias Current (T_J = +25°C)	I_{IB}	-	4.3	8.0	mA
Input Bias Current Change -7.2 Vdc $\geqslant V_I \geqslant$ -25 Vdc 5.0 mA $\leqslant I_O \leqslant$ 1.5 A	$\triangle I_{IB}$	 - -	 - -	 1.3 0.5	mA
Output Noise Voltage (T_A = +25°C, 10 Hz $\leqslant$ f $\leqslant$ 100 kHz)	e_{on}	-	42	-	μV
Long-Term Stability	$\Delta V_O/\Delta t$	-	-	20	mV/1.0k Hrs
Ripple Rejection (I_O = 20 mA, f = 120 Hz)	RR	-	68	-	dB
Input-Output Voltage Differential I_O = 1.0 A, T_J = +25°C	$|V_I - V_O|$	-	2.0	-	Vdc
Average Temperature Coefficient of Output Voltage I_O = 5.0 mA, $0^oC \leqslant T_A \leqslant +125^oC$	$\Delta V_O/\Delta T$	-	-1.0	-	mV/^{o}C

MC7906C ELECTRICAL CHARACTERISTICS (V_I = -11 V, I_O = 500 mA, $0^oC < T_J < +125^oC$ unless otherwise noted.)

Characteristic	Symbol	Min	Typ	Max	Unit
Output Voltage (T_J = +25°C)	V_O	-5.75	-6.0	-6.25	Vdc
Line Regulation	Reg$_{line}$				mV
(T_J = +25°C, I_O = 100 mA)					
-8.0 Vdc $\geqslant V_I \geqslant$ -25 Vdc		–	9.0	60	
-9.0 Vdc $\geqslant V_I \geqslant$ -13 Vdc		–	3.0	30	
(T_J = +25°C, I_O = 500 mA)					
-8.0 Vdc $\geqslant V_I \geqslant$ -25 Vdc		–	43	120	
-9.0 Vdc $\geqslant V_I \geqslant$ -13 Vdc		–	10	60	
Load Regulation	Reg$_{load}$				mV
T_J = +25°C, 5.0 mA $\leqslant I_O \leqslant$ 1.5 A		–	13	120	
250 mA $\leqslant I_O \leqslant$ 750 mA		–	5.0	60	
Output Voltage -8.0 Vdc $\geqslant V_I \geqslant$ -21 Vdc, 5.0 mA $\leqslant I_O \leqslant$ 1.0 A, P $\leqslant$ 15 W)	V_O	-5.7	–	-6.3	Vdc
Input Bias Current (T_J = +25°C)	I_{IB}	–	4.3	8.0	mA
Input Bias Current Change	ΔI_{IB}				mA
-8.0 Vdc $\geqslant V_I \geqslant$ -25 Vdc		–	–	1.3	
5.0 mA $\leqslant I_O \leqslant$ 1.5 A		–	–	0.5	
Output Noise Voltage (T_A = +25°C, 10 Hz $\leqslant f \leqslant$ 100 kHz)	e_{on}	–	45	–	μV
Long-Term Stability	$\Delta V_O/\Delta t$	–	–	24	mV/1.0 k Hrs
Ripple Rejection (I_O = 20 mA, f = 120 Hz)	RR	–	65	–	dB
Input-Output Voltage Differential I_O = 1.0 A, T_J = +25°C	$|V_I-V_O|$	–	2.0	–	Vdc
Average Temperature Coefficient of Output Voltage I_O = 5.0 mA, $0^oC \leqslant T_A \leqslant +125^oC$	$\Delta V_O/\Delta T$	–	-1.0	–	mV/^{o}C

MC7908C ELECTRICAL CHARACTERISTICS (V_I = -14 V, I_O = 500 mA, $0^oC < T_J < +125^oC$ unless otherwise noted.)

Characteristic	Symbol	Min	Typ	Max	Unit
Output Voltage (T_J = +25°C)	V_O	-7.7	-8.0	-8.3	Vdc
Line Regulation	Reg$_{line}$				mV
(T_J = +25°C, I_O = 100 mA)					
-10.5 Vdc $\geqslant V_I \geqslant$ -25 Vdc		–	12	80	
-11 Vdc $\geqslant V_I \geqslant$ -17 Vdc		–	5.0	40	
(T_J = +25°C, I_O = 500 mA)					
-10.5 Vdc $\geqslant V_I \geqslant$ -25 Vdc		–	50	160	
-11 Vdc $\geqslant V_I \geqslant$ -17 Vdc		–	22	80	
Load Regulation	Reg$_{load}$				mV
T_J = +25°C, 5.0 mA $\leqslant I_O \leqslant$ 1.5 A		–	26	160	
250 mA $\leqslant I_O \leqslant$ 750 mA		–	9.0	80	
Output Voltage -10.5 Vdc $\geqslant V_I \geqslant$ -23 Vdc, 5.0 mA $\leqslant I_O \leqslant$ 1.0 A, P $\leqslant$ 15 W	V_O	-7.6	–	-8.4	Vdc
Input Bias Current (T_J = +25°C)	I_{IB}	–	4.3	8.0	mA
Input Bias Current Change	ΔI_{IB}				mA
-10.5 Vdc $\geqslant V_I \geqslant$ -25 Vdc		–	–	1.0	
5.0 mA $\leqslant I_O \leqslant$ 1.5 A		–	–	0.5	
Output Noise Voltage (T_A = +25°C, 10 Hz $\leqslant f \leqslant$ 100 kHz)	e_{on}	–	52	–	μV
Long-Term Stability	$\Delta V_O/\Delta t$	–	–	32	mV/1.0 k Hrs
Ripple Rejection (I_O = 20 mA, f = 120 Hz)	RR	–	62	–	dB
Input-Output Voltage Differential I_O = 1.0 A, T_J = +25°C	$|V_I-V_O|$	–	2.0	–	Vdc
Average Temperature Coefficient of Output Voltage I_O = 5.0 mA, $0^oC \leqslant T_A \leqslant +125^oC$	$\Delta V_O/\Delta T$	–	-1.0	–	mV/^{o}C

MC7912C ELECTRICAL CHARACTERISTICS (V_I = -19 V, I_O = 500 mA, 0°C $<$ T_J $<$ +125°C, unless otherwise noted.)

Characteristic	Symbol	Min	Typ	Max	Unit
Output Voltage (T_J = +25°C)	V_O	-11.5	-12	-12.5	Vdc
Line Regulation (T_J = +25°C, I_O = 100 mA) -14.5 Vdc $\geqslant$ V_I $\geqslant$ -30 Vdc -16 Vdc $\geqslant$ V_I $\geqslant$ -22 Vdc (T_J = +25°C, I_O = 500 mA) -14.5 Vdc $\geqslant$ V_I $\geqslant$ -30 Vdc -16 Vdc $\geqslant$ V_I $\geqslant$ -22 Vdc	Reg$_{line}$	 — — — —	 13 6.0 55 24	 120 60 240 120	mV
Load Regulation T_J = +25°C, 5.0 mA $\leqslant$ I_O $\leqslant$ 1.5 A 250 mA $\leqslant$ I_O $\leqslant$ 750 mA	Reg$_{load}$	 — —	 46 17	 240 120	mV
Output Voltage -14.5 Vdc $\geqslant$ V_I $\geqslant$ -27 Vdc, 5.0 mA $\leqslant$ I_O $\leqslant$ 1.0 A, P $\leqslant$ 15 W	V_O	-11.4	—	-12.6	Vdc
Input Bias Current (T_J = +25°C)	I_{IB}	—	4.4	8.0	mA
Input Bias Current Change -14.5 Vdc $\geqslant$ V_I $\geqslant$ -30 Vdc 5.0 mA $\leqslant$ I_O $\leqslant$ 1.5 A	ΔI_{IB}	 — —	 — —	 1.0 0.5	mA
Output Noise Voltage (T_A = +25°C, 10 Hz $\leqslant$ f $\leqslant$ 100 kHz)	e_{on}	—	75	—	µV
Long-Term Stability	$\Delta V_O/\Delta t$	—	—	48	mV/1.0 k Hrs
Ripple Rejection (I_O = 20 mA, f = 120 Hz)	RR	—	61	—	dB
Input-Output Voltage Differential I_O = 1.0 A, T_J = +25°C	$|V_I - V_O|$	—	2.0	—	Vdc
Average Temperature Coefficient of Output Voltage I_O = 5.0 mA, 0°C $\leqslant$ T_A $\leqslant$ +125°C	$\Delta V_O/\Delta T$	—	-1.0	—	mV/°C

MC7915C ELECTRICAL CHARACTERISTICS (V_I = -23 V, I_O = 500 mA, 0°C $<$ T_J $<$ +125°C, unless otherwise noted.)

Characteristic	Symbol	Min	Typ	Max	Unit
Output Voltage (T_J = +25°C)	V_O	-14.4	-15	-15.6	Vdc
Line Regulation (T_J = +25°C, I_O = 100 mA) -17.5 Vdc $\geqslant$ V_I $\geqslant$ -30 Vdc -20 Vdc $\geqslant$ V_I $\geqslant$ -26 Vdc (T_J = +25°C, I_O = 500 mA) -17.5 Vdc $\geqslant$ V_I $\geqslant$ -30 Vdc -20 Vdc $\geqslant$ V_I $\geqslant$ -26 Vdc	Reg$_{line}$	 — — — —	 14 6.0 57 27	 150 75 300 150	mV
Load Regulation T_J = +25°C, 5.0 mA $\leqslant$ I_O $\leqslant$ 1.5 A 250 mA $\leqslant$ I_O $\leqslant$ 750 mA	Reg$_{load}$	 — —	 68 25	 300 150	mV
Output Voltage -17.5 Vdc $\geqslant$ V_I $\geqslant$ -30 Vdc, 5.0 mA $\leqslant$ I_O $\leqslant$ 1.0 A, P $\leqslant$ 15 W	V_O	-14.25	—	-15.75	Vdc
Input Bias Current (T_J = +25°C)	I_{IB}	—	4.4	8.0	mA
Input Bias Current Change -17.5 Vdc $\geqslant$ V_I $\geqslant$ -30 Vdc 5.0 mA $\leqslant$ I_O $\leqslant$ 1.5 A	ΔI_{IB}	 — —	 — —	 1.0 0.5	mA
Output Noise Voltage (T_A = +25°C, 10 Hz $\leqslant$ f $\leqslant$ 100 kHz)	e_{on}	—	90	—	µV
Long-Term Stability	$\Delta V_O/\Delta t$	—	—	60	mV/1.0 k Hrs
Ripple Rejection (I_O = 20 mA, f = 120 Hz)	RR	—	60	—	dB
Input-Output Voltage Differential I_O = 1.0 A, T_J = +25°C	$|V_I - V_O|$	—	2.0	—	Vdc
Average Temperature Coefficient of Output Voltage I_O = 5.0 mA, 0°C $\leqslant$ T_A $\leqslant$ +125°C	$\Delta V_O/\Delta T$	—	-1.0	—	mV/°C

MC7900C Series

MC7918C ELECTRICAL CHARACTERISTICS (V_I = -27 V, I_O = 500 mA, $0^oC < T_J < +125^oC$, unless otherwise noted.)

Characteristic	Symbol	Min	Typ	Max	Unit
Output Voltage (T_J = +25°C)	V_O	-17.3	-18	-18.7	Vdc
Line Regulation (T_J = +25°C, I_O = 100 mA) -21 Vdc $\geqslant$ V_I $\geqslant$ -33 Vdc -24 Vdc $\geqslant$ V_I $\geqslant$ -30 Vdc (T_J = +25°C, I_O = 500 mA) -21 Vdc $\geqslant$ V_I $\geqslant$ -33 Vdc -24 Vdc $\geqslant$ V_I $\geqslant$ -30 Vdc	Reg$_{line}$	 – – – –	 25 10 90 50	 180 90 360 180	mV
Load Regulation T_J = +25°C, 5.0 mA $\leqslant$ I_O $\leqslant$ 1.0 A 250 mA $\leqslant$ I_O $\leqslant$ 750 mA	Reg$_{load}$	 – –	 110 55	 360 180	mV
Output Voltage -21 Vdc $\geqslant$ V_I $\geqslant$ -33 Vdc, 5.0 mA $\leqslant$ I_O $\leqslant$ 1.0 A, P $\leqslant$ 15 W	V_O	-17.1	–	-18.9	Vdc
Input Bias Current (T_J = +25°C)	I_{IB}	–	4.5	8.0	mA
Input Bias Current Change -21 Vdc $\geqslant$ V_I $\geqslant$ -33 Vdc 5.0 mA $\leqslant$ I_O $\leqslant$ 1.0 A	ΔI_{IB}	 – –	 – –	 1.0 0.5	mA
Output Noise Voltage (T_A = +25°C, 10 Hz $\leqslant$ f $\leqslant$ 100 kHz)	e_{on}	–	110	–	µV
Long-Term Stability	$\Delta V_O/\Delta t$	–	–	72	mV/1.0 k Hrs
Ripple Rejection (I_O = 20 mA, f = 120 Hz)	RR	–	59	–	dB
Input-Output Voltage Differential I_O = 1.0 A, T_J = +25°C	$|V_I - V_O|$	–	2.0	–	Vdc
Average Temperature Coefficient of Output Voltage I_O = 5.0 mA, $0^oC \leqslant T_A \leqslant +125^oC$	$\Delta V_O/\Delta T$	–	-1.0	–	mV/°C

MC7924C ELECTRICAL CHARACTERISTICS (V_I = -33 V, I_O = 500 mA, $0^oC < T_J < +125^oC$, unless otherwise noted.)

Characteristic	Symbol	Min	Typ	Max	Unit
Output Voltage (T_J = +25°C)	V_O	-23	-24	-25	Vdc
Line Regulation (T_J = +25°C, I_O = 100 mA) -27 Vdc $\geqslant$ V_I $\geqslant$ -38 Vdc -30 Vdc $\geqslant$ V_I $\geqslant$ -36 Vdc (T_J = +25°C, I_O = 500 mA) -27 Vdc $\geqslant$ V_I $\geqslant$ -38 Vdc -30 Vdc $\geqslant$ V_I $\geqslant$ -36 Vdc	Reg$_{line}$	 – – – –	 31 14 118 70	 240 120 480 240	mV
Load Regulation T_J = +25°C, 5.0 mA $\leqslant$ I_O $\leqslant$ 1.0 A 250 mA $\leqslant$ I_O $\leqslant$ 750 mA	Reg$_{load}$	 – –	 150 85	 480 240	mV
Output Voltage -27 Vdc $\geqslant$ V_I $\geqslant$ -38 Vdc, 5.0 mA $\leqslant$ I_O $\leqslant$ 1.0 A, P $\leqslant$ 15 W	V_O	-22.8	–	-25.2	Vdc
Input Bias Current (T_J = +25°C)	I_{IB}	–	4.6	8.0	mA
Input Bias Current Change -27 Vdc $\geqslant$ V_I $\geqslant$ -38 Vdc 5.0 mA $\leqslant$ I_O $\leqslant$ 1.0 A	ΔI_{IB}	 – –	 – –	 1.0 0.5	mA
Output Noise Voltage (T_A = +25°C, 10 Hz $\leqslant$ f $\leqslant$ 100 kHz)	e_{on}	–	170	–	µV
Long-Term Stability	$\Delta V_O/\Delta t$	–	–	96	mV/1.0 k Hrs
Ripple Rejection (I_O = 20 mA, f = 120 Hz)	RR	–	56	–	dB
Input-Output Voltage Differential I_O = 1.0 A, T_J = +25°C	$|V_I - V_O|$	–	2.0	–	Vdc
Average Temperature Coefficient of Output Voltage I_O = 5.0 mA, $0^oC \leqslant T_A \leqslant +125^oC$	$\Delta V_O/\Delta T$	–	-1.0	–	mV/°C

MC7900C Series

TYPICAL CHARACTERISTICS
(T_A = +25°C unless otherwise noted.)

FIGURE 1 – WORST CASE POWER DISSIPATION AS A
FUNCTION OF AMBIENT TEMPERATURE (TO-220)

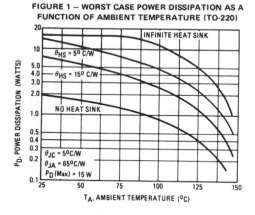

FIGURE 2 – WORST CASE POWER DISSIPATION AS A
FUNCTION OF AMBIENT TEMPERATURE (TO-3)

FIGURE 3 – PEAK OUTPUT CURRENT AS A FUNCTION OF
INPUT-OUTPUT DIFFERENTIAL VOLTAGE

FIGURE 4 – RIPPLE REJECTION AS A FUNCTION
OF FREQUENCY

FIGURE 5 – RIPPLE REJECTION AS A FUNCTION OF
OUTPUT VOLTAGES

FIGURE 6 – OUTPUT VOLTAGE AS A FUNCTION
OF JUNCTION TEMPERATURE

583

TYPICAL CHARACTERISTICS (continued)

FIGURE 7 – QUIESCENT CURRENT AS A FUNCTION OF TEMPERATURE

$V_{in} = -11$ V
$V_O = -6.0$ V
$I_O = 20$ mA

DEFINITIONS

Line Regulation – The change in output voltage for a change in the input voltage. The measurement is made under conditions of low dissipation or by using pulse techniques such that the average chip temperature is not significantly affected.

Load Regulation -- The change in output voltage for a change in load current at constant chip temperature.

Maximum Power Dissipation – The maximum total device dissipation for which the regulator will operate within specifications.

Input Bias Current – That part of the input current that is not delivered to the load.

Output Noise Voltage – The rms ac voltage at the output, with constant load and no input ripple, measured over a specified frequency range.

Long Term Stability - Output voltage stability under accelerated life test conditions with the maximum rated voltage listed in the devices' electrical characteristics and maximum power dissipation.

MC7900C Series

Design Considerations

The MC7900C Series of fixed voltage regulators are designed with Thermal Overload Protection that shuts down the circuit when subjected to an excessive power overload condition, Internal Short-Circuit Protection that limits the maximum current the circuit will pass, and Output Transistor Safe-Area Compensation that reduces the output short-circuit current as the voltage across the pass transistor is increased.

In many low current applications, compensation capacitors are not required. However, it is recommended that the regulator input be bypassed with a capacitor if the regulator is connected to the power supply filter with long wire lengths, or if the output load capacitance is large. An input bypass capacitor should be selected to provide good high-frequency characteristics to insure stable operation under all load conditions. A 0.33 μF or larger tantalum, mylar, or other capacitor having low internal impedance at high frequencies should be chosen. The bypass capacitor should be mounted with the shortest possible leads directly across the regulators input terminals. Normally good construction techniques should be used to minimize ground loops and lead resistance drops since the regulator has no external sense lead. Bypassing the output is also recommended.

FIGURE 8 – CURRENT REGULATOR

The MC7902, -2.0 V regulator can be used as a constant current source when connected as above. The output current is the sum of resistor R current and quiescent bias current as follows:

$$I_O = \frac{2V}{R} + I_B$$

The quiescent current for this regulator is typically 4.3 mA. The 2.0 volt regulator was chosen to minimize dissipation and to allow the output voltage to operate to within 6.0 V below the input voltage.

FIGURE 9 – CURRENT BOOST REGULATOR
(–5.0 V @ 4.0 A, with 5.0 A current limiting)

*Mounted on common heat sink, Motorola MS-10 or equivalent.

When a boost transistor is used, short-circuit currents are equal to the sum of the series pass and regulator limits, which are measured at 3.2 A and 1.8 A respectively in this case. Series pass limiting is approximately equal to 0.6 V/R$_{SC}$. Operation beyond this point to the peak current capability of the MC7905C is possible if the regulator is mounted on a heat sink; otherwise thermal shutdown will occur when the additional load current is picked up by the regulator.

FIGURE 10 – OPERATIONAL AMPLIFIER SUPPLY
(±15 V @ 1.0 A)

The MC7815 and MC7915 positive and negative regulators may be connected as shown to obtain a dual power supply for operational amplifiers. A clamp diode should be used at the output of the MC7815 to prevent potential latch-up problems.

FIGURE 11 – TYPICAL MECL SYSTEM POWER SUPPLY
(–5.2 V @ 4.0 A and –2.0 V @ 2.0 A; for PC Board)

When current-boost power transistors are used, 47-ohm base-to-emitter resistors (R) must be used to bypass the quiescent current at no load. These resistors, in conjunction with the V$_{BE}$ of the NPN transistors, determine when the pass transistors begin conducting. The 1-ohm and 4-ohm dropping resistors were chosen to reduce the power dissipated in the boost transistors but still leave at least 2.0 V across these devices for good regulation.

585

ICL8038
Precision Waveform Generator/Voltage Controlled Oscillator

GENERAL DESCRIPTION

The ICL8038 Waveform Generator is a monolithic integrated circuit capable of producing high accuracy sine, square, triangular, sawtooth and pulse waveforms with a minimum of external components. The frequency (or repetition rate) can be selected externally from .001Hz to more than 300kHz using either resistors or capacitors, and frequency modulation and sweeping can be accomplished with an external voltage. The ICL8038 is fabricated with advanced monolithic technology, using Schottky-barrier diodes and thin film resistors, and the output is stable over a wide range of temperature and supply variations. These devices may be interfaced with phase locked loop circuitry to reduce temperature drift to less than 250ppm/°C.

FEATURES

- **Low Frequency Drift With Temperature — 250ppm/°C**
- **Simultaneous Sine, Square, and Triangle Wave Outputs**
- **Low Distortion — 1% (Sine Wave Output)**
- **High Linearity — 0.1% (Triangle Wave Output)**
- **Wide Operating Frequency Range — 0.001Hz to 300kHz**
- **Variable Duty Cycle — 2% to 98%**
- **High Level Outputs — TTL to 28V**
- **Easy to Use — Just A Handful of External Components Required**

ORDERING INFORMATION

PART NUMBER	STABILITY	TEMP. RANGE	PACKAGE
ICL8038CCJD	250ppm/°C typ	0°C to +70°C	CERDIP
ICL8038BCJD	180ppm/°C typ	0°C to +70°C	CERDIP
ICL8038ACJD	110ppm/°C typ	0°C to +70°C	CERDIP
ICL8038BMJD*	350ppm/°C max	−55°C to +125°C	CERDIP
ICL8038AMJD*	250ppm/°C max	−55°C to +125°C	CERDIP
ICL8038/D	—	—	DICE**

*Add /883B to part number if 883 processing is required.

**Parameter Min/Max Limits guaranteed at 25°C only for DICE orders.

Figure 1: Functional Diagram

Figure 2: Pin Configuration (Outline dwg JD)

302600–002

Note: All typical values have been guaranteed by characterization and are not tested.

ICL8038

TEST CONDITIONS

PARAMETER		R_A	R_B	R_L	C_1	SW_1	MEASURE
Supply Current		10kΩ	10kΩ	10kΩ	3.3nF	Closed	Current into Pin 6
Sweep FM Range[1]		10kΩ	10kΩ	10kΩ	3.3nF	Open	Frequency at Pin 9
Frequency Drift with Temperature		10kΩ	10kΩ	10kΩ	3.3nF	Closed	Frequency at Pin 3
Frequency Drift with Supply Voltage[2]		10kΩ	10kΩ	10kΩ	3.3nF	Closed	Frequency at Pin 9
Output Amplitude:	Sine	10kΩ	10kΩ	10kΩ	3.3nF	Closed	Pk-Pk output at Pin 2
(Note 4)	Triangle	10kΩ	10kΩ	10kΩ	3.3nF	Closed	Pk-Pk output at Pin 3
Leakage Current (off)[3]		10kΩ	10kΩ		3.3nF	Closed	Current into Pin 9
Saturation Voltage (on)[3]		10kΩ	10kΩ		3.3nF	Closed	Output (low) at Pin 9
Rise and Fall Times (Note 5)		10kΩ	10kΩ	4.7kΩ	3.3nF	Closed	Waveform at Pin 9
Duty Cycle Adjust:	MAX	50kΩ	~1.6kΩ	10kΩ	3.3nF	Closed	Waveform at Pin 9
(Note 5)	MIN	~25kΩ	50kΩ	10kΩ	3.3nF	Closed	Waveform at Pin 9
Triangle Waveform Linearity		10kΩ	10kΩ	10kΩ	3.3nF	Closed	Waveform at Pin 3
Total Harmonic Distortion		10kΩ	10kΩ	10kΩ	3.3nF	Closed	Waveform at Pin 2

NOTES: 1. The hi and lo frequencies can be obtained by connecting pin 8 to pin 7 (f_{hi}) and then connecting pin 8 to pin 6 (f_{lo}). Otherwise apply Sweep Voltage at pin 8 (2/3 V_{SUPPLY} + 2V) ≤ V_{SWEEP} ≤ V_{SUPPLY} where V_{SUPPLY} is the total supply voltage. In Figure 7b, pin 8 should vary between 5.3V and 10V with respect to ground.
      2. 10V ≤ V^+ ≤ 30V, or ±5V ≤ V_{SUPPLY} ≤ ±15V.
      3. Oscillation can be halted by forcing pin 10 to +5 volts or −5 volts.
      4. Output Amplitude is tested under static conditions by forcing pin 10 to 5.0V then to −5.0V.
      5. Not tested; for design purposes only.

DEFINITION OF TERMS:

Supply Voltage (V_{SUPPLY}). The total supply voltage from V^+ to V^-.

Supply Current. The supply current required from the power supply to operate the device, excluding load currents and the currents through R_A and R_B.

Frequency Range. The frequency range at the square wave output through which circuit operation is guaranteed.

Sweep FM Range. The ratio of maximum frequency to minimum frequency which can be obtained by applying a sweep voltage to pin 8. For correct operation, the sweep voltage should be within the range

$$(2/3 \ V_{SUPPLY} + 2V) < V_{SWEEP} < V_{SUPPLY}$$

FM Linearity. The percentage deviation from the best-fit straight line on the control voltage versus output frequency curve.

Output Amplitude. The peak-to-peak signal amplitude appearing at the outputs.

Saturation Voltage. The output voltage at the collector of Q_{23} when this transistor is turned on. It is measured for a sink current of 2mA.

Rise and Fall Times. The time required for the square wave output to change from 10% to 90%, or 90% to 10%, of its final value.

Triangle Waveform Linearity. The percentage deviation from the best-fit straight line on the rising and falling triangle waveform.

Total Harmonic Distortion. The total harmonic distortion at the sine-wave output.

Figure 3: Test Circuit

Note: All typical values have been guaranteed by characterization and are not tested.

⌾ **INTERSIL**

ABSOLUTE MAXIMUM RATINGS

Supply Voltage (V^- to V^+) 36V
Power Dissipation[1] 750mW
Input Voltage (any pin) ....V^- to V^+
Input Current (Pins 4 and 5) 25mA
Output Sink Current (Pins 3 and 9) 25mA

Storage Temperature Range .... −65°C to +150°C
Operating Temperature Range:
   8038AM, 8038BM .... −55°C to +125°C
   8038AC, 8038BC, 8038CC .... 0°C to +70°C
Lead Temperature (Soldering, 10sec) ....300°C

Stresses above those listed under Absolute Maximum Ratings may cause permanent damage to the device. These are stress ratings only, and functional operation of the device at these or any other conditions above those indicated in the operational sections of the specifications is not implied. Exposure to absolute maximum rating conditions for extended periods may affect device reliability.

NOTE 1: Derate ceramic package at 12.5mW/°C for ambient temperatures above 100°C.

ELECTRICAL CHARACTERISTICS ($V_{SUPPLY} = \pm 10V$ or $+20V$, $T_A = 25°C$, $R_L = 10k\Omega$, Test Circuit Unless Otherwise Specified)

SYMBOL	GENERAL CHARACTERISTICS	8038CC			8038BC(BM)			8038AC(AM)			UNIT
		MIN	TYP	MAX	MIN	TYP	MAX	MIN	TYP	MAX	
V_{SUPPLY}	Supply Voltage Operating Range										
V^+	Single Supply	+10		+30	+10		30	+10		30	V
V^+, V^-	Dual Supplies	±5		±15	±5		±15	±5		±15	V
I_{SUPPLY}	Supply Current ($V_{SUPPLY} = \pm 10V$)[2]										
	8038AM, 8038BM					12	15		12	15	mA
	8038AC, 8038BC, 8038CC		12	20		12	20		12	20	mA
FREQUENCY CHARACTERISTICS (all waveforms)											
f_{max}	Maximum Frequency of Oscillation	100			100			100			kHz
f_{sweep}	Sweep Frequency of FM Input		10			10			10		kHz
	Sweep FM Range[3]		35:1			35:1			35:1		
	FM Linearity 10:1 Ratio		0.5			0.2			0.2		%
$\Delta f/\Delta T$	Frequency Drift With Temperature[5] 8038 AC, BC, CC 0°C to 70°C		250			180			110		ppm/°C
	8038 AM, BM, −55°C to 125°C						350			250	
$\Delta f/\Delta V$	Frequency Drift With Supply Voltage (Over Supply Voltage Range)		0.05			0.05			0.05		%/V
OUTPUT CHARACTERISTICS											
I_{OLK}	**Square-Wave** Leakage Current ($V_9 = 30V$)			1			1			1	µA
V_{SAT}	Saturation Voltage ($I_{SINK} = 2mA$)		0.2	0.5		0.2	0.4		0.2	0.4	V
t_r	Rise Time ($R_L = 4.7k\Omega$)		180			180			180		ns
t_f	Fall Time ($R_L = 4.7k\Omega$)		40			40			40		ns
ΔD	Typical Duty Cycle Adjust (Note 6)	2		98	2		98	2		98	%
$V_{TRIANGLE}$	**Triangle/Sawtooth/Ramp** Amplitude ($R_{TRI} = 100k\Omega$)	0.30	0.33		0.30	0.33		0.30	0.33		$\times V_{SUPPLY}$
	Linearity		0.1			0.05			0.05		%
Z_{OUT}	Output Impedance ($I_{OUT} = 5mA$)		200			200			200		Ω
V_{SINE}	**Sine-Wave** Amplitude ($R_{SINE} = 100k\Omega$)	0.2	0.22		0.2	0.22		0.2	0.22		$\times V_{SUPPLY}$
THD	THD ($R_S = 1M\Omega$)[4]		2.0	5		1.5	3		1.0	1.5	%
THD	THD Adjusted (Use Figure 6)		1.5			1.0			0.8		%

NOTES: 2. R_A and R_B currents not included.
   3. $V_{SUPPLY} = 20V$; R_A and $R_B = 10k\Omega$, $f \cong 10kHz$ nominal; can be extended 1000 to 1. See Figures 7a and 7b.
   4. 82kΩ connected between pins 11 and 12, Triangle Duty Cycle set at 50%. (Use R_A and R_B.)
   5. Figure 3, pins 7 and 8 connected, $V_{SUPPLY} = \pm 10V$. See Typical Curves for T.C. vs V_{SUPPLY}.

Note: All typical values have been guaranteed by characterization and are not tested.

Figure 7: Connections for Frequency Modulation (a) and Sweep (b)

FREQUENCY MODULATION AND SWEEPING

The frequency of the waveform generator is a direct function of the DC voltage at terminal 8 (measured from V^+). By altering this voltage, frequency modulation is performed. For small deviations (e.g. ±10%) the modulating signal can be applied directly to pin 8, merely providing DC decoupling with a capacitor as shown in Figure 7a. An external resistor between pins 7 and 8 is not necessary, but it can be used to increase input impedance from about 8kΩ (pins 7 and 8 connected together), to about (R + 8kΩ).

The sine wave output has a relatively high output impedance (1kΩ Typ). The circuit of Figure 8 provides buffering, gain and amplitude adjustment. A simple op amp follower could also be used.

Figure 8: Sine Wave Output Buffer Amplifiers

For larger FM deviations or for frequency sweeping, the modulating signal is applied between the positive supply voltage and pin 8 (Figure 7b). In this way the entire bias for the current sources is created by the modulating signal, and a very large (e.g. 1000:1) sweep range is created (f = 0 at V_{sweep} = 0). Care must be taken, however, to regulate the supply voltage; in this configuration the charge current is no longer a function of the supply voltage (yet the trigger thresholds still are) and thus the frequency becomes dependent on the supply voltage. The potential on Pin 8 may be swept down from V^+ by (1/3 V_{SUPPLY} – 2V).

APPLICATIONS

With a dual supply voltage the external capacitor on Pin 10 can be shorted to ground to halt the ICL8038 oscillation. Figure 9 shows a FET switch, diode ANDed with an input strobe signal to allow the output to always start on the same slope.

To obtain a 1000:1 Sweep Range on the ICL8038 the voltage across external resistors R_A and R_B must decrease to nearly zero. This requires that the highest voltage on control Pin 8 exceed the voltage at the top of R_A and R_B by a few hundred millivolts.

The Circuit of Figure 10 achieves this by using a diode to lower the effective supply voltage on the ICL8038. The large resistor on pin 5 helps reduce duty cycle variations with sweep.

Note: All typical values have been guaranteed by characterization and are not tested.

589

Figure 5: Possible Connections for the External Timing Resistors

Neither time nor frequency are dependent on supply voltage, even though none of the voltages are regulated inside the integrated circuit. This is due to the fact that both currents *and* thresholds are direct, linear functions of the supply voltage and thus their effects cancel.

To minimize *sine-wave* distortion the 82kΩ resistor between pins 11 and 12 is best made variable. With this arrangement distortion of less than 1% is achievable. To reduce this even further, two potentiometers can be connected as shown in Figure 6; this configuration allows a typical reduction of sine-wave distortion close to 0.5%.

Figure 6: Connection to Achieve
Minimum Sine-Wave Distortion

SELECTING R_A, R_B and C

For any given output frequency, there is a wide range of RC combinations that will work, however certain constraints are placed upon the magnitude of the charging current for optimum performance. At the low end, currents of less than 1μA are undesirable because circuit leakages will contribute significant errors at high temperatures. At higher currents (I > 5mA), transistor betas and saturation voltages will contribute increasingly larger errors. Optimum performance will, therefore, be obtained with charging currents of 10μA to 1mA. If pins 7 and 8 are shorted together, the magnitude of the charging current due to R_A can be calculated from:

$$I = \frac{R_1 \times (V^+ - V^-)}{(R_1 + R_2)} \times \frac{1}{R_A} = \frac{(V^+ - V^-)}{5R_A}$$

A similar calculation holds for R_B.

The capacitor value should be chosen at the upper end of its possible range.

WAVEFORM OUT LEVEL CONTROL AND POWER SUPPLIES

The waveform generator can be operated either from a single power-supply (10 to 30 Volts) or a dual power-supply (±5 to ±15 Volts). With a single power-supply the average levels of the triangle and sine-wave are at exactly one-half of the supply voltage, while the square-wave alternates between V^+ and ground. A split power supply has the advantage that all waveforms move symmetrically about ground.

The square-wave output is not committed. A load resistor can be connected to a different power-supply, as long as the applied voltage remains within the breakdown capability of the waveform generator (30V). In this way, the square-wave output can be made TTL compatible (load resistor connected to +5 Volts) while the waveform generator itself is powered from a much higher voltage.

Note: All typical values have been guaranteed by characterization and are not tested.

ICL8038

TYPICAL PERFORMANCE CHARACTERISTICS

OP02880I

OP02890I

OP02900I

Performance of the Square-Wave Output

OP02910I

OP02920I

Performance of Triangle-Wave Output

OP02930I

OP02940I

OP02950I

Performance of Sine-Wave Output

OP02960I

OP02970I

Note: All typical values have been guaranteed by characterization and are not tested.

ICL8038

Square-Wave Duty Cycle—50%   Square-Wave Duty Cycle—80%

Figure 4: Phase Relationship of Waveforms

DETAILED DESCRIPTION
(See Figure 1)

An external capacitor C is charged and discharged by two current sources. Current source #2 is switched on and off by a flip-flop, while current source #1 is on continuously. Assuming that the flip-flop is in a state such that current source #2 is off, and the capacitor is charged with a current I, the voltage across the capacitor rises linearly with time. When this voltage reaches the level of comparator #1 (set at 2/3 of the supply voltage), the flip-flop is triggered, changes states, and releases current source #2. This current source normally carries a current 2I, thus the capacitor is discharged with a net-current I and the voltage across it drops linearly with time. When it has reached the level of comparator #2 (set at 1/3 of the supply voltage), the the flip-flop is triggered into its original state and the cycle starts again.

Four waveforms are readily obtainable from this basic generator circuit. With the current sources set at I and 2I respectively, the charge and discharge times are equal. Thus a triangle waveform is created across the capacitor and the flip-flop produces a square-wave. Both waveforms are fed to buffer stages and are available at pins 3 and 9.

The levels of the current sources can, however, be selected over a wide range with two external resistors. Therefore, with the two currents set at values different from I and 2I, an asymmetrical sawtooth appears at terminal 3 and pulses with a duty cycle from less than 1% to greater than 99% are available at terminal 9.

The sine-wave is created by feeding the triangle-wave into a non-linear network (sine-converter). This network provides a decreasing shunt-impedance as the potential of the triangle moves toward the two extremes.

WAVEFORM TIMING

The *symmetry* of all waveforms can be adjusted with the external timing resistors. Two possible ways to accomplish this are shown in Figure 5. Best results are obtained by keeping the timing resistors R_A and R_B separate (a). R_A

controls the rising portion of the triangle and sine-wave and the 1 state of the square-wave.

The magnitude of the triangle-waveform is set at 1/3 V_{SUPPLY}; therefore the rising portion of the triangle is,

$$t_1 = \frac{C \times v}{I} = \frac{C \times \frac{1}{3} \times V_{SUPPLY} \times R_A}{\frac{1}{5} \times V_{SUPPLY}} = \frac{5}{3} R_A \times C$$

The falling portion of the triangle and sine-wave and the 0 state of the square-wave is:

$$t_2 = \frac{C \times V}{I} = \frac{C \times \frac{1}{3} V_{SUPPLY}}{\frac{2}{5} \times \frac{V_{SUPPLY}}{R_B} - \frac{1}{5} \times \frac{V_{SUPPLY}}{R_A}} = \frac{5}{3} \times \frac{R_A R_B C}{2R_A - R_B}$$

Thus a 50% duty cycle is achieved when $R_A = R_B$.

If the duty-cycle is to be varied over a small range about 50% only, the connection shown in Figure 5b is slightly more convenient. If no adjustment of the duty cycle is desired, terminals 4 and 5 can be shorted together, as shown in Figure 5c. This connection, however, causes an inherently larger variation of the duty-cycle, frequency, etc.

With two separate timing resistors, the frequency is given by

$$f = \frac{1}{t_1 + t_2} = \frac{1}{\frac{5}{3} R_A C \left(1 + \frac{R_B}{2R_A - R_B}\right)}$$

or, if $R_A = R_B = R$

$$f = \frac{0.3}{RC} \text{ (for Figure 5a)}$$

If a single timing resistor is used (Figure 5c only), the frequency is

$$f = \frac{0.15}{RC}$$

Note: All typical values have been guaranteed by characterization and are not tested.

Figure 9: Strobe-Tone Burst Generator

Figure 10: Variable Audio Oscillator, 20Hz to 20kHz

Figure 11: Linear Voltage Controlled Oscillator

The linearity of input sweep voltage versus output frequency can be significantly improved by using an op amp as shown in Figure 11.

USE IN PHASE-LOCKED LOOPS

Its high frequency stability makes the ICL8038 an ideal building block for a phase-locked loop as shown in Figure 12. In this application the remaining functional blocks, the phase-detector and the amplifier, can be formed by a number of available IC's (e.g. MC4344, NE562, HA2800, HA2820)

In order to match these building blocks to each other, two steps must be taken. First, two different supply voltages are used and the square wave output is returned to the supply of the phase detector. This assures that the VCO input voltage will not exceed the capabilities of the phase detector. If a smaller VCO signal is required, a simple resistive voltage divider is connected between pin 9 of the waveform generator and the VCO input of the phase-detector.

Second, the DC output level of the amplifier must be made compatible to the DC level required at the FM input of the waveform generator (pin 8, $0.8V^+$). The simplest solution here is to provide a voltage divider to V^+ (R_1, R_2 as shown) if the amplifier has a lower output level, or to ground if its level is higher. The divider can be made part of the low-pass filter.

This application not only provides for a free-running frequency with very low temperature drift, but it also has the

Note: All typical values have been guaranteed by characterization and are not tested.

ICL8038

INTERSIL

unique feature of producing a large reconstituted sinewave signal with a frequency identical to that at the input.

For further information, see Intersil Application Note A013, "Everything You Always Wanted to Know About The ICL8038."

Figure 12: Waveform Generator Used as Stable VCO in a Phase-Locked Loop

Figure 13: Detailed Schematic

Note: All typical values have been guaranteed by characterization and are not tested.

LINEAR
INTEGRATED
CIRCUITS

TYPE TL170C
SILICON HALL-EFFECT SWITCH

D2408, DECEMBER 1977 — REVISED NOVEMBER 1983

- Magnetic-Field Sensing Hall-Effect Input
- On-Off Hysteresis
- Small Size
- Solid-State Technology
- Open-Collector Output

description

The TL170C is a low-cost magnetically operated electronic switch that utilizes the Hall Effect to sense steady-state magnetic fields. Each circuit consists of a Hall-Effect sensor, signal conditioning and hysteresis functions, and an output transistor integrated into a monolithic chip. The outputs of these circuits can be directly connected to many different types of electronic components.

The TL170C is characterized for operation over the temperature range of 0°C to 70°C.

FUNCTION TABLE (T_A = 25°C)

FLUX DENSITY	OUTPUT
≤ -25 mT	Off
-25 mT $< B < 25$ mT	Undefined
≥ 25 mT	On

TOP VIEWS

LP PACKAGE

V_{CC}
GROUND
OUTPUT

1,02 mm (0.040 in)
0,76 mm (0.030 in)
CHIP FACE

LU PACKAGE

V_{CC}
GROUND
OUTPUT

0,48 mm (0.019 in)
0,43 mm (0.017 in)
CHIP FACE

functional block diagram

V_{CC}

V+

SILICON
HALL-EFFECT
SENSOR

SIGNAL
CONDITIONING
AND HYSTERESIS

OUTPUT

TTL
RTL
MOS
TRANSISTORS
SCR'S
TRIACS
RELAYS

GROUND

TEXAS
INSTRUMENTS
POST OFFICE BOX 225012 • DALLAS, TEXAS 75265

TYPE TL170C
SILICON HALL-EFFECT SWITCH

absolute maximum ratings over operating free-air temperature range (unless otherwise noted)

Supply voltage, V_{CC} (see Note 1) . . . . 7 V
Output voltage . . . . 30 V
Output current . . . . 20 mA
Operating free-air temperature range . . . . 0 °C to 70 °C
Storage temperature range . . . . −65 °C to 150 °C
Lead temperature 1,6 mm (1/16 inch) from case for 10 seconds . . . . 260 °C
Magnetic flux density . . . . unlimited

NOTE 1. Voltage values are with respect to network ground terminal.

electrical characteristics at specified free-air temperature, V_{CC} = 5 V ± 5% (unless otherwise noted)

	PARAMETER	TEST CONDITIONS		MIN	TYP	MAX	UNIT
B_{T+}	Threshold of positive-going magnetic flux density[†]		25 °C			25	mT [§]
			0 °C to 70 °C			35	
B_{T-}	Threshold of negative-going magnetic flux density[†]		25 °C	−25[¶]			mT [§]
			0 °C to 70 °C	−35[¶]			
$B_{T+} - B_{T-}$	Hysteresis		0 °C to 70 °C		20		mT [§]
I_{OH}	High-level output current	V_{OH} = 20 V	0 °C to 70 °C			100	μA
V_{OL}	Low-level output voltage	V_{CC} = 4.75 V, I_{OL} = 16 mA	0 °C to 70 °C			0.4	V
I_{CC}	Supply current	V_{CC} = 5.25 V — Output low	0 °C to 70 °C			6	mA
		V_{CC} = 5.25 V — Output high				4	

[†]Threshold values are those levels of magnetic flux density at which the output changes state. For the TL170C, a level more positive than B_{T+} causes the output to go to a low level and a level more negative than B_{T-} causes the output to go to a high level. See Figures 1 and 2.

[§]The unit of magnetic flux density in the International System of Units (SI) is the tesla (T). tesla is equal to one weber per square meter. Values expressed in milliteslas may be converted to gauss by multiplying by ten.

[¶]The algebraic convention, where the most negative limit is designated as minimum, is used in this data sheet for flux-density threshold levels only.

The north pole of a magnet is the pole that is attracted by the geographical north pole. The north pole of a magnet repels the north-seeking pole of a compass. By accepted magnetic convention, lines of flux emanate from the north pole of a magnet and enter the south pole.

**FIGURE 1—DEFINITION OF
MAGNETIC FLUX POLARITY**

The positive-going threshold (B_{T+}) may be a negative or positive B level at which a positive-going (decreasing negative or increasing positive) flux density results in the TL170 output turning on. The negative-going threshold is a positive or negative B level at which a negative-going (decreasing positive or increasing negative) flux density results in the TL170 turning off.

FIGURE 2—REPRESENTATIVE CURVES OF V_O vs B

MOTOROLA

OPTICALLY ISOLATED TRIAC DRIVER

These devices consist of a gallium-arsenide infrared emitting diode, optically coupled to a silicon bilateral switch and are designed for applications requiring isolated triac triggering, low-current isolated ac switching, high electrical isolation (to 7500 V peak), high detector standoff voltage, small size, and low cost.

- UL Recognized File Number 54915
- Output Driver Designed for 115 Vac Line
- Standard 6-Pin DIP

OPTO COUPLER/ISOLATOR

PHOTO TRIAC DRIVER OUTPUT

250 VOLTS

MAXIMUM RATINGS (T_A = 25°C unless otherwise noted)

Rating	Symbol	Value	Unit
INFRARED EMITTING DIODE MAXIMUM RATINGS			
Reverse Voltage	V_R	3.0	Volts
Forward Current — Continuous	I_F	50	mA
Total Power Dissipation @ T_A = 25°C Negligible Power in Transistor Derate above 25°C	P_D	100 1.33	mW mW/°C
OUTPUT DRIVER MAXIMUM RATINGS			
Off-State Output Terminal Voltage	V_{DRM}	250	Volts
On-State RMS Current T_A = 25°C (Full Cycle , 50 to 60 Hz) T_A = 70°C	$I_{T(RMS)}$	100 50	mA mA
Peak Nonrepetitive Surge Current (PW = 10 ms, DC = 10%)	I_{TSM}	1.2	A
Total Power Dissipation @ T_A = 25°C Derate above 25°C	P_D	300 4.0	mW mW/°C
TOTAL DEVICE MAXIMUM RATINGS			
Isolation Surge Voltage (1) (Peak ac Voltage, 60 Hz, 5 Second Duration)	V_{ISO}	7500	Vac
Total Power Dissipation @ T_A = 25°C Derate above 25°C	P_D	330 4.4	mW mW/°C
Junction Temperature Range	T_J	−40 to +100	°C
Ambient Operating Temperature Range	T_A	−40 to +70	°C
Storage Temperature Range	T_{stg}	−40 to +150	°C
Soldering Temperature (10 s)	—	260	°C

(1) Isolation surge voltage, V_{ISO}, is an internal device dielectric breakdown rating.

STYLE 6:
PIN 1. ANODE
2. CATHODE
3. NC
4. MAIN TERMINAL
5. SUBSTRATE
6. MAIN TERMINAL

NOTES:
1. DIMENSIONS A AND B ARE DATUMS.
2. -T- IS SEATING PLANE.
3. POSITIONAL TOLERANCES FOR LEADS:
   $\oplus$ Ø 0.13 (0.005) (M) T | A (M) B (M)
4. DIMENSION L TO CENTER OF LEADS WHEN FORMED PARALLEL.
5. DIMENSIONING AND TOLERANCING PER ANSI Y14.5, 1973.

DIM	MILLIMETERS MIN	MILLIMETERS MAX	INCHES MIN	INCHES MAX
A	8.13	8.89	0.320	0.350
B	6.10	6.60	0.240	0.260
C	2.92	5.08	0.115	0.200
D	0.41	0.51	0.016	0.020
F	1.02	1.78	0.040	0.070
G	2.54 BSC		0.100 BSC	
J	0.20	0.30	0.008	0.012
K	2.54	3.81	0.100	0.150
L	7.62 BSC		0.300 BSC	
M	0°	15°	0°	15°
N	0.38	2.54	0.015	0.100
P	1.27	2.03	0.050	0.080

CASE 730A-01

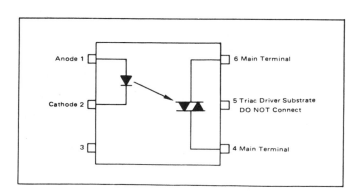

Anode 1

Cathode 2

3

6 Main Terminal

5 Triac Driver Substrate
DO NOT Connect

4 Main Terminal

MOC3009 , MOC3010 , MOC3011

ELECTRICAL CHARACTERISTICS (T_A = 25°C unless otherwise noted)

Characteristic	Symbol	Min	Typ	Max	Unit
LED CHARACTERISTICS					
Reverse Leakage Current (V_R = 3.0 V)	I_R	–	0.05	100	μA
Forward Voltage (I_F = 10 mA)	V_F	–	1.2	1.5	Volts
DETECTOR CHARACTERISTICS (I_F = 0 unless otherwise noted)					
Peak Blocking Current, Either Direction (Rated V_{DRM}, Note 1)	I_{DRM}	–	10	100	nA
Peak On-State Voltage, Either Direction (I_{TM} = 100 mA Peak)	V_{TM}	–	2.5	3.0	Volts
Critical Rate of Rise of Off-State Voltage, Figure 3	dv/dt	–	2.0	–	V/μs
Critical Rate of Rise of Commutation Voltage, Figure 3 (I_{load} = 15 mA)	dv/dt	–	0.15	–	V/μs
COUPLED CHARACTERISTICS					
LED Trigger Current, Current Required to Latch Output (Main Terminal Voltage = 3.0 V) MOC3009 MOC3010 MOC3011	I_{FT}	–	15 8.0 5.0	30 15 10	mA
Holding Current, Either Direction	I_H	–	100	–	μA

Note 1. Test voltage must be applied within dv/dt rating.

2. Additional information on the use of the MOC3009/3010/3011 is available in Application Note AN-780.

TYPICAL ELECTRICAL CHARACTERISTICS
T_A = 25°C

FIGURE 1 – ON-STATE CHARACTERISTICS

FIGURE 2 – TRIGGER CURRENT versus TEMPERATURE

MOC3009 , MOC3010 , MOC3011

FIGURE 3 – dv/dt TEST CIRCUIT

$dv/dt = 8.9 \, f \, V_{in}$

FIGURE 4 – dv/dt versus LOAD RESISTANCE

FIGURE 5 – dv/dt versus TEMPERATURE

FIGURE 6 – COMMUTATING dv/dt versus FREQUENCY

FIGURE 7 – MAXIMUM NONREPETITIVE SURGE CURRENT

599

MOC3009 , MOC3010 , MOC3011

TYPICAL APPLICATION CIRCUITS

FIGURE 8 – RESISTIVE LOAD

FIGURE 9 – INDUCTIVE LOAD WITH SENSITIVE GATE TRIAC
($I_{GT} \leq 15$ mA)

FIGURE 10 – INDUCTIVE LOAD WITH NON-SENSITIVE GATE TRIAC
(15 mA < I_{GT} < 50 mA)

μA78S40
Universal Switching
Regulator Subsystem

Linear Division Voltage Regulators

Description

The μA78S40 is a monolithic regulator subsystem consisting of all the active building blocks necessary for switching
regulator systems. The device consists of a temperature
compensated voltage reference, a duty-cycle controllable
oscillator with an active current limit circuit, an error amplifier, high current, high voltage output switch, a power diode and an uncommitted operational amplifier. The device
can drive external NPN or PNP transistors when currents
in excess of 1.5 A or voltages in excess of 40 V are
required. The device can be used for step-down, step-up
or inverting switching regulators as well as for series pass
regulators. It features wide supply voltage range, low
standby power dissipation, high efficiency and low drift. It
is useful for any stand-alone, low part count switching system and works extremely well in battery operated systems.

- **Step-Up, Step-Down or Inverting Switching Regulators**
- **Output Adjustable From 1.25 V to 40 V**
- **Peak Currents To 1.5 A Without External Transistors**
- **Operation From 2.5 V to 40 V Input**
- **Low Standby Current Drain**
- **80 dB Line And Load Regulation**
- **High Gain, High Current, Independent OP AMP**
- **Pulse Width Modulation With No Double Pulsing**

Connection Diagram
16-Lead DIP
(Top View)

Pin	Signal	Pin	Signal
1	DIODE CATHODE	16	SWITCH COLLECTOR
2	DIODE ANODE	15	DRIVER COLLECTOR
3	SWITCH EMITTER	14	IPK SENSE
4	OP AMP OUT	13	VI
5	OP AMP SUPPLY	12	TIMING CAPACITOR
6	OP AMP +IN	11	GND
7	OP AMP −IN	10	COMPARATOR −IN
8	REFERENCE VOLTAGE	9	COMPARATOR +IN

CD01400F

Order Information

Device Code	Package Code	Package Description
μA78S40DM	7B	Ceramic DIP
μA78S40PV	9B	Molded DIP
μA78S40DC	7B	Ceramic DIP
μA78S40PC	9B	Molded DIP

Block Diagram

EQ00650F

µA78S40

Absolute Maximum Ratings

Storage Temperature Range	
Ceramic DIP	–65°C to +175°C
Molded DIP	–65°C to +150°C
Operating Temperature Range	
Extended (µA78S40M)	–55°C to +125°C
Industrial (µA78S40V)	–40°C to +125°C
Commercial (µA78S40C)	0°C to +70°C
Lead Temperature	
Ceramic DIP (soldering, 60 s)	300°C
Molded DIP (soldering, 10 s)	265°C
Internal Power Dissipation[1,2]	
16L-Ceramic DIP	1.50 W
16L-Molded DIP	1.04 W
Input Voltage from V+ to V–	40 V
Input Voltage from V+ Op Amp to V–	40 V

Common Mode Input Range	
(Error Amplifier and Op Amp)	–0.3 to V+
Differential Input Voltage[3]	± 30 V
Output Short Circuit Duration	
(Op Amp)	Indefinite
Current from V_{REF}	10 mA
Voltage from Switch	
Collectors to GND	40 V
Voltage from Switch	
Emitters to GND	40 V
Voltage from Switch	
Collectors to Emitter	40 V
Voltage from Power Diode to GND	40 V
Reverse Power Diode Voltage	40 V
Current through Power Switch	1.5 A
Current through Power Diode	1.5 A

Notes

1. $T_{J\ Max} = 150°C$ for the Molded DIP, and 175°C for the Ceramic DIP.
2. Ratings apply to ambient temperature at 25°C. Above this temperature, derate the 16L-Ceramic DIP at 10 mW/°C, and the 16L-Molded DIP at 8.3 mW/°C.
3. For supply voltages less than 30 V, the absolute maximum voltage is equal to the supply voltage.

Functional Description

The µA78S40 is a variable frequency, variable duty cycle device. The initial switching frequency is set by the timing capacitor.[1] The initial duty cycle is 6:1. This switching frequency and duty cycle can be modified by two mechanisms — the current limit circuitry ($I_{pk\ sense}$) and the comparator.

The comparator modifies the OFF time. When the output voltage is correct, the comparator output is in the HIGH state and has no effect on the circuit operation. If the output voltage is too high then the comparator output goes LOW. In the LOW state the comparator inhibits the turn-on of the output stage switching transistors. As long as the comparator is LOW the system is in OFF time. As the output current rises the OFF time decreases. As the output current nears its maximum the OFF time approaches its minimum value. The comparator can inhibit several ON cycles, one ON cycle or any portion of an ON cycle. Once the ON cycle has begun the comparator cannot inhibit until the beginning of the next ON cycle.

The current limit modifies the ON time. The current limit is activated when a 300 mV potential appears between lead 13 (V_{CC}) and lead 14 (I_{pk}). This potential is intended to result when designed for peak current flows through R_{SC}. When the peak current is reached the current limit is turned on. The current limit circuitry provides for a quick end to ON time and the immediate start of OFF time.

Note

1. Oscillator frequency is set by a single external capacitor and may be varied over a range of 100 Hz to 100 kHz.

Generally the oscillator is free running but the current limit action tends to reset the timing cycle.

Increasing load results in more current limited ON time and less OFF time. The switching frequency increases with load current.

V_{FD} is the forward voltage drop across the internal power diode. It is listed on the data sheet as 1.25 V typical, 1.5 V maximum. If an external diode is used, then its own forward voltage drop must be used for V_{FD}.

V_{SAT} is the voltage across the switch element (output transistors Q1 and Q2) when the switch is closed or ON. This is listed on the data sheet as output saturation voltage.

Output saturation voltage 1 — defined as the switching element voltage for Q2 and Q1 in the Darlington configuration with collectors tied together. This applies to *Figure 1*, the step down mode.

Output saturation voltage 2 — switching element voltage for Q1 only when used as a transistor switch. This applies to *Figure 2*, the step up mode.

For the inverting mode, *Figure 3*, the saturation voltage of the external transistor should be used for V_{SAT}.

μA78S40

μA78S40
Electrical Characteristics T_A = Operating temperature range, V_I = 5.0 V, $V_{Op\ Amp}$ = 5.0 V, unless otherwise specified.

Symbol	Characteristic	Condition		Min	Typ	Max	Unit
General Characteristics							
I_{CC}	Supply Current (Op Amp Disconnected)	V_I = 5.0 V			1.8	3.5	mA
		V_I = 40 V			2.3	5.0	mA
I_{CC}	Supply Current (Op Amp Connected)	V_I = 5.0 V				4.0	mA
		V_I = 40 V				5.5	mA
Reference Section							
V_{REF}	Reference Voltage[1]	I_{REF} = 1.0 mA	Extend −55°C < T_A < +125°C, Comm 0 < T_A < +70°C, Indus −40°C < T_A < +85°C	1.180	1.245	1.310	V
$V_{R\ LINE}$	Reference Voltage Line Regulation	V_I = 3.0 V to V_I = 40 V, I_{REF} = 1.0 mA, T_A = 25°C			0.04	0.2	mV/V
$V_{R\ LOAD}$	Reference Voltage Load Regulation	I_{REF} = 1.0 mA to I_{REF} = 10 mA, T_A = 25°C			0.2	0.5	mV/mA
Oscillator Section							
I_{CHG}	Charging Current	V_I = 5.0 V, T_A = 25°C		20		50	μA
I_{CHG}	Charging Current	V_I = 40 V, T_A = 25°C		20		70	μA
I_{DISCHG}	Discharge Current	V_I = 5.0 V, T_A = 25°C		150		250	μA
I_{DISCHG}	Discharge Current	V_I = 40 V, T_A = 25°C		150		350	μA
V_{OSC}	Oscillator Voltage Swing	V_I = 5.0 V, T_A = 25°C			0.5		V
t_{on}/t_{off}	Ratio of Charge/Discharge Time				6.0		μs/μs
Current Limit Section							
V_{CLS}	Current Limit Sense Voltage	T_A = 25°C		250		350	mV
Output Switch Section							
$V_{SAT\ 1}$	Output Saturation Voltage 1	I_{SW} = 1.0 A, *Figure 1*			1.1	1.3	V
$V_{SAT\ 2}$	Output Saturation Voltage 2	I_{SW} = 1.0 A, *Figure 2*			0.45	0.7	V
h_{FE}	Output Transistor Current Gain	I_C = 1.0 A, V_{CE} = 5.0 V, T_A = 25°C			70		
I_L	Output Leakage Current	V_O = 40 V, T_A = 25°C			10		nA
Power Diode							
V_{FD}	Forward Voltage Drop	I_D = 1.0 A			1.25	1.5	V
I_{DR}	Diode Leakage Current	V_D = 40 V, T_A = 25°C			10		nA
Comparator							
V_{IO}	Input Offset Voltage	V_{CM} = V_{REF}			1.5	15	mV
I_{IB}	Input Bias Current	V_{CM} = V_{REF}			35	200	nA
I_{IO}	Input Offset Current	V_{CM} = V_{REF}			5.0	75	nA

μA78S40

μA78S40 (Cont.)

Electrical Characteristics T_A = Operating temperature range, V_I = 5.0 V, $V_{Op\ Amp}$ = 5.0 V, unless otherwise specified.

Symbol	Characteristic	Condition	Min	Typ	Max	Unit
V_{CM}	Common Mode Voltage Range	T_A = 25°C	0		V1 − 2	V
PSRR	Power Supply Rejection Ratio	V_I = 3.0 V to 40 V, T_A = 25°C	70	96		dB
Output Operational Amplifier						
V_{IO}	Input Offset Voltage	V_{CM} = 2.5 V		4.0	15	mV
I_{IB}	Input Bias Current	V_{CM} = 2.5 V		30	200	nA
I_{IO}	Input Offset Current	V_{CM} = 2.5 V		5.0	75	nA
$A_{VS}+$	Voltage Gain +	R_L = 2.0 kΩ to GND; V_O = 1.0 V to 2.5 V, T_A = 25°C	25	250		V/mV
$A_{VS}-$	Voltage Gain −	R_L = 2.0 kΩ to V+ (Op Amp) V_O = 1.0 V to 2.5 V, T_A = 25°C	25	250		V/mV
V_{CM}	Common Mode Voltage Range	T_A = 25°C	0		V_{CC} − 2	V
CMR	Common Mode Rejection	V_{CM} = 0 V to 3.0 V, T_A = 25°C	76	100		dB
PSRR	Power Supply Rejection Ratio	V+ Op Amp = 3.0 to 40 V, T_A = 25°C	76	100		dB
I_O+	Output Source Current	T_A = 25°C	75	150		mA
I_O-	Output Sink Current	T_A = 25°C	10	35		mA
SR	Slew Rate	T_A = 25°C		0.6		V/μs
V_{OL}	Output Voltage LOW	I_L = −5.0 mA, T_A = 25°C			1.0	V
V_{OH}	Output Voltage HIGH	I_L = 50 mA, T_A = 25°C	V + OP Amp −3.0 V			V

Note
1. Selected devices with tightened tolerance reference voltage available.

Design Formulas

CHARACTERISTIC	STEP-DOWN	STEP-UP	INVERTING	UNIT
$\dfrac{t_{on}}{t_{off}}$	$\dfrac{V_O + V_D}{V_I - V_{SAT} - V_O}$	$\dfrac{V_O + V_D - V_I}{V_I - V_{SAT}}$	$\dfrac{|V_O| + V_D}{V_I - V_{SAT}}$	
$(t_{on} + t_{off})$ Max	$\dfrac{1}{f_{Min}}$	$\dfrac{1}{f_{Min}}$	$\dfrac{1}{f_{Min}}$	μs
C_T	$4 \times 10^{-5} \ t_{on}$	$4 \times 10^{-5} \ t_{on}$	$4 \times 10^{-5} \ t_{on}$	μF
I_{pk}	$2 \ I_{O \ Max}$	$2 \ I_{O \ Max} \cdot \dfrac{t_{on} + t_{off}}{t_{off}}$	$2 \ I_{O \ Max} \cdot \dfrac{t_{on} + t_{off}}{t_{off}}$	A
L_{Min}	$\left(\dfrac{V_I - V_{SAT} - V_O}{I_{pk}} \right) t_{on \ Max}$	$\left(\dfrac{V_I - V_{SAT}}{I_{pk}} \right) t_{on \ Max}$	$\left(\dfrac{V_I - V_{SAT}}{I_{pk}} \right) t_{on \ Max}$	μH
R_{SC}	$0.33/I_{pk}$	$0.33/I_{pk}$	$0.33/I_{pk}$	Ω
C_O	$\dfrac{I_{pk} \ (t_{on} + t_{off})}{8 \ V_{ripple}}$	$\approx \dfrac{I_o}{V_{ripple}} \cdot t_{on}$	$\approx \dfrac{I_o}{V_{ripple}} \cdot t_{on}$	μF

Note
V_{SAT} = Saturation voltage of the switching element
V_D = Forward voltage of the flyback diode

Figure 1 Typical Step-Down Operational Performance
(T$_A$ = 25°C)

Figure 2 Typical Step-Up Operational Performance
(T$_A$ = 25°C)

Characteristic	Condition	Typical Value
Output Voltage	I$_O$ = 200 mA	10 V
Line Regulation	20 ≤ V$_I$ ≤ 30 V	1.5 mV
Load Regulation	5.0 mA ≤ I$_O$	
	I$_O$ ≤ 300 mA	3.0 mV
Max Output Current	V$_O$ = 9.5 V	500 mA
Output Ripple	I$_O$ = 200 mA	50 mV
Efficiency	I$_O$ = 200 mA	74%
Standby Current	I$_O$ = 200 mA	2.8 mA

Characteristic	Condition	Typical Value
Output Voltage	I$_O$ = 50 mA	25 V
Line Regulation	5.0 V ≤ V$_I$ ≤ 15 V	4.0 mV
Load Regulation	5.0 mA ≤ I$_O$	
	I$_O$ ≤ 100 mA	2.0 mV
Max Output Current	V$_O$ = 23.75 V	160 mA
Output Ripple	I$_O$ = 50 mA	30 mV
Efficiency	I$_O$ = 50 mA	79%
Standby Current	I$_O$ = 50 mA	2.6 mA

Notes
1. For I$_O$ ≥ 200 mA use external diode to limit on-chip power dissipation.
2. It is recommended that the internal reference (lead 8) be bypassed by a
   0.1 µF capacitor directly to (lead 11) the ground point of the µA78S40.

µA78S40

Figure 3 Typical Inversion Operational Performance ($T_A = 25°C$)

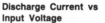

Characteristic	Condition	Typical Value
Output Voltage	$I_O = 100$ mA	−15 V
Line Regulation	8.0 V $\leqslant V_I < 18$ V	5.0 mV
Load Regulation	5.0 mA $\leqslant I_O$	
	$I_O \leqslant 150$ mA	3.0 mV
Max Output Current	$V_O = 14.25$ V	160 mA
Output Ripple	$I_O = 100$ mA	20 mV
Efficiency	$I_O = 100$ mA	70%
Standby Current	$I_O = 100$ mA	2.3 mA

Typical Performance Curves

Capacitance vs OFF Time

Reference Voltage vs Junction Temperature

Discharge Current vs Input Voltage

607

Typical Performance Curves (Cont.)

Current Limit Sense Voltage vs Input Voltage

PC09630F

Typical Pulse Width Modulator Application

CR03971F

APPENDIX E

ANSWERS TO SELECTED PROBLEMS

Chapter 1

2. (a) $I_{E1} = I_{E2} = 1.19$ mA
   $V_{B1} = V_{B2} = 0$ V (assuming $B_1 + B_2$ are connected to ground)
   $V_{E1} = V_{E2} = -0.715$ V
   $V_{C1} = V_{C2} = 6.79$ V
   (b) $A_d = 64.29$
   (c) $R_{i1} = R_{i2} = 4.2$ kΩ

4. (a) $I_{E1} = I_{E2} = 1.19$ mA
   $V_{B1} = V_{B2} = 0$ V (assuming B_1 and B_2 are connected to ground)
   $V_{E1} = V_{E2} = -0.715$ V
   $V_{C1} = V_{C2} = 3.34$ V
   (b) $A_d = 266.67$
   (c) $R_{i1} = R_{i2} = 4.2$ kΩ

6. (a) $I_{E1} = I_{E2} = 1.19$ mA
   $V_{B1} = V_{B2} = 0$ V (assuming B_1 and B_2 are connected to ground)
   $V_{E1} = V_{E2} = -0.715$ V
   $V_{C1} = V_{C2} = 3.34$ V
   (b) $A_d = 133.33$
   (c) $R_{i1} = R_{i2} = 4.2$ kΩ

8. (a) $v_o = 385.5$ mV pp
   (b) $v_{o\,max} = 12.84$ V pp

10. (a) $v_o = 40.2$ mV pp
   (b) $v_{o\,max} = 12.53$ V pp (ideal)

12. $A_d = 15.6$
   $R_i = 1$ MΩ

14. Q_1, Q_2, Q_3, Q_4: CA3086
   $R_2 = 4.7$ kΩ
   $R_C = 1$ kΩ

16. $I_{E3} = 0.988$ mA

18. (a) $I_{CQ1} = I_{CQ2} = 1.15$ mA
   $V_{CEQ1} = V_{CEQ2} = 5.31$ V
   (b) $A_d = 216.2$
   (c) $R_i = 4.35$ kΩ
   (d) $v_{o\,max} = 20$ V (ideal)

20. (a) $I_{E1} = I_{E2} = 0.988$ mA
   $I_{E3} = 1.95$ mA
   $I_{E4} = I_{E6} = 0.974$ mA
   $I_{E5} = I_{E7} = 9.74$ μA
   $V_{B1} = V_{B2} = 0$ V (assuming B_1 and B_2 are connected to ground)
   $V_{E1} = V_{E2} = -0.715$ V
   $V_{C1} = V_{C2} = V_{B5} = V_{B7} = 7.33$ V
   $V_{E4} = V_{E6} = V_{C3} = 5.9$ V
   $V_{C4} = V_{C5} = 10$ V
   $V_{C6} = V_{C7} = 8.25$ V
   $V_{B3} = -5$ V, and $V_{E3} = -5.715$ V
   (b) $A_{dT} = 1865.2$
   (c) $v_{o\,max} = 3.5$ V pp

22. (a) $I_{E3} = I_{E5} = 2.11$ mA
$I_{E1} = I_{E2} = I_{E6} = I_{E7} = 1.055$ mA
$V_{B1} = V_{B2} = 0$ V with B_1 and B_2 connected to ground.
$V_{E1} = V_{E2} = V_{C3} = -0.715$ V
$V_{C1} = V_{C2} = V_{B6} = V_{B7} = 4.04$ V
$V_{E6} = V_{E7} = V_{C5} = 3.325$ V
$V_{C6} = V_{C7} = 5.52$ V
$V_{E3} = V_{E5} = -9$ V
(b) $A_{dT} = 6{,}932.2$
(c) $v_{o\,max} = 4.39$ V pp

24. (a) $I_{E1} = I_{E2} = 0.368$ mA
$I_{E3} = I_{E4} = 3.8$ mA
$I_{E5} = 0.736$ mA
$V_{B1} = V_{B2} = V_{CM} = 3.5$ V
$V_{E1} = V_{E2} = V_{C5} = 2.785$ V
$V_{C1} = V_{C2} = V_{B3} = V_{B4} = 8.32$ V
$V_{B5} = 2.26$ V
$V_{E5} = 1.545$ V
$V_{C3} = V_{C4} = 12$ V
$V_{E3} = V_{E4} = 7.61$ V
(b) $A_{dT} = 143.63$
(c) $v_{o\,max} \cong 7.36$ V pp (unbalanced output riding on 7.61 V dc.)

26. D_1, D_2, Q_5, Q_6: CA3086 transistor array
$R_2 = 6.2$ kΩ
$R_E = 680$ Ω
$R = 4.39$ kΩ (5 kΩ pot)

28. Q_6, Q_7, Q_8: CA3086 transistor array
$R_2 = 5.6$ kΩ
$R = 3.51$ kΩ (5 kΩ pot)

30. (a) $I_{E1} = I_{E2} = 1.02$ mA
$V_{B1} = 2.87$ V, $V_{B2} = 7.01$ V
$V_{E1} = 2.24$ V
$V_{E2} = V_{C1} = 6.38$ V
$V_{C2} = 10.16$ V
(b) $A_v = -73.44$
(c) $A_i = 104.86$
(d) $R_i = 2.57$ kΩ

Chapter 2

2. (a) $A_{dT} = 4{,}812.8$
(b) $R_i = 2.79$ MΩ
(c) $v_{o\,max} = 16.96$ V pp

Chapter 3

2. $A_{cm} = 1$

4. $V_o \cong +15$ V (ideal)
6. $v_o = 3$ V dc

Chapter 4

2. (a) $A_F = 10.99$
(b) $A_F = 11$
(c) Results in a and b are the same because AB >> 1.

4. $A_F = 10.99$
$R_{iF} = (1.2)(10^{12})$ Ω
$R_{oF} = 0.0016$ Ω
$f_F = 54.60$ kHz
$V_{ooT} = \pm 0.357$ mV

6. $v_o = 1.1$ V pp

8. (a) $A_F = -9.99$
(b) $A_F = -10$
(c) Results in a and b are the same because AB >> 1

10. $A_F = -4.6$
$R_{iF} \cong 1$ kΩ
$R_{oF} = 0.00083$ Ω
$f_F = 106.96$ kHz
$V_{ooT} = \pm 180.56$ μV

12. (a) $v_o = -4.6$ V pp

14. (a) $A_D = -39$
(b) $R_{iFy} = 4$ kΩ, $R_{iFx} = 100$ Ω.

16. (a) $A_D = -44$
(b) $A_D = -66$

18. (a) $A_D = -14$
(b) $R_{iF_x} = R_{iFy} = 342.86$ GΩ
(c) $R_{oF} = 5.25$ mΩ
(d) $f_F = 71.43$ kHz

Chapter 5

2. $V_{oo} = \pm 240$ mV or -240 mV
4. $v_o \cong 1$ V pp
6. $V_{ol_{io}} = 7$ mV
8. $V_{ooT} = 71$ mV
$V_{ooT} = 68$ mV
10. (a) $E_v = \pm 30.6$ mV
$V_o = 101$ mV ± 30.6 mV
(b) $E_v = \pm 30.6$ mV
$V_o = 1.01$ V ± 30.6 mV
12. (a) $E_v = \pm 51.44$ mV
$v_o = 1.66$ V peak ± 51.44 mV

14. (a) $\Delta V_{oo} = \pm 60.6$ mV
    (b) $v_o = -1.0$ V peak ± 60.6 mV
16. $\Delta V_{oo} = 7.61$ μV rms
18. $\Delta V_{oo} = 12.51$ mV
20. (a) $v_o = 470$ mV sine wave at 1 kHz
    (b) $V_{no} = 4.7$ μV at 60 Hz
22. $V_{cm} = 10.64$ V rms

Chapter 6

2. $BW \cong 80$ kHz
4. $BW \cong 5$ kHz
6. $A_{oL} =$

$$\frac{10^4}{\left(1 + j\dfrac{f}{0.7 \text{ M}}\right)\left(1 + j\dfrac{f}{3 \text{ M}}\right)\left(1 + j\dfrac{f}{18 \text{ M}}\right)}$$

8. $UGB = 72.3$ MHz
10. (a) $-97.46°$
    (b) $-260.95°$
12. $C_1 = 10$ pF, $R_1 = 0$ Ω, $C_2 = 3$ pF
14. $f_{max} = 79.6$ MHz
16. $v_{in\,max} = 0.398$ V pp
18. $SR = 36.10$ V/μS
20. $v_{o\,max} = 3.02$ V

Chapter 7

2. $V_o = 4.8$ V
4. (a) $BW = 90.74$ kHz
   (b) $v_{o\,max} = 15$ V
   (c) $v_o' = 7.5$ V dc $- 2$ V peak sine wave
       $v_o = -2$ V peak sine wave
6. (a) $V_o \cong -1.14$ V
   (b) Scaling amplifier or weighted amplifier
8. $R_F = 2.4$ kΩ
10. (a) V_{ab} varies from -0.122 V to 0.236 V
    (b) V_o varies from 0.326 V to -0.587 V
14. $v_{in} = 11.34$ mV
16. $v_{in} \cong 0.52$ V rms
18. (a) $I_o = 3.33$ mA
    (b) $V_o = 10.1$ V
20. $V_o = 5.06$ V
22. $v_o = 3.98 \cos \omega t - 3.98$

24. $R_1 = 7.9$ kΩ (10 kΩ pot)
    $C_1 = 0.01$ μF
    $C_F = 500$ pF
    $R_F = 159$ kΩ (200 kΩ pot)
    $R_{OM} = 8.2$ kΩ

Chapter 8

2. $R = 5.3$ kΩ (10 kΩ pot)
   $C = 0.01$ μF
   Use a voltage follower.
4. $R = 13.26$ kΩ (use 20 kΩ pot)
   $C = 0.01$ μF
   $R_1 = 4.7$ kΩ and $R_F = 2.7$ kΩ
8. For $f_H = 2$ kHz: $R' = 7.95$ kΩ (use 10 kΩ pot)
   $C' = 0.01$ μF
   For $f_L = 400$ Hz: $R \cong 39$ kΩ and $C = 0.01$ μF
   For $A_{FT} = 4$: $R_1 = R_1' = R_F = R_F' = 10$ kΩ
10. For $f_L = 2$ kHz: $R = 7.95$ kΩ (use 10 kΩ pot), $C = 0.01$ μF
    For $f_H = 400$ Hz: $R' \cong 39$ kΩ, and $C' = 0.01$ μF
    For $A_{FT} = 2$: $R_1 = R_F = R_1' = R_F' = R_2 = R_3 = R_4 = 10$ kΩ
12. $\phi = -126.83°$
    To obtain positive phase shift interchange resistor R and capacitor C.
14. $R = 7.95$ kΩ (use 10 kΩ pot)
    $C = 0.01$ μF
    $R_1 = 10$ kΩ
    $R_F = 20$ kΩ
16. $f_o = 500$ Hz
18. (a) $V_o = 4.94$ V pp
    (b) $f_o = 1.18$ kHz
20. f_o changes from 9.8 kHz to 2.18 kHz if R_1 is varied from 4 kΩ to 18 kΩ.

Chapter 9

2. (a) $V_{ut} = 30.8$ mV
       $V_{lt} = -30.8$ mV
   (b) $V_{hy} = 61.6$ mV
4. The output voltage swings between -12 V and $+6.9$ V

6. $V_o = 700$ mV

    Change R_{int} from 1 MΩ to 100 kΩ if F_{in} is changed from 10 kHz to 100 kHz.

8. (a) Size of each step = 0.84375 V
    (b) $V_o = -12.66$ V
    (c) The advantage of $R/2R$ DAC is that it requires only two sets of resistance values.

10. The output voltage is a negative going half wave rectified waveform having a peak value of -300 mV.

12. $v_o \cong 1$ V

Chapter 10

2. $R_2 = 6.32$ kΩ (use 10 kΩ pot)
    $R_3 = 172.41$ kΩ (use 200 kΩ pot)
    $R_4 = R_5 = 5.03$ kΩ (use 5.1 kΩ)
    $R_6 = 1$ kΩ
    $R_7 = 9$ kΩ (use 10 kΩ pot)
    $R_8 = 900$ Ω (use 820 Ω)

4. $R_6 = R_7 = R_8 = 10$ kΩ
    $R_9 \cong 3.3$ kΩ

6. $R_1 = 10$ kΩ
    $R_2 = 20$ kΩ
    $R_3 = 20$ kΩ
    $f_{CLK} = 50$ kHz

8. $R_A = 909.09$ kΩ (use 1 MΩ pot)
    $C = 0.1$ μF, and NE555 timer

10. $R_A = 3.33$ kΩ (use 3.3 kΩ)

12. $R_A = 3.6$ kΩ (use 5 kΩ pot)
    $R_B = 1.8$ kΩ
    $C = 0.02$ μF, and NE555 timer

14. 1N914 small signal diode
    2N404 PNP transistor
    NE555 timer
    $R = 516.2$ Ω (use 1 kΩ pot)
    $C = 1$ μF
    $C_1 = 0.01$ μF, and $C_3 = 10$ μF

16. $f_{OUT} = 53.19$ kHz
    $f_L = \pm21.28$ kHz
    $f_C = \pm306.7$ Hz

18. $R_1 = 12$ kΩ

20. $R = 34.3$ Ω, and $V_{in} \geq 8.2$ V

22. $R_1 = 240$ Ω
    $R_2 = 3$ kΩ pot
    $C_2 = 1$ μF
    $C_3 = 10$ μF

24. $R_{SC} = 0.56$ Ω
    $L = 100$ μH
    $C_T = 0.01$ μF
    $C_o = 100$ μF
    $R_2 = 12$ kΩ
    $R_1 = 130$ kΩ

26. $V_o = -5$ V
    $V_o = -8.3$ V

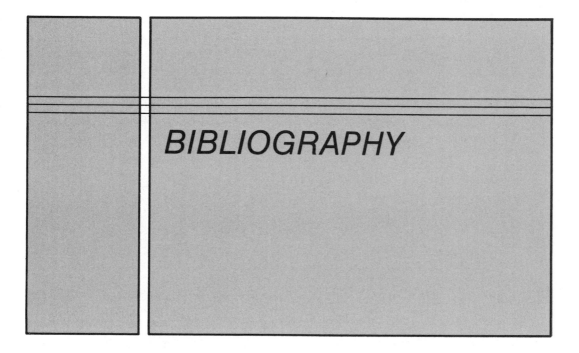

BIBLIOGRAPHY

Burr-Brown General Catalog. Tucson, Ariz.: Burr-Brown, 1979.

Datal-Intersil Data Acquisition Component Handbook. Mansfield, Mass.: Datel-Intersil, Inc., 1980.

Fairchild Linear Op-Amp Data Book. Mountain View, Calif.: Fairchild Camera and Instrument Corporation, 1979.

Intersil Data Book. Cupertino, Calif.: Intersil Inc., 1979.

Motorola Linear Integrated Circuit. Motorola CMOS Data. Phoenix, Ariz.; Motorola Semiconductor Products, 1979.

National Linear Data Book. Santa Clara, Calif.: National Semiconductor Corporation, 1978.

RCA Linear Integrated Circuits. Somerville, N.J.: RCA Solid State, 1978.

Signetics Analog Data Manual. Sunnyvale, Calif.: Signetics Corporation, 1979.

Teledyne Data Conversion Design Manual. Mountain View, Calif.: Teledyne Semiconductor, 1980.

The TTL Data Book for Design Engineers. Texas Instruments, 1976.

BERLIN, H. M., *The Design of Active Filters with Experiments.* Derby, Conn.: E and L Instruments, Inc., 1977.

BYLERLY, J. E., and M. VANDER KOOI, LM380 Power Audio Amplifier. *National Semiconductor Linear Applications,* Vol. 1, 1976.

KAHN, M., *The Versatile Op-Amp.* New York: Holt, Rinehart and Winston, Inc., 1970.

MALVINO, A. P., *Electronic Principles*. New York: McGraw-Hill Book Company, 1979.

Phase-Locked Loops. Benton Harbor, Mich.: Heathkit Continuing Education, 1980.

PRENSKY, S. D., *Manual of Linear Integrated Circuits: Operational Amplifiers and Analog ICs*. Reston, Va.: Reston Publishing Company, Inc., 1974.

RUTKOWSKI, G. B., *Handbook of Integrated Circuit Operational Amplifier*. Englewood Cliffs, N. J.: Prentice-Hall, Inc., 1975.

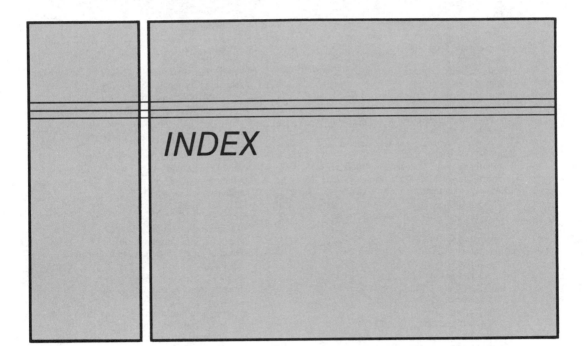

INDEX